ANNUAL REPORTS ON THE PROGRESS OF CHEMISTRY

SECTION B

Annual Reports on the Progress of Chemistry

Volume 90, 1993

SECTION B
Organic Chemistry

The Royal Society of Chemistry
Thomas Graham House, Cambridge CB4 4WF

ISBN 0-85186-251-9
ISSN 0069-3030

Annual Reports on the Progress of Chemistry Section 'B' is published annually by the Royal Society of Chemistry, Thomas Graham House, Science Park, Milton Road, Cambridge CB4 4WF, U.K. 1994 Annual subscription rate (for Volume 90) E.C. £95.00, Overseas £102.00, U.S.A. $179.00, including air speeded delivery. (Combined subscription package to *Annual Reports on the Progress of Chemistry Sections 'A, B and C'* details available from The Distribution Centre.) Change of address and orders with payment in advance to The Royal Society of Chemistry, The Distribution Centre, Blackhorse Road, Letchworth, Herts SG6 1HN, England. Air Freight and mailing in the U.S.A. by Publications Expediting Inc., 200 Meacham Avenue, Elmont, New York 11003. Second class postage paid at Jamaica, New York 11431. All other despatches outside the U.K. by Bulk Airmail, and Accelerated Surface Post outside Europe. PRINTED IN THE U.K.

Typeset by Vision Typesetting, Manchester
Printed and Bound in Great Britain by Bookcraft (Bath) Ltd.

Contributors

Senior Reporters

J. A. Joule, B.Sc., Ph.D., C.Chem, F.R.S.C., *University of Manchester*
C. I. F. Watt, B.Sc., Ph.D., *University of Manchester*

Reporters

C. P. Ball, B.Sc., *University of Warwick*
S. Caddick, B.Sc., G.R.S.C., Ph.D., *University of Sussex*
A. M. Chippendale, M.Sc., Ph.D., C.Chem., M.R.S.C., *Zeneca Specialties, Blackley*
A. P. Chorlton, B.Sc., M.Sc., Ph.D., *Zeneca Specialties, Blackley*
P. J. Derrick, B.Sc., Ph.D., F.R.S.C., F.R.A.C.I., F. Inst. Phys., *University of Warwick*
R. A. Hill, B.A., M.A., Ph.D., *University of Glasgow*
N. J. Lawrence, M.A., Ph.D., C.Chem., M.R.S.C., *UMIST*
P. Quayle, B.Sc., Ph.D., A.R.C.S., D.I.C., *University of Manchester*
N. Ramsden, D.Phil., *Glaxo Group Research, Greenford*
C. A. Reynolds, B.Sc., Ph.D., M.R.S.C., C.Chem., *University of Essex*
P. W. Sheldrake, M.A., Ph.D., C.Chem., F.R.S.C., *SmithKline Beecham Pharmaceuticals, Tonbridge*
G. R. Stephenson, B.A., M.A., Ph.D., C.Chem., M.R.S.C., A.R.A.C.I., *University of East Anglia*
A. G. Sutherland, B.Sc., M.Sc., Ph.D., C.Chem., M.R.S.C., *University of North London*

Contents

Chapter 1 Introduction 1
By J. A. Joule and C. I. F. Watt

Chapter 2 Physical Methods and Techniques 3
Part (i) NMR Spectroscopy
By A. M. Chippendale

1 Introduction 3

2 Solution NMR Techniques and Applications 3
2D NMR 3
3D and 4D NMR Chemical Shift Correlation 6
Pseudomultidimensional NMR 7
Coupling Constant Determination 8
Gradient-enhanced NMR Spectroscopy 11
Analysis of Biological Fluids 12

3 Solid-state NMR 12
High Resolution Techniques and Applications 12
Spin-1/2 Nuclei 12
Quadrupolar Nuclei 13
CP/MAS Experiments 13
Spin–Spin Coupling Constants 14
Spinning Sidebands 15
Multidimensional NMR 16
Measurement of Internuclear Distances 18
Study of Exchange Processes 18

Part (ii) Mass Spectrometry
By C. P. Ball and P. J. Derrick

1 Introduction 21

2 Physical Organic Chemistry 21

3 Fullerenes and Clusters 25

4 Organometallics 27

5 Polymers 28

6 Bioorganic compounds 29

Peptides and Proteins 30
Carbohydrates 34
Lipids and Steroids 36
DNA and Nucleotides 36
Narcotics and Drugs of Abuse 38

7 **Electrospray** 38

8 **Matrix-assisted Laser Desorption/Ionization** 40

9 **Fourier-transform Ion Cyclotron Resonance** 42

10 **Tandem Mass Spectrometry** 45

11 **Ion Trap** 47

Chapter 3 Theoretical Organic Chemistry 51
By C. A. Reynolds

1 **Introduction** 51

2 **Density Functional Theory** 51
Background 51
Density Functional Methods 52
Model Studies on Small Molecules 53
Applications 54
Carr–Parrinello Methods 56
Excited States 56

3 **Solvation** 56
Background 56
Computer Simulation Methods 58
Combining Quantum Mechanical Calculations with Computer Simulations 58
Tautomeric Equilibria 58
Computational Cost 59
Continuum Methods 59
Semiempirical and Classical Implementations of Continuum Methods 61
Applications 61
Effect of Hydration on Cycloadditions 61
Nucleophilic Additions and the Supermolecule Approach to Solvation 62
Potential of Mean Force Calculations 62

4 **Transition Structures** 62
Diels–Alder Reactions 62
Competition Between S_N2 and $E2$ Mechanisms 65
Hybrid Molecular Mechanical–Quantum Mechanical Models of the Transition State 66
Photochemistry 66

Tunnelling 67

5 Neural Networks 68

6 Conclusions 70

Chapter 4 Reaction Mechanisms 71
Part (i) Pericyclic Reactions
By N. G. Ramsden

1 Cycloadditions 71

2 Sigmatropic Reactions 77

3 Other Reactions 80

Part (ii) Free-radical Reactions
By S. Caddick

1 Introduction 81

2 Initiators, Promoters, and Mediators 81

3 Intramolecular Reactions 84
General 84
Ipso Substitution 86
Rearrangements 86
Hydrogen-atom Transfer 86
Group Transfer 89
Ring Expansions 90
Tandem Cyclizations 93

4 Intermolecular Reactions 94
General 94
Acyclic Stereocontrol 97
π-Substituted Radicals 97
Heteroatom-substituted Radicals 98
Reagent Controlled Stereoselectivity 100
Radical Sequences 100

5 Applications of Radical Processes to Synthetic and Biological Chemistry 101
Synthetic Chemistry 101
Biological Chemistry 102

Chapter 5 Aliphatic and Alicyclic Chemistry 105
By P. Quayle

1 Introduction 105

2 Aliphatic Chemistry 107
General 107
Coupling Reactions 108

Asymmetric Catalytic Coupling Reactions 113
Alkenes and Polyenes 115
Epoxides, Diols, and Related Compounds 119
Alkynes 123
Enolate, Aldol, and Related Reactions 125
Radical Reactions 128
Reductions 130
Asymmetric Hydrogenation and Related Reactions 131
Pot Pourri 132

3 Alicyclic Chemistry 133
Cyclopropanes 133
Cyclobutanes 135
Cyclopentanes 137
Cyclohexanes 141
Medium Rings 144

Chapter 6 Aromatic Compounds 145
By A. P. Chorlton

1 General and Theoretical Studies 145

2 Preparation of Benzenes from Non-aromatic Precursors 148

3 Non-aromatic Compounds from Benzene Precursors 155

4 Substitution in the Benzene Ring 158
Electrophilic Substitution 158
Nucleophilic Substitution 162
Substitution *via* Organometallic Intermediates 165
Substitution *via* Aryl Radicals 168

5 Condensed Polycyclic Aromatic Compounds 169
Benzene Aromatics 169
Non-benzenoid Aromatics 171

6 Cyclophanes 174

Chapter 7 Heterocyclic Compounds 179
By P. W. Sheldrake

1 Three-membered Rings 179

2 Four-membered Rings 182

3 Five-membered Rings 187

4 Six-membered Rings 200

5 Seven-membered Rings 210

6 Larger Rings 214

Chapter 8 Organometallic Chemistry. The Transition Elements 217
By G. R. Stephenson

1 **Introduction** 217

2 **Alkynes in C–C Bond Formation** 218
Enyne Cyclization 218

3 **Palladium-catalysed Cross Coupling** 223
Coupling with Vinyl Halides and Triflates 230
Nucleophilic Reagents in Palladium-catalysed Cross Coupling 232
Double Heteroatom Attachment 235
Multi-step Coupling Reactions 236
Unusual Coupling Processes 240
Asymmetric Palladium-catalysed Cross Coupling Reactions 242

4 **Use of Other Metals in Coupling Reactions** 243

5 **The Nicholas Reaction** 245

6 **Transition Metal Alkene and Arene π-Complexes** 247
η^2-Complexes 247
η^3-Complexes 248
Enantioselective Palladium-catalysed Allylic Displacement 252
η^4-Complexes 254
η^5-Complexes 256
η^6-Complexes 259

7 **Other Aspects** 265

Chapter 9 Synthetic Methods 269
By N. J. Lawrence

1 **Introduction** 269

2 **Carbon–Carbon Bond Formation** 270
Alkene Synthesis 274
Cycloadditions 278
Radical-based Methods 279

3 **Reduction** 280

4 **Oxidation** 284

5 **Protection** 290

6 **Miscellaneous Preparations** 292

Chapter 10 Enzyme Chemistry 299
By A. G. Sutherland

1 **Introduction** 299

2 **Hydrolysis and Condensation Reactions** 299

Alcohols, Carboxylic Acids, and Esters 299
Amines, Amides, and Lactams 303
Epoxides 304

3 Oxidation Reactions 305

4 Reduction Reactions 307

5 Carbon–Carbon Bond-forming and -cleaving Reactions 309

Chapter 11 Biosynthesis 311
By R. A. Hill

1 Introduction 311

2 Fatty Acid and Polyketide Biosynthesis 311

3 Terpenoids 318

4 Shikimate and Related Metabolites 321

5 Alkaloids and Other Amino-acid-derived Metabolites 325

6 Porphyrins 325

7 Miscellaneous Metabolites 336

Author Index 337

Section A, Inorganic Chemistry, contains the following items

1 Introduction. By F. J. Berry and N. G. Connelly

p-Block Element Surveys

2 Boron. By M. A. Beckett
3 Al, Ga, In, and Tl. By J. P. Maher
4 C, Si, Ge, Sn, and Pb. By D. A. Armitage
5 N, P, As, Sb, and Bi. By D. A. Armitage
6 O, S, Se, and Te. By J. D. Woollins
7 The Halogens and Noble Gases. By D. A. Armitage
8 Zn, Cd, and Hg. By H. Sloan
9 Inorganic and Organometallic Polymers. By I. Manners

d- and f-Block Element Surveys

10 Ti, Zr, and Hf. By S. A. Cotton
11 V, Nb, and Ta. By H. Sloan
12 Cr, Mo, and W. By H. Sloan
13 Mn, Tc, and Re. By J. D. Crane
14 Fe, Co, and Ni. By S. A. Cotton
15 Copper. By D. W. Smith
16 The Noble Metals. By G. Reid
17 Sc, Y, the Lanthanides, and the Actinides. By S. A. Cotton

Coordination and Organometallic Chemistry

18 The Coordination Chemistry of Open-chain Polydentate Ligands. By M. D. Ward
19 The Coordination Chemistry of Macrocyclic Ligands. By S. L. W. McWhinnie
20 Organometallic Chemistry of Monometallic Species. By P. K. Baker
21 Organometallic Chemistry of Bi- and Polynuclear Complexes. By G. Hogarth

Solid State Chemistry

22 New Compounds and Structures. By J. T. S. Irvine
23 Magnetism. By A. Harrison and A. S. Wills
24 Conducting Solids. By P. Slater

Other Topics

25 Radiochemistry. By D. S. Urch
26 Inorganic Mechanisms. By N. Winterton
27 Bioinorganic Chemistry. By J. D. Crane

Section C, Physical Chemistry, contains the following items

1 Introduction. By G. A. Webb
2 Low-frequency Spectroscopic Studies of Interactions in Liquids. By O. Faurskov Nielsen
3 Molecular Modelling. By B. J. Howlin
4 Electron Spin Resonance. By C. J. Rhodes
5 Photoemission Studies of Electronic Structure in Low-dimensional Conductors. By K. E. Smith
6 Carbon-based Catalysts for the Low-temperature Reduction of NO with NH_3. By J. J. Spivey
7 Intercalation Compounds of Low-dimensional Transition Metal Chalcogenides. By A. V. Powell

Table *Synopsis of Reports* 1986–1993

	1986	1987	1988	1989	1990	1991	1992	1993
Physical Methods:								
Mass spectrometry	×	√	×	√	×	√	×	√
NMR spectoscopy	√	√	√	√	√	√	√	√
Computer graphics	×	×	×	×	√	×	√	×
Theoretical chemistry	√	√	√	√	√	√	√	√
Reaction Mechanisms:								
Pericyclic	√	√	√	√	√	√	√	√
Polar reactions	√	√	√	√	√	√	√	×
Free-radical reactions	√	√	√	√	√	√	√	√
Aliphatic compounds	√	√	√	√	√	√	√	√
Alicyclic compounds	√	√	√	√	√	√	√	√
Aromatic compounds	√	√	√	√	√	√	√	√
Heterocyclic compounds	√	√	√	√	√	√	√	√
Organometallic compounds	√	√	√	√	√	√	√	√
Synthetic methods	√	√	√	√	√	√	√	√
Biological Chemistry:								
Enzymes	√	√	√	√	√	√	√	√
Biosynthesis	√	×	√	×	×	√	×	√
Other chapters*	*a*, *b*,	*c*, *d*,	*e*	*f*				

* (*a*) Terpenoid chemistry. (*b*) Organic chemistry related to molecular devices. (*c*) Arachidonic acid metabolites. (*d*) Polymer-supported reagents (*e*) Host-guest. (*f*) Carbohydrates.

1 Introduction

By J. A. JOULE and C. I. F. WATT
Department of Chemistry, University of Manchester, Manchester M13 9PL, UK

The topics covered in the 1992 *Annual Report*, *Section B* include discussions of Biosynthesis and Mass Spectrometry (biennial reviews), Theoretical Organic Chemistry, Pericyclic Chemistry, Aliphatic and Alicyclic Chemistry, Aromatic Compounds, Organometallic Chemistry (Transition Elements), Heterocyclic Compounds, Synthetic Methods, and Enzyme Chemistry. This year we welcome Drs. S. Caddick and C. Reynolds to the reporting team; we regret that Dr. Percy was unable to supply the planned chapter on Polar Reactions. Readers will find that some work is reviewed more than once where, for example, it is relevant to both Alicyclic Chemistry and Synthetic Methods; the time scale in which these Reports are assembled does not allow for editorial elimination of such repetitions nor, it can be argued, would that be desirable. The reader is referred to the Society's bimonthly journal *Natural Product Reports* for authoritative reviews of the chemistry and biosynthesis of particular groups of natural products. Some highlights, which tend to reflect the Senior Reporters' inclinations and cannot claim to be a measured assessment of the work of 1993, are given below.

The sophistication of methods for the determination of the biosynthesis (Chapter 11) of natural products has burgeoned over the last twenty years but doubts have now been expressed over the use of natural abundance deuterium NMR in such work. The days have long gone when it seemed that structural inspection alone could usually predict biosynthetic precursors, for who would have been brave enough to wager that thiotropocin is derived from phenylacetic acid.

Furans can be converted by dimethyldioxirane into malealdehydes; methyl(trifluoromethyl)dioxirane oxidizes unactivated C–H bonds in amines, leading to piperideines (Chapter 7). In a remarkable seven-component condensation, NaHS, NH_3, CO_2, MeOH, Bu^tNC, MeCH(Me)CH=O, and MeC(Me)(Br)CH=O produce a l-aza-3-thia five-membered heterocycle.

Notable in the area of enzyme chemistry (Chapter 10) are the enantioselective epoxidation of *cis* alkenes with H_2O_2 in the presence of *Caldariomyces fumago* and the quite delightful conversion of (±)-styrene epoxide into the corresponding ring-opened *R* diol, in 92% yield, with 89% e.e., using a mixture of two organisms.

The lengths to which aromatic chemistry (Chapter 6) can now be pushed are fascinating. For example cyclohexa-1,2,3-triene and cyclohex-1-en-3-yne, almost unbelievable isomers of benzene, have been synthesized in the form of zirconocene complexes.

The use of organometallic chemistry (Chapter 8) and new synthetic methods

(Chapter 9) are often intertwined, and often of relevance to aliphatic and alicyclic chemistry (Chapter 5). The use of supercritical fluids as solvents for organometallic synthesis has been described by several groups and will certainly have an important impact as it is developed.

There is considerable evidence that the addition of copper(I) salts can have a pronounced effect on the rate-determining transmetallation step of Heck couplings. Palladium-catalysed cross coupling processes allow the assembly of incredible arrays (Chapter 8).

Tris(pentafluorophenyl)boron and lanthanum triflate are air-stable, water-tolerant Lewis-acid catalysts for the aldol and Michael reactions of silyl enol ethers and can be used for the reactions of aldehydes in aqueous solution! Scandium(III) triflate is an excellent catalyst for Diels–Alder reactions. In 'chiral poisoning' processes a cheap homochiral impurity is added to an expensive catalyst. Of the many fascinating aspects of C_{60} chemistry, none is more bizarre than its catalysis of acetal hydrolysis! Microwaves continue to find diverse uses, for example in the deprotection of esters and ethers on alumina. The selective generation of *E* enolates can be achieved using the very hindered base lithium bis(2-adamantyl)amide; highly enantioselective kinetic deprotonation of cyclohexanones results from the use of lithium 1-phenyl-2-(piperidin-1-yl)ethylamide.

The development of multidimensional NMR techniques (Chapter 2i) that reduce spectral overlap, enhance sensitivity, and minimize acquisition times continues. Reports of the use of pulse field gradients have matched the availability of appropriately equipped spectrometers; in heteronuclear correlation experiments, an important benefit is elimination of signals for hydrogens not attached directly to ^{13}C and ^{15}N, including solvent H_2O. In solid state NMR, the REDOR technique has now been applied to determination of the ^{31}P–^{13}C distance in shikimate-3-phosphate.

Chapter 3 on Theoretical Organic Chemistry again focuses heavily on Density Functional Theory, which is potentially exact, and scales very favourably with size. Results compare favourably with those from high level *ab initio* methods, and such methods are now available within traditional quantum chemical codes such as GAUSSIAN and CADPAC.

Thermolysis of bis(tri-n-butylstannyl)benzopinacolinate affords a new method for the generation of butyltin radicals (Chapter 4ii). Tetrathiofulvene has been shown to promote intramolecular alkene addition of an aryl residue from a diazonium salt, presumably by donating an electron to the diazonium ion to generate the radical and nitrogen.

The expanding application of mass spectrometry (Chapter 2ii) to the studies of biological molecules and their interactions follows directly from the development of new ways of creating gaseous ions from heavy, complex molecules. Electrospray and MALDI are also revolutionizing the mass spectrometrical study of polymers. With these exciting developments, it is often easy to forget that MS remains a widely used method for the study of reaction mechanisms and solvent effects. Mass spectrometry has always been particularly suited to the study of fullerene chemistry, and C_{60} has, for example, been shown to add a variety of charged and neutral nucleophiles, and to react with methane under both positive and negative ionization.

2 Physical Methods and Techniques
Part (i) NMR Spectroscopy

By A. M. CHIPPENDALE

ZENECA Specialties, ZENECA plc, Blackley, Manchester, M9 8ZS, UK

1 Introduction

This review will follow essentially the same format as that employed last year. Technique development and applications will be reviewed for both solution and solid-state NMR. Once again technique development in solution NMR has continued to be driven by ever more demanding applications in the structural determination of proteins and oligonucleotides. Coverage of solution-state applications is therefore dominated by this area. The solid-state section will again focus on high resolution, solid-state NMR, which at present is a very active, rapidly developing area of research. The reader should bear in mind that the coverage will be selective rather than comprehensive.

2 Solution NMR Techniques and Applications

This section contains a review of a selection of new techniques and applications. They are mainly, but not exclusively, targeted at biochemical applications. Structure determination of biological macromolecules is based on the determination of *J* coupling connectivities, nuclear Overhauser enhancements (NOEs) and the magnitude of *J* couplings. In 1992 the focus was on the development of increasingly sophisticated 3D and 4D multidimensional NMR tools aimed mainly at the determination of chemical shift correlations and NOEs. This year several of the most elegant experiments have been concerned with methods for the accurate measurement of *J* couplings. In addition the use of pulsed field gradients in a variety of experiments has increased.

2D NMR.—In this section a selection of new 2D experiments and/or applications are briefly described. Several of them are concerned with chemical shift correlation for specific applications. Chary *et al.*[1] report a novel proton detected experiment, HELCO (heteronuclear long-range correlation), for the sequential assignment of ^{31}P and ^{1}H resonances in DNA. The technique has the same basic principle as that of 2D heteronuclear correlation used for the selection of ^{13}C–^{1}H and ^{15}N–^{1}H satellites in a ^{1}H NMR spectrum. Its application to the direct sequential assignment of ^{31}P spins and sugar ring protons is demonstrated for a duplex dodecanucleotide. Kayser *et al.*[2] have

[1] K. V. R. Chary, V. K. Rastogi, and G. Govil, *J. Magn. Reson., Series B*, 1993, **102**, 81.
[2] F. Kayser, M. Biesemans, M. Gielen, and R. Willem, *J. Magn. Reson., Series A*, 1993, **102**, 249.

demonstrated the use of proton detected 1D and 2D heteronuclear multiple-bond correlation (HMBC) experiments for the detection and measurement of long-range and/or small ^{1}H–^{119}Sn couplings. Berger[3] has designed a new pulse sequence, INEPT-HMQC, for ^{13}C–^{29}Si correlation. When HMQC is used alone for this type of correlation it is very time consuming due to long ^{29}Si relaxation times. When a third radiofrequency channel is available the INEPT proton polarization transfer technique can be combined with the HMQC method so that the repetition time of the experiment is governed by the relaxation time of the protons and not the ^{29}Si nuclei. A correlation spectrum for a polymeric silicon oil demonstrates the use of the technique. The experiment enables badly overlapping 1D carbon resonances to be correlated and assigned.

In 1992 Günther's group[4] reported the extension of the 1D and 2D INADEQUATE experiments to the spin-1 ^{6}Li nucleus. In this work chemically non-equivalent scalar coupled ^{6}Li nuclei in organolithium clusters were detected *via* the observation of NMR signals arising from two-spin double quantum coherences. The group have extended their work[5] on spin-1 nuclei to deuterion spin systems of the AX, AA′XX′, and AA′MM′X type. This time, in addition to two-spin double quantum coherences, one-spin double quantum coherences were also observed. Inspection of the relevant energy level diagrams confirm that such coherences are to be expected and it is unclear why they were not observed in the ^{6}Li case.

Many proteins are only biologically active when they are combined into symmetrical dimers. However, structure determination of such dimers by NMR is obstructed because it is impossible to distinguish between their intra- and intersubunit NOEs by a normal NOESY spectrum. This problem can be resolved by the use of asymmetric isotope labelling. In the past this has been achieved by the use of heterodimers composed of different selectively deuteriated proteins. Folkers *et al.*[6] have used an alternative approach in which heterodimers are formed from uniformly ^{13}C-labelled and unlabelled protein. The mixture thus generated contains a proportion of asymmetrically labelled dimer. 2D heteronuclear double-half-filter NOESY methodology is used to detect selectively NOE cross peaks from such molecules. Two sets of data are recorded, one which uses ^{13}C editing pulses and one in which they are omitted. Addition of the two recordings yields a spectrum in which NOE cross peaks are observed for both ^{13}C and ^{12}C-bound protons. Subtraction gives a spectrum in which NOE cross-peaks are only observed between, on the one hand, ^{13}C-bound protons and, on the other, ^{12}C-bound protons.

Batta *et al.*[7] have used a combination of ^{13}C and ^{205}Tl NMR 2D exchange spectroscopy (EXSY) to study cyanide exchange reactions in an aqueous solution containing the complexes $Tl(^{13}CN)_3$, $Ti(^{13}CN)_4^-$, and $H^{13}CN$. The well-resolved heteronuclear spin–spin couplings result in six exchanging sites for ^{13}C NMR and nine for ^{205}Tl NMR. 2D exchange spectroscopy provides an overview of the exchange reactions and gives their rates. In some cases information not otherwise accessible about the equilibrium dynamics can be obtained by quantitative analysis of the cross and auto peaks in the 2D spectra.

[3] S. Berger, *J. Magn. Reson., Series A*, 1993, **101**, 329.
[4] O. Eppers, T. Fox, and H. Günther, *Helv. Chim. Acta*, 1992, **75**, 883.
[5] H.-E. Mons, K. Bergander, and H. Günther, *Helv. Chim. Acta*, 1993, **76**, 1216.
[6] P. J. M. Folkers, R. H. A. Folmer, R. N. H. Konigs, and C. W. Hilbers, *J. Am. Chem. Soc.*, 1993, **115**, 3798.
[7] G. Batta, I. Bányai, and J. Glaser, *J. Am. Chem. Soc.*, 1993, **115**, 6782.

A completely different type of 2D NMR spectroscopy has been reported by Morris and Johnson.[8] They describe improvements in the electrophoretic or mobility-ordered experiment (ENMR) for the analysis of ionic mixtures. In 2D ENMR experiments, the chemical shift dimension is always obtained by Fourier transformation of the FID or half-echo with respect to the acquisition time. The second dimension in this experiment was obtained by incrementing the amplitude (I) of the electric current pulse. Transformation with respect to I then yielded peaks in the mobility dimension, the resulting method being named mobility ordered NMR spectroscopy (MOSY). The method was demonstrated with a mixture of 1.00 mM sodium dodecyl sulfate (SDS), 1.00 mM EDTA, and 1.00 mM 2,4,6-trimethylbenzoic acid (TMBA). Figure 1 shows

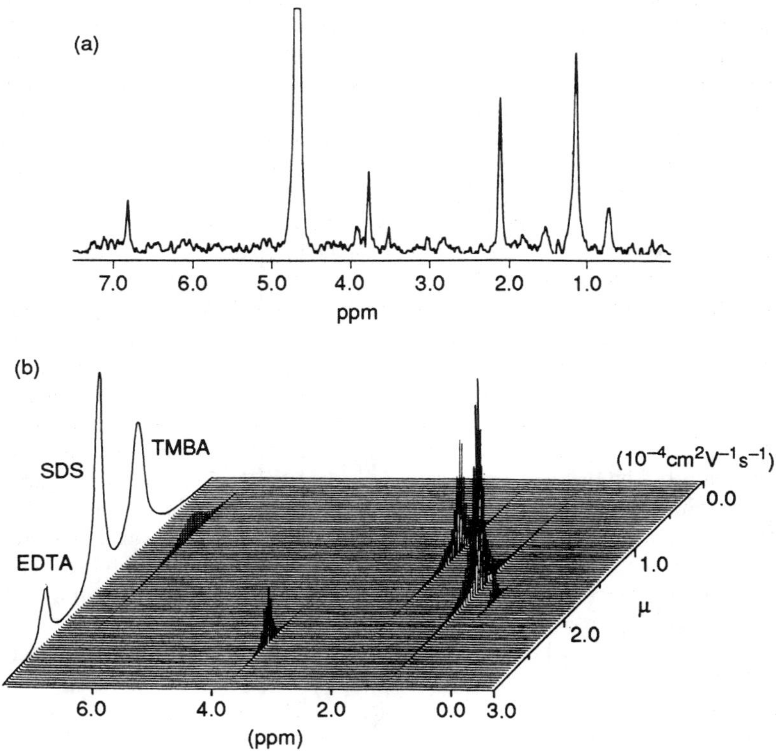

Figure 1 *(a) The ^1H NMR spectrum of a mixture containing SDS, EDTA, and TMBA. (b) The mobility-ordered 2D NMR spectrum for the same sample*
(Reproduced with permission from *J. Magn. Reson., Series A*, 1993, **101**, 67)

both the 1D ^1H NMR and 2D MOSY spectra obtained from the sample. The authors propose to use MOSY to study the binding of small ions to macromolecules and colloids, and to the study of phospholipid vesicles.

[8] K. F. Morris and C. S. Johnson, *J. Magn. Reson., Series A*, 1993, **101**, 67.

3D and 4D NMR Chemical Shift Correlation.—The last few years have seen the development of a relatively large number of multidimensional NMR experiments, which have facilitated the structure determination of biological macromolecules of increasing size and complexity. New techniques continue to be developed and others refined. Of particular interest are methods that further reduce spectral overlap, increase sensitivity, minimize acquisition times, or reduce the overall number of spectra required for an identification. It is worth noting at this stage that Bax's group at NIH has been particularly active and inventive in such developments.

In one of their publications[9] they describe two new 3D experiments, named H(CCO)NH and C(CO)NH, that can correlate all the aliphatic ^{1}H or ^{13}C resonances of a given amino acid directly with the amide of the next residue in ^{13}C/^{15}N-enriched proteins. The new methods are far more convenient for obtaining assignments than the previous combinations of H_2O and D_2O experiments as they provide a direct linkage between the backbone and entire side chains. The power of the technique is demonstrated in Figure 2, which shows strip plots observed for the amides of residues Gly-121 to Leu-129 of calcineurin B. The strips are taken parallel to the F_1 axis at the F_2/F_3 (^{15}N/^{1}H) position for each amide which correlates with (a) the alphatic

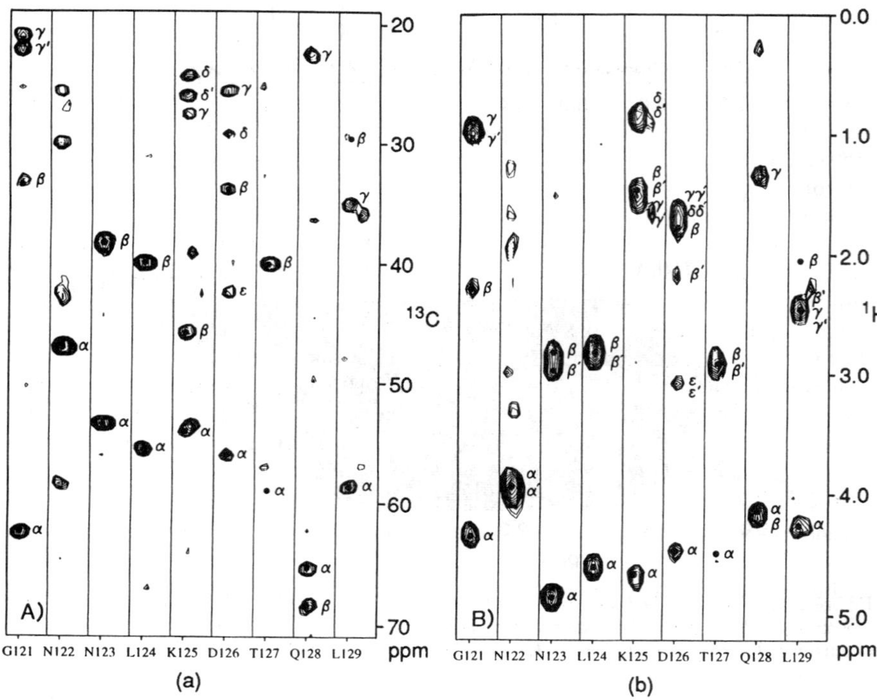

Figure 2 *Strip plots of the correlations observed for the amides of residues Gly-*121 *to Leu-*129 *of calcineurin B taken from (a) the C(CO)NH and (b) the H(CCO)NH spectrum*
(Reproduced with permission from *J. Magn. Reson., Series B*, 1993, **101**, 114)

[9] S. Grzesiek, J. Anglister, and A. Bax, *J. Magn. Reson., Series B*, 1993, **101**, 114.

side-chain carbon resonances of the preceding residue and (b) the side-chain protons of the preceding residue. Resonances that are not marked by Greek letters correspond to correlations to amide ^{1}H–^{15}N pairs that are close in frequency to the one for which the strip has been selected. Grzesiek and Bax[10] have also refined the 3D HCACO experiment, which correlates the chemical shifts of the intraresidue H^{α}, C^{α}, and carbonyl resonances in proteins uniformly enriched with ^{13}C. The source of a large number of artifacts in such spectra was investigated and an effective, simple remedy found.

Kay[11] has reported a new pulse sequence for the separation of aliphatic carbon chemical shifts *via* the carbonyl (C′) chemical shift in ^{15}N, ^{13}C-labelled proteins. The experiment is described as C′-edited ^{13}C–^{13}C TOCSY. In a search for increased sensitivity Wittekind and Mueller[12] report a new 3D experiment, HNCACB, to correlate the resonances of the amide proton and nitrogen frequencies with those of the intra- and interresidue α- and β-carbon resonances of proteins. It provides the same correlations as the CBCANH experiment reported in 1992 by Grzesiek and Bax.[13] However, the new experiment is claimed to be more sensitive because it exploits the long T_2 relaxation properties of the ^{15}N nuclei relative to those of the $^{13}C_{\alpha/\beta}$ spins and is expected to provide higher signal-to-noise ratios for large proteins where sensitivity losses due to relaxation become problematic.

Bax's group[14] has devised an alternative approach to the problems associated with short T_2 ^{13}C relaxation times in correlation experiments with larger proteins. They have achieved ^{13}C line narrowing by ^{2}H decoupling in ^{2}H, ^{13}C, ^{15}N-enriched proteins. The ^{13}C T_2 is normally dominated by strong dipolar interaction with its attached protons. As the gyromagnetic ratio of ^{2}H is ~6.5 times lower than that of ^{1}H, the heteronuclear dipolar interaction is greatly reduced by deuteriation. High power ^{2}H decoupling with an RF field strength much stronger than the inverse ^{2}H T_1 results in a ^{13}C linewidth that is much narrower than for the a protonated ^{13}C. The success of the method is demonstrated for a sample of the protein calcineurin B, uniformly enriched with ^{2}H, ^{15}N, and ^{13}C to levels of 50%, 98%, and 88% respectively. The ^{13}C resolution enhancement obtained by this technique is illustrated for a small region of the 2D H(N)CA correlation spectrum displaying connectivities between amide protons and their intraresidue C^{α}. A 4D experiment, HN(COCA)NH, is also described which allows J correlation between ^{1}H–^{15}N resonances of sequential residues. Efficient transfer of magnetization from $^{13}C^{\alpha}$ to ^{15}N is possible because of the ^{13}C line narrowing afforded by the deuteriation and ^{2}H decoupling.

Pseudo Multidimensional NMR.—Multidimensional NMR experiments are powerful tools for the extraction and correlation of spectral parameters. However, for small and medium-sized molecules, where only a limited amount of the information obtained is required, selective 1D analogues often become more attractive. Kövér *et al.*[15] have described several new 1D variants of 2D inverse ^{1}H–^{13}C correlation experiments for mapping of direct, long-range, relayed, and NOE connectivities. In all the experiments

[10] S. Grzesiek and A. Bax, *J. Magn. Reson., Series B*, 1993, **102**, 103.
[11] L. E. Kay, *J. Magn. Reson., Series B*, 1993, **101**, 110.
[12] M. Wittekind and L. Mueller, *J. Magn. Reson., Series B*, 1993, **101**, 201.
[13] S. Grzesiek and A. Bax, *J. Magn. Reson.*, 1992, **99**, 201.
[14] S. Grzesiek, J. Anglister, H. Ren, and A. Bax, *J. Am. Chem. Soc.*, 1993, **115**, 4369.
[15] K. E. Kövér, D. Uhrin, G. Szalontai, and G. Batta, *J. Magn. Reson., Series B*, 1993, **101**, 1.

the required selectivity is achieved by the incorporation of a chemical-shift-selective filter in the carbon frequency domain.

Mons *et al.*[16] have reported a selective ^{2}H, ^{2}H TOCSY experiment for signal assignment in perdeuteriated compounds. In such compounds the small homonuclear ^{2}H–^{2}H spin–spin coupling constants are not usually resolved and therefore give no indication of connectivities. The experiment was demonstrated by its application to pyridine-d_5, iodocyclohexane-d_{11}, and chlorocyclohexane-d_{11}.

Mueller and Bigler[17] have designed two 1D pulse sequences for the simultaneous selective detection and measurement of $^1J_{cc}$ and $^nJ_{cc}$ coupling interactions among carbons. Berger[18] had previously reported a modified 1D INADEQUATE experiment (SELINQUATE) by which a single connectivity could be detected by the application of a final selective 90° pulse. In these new experiments a series of selective 180° pulses are applied to enable one sequentially to perturb several carbon spins and simultaneously detect the corresponding responses. Such experiments represent valuable alternatives to 2D INADEQUATE experiments when only part of the full carbon–carbon connectivity matrix is to be studied and as long as the corresponding carbon resonances are well resolved.

Coupling Constant Determination.—Long range homo- and heteronuclear *J* couplings provide valuable conformational information in the structure determination of proteins and oligonucleotides. However, to date such couplings have not been very widely used mainly because of difficulties associated with their accurate measurement in many systems. Much effort is now being expended in devising experiments and data processing procedures for such measurements.

Griesinger's group[19] have presented a 3D experiment to measure three-bond ^{13}C–^{13}C couplings between carbonyl carbon atoms (C′) and C_γ atoms for the stereospecific assignment of valine methyl groups in proteins that are uniformly ^{13}C labelled. The method relies on the relatively large $^1J_{C'C_\alpha}$ coupling of about 55 Hz in amino acids. Correlating C_α with C_γ without touching the carbonyl allows one to measure $^3J_{C'C_\gamma}$. For sensitivity reasons the magnetization on C_γ is transferred to H_γ for detection.

Emsley *et al.*[20] report the application of selective 2D NMR correlation experiments to obtain quantitative three-bond coupling constants in DNA. A soft-COSY experiment is used, which results in a cross peak for a selected pair of *active* spins. The cross-peak structure is the same as that observed in E.COSY spectra. Compared with a non-selective COSY experiment, the very small spectral widths excited in the selective experiment enable a high digital resolution to be achieved. The technique was demonstrated by obtaining cross peaks between protons in the sugar rings in DNA fragments, which were then iteratively fitted to simulations derived from trial coupling topologies to determine the coupling constants that contribute to the cross-peak patterns.

Bax's group has devised a series of ingenious multidimensional experiments for the

16 H.-E. Mons, K. Bergander, and H. Günther, *Magn. Reson. Chem.*, 1993, **31**, 509.
17 C. Mueller and P. Bigler, *J. Magn. Reson., Series A*, 1993, **102**, 42.
18 S. Berger, *Angew. Chem., Int. Ed. Engl.*, 1988, **27**, 1196.
19 H. Schwalbe, A. Rexroth, U. Eggenberger, T. Geppert, and C. Griesinger, *J. Am. Chem. Soc.*, 1993, **115**, 7878.
20 L. Emsley, T. J. Dwyer, H. P. Spielmann, and D. E. Wemmer, *J. Am. Chem. Soc.*, 1993, **115**, 7765.

measurement of J couplings. In their approach, referred to as quantitative J correlation, the J value is derived from the fraction of magnetization transferred 'out and back' to its J-coupled partner relative to the fraction which does not get transferred. In one[21] such experiment, measurement of two and three-bond proton to methyl-carbon J couplings in proteins uniformly enriched with ^{13}C is described. A long-range ^{13}C–H 3D correlation experiment (LRCH) was used. Figure 3 shows six F_2 strips from the resulting 3D ^1H-detected (^{13}C–^1H) long-range spectrum of calmodulin complexed with a 26-residue peptide. The strips are taken at the $^{13}C/^1H$ (F_1/F_2) resonance frequencies of $C^{\gamma 2}H_3$ and $C^{\delta}H_3$ of Ile52 (a), $C^{\delta 2}H_3$ and $C^{\delta 1}H_3$ of Leu48 (b), and $C^{\gamma 2}H_3$ and $C^{\gamma 1}H_3$ of Val142 (c). The magnitudes of the long-range coupling constants to each methyl group are reflected in the intensities of the corresponding cross peaks. In addition to the 3D spectrum, a 2D reference spectrum is also recorded using the same pulse scheme but with t_2 fixed at zero. The methyl-group ^{13}C–^1H cross-peak volumes in the 2D spectrum are a function of the magnitude of the J couplings of *all* the protons coupled to each methyl carbon. The value of the long-range coupling of a methyl carbon atom to proton k, J_{CHk}, can be extracted from the ratio of the relevant 3D cross-peak volume to the 2D reference peak volume.

Another functionally similar experiment has been reported by Zhu and Bax[22] for the measurement of long-range ^1H–^{13}C coupling constants from quantitative 2D heteronuclear multiple-quantum correlation spectra. The pulse scheme used for correlating the long-range-coupled ^1H and ^{13}C nuclei is equivalent to the well-known HMQC one-bond correlation scheme with the dephasing and rephasing intervals adjusted to a suitably long value. A reference spectrum is acquired with the same ^1H pulse scheme, but with the 90° ^{13}C pulses replaced by short delays equal to the pulse widths. This results in a 'reference spectrum', which is symmetrical about the carrier in the F_1 dimension of the 2D spectrum. The shapes of the corresponding peaks in the correlation spectrum and the 'reference spectrum' are identical and their relative intensities are related in a straightforward manner to the size of the long-range coupling involved.

In another publication Vuister and Bax[23] report a further type of quantitative J correlation experiment, this time used to measurement homonuclear three-bond $J_{HNH\alpha}$ coupling constants in ^{15}N-enriched proteins. The method relies on the quantitative analysis of the diagonal-peak to cross-peak intensity ratio in a ^{15}N-separated H^N–H^α homonuclear J correlation experiment. The experiment was demonstrated for the protein staphylococcal nuclease, uniformly enriched with ^{15}N, ligated with thymidine 3′,5′-bisphosphate and Ca^{2+}. J couplings were measured for 96 residues which had sufficiently resolved ^1H–^{15}N correlations. The same group[24] also report a method which is essentially a 2D difference experiment used to measure small unresolvable $^3J_{CN}$ couplings in proteins uniformly enriched in ^{15}N and ^{13}C.

A variety of numerical techniques are being developed for the analysis of multiplets in 1D and 2D spectra. Although space does not permit these to be discussed at any length they deserve a mention. Jones *et al.*[25] have used the following strategy for the

[21] G. W. Vuister and A. Bax, *J. Magn. Reson., Series B*, 1993, **102**, 228.
[22] G. Zhu and A. Bax, *J. Magn. Reson., Series A*, 1993, **104**, 353.
[23] G. W. Vuister and A. Bax, *J. Am. Chem. Soc.*, 1993, **115**, 7772.
[24] G. W. Vuister, A. C. Wang, and A. Bax, *J. Am. Chem. Soc.*, 1993, **115**, 5334.
[25] J. A. Jones, D. S. Grainger, P. J. Hore, and G. J. Daniell, *J. Magn. Reson., Series A*, 1993, **101**, 162.

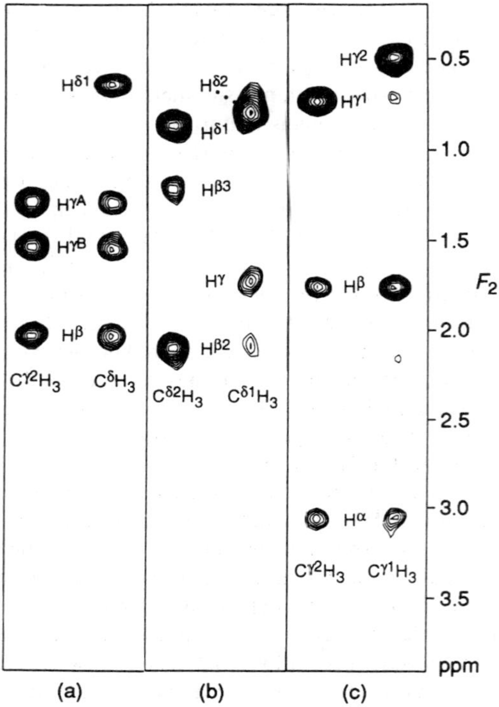

Figure 3 *Six* F_2 *strips from the 3D* ^1H*-detected* [^{13}C–^1H] *long-range spectrum of calmodulin complexed with a 26-residue peptide*
(Reproduced with permission from *J. Magn. Reson., Series B*, 1993, **102**, 228)

analysis of cross sections through COSY cross peaks. The active coupling constant has been determined by the *J* doubling method described by McIntyre and Freeman.[26] *J* convolution is then used to remove the active splitting by the maximum entropy method. Estimates of the passive couplings are next obtained from the deconvolved spectrum. Finally the values of all the coupling constants are refined by model fitting to the original spectrum. The approach was shown to be successful for complex, congested spectra.

Portilla and Freeman[27] have measured spin-coupling constants using decoupling followed by reconvolution. A well-digitized spin multiplet (a) is extracted from a conventional spectrum. The splitting of interest, J_{IS}, is then decoupled to give spectrum (b). Reconvolution introduces a trial splitting, J^*, giving a multiplet (c) that is a good match to (a) only when $J^* = J_{IS}$. Huber and Bodenhausen[28] have also reported the use of deconvolution methodology with 1D and 2D spectra.

[26] L. McIntyre and R. Freeman, *J. Magn. Reson.*, 1992, **96**, 425.
[27] F. del Río Portilla and R. Freeman, *J. Magn. Reson., Series A*, 1993, **104**, 358.
[28] P. Huber and G. Bodenhausen, *J. Magn. Reson., Series A*, 1993, **102**, 81.

Gradient-enhanced NMR Spectroscopy.—The increasing availability of commercial spectrometers equipped for the use of pulsed field gradients (PFGs) has led to a significant increase in the number of reports of their use. Applications are mainly in the multidimensional area. In heteronuclear correlation experiments an important benefit has been the elimination of signals from protons not attached to observed ^{13}C and ^{15}N nuclei, *e.g.* from ^{12}C nuclei and, most importantly for protein studies, H_2O solvent signals. In relatively sensitive ^{1}H homonuclear correlation experiments they can save experimental time by affording coherence selection and quadrature detection in the F_1 dimension in a single acquisition per t_1 increment.

Willker *et al.*[29] have reported several new gradient-selected heteronuclear correlation experiments. Coherence pathway selection in the presence of magnetic field gradients is described in terms of product operator formalism. A new approach to obtain pure absorption line shapes in gradient-selected spectroscopy is used to measure phase-sensitive gs-HMQC spectra, to carry out multiplicity editing in HSQC spectra, and to distinguish direct and long-range correlations in HMQC/HSQC-TOCSY spectra. In many PFG multidimensional heteronuclear correlation experiments some of the benefits are offset by the loss of $\sqrt{2}$ in signal to noise that results from the selection of a single or limited number of coherence pathways.

During 1993 several groups[30–32] have developed experiments using methods aimed at circumventing this problem, *i.e.* to combine full sensitivity with efficient solvent suppression. In one of these,[32] from Wüthrich's group, a scheme was developed for a phase-sensitive, proton-detected, heteronuclear COSY. Loss of sensitivity was avoided by rejecting unwanted coherences instead of selecting desired ones, a strategy originating from Bax's group,[33] which has also used it to develop a gradient version of the 4D ^{13}C–^{13}C-separated HMQC-NOESY-HMQC experiment.[34] The latter experiment is demonstrated for the protein interleukin-4 uniformly enriched (>95%) with ^{13}C and ^{15}N. A major benefit in this application is that as a result of reduced-phase cycling requirements it is possible to increase the number of increments, and therefore the resolution, in the indirectly detected dimensions without increasing the total measuring time. Chung *et al.*[35] noted similar benefits from the use of PFGs in a 3D ^{1}H-detected ^{13}C single-quantum (SQ)–double-quantum (DQ) coupling-correlated experiment used for assigning ^{1}H and ^{13}C resonances in ^{13}C-enriched oligosaccharides.

At the opposite end of the scale of molecular and NMR complexity, gradient-enhanced NMR has been used in a study[36] to determine the conformation of ethane-1,2-diol. Conformations to substituted ethanes, which are usually described as mixtures of staggered rotamers, can usually be determined from *vicinal* spin-coupling constants obtained by an analysis of their ^{1}H NMR spectra. For symmetrical compounds of the type XCH_2CH_2X this is not possible because the ^{1}H spectrum is a single line. Under such circumstances an analysis of the ^{13}C satellite lines, which form

[29] W. Willker, D. Leibfritz, R. Kerssebaum, and W. Bermel, *Magn. Reson. Chem.*, 1993, **31**, 287.
[30] A. Majumdar and E. R. P. Zuiderweg, *J. Magn. Reson., Series B*, 1993, **102**, 242.
[31] B. K. John, D. Plant, and R. E. Hurd, *J. Magn. Reson., Series A*, 1993, **101**, 113.
[32] G. Wider and K. Wüthrich, *J. Magn. Reson., Series B*, 1993, **102**, 239.
[33] A. Bax and S. Pochapsky, *J. Magn. Reson.*, 1992, **99**, 638.
[34] G. W. Vuister, G. M. Clore, A. M. Gronenborn, R. Powers, D. S. Garrett, R. Tschudin, and A. Bax, *J. Magn. Reson., Series B*, 1993, **101**, 210.
[35] J. Chung, J. R. Tolman, K. P. Howard, and J. H. Prestegard, *J. Magn. Reson., Series B*, 1993, **102**, 137.
[36] S. R. Salman, R. D. Farrant, P. N. Sanderson, and J. C. Lindon, *Magn. Reson. Chem.*, 1993, **31**, 585.

part of an AA′BB′X system, is required. A full analysis for the spin system requires observation of both the inner and outer satellite lines. No previous studies had managed to observe the inner lines because of interference from the large singlet from the ^{12}C-containing molecules. Suppression of this singlet in inverse detected 1D ^{1}H–^{13}C correlation experiments using phase-cycling methods has been inadequate. In the present study PFG methodology enabled these inner satellites to be observed.

Analysis of Biological Fluids.—High resolution NMR is now widely used for the analysis of metabolites in biofluids such as urine and blood plasma. The samples are complex mixtures in which the species of interest are usually of relatively low molecular weight. In the case of blood plasma the spectra also contain broad overlapping resonances from macromolecules such as proteins. A wide range of experimental and data processing techniques are being used in such analyses. Spraul *et al.*[37] have demonstrated the capabilities of high-performance liquid chromatography coupled to high-field proton NMR spectroscopy for the efficient separation, detection, and identification of urinary metabolites of ibuprofen. Nicholson's group[38,39] has reported the use of 600 and 750 MHz 2D *J*-resolved ^{1}H NMR spectroscopy in the analysis of human urine and plasma. In the case of plasma the broad signals from proteins are selectively attenuated as a consequence of their short T_2 relaxation times.

3 Solid-state NMR

1993 has seen the development of many new experiments and applications of other relatively new techniques, particularly in the field of high resolution solid-state NMR.

High Resolution Techniques and Applications.—There appears to have been an increasing number of publications in this area. In part this is due to the exploitation of relatively new methodologies such as those which enable better resolved spectra to be obtained from quadrupolar nuclei and those which facilitate the measurement of internuclear distances. There are also reports of interesting *J* coupling effects in some magic angle spinning (MAS) spectra as well as new experiments for purposes such as spectral editing.

Spin-$\frac{1}{2}$ Nuclei. Applications concerned with a few less-common nuclei will be briefly mentioned here. ^{89}Y is a low γ nucleus from which MAS spectra traditionally have been obtained using the single-pulse technique, which suffers from low sensitivity making it unsuitable for the study of air-sensitive compounds. However, Wu *et al.*[40] have demonstrated that when proton-containing ligands are present CP/MAS techniques allow many spectra to be obtained in minutes enabling less-stable materials to be studied and characterized. Dobson's group has found[41] the ^{119}Sn MAS spectra of the paramagnetic stannates $Nd_2Sn_2O_7$, $Sm_2Sn_2O_7$, and $Y_{1.8}Sm_{0.2}Sn_2O_7$ to be very temperature dependent with isotropic resonance positions proportional to T^{-1}. In the case of $Nd_2Sn_2O_7$ changing the temperature from 254 K to 350 K results in a chemical

37 M. Spraul, M. Hofmann, P. Dvortsak, J. K. Nicholson, and I. D. Wilson, *Anal. Chem.*, 1993, **65**, 327.
38 P. J. D. Foxall, J. A. Parkinson, I. H. Sadler, J. C. Lindon, and J. K. Nicholson, *J. Pharm. Biomed. Anal.*, 1993, **11**, 21.
39 P. J. D. Foxall, M. Spraul, R. D. Farrant, J. C. Lindon, G. H. Neild, and J. K. Nicholson, *J. Pharm. Biomed. Anal.*, 1993, **11**, 267.
40 J. Wu, T. J. Boyle, J. L. Shreeve, J. W. Ziller, and W. J. Evans, *Inorg. Chem.*, 1993, **32**, 1130.
41 C. P. Grey, A. K. Cheetham, and C. M. Dobson, *J. Magn. Reson., Series A*, 1993, **101**, 299.

shift change of ~1500 ppm. It is suggested that the latter compound should be used for calibrating high temperature probes.

Quadrupolar Nuclei. Wu and Wasylishen[42] report the first observation of ^{201}Hg NMR signals in the solid state. Both static and MAS spectra were reported for $K_2Hg(CN)_4$. Dec *et al.*[43] report high-field ^{47}Ti, ^{49}Ti, ^{67}Zn, ^{91}Zr, and ^{137}Ba MAS NMR spectra for a series of ferroelectric, piezoelectric, and related materials. They conclude that the techniques are not limited to the study of materials that have cubic symmetry, that useful structural information can be obtained from samples with lower symmetry, and that quadrupolar coupling constants and asymmetry parameters should be useful for obtaining structural information on the metal cation sites in electrical materials.

Hayashi and Hayamizu[44] have performed CP experiments under MAS conditions from ^{1}H to the $S = \frac{3}{2}$ quadrupolar nuclei ^{23}Na and ^{11}B for powder samples of $Na_2B_4O_7{\cdot}10H_2O$ and H_3BO_3. The line shapes obtained in the CP experiments were different from those obtained in analogous single-pulse experiments. Moreover, it was demonstrated that the line shapes in the CP/MAS NMR spectra are largely dependent on the spin-locking efficiency.

Skibsted *et al.*[45] have used ^{27}Al MAS NMR spectroscopy in the characterization of calcium aluminate phases in cements. For a series of synthetic samples ^{27}Al quadrupole coupling constants, asymmetry parameters of the electric field gradients tensors, and isotropic chemical shifts were determined from computer simulations of the complete spinning side-band pattern for the satellite transitions or the central transition. The quadrupole coupling parameters are shown to be an important supplement to the isotropic chemical shift in the characterization of structurally different A1 sites. The recently introduced double rotation (DOR) technique, which averages out line broadening due to second order quadrupolar interactions, has been used by Haddix *et al.*[46] in a ^{27}Al study of $Al_2(MoO_4)_3$. DOR enabled four inequivalent aluminium resonances to be resolved compared with the MAS spectrum, which consists of two poorly resolved peaks.

CP/MAS Experiments. Hagaman[47] has reported the use of a ^{13}C–^{19}F dipolar dephasing experiment for monofluorinated organic substances. ^{1}H dipolar decoupled CP/MAS NMR spectra of monofluorinated organic compounds exhibit dramatic homogeneous broadening of those carbons in close spatial proximity to the magnetic dipole of the ^{19}F nucleus. The ^{1}H–^{13}C–^{19}F triple resonance experiments described employ full ^{1}H–^{19}F double dipolar decoupling. ^{13}C–^{19}F dipolar dephasing is used to produce edited spectra in which resonances from carbon atoms close to the fluorine nucleus are eliminated or strongly attenuated. Wu *et al.*[48] describe a method for editing ^{13}C CPMAS spectra in which cross polarization is used in combination with polarization inversion (PI). The appearance of the spectrum in terms of the phase and amplitude of signals from different types of carbon atoms is dependent on the PI time.

[42] G. Wu and R. E. Wasylishen, *Magn. Reson. Chem.*, 1993, **31**, 537.
[43] S. F. Dec, M. F. Davis, G. E. Maciel, C. E. Bronnimann, J. J. Fitzgerald, and S. Han, *Inorg. Chem.*, 1993, **32**, 955.
[44] S. Hayashi and K. Hayamizu, *Chem. Phys. Lett.*, 1993, **203**, 319.
[45] J. Skibsted, E. Henderson, and H. J. Jakobsen, *Inorg. Chem.*, 1993, **32**, 1013.
[46] G. W. Haddix, M. Narayana, S. C. Tang, and Y. Wu, *J. Phys. Chem.*, 1993, **97**, 4624.
[47] E. W. Hagaman, *J. Magn. Reson., Series A*, 1993, **104**, 125.
[48] X. Wu and K. W. Zilm, *J. Magn. Reson., Series A*, 1993, **102**, 205.

In a subsequent publication[49] they demonstrate that by careful selection of CP and PI times it is possible to obtain a methylene-only subspectrum.

Zumbulyadis *et al.*[50] have used ^1H–^2H CP/MAS NMR to study polymer miscibility. The technique is demonstrated using homogeneous and phase-separated blends of perdeuteriated syntactic poly(methylmethacrylate) in poly(vinylphenol). Under the influence of cross-polarization only those deuterium nuclei close to protons are detected thus making the technique interface selective.

Cross polarization from abundant spin nuclei (I), usually ^1H, to dilute spins (S) is achieved *via* a pair of spin-lock pulses whose amplitudes are adjusted such that the I and S spins have equal precession frequencies about their respective RF fields. In rigid proton-rich solids under static or low-field MAS conditions the CP matching width is generally several kilohertz wide and as a result it is relatively straightforward to obtain the maximal S signal. However this profile breaks down when combined with MAS at a spinning frequency comparable to dipolar I–I and I–S interactions in the sample. Peersen *et al.*[51] have reported that variable-amplitude cross polarization (VACP) can be used to overcome the oscillatory behaviour of cross polarization at high speed and eliminate the burden of finding and maintaining an exact CP match.

Spin–Spin Coupling Constants. Eichele *et al.*[52] report the first observation of $^1J_{101,99\mathrm{Ru,P}}$ in the ^{31}P CP/MAS NMR spectra of some solid ruthenium–phosphorus complexes. Both ^{99}Ru and ^{101}Ru are quadrupolar nuclei with nuclear spins of $\frac{5}{2}$. As a result the dipolar interactions between the ruthenium and phosphorus nuclei are not completely eliminated by magic angle spinning. Therefore the multiplets observed in the ^{31}P spectra had to be analysed in terms of both J and the residual dipolar coupling.

Eichele, Wu, and Wasylishen[53] have observed some unusual 'AB' patterns in 81.03 MHz ^{31}P CP/MAS spectra of $Cd(NO_3)_2(PPh_3)_2$, which have been recorded at several different spinning rates and are shown in Figure 4. Although the two ^{31}P nuclei are crystallographically equivalent and have the same isotropic chemical shift in the solid state, they exhibit spinning rate-dependent MAS spectra, which have been analysed to obtain the value of $^2J_{PP}$. At high spinning rates the spectra are analogous to 'A_2' spectra in isotropic solutions, whilst at slower spinning rates the spectra resemble 'AB' solution spectra. The AB spectra are unusual in that $\delta_A = \delta_B$ and J_{AB} is given by the splitting between alternate peaks in the four-peak multiplet. The spinning-rate-dependent MAS lineshapes result from recoupling of the J interaction between the two crystallographically equivalent nuclei *via* anisotropic interactions, *i.e.* weak homonuclear dipolar coupling and differences in the orientation dependence of the chemical shift tensors. Such effects are predicted to be a more frequent observation at higher magnetic fields. In a subsequent publication Wu and Wasylichen[54] describe another example of a J-recoupling pattern and a theoretical treatment of the phenomenon.

Challoner *et al.*[55] report the observation of a septet resulting from J coupling to six fluorine atoms in the ^{31}P MAS NMR spectrum of the PF_6^- moiety in KPF_6. In solids

[49] X. Wu and K. W. Zilm, *J. Magn. Reson., Series A*, 1993, **104**, 119.
[50] N. Zumbulyadis, C. J. T. Landry, and T. E. Long, *Macromolecules*, 1993, **26**, 2647.
[51] O. B. Peersen, X. Wu, I. Kustanovich, and S. O. Smith, *J. Magn. Reson., Series A*, 1993, **104**, 334.
[52] K. Eichele, R. E. Wasylishen, J. F. Corrigan, S. Doherty, Y. Sun, and A. J. Carty, *Inorg. Chem.*, 1993, **32**, 121.
[53] K. Eichele, G. Wu, and R. E. Wasylishen, *J. Magn. Reson., Series A*, 1993, **101**, 157.
[54] G. Wu and R. E. Wasylishen, *J. Chem. Phys.*, 1993, **98**, 6138.
[55] R. Challoner, T. Schaller, and A. Sebald, *J. Magn. Reson., Series A*, 1993, **101**, 106.

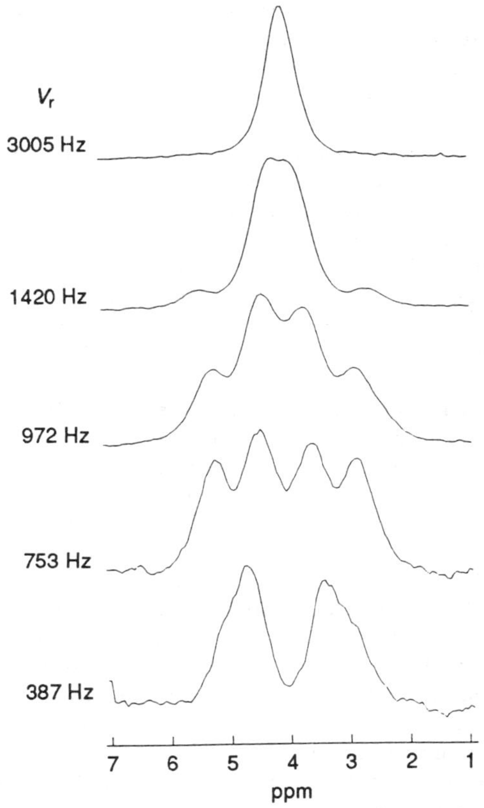

Figure 4 *Isotropic region of* ^{31}P *CP/MAS NMR spectra of* $Cd(NO_3)_2(PPh_3)_2$ *obtained at 81.03 MHz showing the spinning-rate-dependent splittings*
(Reproduced with permission from *J. Magn. Reson., Series A*, 1993, **101**, 157)

the observation of heteronuclear *J* coupling in MAS experiments is typically obscured by the influence of the homonuclear direct coupling interaction in systems containing nuclei that are both isotopically abundant and ubiquitous. Indeed the conventional MAS ^{19}F-coupled ^{31}P spectrum of KPF_6 is significantly broadened by such ^{19}F–^{19}F interactions. This paper reports the use of multiple pulse decoupling methodology to minimize such broadening, yielding a *J*-resolved spectrum in which the observed coupling is scaled by a factor, which depends on the pulse sequence used.

Spinning Sidebands. MAS is the principal method for obtaining high-resolution NMR spectra of dilute spins in solids. However, intense spinning side bands can be present in the resulting spectra. TOSS (total side-band suppression) sequences are therefore used frequently to achieve side-band cancellation for isotropic powder samples. However, TOSS also attenuates the centre band. Green and Bodenhausen[56] report a modified

[56] H. Green and G. Bodenhausen, *J. Am. Chem. Soc.*, 1993, **115**, 1579.

TOSS experiment which not only suppresses the side bands completely but also transfers the intensity from each side-band family into the corresponding centre band.

The dynamic angle spinning (DAS) NMR technique, which involves a 2D echo experiment, was developed by Pines to obtain high resolution spectra from half odd-integer quadrupolar nuclei. Spin evolution must occur at two different spinning angles of the sample rotor with respect to the magnetic field in order for the broadening from both the quadrupolar and chemical shift anisotropies to disappear. DAS can be achieved with a variety of complementary angles (θ_1,θ_2). However, the DAS spectra may display complex spinning side-band patterns. Pines' group[57] describes a theory of the side-band pattern, which is presented with illustrative experimental results over a range of DAS angles. In another publication[58] from this group a technique, dynamic-angle hopping (DAH), is described which produces DAS spectra without spinning side bands.

Multidimensional NMR.—Eck and Veeman[59] have reported a new 2D heteronuclear correlation pulse sequence for solids under MAS conditions which enables the correlation of spectra of two different nuclei through their weak heteronuclear dipolar coupling. The sequence provides an alternative to other 2D-correlation techniques for establishing through-space correlations. ^{27}Al–^{31}P correlation is demonstrated for the hydrated form of aluminophosphate molecular sieve VPI-5.

Frydman *et al.*[60] have reported a very elegant 3D NMR experiment for the characterization of ultraslow molecular motions in complex solid systems. The method is illustrated with an application to the analysis of motions in isotactic polypropylene (iPP). In a recent publication[61] the same group described a 2D variable-angle correlation spectroscopy (VACSY) technique which yields isotropic–anisotropic correlation spectra using a series of variable-angle spinning experiments. In the present paper this approach is incorporated into exchange experiments to produce 3D VACSY exchange NMR spectra. In these a spectral distribution $I(\omega_1^a,\omega_2^a,\omega^i)$ is obtained in which chemical shift anisotropy (CSA) frequencies ω^a at times t_1 and t_2 are correlated as well as separated along a third dimension according to the isotropic chemical shifts ω^i of individual sites. Results from such an experiment are illustrated in Figure 5, which shows a plot of the 1D spectrum obtained after projecting the 3D set on to the ω^i axis, together with 2D exchange distributions corresponding to anisotropic slices extracted from the 3D spectrum at the isotropic shifts of individual iPP resonances. Off-diagonal peaks and broadened diagonal ridges reveal the presence of molecular motions for all three sites of iPP.

Spiess' group recently reported[62] an experiment for the separation of overlapping side-band resonances in powder NMR spectra into two frequency dimensions. The group has extended the work[63] to a 3D MAS method which facilitates the extraction of

57 P. J. Grandinetti, Y. K. Lee, J. H. Baltisberger, B. Q. Sun, and A. Pines, *J. Magn. Reson., Series A*, 1993, **102**, 195.
58 S. L. Gann, J. H. Baltisberger, and A. Pines, *Chem. Phys. Lett.*, 1993, **210**, 405.
59 E. R. H. van Eck and W. S. Veeman, *J. Am. Chem. Soc.*, 1993, **115**, 1168.
60 L. Frydman, Y. K. Lee, L. Emsley, G. C. Chingas, and A. Pines, *J. Am. Chem. Soc.*, 1993, **115**, 4825.
61 L. Frydman, G. C. Chingas, Y. K. Lee, P. J. Grandinetti, M. A. Eastman, G. A. Barrall, and A. Pines, *J. Phys. Chem.*, 1992, **97**, 4800.
62 S. Féaux de Lacroix, J. J. Titman, A. Hagemeyer, and H. W. Spiess, *J. Magn. Reson.*, 1992, **97**, 435.
63 J. J. Titman, S. Féaux de Lacroix, and H. W. Spiess, *J. Chem. Phys.*, 1993, **98**, 3816.

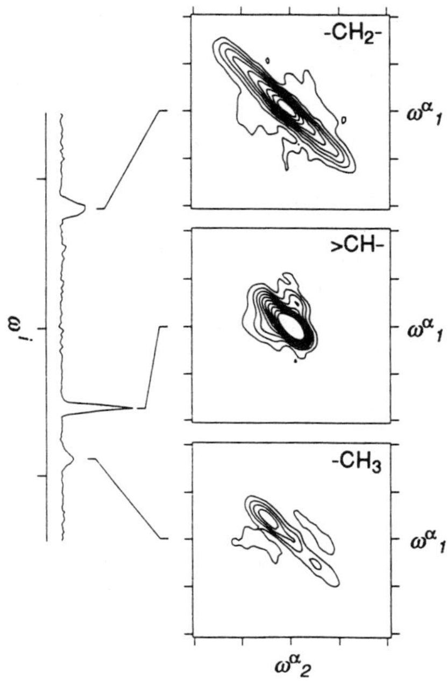

Figure 5 *Results of the high resolution 3D VACSY exchange NMR experiment applied to isotactic polypropylene.*
(Reproduced with permission from *J. Am. Chem. Soc.*, 1993, **115**, 4825)

orientational distribution functions from partially ordered solid materials. The experiment allows the separation of all the 2D side-band patterns arising from carbon sites whose isotropic shifts are resolved. This enables the degree of order for each of the carbon sites to be determined. The experiment is demonstrated for a liquid-crystalline side-group polymer, PA6.

Kumar and Opella[64] report a 3D separated-local-field/dilute-spin-exchange solid state NMR experiment. Homonuclear spin-exchange cross peaks provide correlation in the ω_2/ω_3 plane. Resonances are resolved along the ω_1 axis by their heteronuclear dipolar couplings. The experiment is demonstrated experimentally with the ^1H–^{15}N heteronuclear dipole–dipole interactions and ^{15}N–^{15}N homonuclear dipole–dipole interactions in a single-crystal sample of uniformly ^{15}N-labelled glycylglycine at an arbitrary orientation. Glycylglycine has two magnetically inequivalent molecules per unit cell giving rise to four nitrogen resonances. However, the resonances from the two amino acid groups in the unit cell overlap at this orientation of the crystal. In the 3D experiment the resonances are nevertheless resolved by their heteronuclear dipole–dipole couplings. This type of experiment is expected to be particularly important for studying proteins in the solid state.

[64] B. S. A. Kumar and S. J. Opella, *J. Magn. Reson., Series A*, 1993, **101**, 333.

Measurement of Internuclear Distances. In 1989 Gullion and Schaefer[65] described the rotational-echo, double-resonance (REDOR) technique. In this experiment transverse, *S*-spin magnetization is dephased by rotor synchronized *I*-spin π-pulses, where *I* and *S* refer to different types of dipolar-coupled rare spins. When interactions with all other spins can be ignored or suppressed, the comparison of *S*-spin echo intensities with and without *I*-spin dephasing pulses leads directly to the strength of the *I–S* dipolar coupling and hence *I–S* distances.

Structural studies of ligand–receptor complexes can contribute to the development of new drugs. Distances measured by the REDOR technique may be used in the determination of bound-ligand conformations without the limitations imposed by crystallization, molecular weight, or solubility. With this type of application in mind Garbow and McWherter[66] report their work aimed at defining the methods and requirements for constructing structural models from REDOR data. As a model system they used the tripeptide melanostatin (Pro-Leu-Gly-NH_2). REDOR experiments were performed on a series of selectively $^{13}C,^{15}N$-labelled melanostatins allowing the distances between the labelled carbon and nitrogen atom in each case to be determined. A comparison of these with those from the *X*-ray structure showed excellent agreement. The REDOR distances provided effective constraints on the values of the backbone dihedral angles of melanostatin, therefore providing a direct determination of the solid-state conformation.

Christensen and Schaefer[67] have used ^{31}P–^{13}C REDOR to obtain intra- and intermolecular ^{31}P–^{13}C distances for shikimate 3-phosphate (S3P) and [1-^{13}C] glyphosate bound to EPSP synthase, a 46 kDa enzyme essential for the synthesis of aromatic amino acids in plants and microorganisms. Because of the inherent accuracy of the REDOR experiment, just a few distances provide important constraints in determining the geometry of the S3P–glyphosate complex in the binding site. In another publication Schaefer's group[68] reports the use of ^{13}C–^{2}H REDOR NMR to study local packing in glassy polycarbonate. A homogeneous mixture of 5% [*carbonyl*-^{13}C]polycarbonate and 95% [*methyl*-d_6]polycarbonate was used for this initial work.

Study of Exchange Processes. The use of high resolution NMR as a technique for studying exchange processes in solid samples seems to be increasing. Riddell *et al.*[69] have used ^{13}C CP/MAS techniques to study the rates of rotation of the t-butyl groups in *cis*-4-t-butyl-1-(t-butylsulfonyl)cyclohexane. Line-shape analysis and T_1 measurements were used to obtain rates of between 6×10^3 and $2 \times 10^7 \, s^{-1}$ over the temperature range 202–320 K. Kümmerlen and Sebald[70] have studied the rate of Me_3Sn reorientation for a series of six crystalline trimethyltin compounds. A combination of line-shape analysis of variable temperature 1D ^{13}C CP/MAS spectra and 2D ^{13}C EXSY spectra yields activation energies in the range 38.6–72.5 kJ mol^{-1} for the series of compounds. The 2D experiments are used to obtain exchange rates

[65] T. Gullion and J. Schaefer, *J. Magn. Reson.*, 1989, **81**, 196.
[66] J. R. Garbow and C. A. McWherter, *J. Am. Chem. Soc.*, 1993, **115**, 238.
[67] A. M. Christensen and J. Schaefer, *Biochemistry*, 1993, **32**, 2868.
[68] A. Schmidt, T. Kowalewski, and J. Schaefer, *Macromolecules*, 1993, **26**, 1729.
[69] F. G. Riddell, S. Arumugan, K. D. M. Harris, M. Rogerson, and J. H. Strange, *J. Am. Chem. Soc.*, 1993, **115**, 1881.
[70] J. Kümmerlen and A. Sebald, *J. Am. Chem. Soc.*, 1993, **115**, 1134.

where the exchange-broadened regime is not accessible by temperature variation and also to extend the range of available jump rates for those compounds where a variable temperature 1D study *is* feasible. Comparison of the NMR results with information from *X*-ray diffraction measurements led the authors to postulate a $2\pi/3$ propeller-like jump around Sn–O or Sn–N bonds as the reorientation mechanism.

Aliev and Harris[71] have studied the conformational properties of monosubstituted cyclohexanes, $C_6H_{11}X$, in their thiourea inclusion compounds, again using both variable-temperature 1D and 2D ^{13}C exchange NMR. The variable-temperature MAS spectra demonstrate that a 'chair–chair' ring inversion process occurs in the thiourea tunnel, in which the axial and equatorial conformers are interconverted. Guest molecules with X = Cl, Br, and I have a predominance of the axial conformer whereas those with X = CH_3, NH_2, and OH have a predominance of the equatorial conformer. 2D ^{13}C CP/EXSY spectroscopy also provides direct evidence for the exchange process and enables minor conformers with relative populations as low as 0.03 to be identified.

Zheng *et al.*[72] report the use of CP/MAS techniques combined with deuterium spin dilution to obtain high resolution solid-state ^1H NMR spectra. Deuteriation results in a dramatic increase in the proton T_1 relaxation time. However, by reverse cross polarization (RCP) from deuterions to protons well-resolved 1D and 2D spectra can be obtained in a reasonable amount of time by taking advantage of the relatively short T_1 of deuterium. 2D-RCP NOESY exchange spectroscopy was used to study proton exchange in perdeuteriated oxalic acid dihydrate, aspartic acid, and ammonium hydrogen oxalate hemihydrate.

McGrath and Weiss[73] report the use of ^2H MAS NMR to study the rate of chair-to-chair interconversion of cyclohexane-d_{12} in its neat plastic crystalline phase. Well resolved resonances are obtained in this way at 193 K from the axial and equatorial deuterions. A rate constant of 51 s^{-1} was calculated at 222 K, and the NMR coalescence temperature at 7.04 T.

[71] A. E. Aliev and K. D. M. Harris, *J. Am. Chem. Soc.*, 1993, **115**, 6369.
[72] L. Zheng, K. W. Fishbein, R. G. Griffin, and J. Herzfeld, *J. Am. Chem. Soc.*, 1993, **115**, 6254.
[73] K. J. McGrath and R. G. Weiss, *J. Phys. Chem.*, 1993, **97**, 2497.

2 Physical Methods and Techniques
Part (ii) Mass Spectrometry

By C. P. BALL and P. J. DERRICK

Department of Chemistry, University of Warwick, Coventry, CV4 7AL, UK

1 Introduction

Mass spectrometry (MS) is currently enjoying a period of rapid expansion, due in large part to the developments in methods of creating gaseous ions from complex molecules and molecular assemblies. Applications of mass spectrometry in structural biochemistry are attracting a great deal of attention, in particular those made possible by the advances with electrospray and Fourier transform ion cyclotron resonance (FT-ICR). The possibilities of studying non-covalent interactions between proteins and substrates are widely discussed, and there have been examples of such interactions apparently surviving intact the transition from solution to vacuum (*vide infra*). Matrix-assisted laser desorption/ionization (MALDI) is of great current interest, and there is again an exciting possibility that through MALDI mass spectrometry will come to make significant contributions in the field of polymer science. On the more physical side of the subject, the advances are in the sophistication of experiments to study the behaviour of gaseous ions. The complexity of the systems which can be studied increases steadily, and the level of detail provided by spectroscopic and ion–molecule experiments steadily rises. Meanwhile, gas-phase ion chemistry has truly come of age, and knowledge from mass spectrometry on organic mechanisms and metal–organic interactions does now have relevance far beyond the bounds of mass spectrometry.

2 Physical Organic Chemistry

Mass spectrometry remains a widely-used method for the elucidation of organic reaction mechanisms, particularly if short-lived intermediates are involved. Increasingly, isotopically labelled compounds are studied, not simply to 'tag' particular atoms in an ion but so as to make use of kinetic isotope effects to elucidate reaction mechanism.[1] Distonic ions and ion–molecule complexes have become more firmly established as important concepts in gas-phase ion chemistry over the last two years. Zeller *et al.*[2] have found that long-lived radical cations of simple organophosphate esters isomerize spontaneously to more stable distonic structures in the gas phase. On the basis of collision-induced dissociation experiments, Bowen and Wright[3] proposed that the

[1] P. C. Burgers, C. A. Kingsmill, G. A. McGibbon, and J. K. Terlouw, *Org. Mass. Spectrom.*, 1992, **27**, 398.
[2] L. Zeller, J. Farrell, Jr., H. I. Kenttamaa, and P. Vainiotalo, *J. Am. Chem. Soc.*, 1992, **114**, 1205.
[3] R. D. Bowen and A. D. Wright, *J. Chem. Soc., Chem. Commun.*, 1992, 96.

crucial step in alkyl radical loss from ionized pentenyl and hexenyl methyl esters involves a 1,2-hydrogen shift at the radical site of a distonic ion, followed by γ-cleavage of the resultant ionized enol ether. The loss of water from ionized ethyl formate produced[4] the rarely seen ion $\cdot CH_2CH_2CO^+$. The distonic ion $\cdot CH_2OSi^+$ was shown to be stable by Srinivas *et al.*,[5] the ions being generated by electron-impact ionization of $Si(OCH_3)_4$. Holmes and coworkers[6] have shown that intramolecular hydrogen bonds in β-diols and β-hydroxy ethers are not preserved in their molecular ions. The fragment ions of metastable formyl loss from ionized ethylene glycol and 2-methoxyethanol have been shown to be the conjugate acids of methanol and dimethyl ether respectively.[7] Mechanisms involving ion–neutral complexes were ruled out, and it was concluded that distonic ions were the intermediates. A new isomer of dimethyl propyl phosphate was synthesized by transfer of ionized ethylene to trimethylphosphate in the gas phase,[8] and it was proposed that this ion has separate radical and charge sites. The distonic radical cation $C_5H_5N^+–C^{\bullet}CH_2$ has been generated by reacting pyridine with the radical cations of cyclopropane, ethylene oxide, and ketene. This species transfers $(CH_2)^{+\bullet}$ to more nucleophilic, substituted pyridines.[9] The β-distonic ion formed by the reaction between pyridine radical cation and ethene behaves differently. It can lose C_2H_4 by collisional activation, or can ionize again, giving a doubly-charged ion. This β-distonic ion requires collisional stabilization to be visible.[10]

The *ortho* effects in collision-induced dissociation of closed-shell aromatic ions has been examined using both quadrupole ion traps and triple quadrupole mass spectrometers.[11] The *meta* and *para* isomers were found to have qualitatively similar dissociation pathways, whereas the *ortho* isomers dissociated differently. Several isomers of $C_2H_3O^+$ and $C_2H_2S^+$ ions were distinguished, using ion–molecule reactions with isoprene and methylanisoles, to show that the ions have different reactivities depending on the precursor molecule from which they are generated.[12] Epimers of methylcyclohexanol were distinguished on the basis of their relative gas-phase acidities, which were obtained using a kinetic method based on the relative rates of competitive fragmentation of cluster ions.[13] McCarley and Brodbelt have studied the behaviour of 1,4-benzodiazepins in ion–molecule reactions using a quadrupole ion trap. The reactions with dimethyl ether ions give selectively formed adducts which can be used to diagnose the benzodiazepin structure.[14]

Huang and Dunbar[15] studied the breakdown pattern of n-heptane ions by photodissociation, and discovered that the *m/z* 71 and 43 ion abundances were strongly dependent upon the internal energy of the parent ion. Photodissociation and photoionization techniques have been employed very successfully to elucidate not only

[4] C. E. Hudson and D. J. McAdoo, *Org. Mass. Spectrom.*, 1992, **27**, 1384.
[5] R. Srinivas, D. K. Bohme, J. Hrusak, D. Schroeder, and H. Schwarz, *J. Am. Chem. Soc.*, 1992, **114**, 1939.
[6] J. R. Cao, M. George, J. L. Holmes, M. Sirios, J. K. Terlouw, and P. C. Burgers, *J. Am. Chem. Soc.*, 1992, **114**, 2017.
[7] H. E. Audier, A. Milliet, D. Leblanc, and T. H. Morton, *J. Am. Chem. Soc.*, 1992, **114**, 2020.
[8] L. K. M. Kiminkinen, K. G. Stirk, and H. I. Kenttamaa, *J. Am. Chem. Soc.*, 1992, **114**, 2027.
[9] S. J. Yu and M. L. Gross, *Org. Mass. Spectrom.*, 1993, **4**, 117.
[10] S. J. Yu, C. L. Holliman, D. L. Rempel, and M. L. Gross, *J. Am. Chem. Soc.*, 1993, **115**, 9676.
[11] T. Donovan and J. Brodbelt, *Org. Mass. Spectrom.*, 1992, **27**, 9.
[12] M. N. Eberlin, T. K. Majumdar, and R. G. Cooks, *J. Am. Chem. Soc.*, 1992, **114**, 2884.
[13] T. K. Majumdar, F. Clairet, J. C. Tabet, and R. G. Cooks, *J. Am. Chem. Soc.*, 1992, **114**, 2897.
[14] T. D. McCarley and J. Brodbelt, *Anal. Chem.*, 1993, **65**, 2380.
[15] S. F. Huang and R. C. Dunbar, *Int. J. Mass Spectrom. Ion. Processes*, 1992, **112**, 101.

fragmentations of ions but also of neutrals such as alkenes and dienes.[16,17] Energy randomization in the benzene dimer ion has been studied by Boernson and Selzle.[18] Employing isotopically mixed benzene dimers, it was demonstrated that charge and energy can be totally shared. At higher energies, fragmentation can, however, occur without communication between the two dimer halves.

The kinetic energy releases of the photodissociation processes $C_6H_5SH + h\nu$ (193 nm) $\rightarrow C_6H_5 + SH$ and $C_6H_5SCH_3 + h\nu$ (193 nm) $\rightarrow C_6H_5 + SCH_3$ and $C_6H_5S + CH_3$ were measured by a time-of-flight mass spectrometric method; values for the dissociation energies of the C_6H_5–SH, C_6H_5–SCH_3, and C_6H_5S–CH_3 bonds at 0 K were obtained.[19] The borylium ions $(CH_3OBOCH_3)^+$ and $(CH_3BCH_3)^+$ studied by dual-cell FT-ICR showed an unusual bimolecular reaction, involving the facile abstraction of a water molecule from an organic ether.[20] Ruscic and Berkowitz have studied the products of the reactions of fluorine atoms with MeSH, by photoionization mass spectrometry. They found adiabatic thresholds at *m/z* 47, of 7.536 ± 0.003 eV and 9.262 ± 0.005 eV, identified as the adiabatic ionization energies of CH_2SH and MeS respectively.[21]

Gable, Phan, and Griffin have generated the iodiranium ion by gas-phase reaction of ethylene with positive ions generated from I_2.[22] The unimolecular chemistry of methyl formate cation radical has been studied by Heinrich *et al.*[23] The mobility of t-butyl ion within t-butyl substituted diphenylalkanes has been investigated by Kuck and Matthias.[24] Metastable protonated 4-$Bu^tC_6H_4CH_2Ph$ lost isobutane following the abstraction of a hydride by the t-butyl ion from the benzylic C–H bonds in a π complex. Protonated 4-$Bu^tC_6H_4(CH_2)_3Ph$ did not involve free t-butyl ions, but did involve an ion–neutral complex. McGibbon *et al.*[25] have determined stabilization energies of α- and β-silyl substituents of vinyl cations, by measuring the appearance energies for metastable peaks for loss of $I^\bullet$ from $Me_3SiCI{=}CH_2$ and (*E*)-$Me_3SiCH{=}CHI$.

Enantioselectivity in fast atom bombardment (FAB) MS was established by using a complex of a selected aldohexose derivative with an enantiomeric pair of naphthyl ether ammonium ions.[26] The selectivity was detected and evaluated from the relative intensities of the 1 : 1 adduct ions (formed with an internal standard of 12-crown-4). Similar results were seen using a crown ether host with an ammonium ion guest. It has been claimed that the absolute configurations of a sample molecule can be predicted by examination of the FAB mass spectra of the sample measured with reagents of *R* and *S* configuration.[27] The proton affinities of 21 amines have been calculated using data from collision experiments using a hybrid tandem mass spectrometer. These results can be used to estimate the proton affinities of amino acids, aside from those containing a

[16] S.E. Van Bramer and M.V. Johnston, *Org. Mass. Spectrom.*, 1992, **27**, 949.
[17] S.E. Van Bramer, P.L. Ross, and M.V. Johnston, *J. Am. Soc. Mass Spectrom.*, 1993, **4**, 65.
[18] K.O. Boernsen and H.L. Selzle, *Chem. Phys. Lett.*, 1992, **190**, 497.
[19] S. Nourbakhsh, H.M. Yin, C.L. Liao, and C.Y. Ng, *Chem. Phys. Lett.*, 1992, **190**, 469.
[20] T.D. Ranatunga and H.I. Kenttamaa, *J. Am. Chem. Soc.*, 1992, **114**, 8600.
[21] B. Ruscic and J. Berkowitz, *J. Chem. Phys.*, 1992, **97**, 1818.
[22] K.P. Gable, T.N. Phan, and D. Griffin, *Org. Mass. Spectrom.*, 1993, **28**, 397.
[23] N. Heinrich, T. Drewello, P.C. Burgers, J.C. Morrow, J. Schmidt, W. Kulik, J.K. Terlouw, and H. Schwarz, *J. Am. Chem. Soc.*, 1992, **114**, 3776.
[24] D. Kuck and C. Matthias, *J. Am. Chem. Soc.*, 1992, **114**, 1901.
[25] G.A. McGibbon, M.A. Brook, and J.K. Terlouw, *J. Chem. Soc., Chem. Commun.*, 1992, 360.
[26] M. Sawada, M. Shizuma, Y. Takai, H. Yamada, T. Kaneda, and T. Hanafusa, *J. Am. Chem. Soc.*, 1992, **114**, 4405.
[27] H. Yang and Y. Chen, *Org. Mass. Spectrom.*, 1992, **27**, 736.

functional group capable of stabilizing a protonated species by hydrogen bonding.[28]

The question of whether the McLafferty rearrangement of ketones is concerted or stepwise has been addressed by Bowie, Derrick, and coworkers,[29] using intramolecular ^{13}C and ^{2}H effects. They conclude that the isotope effects observed are consistent with a stepwise mechanism with more than one rate-determining step. Ionized 4-methoxyheptane exhibits just two significant fragmentations, which consist of hydrogen transfer steps with interesting site selectivity.[30] Ketene has been discovered to dimerize in the gas phase when ionized by low-energy charge exchange.[31]

Morton has addressed the reorientation criterion with regards to positive ion–neutral complexes.[32] Ion–neutral complexes have been convincingly invoked in quite a number of mechanistic studies.[33,34] Morton postulates two categories of gaseous ion–neutral complexes: potential energy barriers prevent the separation and collapse of type 1 complexes, but only an entropic barrier prevents the collapse of a type 2 complex. Isotopic labelling has shown that the dissociation of the metastable ion $AcCH_2CO^+$ to produce the Ac^+ ion is preceded by an atom exchange involving an ion–neutral complex.[35] An ion–neutral complex intermediate has been postulated for the loss of water from fatty acid carboxylates by Jensen, Haas, and Gross.[36] Fatty acid carboxylates formed by FAB have been found to undergo both metastable ion and collisionally activated dissociation (CAD) to eliminate water. Upon CAD, the water elimination competes with charge-remote fragmentation, and deuterium labelling suggested that the elimination proceeds *via* an ion–neutral complex. Mass-analysed ion kinetic energy spectroscopy (MIKES) has been used to examine the unimolecular reactivity of the protonated forms of EtCN and EtNC in the gas phase. Both species were found to isomerize into weak π complexes between ethene and $CH{=}NH^+$, which may then dissociate or interconvert. The 1,2 elimination processes proposed may require the intermediacy of loosely bound ion–neutral complexes.[37] $[M - H]^-$ ions prepared from 17β-oestradiol-17 fatty acid esters by negative-ion chemical ionization have exhibited interesting low collisional energy processes.[38] Both phenoxide and enolate forms of the $[M - H]^-$ were seen. Evidence was produced for molecular isomerizations into ion–dipole complexes.

The mechanism of [H,C,N] loss from metastable molecular ions of benzene cyanides containing F, Cl, CF_3, CH_3, OH, or CH_3O substituents has been investigated by deuterium and ^{13}C labelling.[39] The eliminations from the F, Cl, or CF_3-substituted species was found to involve a randomization of the hydrogen atoms and a minor rearrangement of the CH_2CN function, causing expulsion of a benzylic carbon 20–30% of the time. When the substituents are CH_3, OH, or CH_3O, there appears to

[28] X. Li and A. G. Harrison, *Org. Mass. Spectrom.*, 1993, **28**, 366.
[29] M. B. Stringer, D. J. Underwood, J. H. Bowie, C. E. Allison, K. F. Donchi, and P. J. Derrick, *Org. Mass. Spectrom.*, 1992, **27**, 270.
[30] R. D. Bowen and P. J. Derrick, *J. Chem. Soc., Perkin Trans. 2*, 1992, 1041.
[31] C. H. Dass, *Rapid Commun. Mass Spectrom.*, 1993, **7**, 95.
[32] T. H. Morton, *Org. Mass. Spectrom.*, 1992, **27**, 353.
[33] V. Nguyen, X. Cheng, and T. H. Morton, *J. Am. Chem. Soc.*, 1992, **114**, 7127.
[34] M. Gu and F. Turecek, *J. Am. Chem. Soc.*, 1992, **114**, 7146.
[35] J. Tortajorda, D. Berthomieu, J. P. Morizur, and H. E. Audier, *J. Am. Chem. Soc.*, 1992, **114**, 10 874.
[36] N. J. Jensen, G. W. Haas, and M. L. Gross, *Org. Mass. Spectrom.*, 1992, **27**, 423.
[37] G. Bouchoux, M. T. Nguyen, and P. Longvialle, *J. Am. Chem. Soc.*, 1992, **114**, 10 000.
[38] L. Debrauwer, A. Paris, D. Rao, F. Fournier, and J. C. Tabet, *Org. Mass. Spectrom.*, 1992, **27**, 709.
[39] T. A. Molenaar-Langeveld, R. H. Fokkens, S. Ingemann, and N. M. M. Nibbering, *Org. Mass. Spectrom.*, 1992, **27**, 390.

be a rearrangement of the *meta* and *para* isomers to the *ortho* isomers. Tkaczyk and Harrison have developed a method for the identification of esters by collisional charge inversion of their enolate ions.[40] The enolate ions of *m/z* 115 derived from the four butyl acetates have been distinguished by their collisional charge inversion and neutralization–reionization mass spectra. The reactions of the ion $HC_2{=}CHCH{=}O^+CH_3$, generated by radical loss from allylic alkenyl methyl ethers, have been probed using ^2H and ^{13}C labelling experiments.[41]

3 Fullerenes and Clusters

Mass spectrometry has always had a connection with the field of fullerenes, and these compounds have continued to generate a huge amount of interest.[42] Gaseous C_{60} was studied by photoionization mass spectrometry between the ionization threshold and 40.8 eV.[43] Fragmentation to form C_{58}^+ was only observed at the highest energy, and C_{60}^{2+} also became more prominent with increasing energy. An adiabatic threshold of 7.57 ± 0.01 eV was observed for C_{60}^+. An Fe adduct with C_{60} was obtained by carrying out contact-arc vaporization of graphite in a partial atmosphere of $[Fe(CO)_6]$.[44] The properties of the molecule differ from those of a molecule prepared with Fe outside the cage by solution methods. Wu, Christian, and Anderson examined the effects of colliding a beam of Ne^+ with C_{60} vapour.[45] C_{60} has been found to add to a large variety of neutral and charged nucleophiles,[46] and to exhibit facile reactivity with CH_4 in both positive and negative ionization. The reaction of fullerenes with fluorine gives a variety of products which may well have the fluorine externally attached to the skeleton.[47]

C_{60} and C_{70} in 3-nitrobenzyl alcohol and 2-nitrophenyl octyl ether have been investigated by both positive and negative-ion FAB.[48] Many protonated and hydroxylated ions were observed. Collision-induced dissociation of the fullerene-adduct ions indicated that the oxygen and hydroxyl were connected to the outer surface of the fullerene. A four-sector tandem mass spectrometer was used to prove that, as a result of collisions, small target gases such as helium and deuterium are incorporated into intact fullerene radical cations.[49] The internal energy of these complexes was varied by changing the centre-of-mass collision energy, which is converted into the internal energy of the complex when the target gas is captured. The formation of endohedral $(C_{60}He)^+$ complexes in mass spectrometric experiments has been modelled with molecular dynamic simulations.[50,51] These simulations predicted optimal complex formation at a C_{60}^+ incident energy of 8000 eV. The results were consistent

40 M. Tkaczyk and A. G. Harrison, *Org. Mass. Spectrom.*, 1992, **27**, 585.
41 R. D. Bowen, A. W. Colburn, and P. J. Derrick, *Org. Mass. Spectrom.*, 1992, **27**, 625.
42 S. W. McElvany and M. M. Ross, *J. Am. Soc. Mass Spectrom.*, 1992, **3**, 268.
43 R. K. Yoo, B. Ruscic, and J. Berkowitz, *J. Chem. Phys.*, 1992, **96**, 911.
44 T. Pradeep, G. U. Kulkarni, K. R. Kannan, T. N. Row, and C. N. R. Rao, *J. Am. Chem. Soc.*, 1992, **114**, 2272.
45 Z. Wu, J. F. Christian, and S. L. Anderson, *J. Chem. Phys.*, 1992, **96**, 3344.
46 F. Wudl, A. Hirsch, K. C. Khemani, T. Suzuki, P. M. Allemand, A. Koch, H. Eckert, G. Srdanov, and H. M. Webb, in 'Fullerenes', *ACS Symposium Series no. 481*, ACS, Washington DC, 1992, p. 161.
47 K. Kniaz, J. E. Fischer, H. Selig, G. B. Vaughan, W. J. Romanow, D. M. Cox, S. K. Chowdhury, J. P. McCauley, R. M. Strongin, and A. B. Smith, III, *J. Am. Chem. Soc.*, 1993, **115**, 6060.
48 J. M. Miller and L. Z. Chen, *Rapid Commun. Mass Spectrom.*, 1993, **6**, 184.
49 K. A. Caldwell, D. E. Giblin, and M. L. Gross, *J. Am. Chem. Soc.*, 1992, **114**, 3743.
50 S. W. McElvaney, M. M. Ross, and J. H. Callahan, *Acc. Chem. Res.*, 1992, **96**, 3344.
51 R. C. Mowrey, M. M. Ross, and J. H. Callahan, *J. Phys. Chem.*, 1992, **96**, 4755.

with a collision involving momentum transfer between He and an isolated portion of C_{60}^{+}. Lykke and Wurtz have reported neutral products from photodissociated C_{60} for the first time.[52]

The electron impact cross section of C_{60} was determined by Baba and coworkers[53] using Knudsen cell mass spectrometric techniques to be $(5.35 \pm 5.6) \times 10^{-16}\,cm^2$ at 38 eV. Doubly-charged negative ions of fullerenes have been produced by chemical ionization using isobutane as the reagent gas. The efficiency of the attachment increases with the size of the fullerene. Under low-energy collision conditions an electron is ejected[54] from the doubly-charged C_{70}^{2-} ion to form $C_{70}^{-\bullet}$. Knudsen cell mass spectrometry of pure C_{60} was carried out at 600–800 K to determine[55] the vapour pressure and sublimation energies of C_{60}. Evidence of a 'radical sponge' effect of C_{60} has been presented by McEwen, McKay, and Larsen, when additions of H, CH_3, and C_2H_5 to C_{60} were observed using methane as the source under electron attachment conditions.[56] The results are explained by a mechanism in which radicals produced in the gas phase react with C_{60} at a surface.

Multiphoton ionization of fullerenes has shown prompt electrons resulting from direct two-photon ionization, as well as a low-energy delayed electron signal following a thermal energy distribution with average energy of $\varepsilon^- = 0.3 \pm 0.05$ eV. The results lead to an estimate[57] of the photoionization cross-section for C_{60} in the lowest triplet state of $(5.1 \pm 2.0) \times 10^{-17}\,cm^{-2}$. Delayed ionization of fullerene over a timescale of 16 μs was reported, in good agreement with Klots' model for thermionic emission.[58] FAB of fullerenes using helium, argon, or xenon beams gives characteristic patterns, with helium showing abundant triply-charged peaks.[59]

Soluble yttrium-containing metallofullerenes, YC_{82} and Y_2C_{82}, were obtained from yttrium–carbon soot, and detected using laser desorption time-of-flight (TOF) mass spectrometry, which showed a series of peaks Y_2C_{2n} ($2n = 84, 90 \leq 2n \leq 140$).[60] The metallofullerenes YC_{60} and YC_{70} were not observed. The YC_{82} compound was shown by *X*-ray crystallography to have a Y–Y interacton with the Y–Y neighbour distance at 4.05 ± 0.05 Å,[61] which might indicate that Y is not endohedral. Gas-phase Y–fullerene adduct ions, $(Y_xC_n)^+$ ($x = 1,2$) with even numbered n of 60–100, were formed in a FT-ICR mass spectrometer by direct laser ionization of graphite and Y_2O_3 mixtures.[62] These compounds were found to be extremely stable towards fragmentation and oxidation, and hence it was proposed that they could be endohedral. Endohedral scandium fullerenes have been produced by arc burning scandium oxide/carbon graphite rods.[63] The fullerenes were confirmed as discandium fullerenes by MALDI,

52 K. R. Lykke and P. Wurtz, *J. Phys. Chem.*, 1992, **96**, 3191.
53 M.Sai. Baba, T. S. L. Narasimhan, R. Balasubramanian, and C. K. Mathews, *Int. J. Mass Spectrom. Ion. Processes*, 1992, **114**, R1.
54 A. Mandelbaum and A. Etinger, *Org. Mass. Spectrom.*, 1993, **28**, 487.
55 C. K. Mathews, M. S. Baba, T. S. L. Narasimhan, R. Balasubramanian, N. Sivaraman, T. G. Srinivasan, and P. R. V. Rao, *J. Chem. Phys.*, 1992, **96**, 3566.
56 C. N. McEwen, R. G. McKay, and B. S. Larsen, *J. Am. Chem. Soc.*, 1992, **114**, 4412.
57 D. Ding, R. N. Compton, R. E. Haufler, and C. E. Klots, *J. Phys. Chem.*, 1993, **97**, 2500.
58 E. E. B. Campbell, G. Ulmer, and I. V. Hertel, *Z. Phys. D: At. Mol. Clusters*, 1992, **24**, 81.
59 M. Takayama and H. Shinohara, *Int. J. Mass Spectrom. Ion Processes*, 1993, **123**, R7.
60 H. Shinohara, H. Sato, Y. Saito, M. Ohkohchi, and Y. Ando, *J. Phys. Chem.*, 1992, **96**, 3571.
61 L. Sonderholm, P. Wurz, K. R. Lykke, D. H. Parker, and F. W. Lytle, *J. Phys. Chem.*, 1992, **96**, 7153.
62 S. W. McElvany, *J. Phys. Chem.*, 1992, **96**, 4935.
63 H. Shinohara, H. Yamaguchi, N. Hayashi, H. Sato, M. Obkohchi, Y. Ando, and Y. Saito, *J. Phys. Chem.*, 1993, **97**, 4259.

and have the formulae $Sc_2@C_{74}$, $Sc_2@C_{82}$, and $Sc_2@C_{84}$.

Propane(1 +)-propane $[(C_3H_7^+)(C_3H_8)_n]$ cluster ions induced by isomerization of the propyl ion have been observed by Foltin *et al.*[64] to decay by the evaporation of up to four propane monomers, as well as the normal single-monomer evaporation. This additional fragmentation is described by a new decay mechanism.

Dissociative ionization of clusters has been studied using photoionization by vacuum UV laser, dispersed light from synchrotron sources, or discharge light sources.[65]

4 Organometallics

Mass spectrometry is being used increasingly to study reactions of metal ions with organic molecules. A review has been written by Teesch and Adams.[66] A review has appeared comparing the primary processes undergone by transition metal complexes in photochemical excitation in solution, and FAB in a matrix.[67]

Schroeder *et al.* have studied the gas-phase oxidation of methane by FeO^+. There was evidence[68] of four distinguishable isomers of $[Fe,C,H_4,O]^+$. Fisher and Armentrout[69] investigated the activation of alkanes by Cr^+, employing ion-beam mass spectrometry to study the reactions of ground-state Cr^+ (^6S) with propane, butane, methylpropane, dimethylpropane, and selectively deuteriated propane and methylpropane. Ground state Cr^+ ions were found to undergo no bimolecular reactions at thermal energies, but did react at elevated kinetic energies. The structure and mechanisms of formation of $(C_6H_6Fe)^+$ produced by electron impact on η-$C_5H_5Fe(CO)_2CH_3$ were examined by Uggerud and coworkers[70] using collisional activation. The ions consist of a 1:1 mixture of benzene–iron and fulvene–iron structures, and two different fragmentation pathways were observed. Fe^+ was shown to mediate in [4 + 2] additions of buta-1,3-diene with ethyne and propyne. The proposed mechanisms involve formation of $[Fe(benzene)]^+$ for the ethyne reaction, and $[Fe(toluene)]^+$ for the reaction with propyne. $[FeCp]^+$ was generated at low pressure and its sequential reactions with ethylene oxide and related heterocycles monitored.[71]

Molecular mass determination of ferrocenyl–aryl–plumbanes has been carried out by californium-252 plasma desorption, combined with TOF analysis.[72] Positive and negative plasma desorption mass spectra of methylbenzenethiol cadmium salts, which are three-dimensional non-molecular materials, were found to contain fragment ions directly related to the compound lattice structure.[73]

A reverse-geometry tandem mass spectrometer was used to analyse the unimolecular and collision-induced dissociation of doubly-charged porphyrins containing nickel and vanadyl ions. The major reactions were classified and illustrated with appropriate

[64] M. Foltin, V. Grill, T. Rauth, Z. Herman, and T. D. Maerk, *Phys. Rev. Lett.*, 1992, **68**, 2019–22.
[65] J. A. Booze and T. Baer, *J. Chem. Phys.*, 1992, **96**, 5541.
[66] L. M. Teesch and J. Adams, *Org. Mass. Spectrom.*, 1992, **27**, 931.
[67] T. J. Kemp, *Coord. Chem. Rev.*, 1993, **125**, 333.
[68] D. Schroeder, A. Fieldler, J. Hrusak, and H. Schwarz, *J. Am. Chem. Soc.*, 1992, **114**, 1215.
[69] E. R. Fisher and P. B. Armentrout, *J. Am. Chem. Soc.*, 1992, **114**, 2039.
[70] D. Ekeberg, G. Hvistendahl, Y. Stenstroem, and E. Uggerud, *Acta Chem. Scand.*, 1992, **46**, 92.
[71] R. Bahktiar, J. J. Drader, and D. B. Jacobson, *J. Am. Chem. Soc.*, 1992, **114**, 8304.
[72] G. Allmaier, O. Roepstroff, and K. E. Schwarzhans, *Rapid Commun. Mass Spectrom.*, 1992, **6**, 58.
[73] A. G. Craig, D. Kinny, R. Garbutt, I. G. Dance, and P. J. Derrick, *Org. Mass. Spectrom.*, 1992, **27**, 1019.

spectra, and information from tandem quadrupole experiments was presented.[74]

Electrospray (ES) and four-sector tandem mass spectrometry has been used to study complexes of lanthanides and trifluoracetylcamphorates.[75] Electrospray mass spectra of mercury phosphine complexes has been used to examine the cation exchange that readily takes place with these species.[76]

5 Polymers

Developments in electrospray and, more particularly, MALDI are revolutionizing the mass spectrometry of polymers. Mass discrimination has been reported, however, in the laser desorption–FT-ICR mass spectrometry cation-attachment spectra of polymers.[77] It is unclear whether mass discrimination occurs with laser desorption and FT-ICR of polymers in general. The interactions by collision-induced decomposition and ion–molecule reactions of poly(ethylene glycol) and ethoxylated alcohols have been studied by FT-ICR. Poly(ethylene glycol), when protonated, can transfer protons to both polymer and alcohol, and measurements of the proton affinity and entropy of protonation of the ethoxylated alcohols give evidence for a strong intramolecular hydrogen bond.[78]

Plasma desorption mass spectrometry has become less popular in recent years, but work is still being carried out using this technique on polymers. Coincidence-counting techniques were used with TOF mass spectrometry to study the impact of MeV/amu ions on surfaces.[79] Correlations were observed between proton emission and certain hydrocarbon-ion emissions from polymer samples. ^{252}Cf fragments were used to probe the surface of polystyrene/PVME blends by coincidence-counting TOF mass spectrometry.[80]

Nohmi and Fenn have shown that ions formed from electrosprayed poly(ethylene oxide) with molecular weights from 200 to 5 000 000 can be mass analysed.[81] The number of charges per unit length decreased with increasing molecular weight for molecular weight below 20 000. Above this figure, the number of charges per unit length increased. Above 5 000 000, the number exceeds that allowed by the electrostatic model. Two different mechanisms were proposed for formation of ions from charged droplets.

MALDI-TOF has been used to examine the absolute masses and mass distribution of poly(styrene sulfonic acid) and poly(acrylic acid).[82] MALDI was used successfully to characterize polyethylenes, PMMA, polystyrene, and poly(ethylene glycols), by the use of appropriate matrices and the addition of metal salts. Molecular masses up to 70 000 Da are observed, but no fragmentation.[83] It has also been discovered that Al^+, Cr^+, Fe^+, and Cu^+ all attach efficiently to polystyrene and poly(ethylene glycol) in

74 G. J. Van Berkel, S. A. McLuckey, and G. L. Glish, *J. Am. Soc. Mass Spectrom.*, 1992, **3**, 235.
75 J. M. Curtis, P. J. Derrick, A. Schnell, E. Constantin, R. T. Gallagher, and J. R. Chapman, *Org. Mass. Spectrom.*, 1992, **27**, 1176.
76 R. Colton and D. Dakternieks, *Inorg. Chim. Acta*, 1993, **208**, 173.
77 J. D. Hogan and D. A. Laude, Jr., *Anal. Chem.*, 1992, **64**, 763.
78 H. Y. Lin, A. Rockwood, M. S. B. Munson, and D. P. Ridge, *Anal. Chem.*, 1993, **65**, 2119.
79 M. A. Park, E. A. Schweikert, and E. F. Da Silveira, *J. Chem. Phys.*, 1992, **96**, 3206.
80 D. B. Cox, M. A. Park, G. R. Kaercher, and E. A. Schweikert, *Anal. Chem.*, 1992, **64**, 843.
81 T. Nohmi and J. B. Fenn, *J. Am. Chem. Soc.*, 1992, **114**, 3241.
82 P. O. Danis, D. E. Karr, F. Mayer, A. Holle, and C. H. Watson, *Org. Mass. Spectrom.*, 1992, **27**, 843.
83 U. Bahr, A. Deppe, M. Karas, and F. Hillenkamp, *Anal. Chem.*, 1992, **64**, 2866.

laser desorption TOF mass spectrometry. The metals can be added either by applying the salt to the matrix, or by desorption from a metal support.[84]

6 Bioorganic Compounds

Using ES mass spectrometry it has been possible to demonstrate the attachment of coenzyme A to a flavodoxin by a modified strain of *Escherichia coli*.[85] Ion-spray mass spectrometry has been used to characterize lipopeptides, with high sensitivity and simple sample preparation.[86] Molecular mass determination up to 6000 Da, with an accuracy of 1 Da, was routinely possible.

Using 266 nm laser desorption followed by 255 nm multiphoton ionization, a variety of structurally significant sequence fragment ions have been obtained for peptides.[87] The fragment ions were mostly obtained during the 266 nm laser desorption. A variety of immonium ions were also formed at higher ionizing power densities, which gave an overview of the amino acid residues present. Fragment ions observed in mass spectra produced during resonant 266 nm laser desorption may originate as neutral fragments formed from the intact molecule during laser desorption and subsequently post-ionized.[88]

MALDI-TOF and ES ionization were used to investigate the interactions of copper(II) ions with a 26-residue sequence of amino acids found on the surface of the human plasma metal-transport protein, histidine-rich glycoprotein.[89] The peptide $(GHHPH)_5G$ was reacted with copper and manganese, and was observed to bind one copper ion per repeat unit. Manganese showed no reaction.

New reagents are frequently proposed for the Edmann degradation, and one such has been heavily studied by mass spectrometric methods.[90] 3-[4′-(ethylene-*N*,*N*,*N*-trimethylamino)phenyl]-2-isothiocyanate was detected as amino acid derivatives with high sensitivity by ion-evaporation mass spectrometry. Sensitivity was subfemtomolar, and amino acids of all types of side chain were detected with equal facility. The polar nature of the reagent required covalent immobilization of polypeptides prior to automatic sequence analysis.

The veterinary drugs dimetridazole and ipronidazole and their metabolites were detected in turkey muscle by HPLC followed by thermospray tandem mass spectrometry[91] at the ppb level. Tandem mass spectrometry confirmed the identity of the drugs and their metabolites. Taxol®, cephalomannine, and baccatin III® have been detected by tandem mass spectrometry in crude plant extracts at less than 500 pg quantities.[92] No interferences were observed. Diclofenac®, an antiinflammatory drug, has been investigated by means of electron impact, chemical ionization, and FAB tandem mass spectrometry, along with hydroxylated metabolites.[93] GC-MS with

[84] C. F. Llenes and R. M. O'Malley, *Rapid Commun. Mass Spectrom.*, 1992, **6**, 564.
[85] R. N. F. Thorneley, C. Abell, G. A. Ashby, M. H. Drummond, R. R. Eady, S. Huff, C. J. Macdonald, and A. Schneier, *Biochemistry*, 1992, **31**, 1216.
[86] J. W. Metzger, W. Beck, and G. Jung, *Angew. Chem.*, 1992, **104**, 235.
[87] G. R. Kinsel, J. Lindner, and J. Grotemeyer, *J. Chem. Phys.*, 1992, **96**, 3157.
[88] G. R. Kinsel, J. Lindner, and J. Grotemeyer, *J. Chem. Phys.*, 1992, **96**, 3162.
[89] W. T. Hutchens, R. W. Nelson, M. H. Allen, C. M. Li, and T. T. Yip, *Biol. Mass Spectrom.*, 1992, **21**, 151.
[90] R. Aebersold, E. J. Bures, M. Namchuk, M. H. Goghari, B. Shushan, and T. C. Covey, *Protein Sci.*, 1992, **1**, 494.
[91] J. E. Matusik, M. G. Leadbetter, C. J. Barnes, and J. A. Sphon, *J. Agric. Food Chem.*, 1992, **40**, 439.
[92] S. H. Hoke, II, J. M. Wood, R. G. Cooks, X. H. Li, and C. J. Chang, *Anal. Chem.*, 1992, **64**, 2313.
[93] F. M. Rubino and L. Zecca, *Biol. Mass Spectrom.*, 1992, **21**, 109.

selected ion monitoring has been employed to detect cortisol and cortisone in human plasma. [$^2H_2$]cortisol and [$^2H_5$]cortisone were used as internal standards, and the determination was based on the $[M-31]$ fragment ions of the methoxine–trimethylsilyl derivatives.[94]

Macrolide antibiotics have been examined by capillary zone electrophoresis, combined with ES. Increasing the ES skimmer voltage gave structurally significant fragment ions. This method gave a better limit of detection than any other currently in use.[95] Tandem mass spectrometry has been used to demonstrate that halopridoe®, a neuroleptic drug, is metabolized into at least eight different compounds.[96]

The purification of saxitoxin, the parent compound in the family of toxins responsible for paralytic shellfish poisoning (PSP), has been monitored using flow-injection analysis, and capillary electrophoresis–mass spectrometry.[97] Tandem mass spectrometry was used to obtain structural information. Quilliam *et al.*[98] have analysed okadaic acid and dinophysistoxin-1, the principle toxins implicated in cases of diarrhetic shellfish poisoning, by a liquid chromatography/mass spectrometric method, which provided a mass detection limit of 0.4 ng for each toxin. The trace level analysis[99] of sulfonylurea herbicides has been reported using thermospray. Enantiomers of chiral octachloranes in tissue samples from aquatic vertebrates in Antarctica[100] were separated using chiral high resolution GC and detected by electron capture negative ionization mass spectrometry.

Taxol® and related diterpenoids have been analysed by microbore HPLC coupled with electrospray.[101] 100 pg of all taxanes were readily detected using secondary ion monitoring. Vitamin D_3 metabolites have been characterized by cycloaddition with 4-phenyl-1,2,4-triazoline-3,5-dione, followed by continuous-flow FAB-MS.[102] The spectra are dominated by peaks corresponding to a protonated molecule, and by side-chain fragments. The siderophores of the fire blight pathogen *Erwinia amylovoria*, a group of proferrioxamines, have been detected and characterized by LC-ES. In addition, six hydroxamate compounds were characterized and another five detected.[103] Tomatine, a glycoalkaloid, has been identified at the femtomole level by four-sector tandem mass spectrometry and scanning-array detection.[104] Capillary electrophoresis combined with electrospray was used to make the intriguing discovery that the venom of the black mamba contains at least 70 peptides and has a molecular mass range of 6000–9000 Da.[105]

Peptides and Proteins.—Mass spectrometry is now fairly well established as a reliable

[94] H. Shibasaki, I. Arai, T. Furuta, and Y. Kasuya, *J. Chromatogr.*, 1992, **576**, 47.
[95] C. E. Parker, J. R. Perkins, K. B. Tomer, Y. Shida, K. O'Hara, and M. Kono, *J. Am. Soc. Mass Spectrom.*, 1992, **3**, 263.
[96] J. Fang, J. W. Gorrod, M. Kajbaf, J. H. Lamb, and S. Naylor, *J. Mass Spectrom. Ion Processes*, 1992, **122**, 121.
[97] S. Pleasance, S. W. Ayer, M. V. Laycock, and P. Thibault, *Rapid Commun. Mass Spectrom.*, 1992, **6**, 14.
[98] M. A. Quilliam, S. Pleasance, and J. C. Marr, *Rapid Commun. Mass Spectrom.*, 1992, **6**, 121.
[99] M. L. Shalaby, F. Q. Bramble, Jr., and P. W. Lee, *J. Agric. Food Chem.*, 1992, **40**, 513.
[100] H. R. Buser, M. D. Mueller, and C. Rappe, *Environ. Sci. Technol.*, 1992, **26**, 1533.
[101] F. Bitsch, W. Ma, F. Macdonald, M. Nieder, and C. H. L. Shackleton, *J. Chromatogr. Biomedical Appl.*, 1993, **615**, 273.
[102] B. Yeung, P. Vouros, and G. Satanarayana Reddy, *J. Chromatogr.*, 1993, **645**, 115.
[103] G. J. Feistner, D. C. Stahl, and A. H. Gabrik, *Org. Mass. Spectrom.*, 1993, **28**, 163.
[104] S. Evans, R. Buchanan, A. Hoffman, F. A. Mellon, K. R. Price, S. Hall, F. C. Walls, A. L. Burlingame, S. Chen, and P. J. Derrick, *Org. Mass. Spectrom.*, 1993, **28**, 289.
[105] J. R. Perkins, C. E. Parker, and K. B. Tomer, *Electrophoresis (Weinheim)*, 1993, 14.

and sensitive method of sequencing peptides. The rapid development of ES and MALDI of proteins, and advances in techniques for interfacing mass spectrometers with separation methods such as capillary zone electrophoresis (CZE) has led to more and more mass spectrometry being undertaken in the protein field.

Antigens recognized by T-cells are expressed as peptides bound to major histocompatibility complex (MHC) molecules. Microcapillary HPLC-ES tandem mass spectrometry has been used to sequence subpicomolar amounts of peptides isolated from the MHC molecule HLA-A2.1.[106] Of 200 species quantitated, eight were sequenced, and four found in cellular proteins. All were nine residues long, and shared a distinct structural motif. The enzyme 6-pyruvoyl tetrahydropterin synthase, which occurs at very low levels in the metabolic disorder atypical phenylketonuria, was purified and characterized from salmon liver, digested, and sequenced by tandem mass spectrometry.[107]

Biemann and coworkers[108] showed, from the m/z ratio of ions of the tryptic peptide of myoglobin encompassing the disputed amino acid 122 (previously commonly held to be asparagine), derived from horse heart and horse skeletal myoglobins, the apomyoglobin of the latter, and the tryptic and chymotryptic peptide of sperm whale myoglobin, that amino acid 122 of the myoglobin tryptic peptide is aspartic acid rather than asparagine. This finding was confirmed by the collision-induced dissociation spectra of the $[M + H]^+$ ions of the tryptic peptides. Thus, the correct relative molecular mass of horse myoglobin is 16 951.49 (not 16 950.5 as previously reported), and that of the sperm whale protein is 17 199.91 (not 17 199 as previously reported). Since myoglobin is used as a common mass calibrant, this finding is of importance.

Destruxins (cyclodepsipeptides) were sequenced using FAB by Lange *et al.*,[109] with deuteriation used to overcome problems caused by isobaric fragment ions. Lumb *et al.*[110] have studied D52S hen lysozome oligosaccharide complexes by ES. They found that in the spectrum of a mutant form, with a catalytically important residue, Asp-52, replaced by serine, the most intense peaks arise from uncomplexed protein, indicating that dissociation takes place in the mass spectrometer. Peaks from minor species corresponding to complexes between the protein and oligosaccharide were also observed. The leucine zipper, found in several DNA-binding proteins, has been shown by HPLC-ES and tandem mass spectrometry to dimerize both in solution and in the gas phase. These dimers are stable in the gas phase for minutes.[111] The infectious scrapie prion, PrP^{Sc}, has been sequenced from endoprotease digests by mass spectrometry and Edman degradation data.[112] Species-selective effects have been observed in the production of peptide ions by ES. For Arg-Arg-Arg and Lys-Lys in methanol–water (80:20 by volume), the relative intensities of $(CH_3OH_2)^+$ ions were found to decrease faster than those of other ions with distance from the capillary tip to

106 D. F. Hunt, R. A. Henderson, J. Shabanowitz, K. Sakaguchi, H. Michel, N. Sevilir, A. L. Cox, E. Appella, and V. H. Engelhard, *Science* (*Washington, D.C.*), 1992, **255**, 1261.

107 C. R. Hauer, W. Leimbacher, P. Hunziker, F. Neukeiser, N. Blau, and C. W. Heizmann, *Biochem. Biophys. Res. Commun.*, 1992, **182**, 953.

108 J. Zaia, R. S. Annan, and K. Biemann, *Rapid Commun. Mass Spectrom.*, 1992, **6**, 32.

109 C. Lange, C. Mulheim, A. Vey, and M. Pais, *Biol. Mass Spectrom.*, 1992, **21**, 33.

110 K. J. Lumb, R. T. Aplin, S. E. Radford, D. B. Archer, D. J. Jeenes, N. Lambert, D. A. MacKenzie, C. M. Dobson, and G. Lowe, *FEBS Letters*, 1992, **296**, 153.

111 G. S. Gorman and J. I. Amster, *Org. Mass. Spectrom.*, 1993, **28**, 437.

112 N. Stahl, M. A. Baldwin, D. B. Teplow, L. Hood, B. W. Gibson, A. L. Burlingame, and S. B. Prusiner, *Biochemistry*, 1993, **32**, 1991.

interface plate. These ions are believed to have maximum intensities in the central region of the spray cone, whilst ammonium and singly charged oligopeptide ions are found in the peripheral region. The results are interpreted in terms of the volatility of ions in the charged liquid droplets.[113]

A detailed study of the field desorption of valinomycin found that the ratio of the molecular ion to the protonated molecule was independent of the emitter heating current, and the nature of particular additives to the solution. In contrast, the presence of the $[M + Na]^+$ ions was very strongly dependant on the heater current. There was also found to be no relationship between the abundance of cationized species and the selectivity of valinomycin for monovalent species in solution.[114] Tandem mass spectrometry of valinomycin has been studied with a number of ionization methods; these show a marked difference between $[M + Na]^+$ spectra following different methods of ionization.[115] Small peptides react with alkaline earth metals under FAB conditions to form metal–bis(peptide) complexes, $[2\text{ peptide} + \text{metal} - 3\text{ H}]^-$. The metal ion binds to the deprotonated C-terminal carboxylate group of both peptides and to a deprotonated amide nitrogen of the constituent peptides. This gives rise to fragments containing metal ions tightly bound to the C-terminal end of the peptide, with few N-terminal fragment ions.[116] A similar study has been described for tripeptides.[117] Fragmentation mechanisms for the formation of several commonly occurring product ions in high-energy collision-induced decomposition spectra of $[M + Ca{-}H]^+$ ions have been evaluated by using deuterium-labelled peptides.[118] The $[M - 17]$ peak observed in the multiphoton ionization mass spectra of dipeptides has been explained as the elimination of ammonia from the molecular ion. The $[M - 18]$ peak is generated from the ionization of the thermally decomposed dipeptide.[119]

Tripeptide stereoisomers have been distinguished successfully by tandem mass spectrometry on a hybrid instrument by using CAD. The tripeptides were of the form X-L-Pro-L-Phe, where X is L-Ala, D-Ala, L-Asp, or D-Asp. The differences in fragmentation were correlated with force-field calculations on hydrogen-bonded tautomers.[120] Abundant ions due to the cleavage of Asp-Pro and Asp-Xxx ions following MALDI have been observed by TOF. Using four-sector tandem mass spectrometry, abundant metastable decomposition of Asp-Pro bonds takes place in larger peptides and proteins. The implication is that the Asp-Pro bond is especially labile.[121] The gas-phase basicities of 22 peptides have been estimated by using MALDI to generate protonated peptides, which were then reacted with reference compounds to bracket the gas-phase basicity. Neutral peptides were desorbed using substrate-assisted laser desorption and reacted with protonated references to confirm proton affinity values. The results are consistent with an intramolecular hydrogen bond between the N-terminus nitrogen and the amide carbonyl oxygen of a dipeptide, which

[113] K. Hiraoka, K. Murata, and I. Kudaka, *Rapid Commun. Mass Spectrom.*, 1993, **7**, 363.
[114] M. M. Sheil and P. J. Derrick, *Org. Mass. Spectrom.*, 1992, **27**, 1000.
[115] M. M. Sheil, G. W. Kilby, J. M. Curtis, C. D. Bradley, and P. J. Derrick, *Org. Mass. Spectrom.*, 1993, **28**, 574.
[116] P. Hu and M. L. Gross, *J. Am. Chem. Soc.*, 1992, **114**, 9161.
[117] P. Hu and M. L. Gross, *J. Am. Chem. Soc.*, 1992, **114**, 9153.
[118] H. Zhoa and J. Adams, *Int. J. Mass Spectrom. Ion Processes*, 1992, **125**, 195.
[119] D. S. Nagra, J. Y. Zhang, A. P. L. Wang, and L. Li, *Int. J. Mass Spectrom. Ion Processes*, 1992, **116**, 127.
[120] B. L. Schwartz, R. D. McClain, B. W. Erickson, and M. M. Bursey, *Rapid Commun. Mass Spectrom.*, 1993, **7**, 339.
[121] W. Yu, J. E. Vath, M. C. Huberty, and S. A. Martin, *Anal. Chem.*, 1993, **65**, 3015.

increases in strength with the strength of the basicity of the C-terminal amino acid.[122]

Bean and Carr used tandem mass spectrometry and high-energy CAD to sequence through the cystine bridge of intermolecularly disulfide-bonded peptides.[123] Using this method, they found that a characteristically-defined triplet of intense fragment ions is observed.

FAB tandem mass spectrometry of protonated and metal-cationized peptides was found[124] to give evidence of intramolecular rearrangements in the gas phase. ^{18}O was used to label the C-terminal carboxyl group, and suggested a mechanism where the C-terminal amino acid group is lost with retention of the original carboxyl residue. Using two steps of tandem mass spectrometry (MS/MS/MS) analysis, the intermediate formed in the process was found to be the same for both the [$^{18}O_2$] and the [^{18}O]-labelled rearrangement products. The probable intermediate was concluded to be almost certainly cyclic. The fragmentation of the peptide luteinizing of hormone releasing-hormone measured in plasma desorption mass spectrometry (PDMS) and keV-atom bombardment has been compared with the field desorption (FD) mass spectrum.[125] Kenny has proposed a mechanism for the formation of *y* ions in the collision-induced dissociation spectra of protonated peptides produced by FAB,[126] in which a hydrogen atom attached to nitrogen migrates during the amide bond cleavage.

Evidence was obtained by Ackermann *et al.*[127] for a lysine-specific fragmentation in FAB of peptides. Fragment ions from peptides containing lysine (or other amino acids with a free amino group on a side chain) were found 16 Da below the *b* ions normally resulting from C-terminal fragmentation. Using liquid secondary ion mass spectrometry (LSIMS), the peptide somatostatin-14 (SS-14) was found to undergo unsymmetrical disulfide bridge cleavage.

The mechanism for c'' ions from protonated peptides produced during fast atom bombardment and collisonal activation has been elucidated. Neither the carboxylic proton nor a hydrogen atom from a nitrogen is involved in the process of hydrogen migration that gives rise to the c'' ion. The most feasible position from which the hydrogen originates is that at the first C-atom in the adjacent amino acid side chain.[128] FAB-MS/MS of a disulfide-containing enkephalin, [Pen]-enkephalin, gives a large number of fragment ions, in contrast to other disulfide peptides. Most of the fragments are due to backbone cleavages, with unsymmetric cleavage of the disulfide bond giving charge retention at the N-terminus; the fragment ions rarely include a sulfur atom. The N-terminal ethyltriphenylphosphonium derivative, however, shows very little fragmentation, due to its suggested charge localization at one end of the peptide.[129] The disulfide bridges in recombinant human macrophage-stimulating factor, a 49 kDa homodimeric protein, have been assigned through FAB after cleavage using BrCN.[130]

Papayannopoulos and Biemann have suggested a reason for the non-detection of

122 G. S. Gorman and J. I. Amster, *J. Am. Chem. Soc.*, 1993, **115**, 5729.
123 M. F. Bean and S. A. Carr, *Anal. Biochem.*, 1992, **201**, 216.
124 K. D. Ballard and S. J. Gaskell, *J. Am. Chem. Soc.*, 1992, **114**, 64.
125 A. G. Grey, H. Bennich, and P. J. Derrick, *Aust. J. Chem.*, 1992, **45**, 403.
126 P. T. M. Kenny, *Rapid Commun. Mass Spectrom.*, 1992, **6**, 95.
127 B. L. Ackermann, R. J. Barbuch, J. E. Coutant, J. L. Krstenansky, and T. J. Owen, *Rapid Commun. Mass Spectrom.*, 1992, **6**, 257.
128 W. D. Van Dongen, C. G. De Koster, W. Heerma, and J. Haverkamp, *Rapid Commun. Mass Spectrom.*, 1993, **7**, 241.
129 Y. S. Chang, D. A. Gage, and J. T. Watson, *Biol. Mass Spectrom.*, 1993, **22**, 176.
130 M. O. Glocker, B. Arbogast, J. Schreurs, and M. L. Deinzer, *Biochemistry*, 1993, **32**, 482.

components in peptide mixtures in FAB-MS. This phenomenon has always been attributed to the suppression of ionization of some of the mixture components by other components; a change in the sample probe temperature has been suggested as a method of overcoming this effect.[131] Derivation has been used to increase negative ion sensitivity and positive ion sequence information for LSIMS of peptides. The analyte is acylated at unprotected N-termini, tyrosine and lysine, by application of pentafluorobenzoyl fluoride.[132] The obervation that certain preparations for FAB yield an intense pulse of peptide ions at the start of ionization has prompted the development of a source incorporating a probe that permits mechanical movement of the sample stage and mimics a pulsed atom-beam. Following bombardment initiation, sample ions are preferentially desorbed with respect to matrix ions. Using an array detector, the system allows significantly enhanced sensitivity, up to an improvement in the detection limit of three to four orders of magnitude.[133]

A computer program has been developed specifically for interpretation of four-sector tandem mass spectra of peptides.[134] The sequencing algorithm uses a pattern based on the polymeric nature of peptides to classify spectral peaks. An algorithm has been described which uses both Edman and mass spectrometric data for the simultaneous detection of the amino acid sequences of several peptides in a mixture. The algorithm utilizes Edman data to detect all hypothetical sequences with a calculated mass equal to the observed mass of one of the peptides present in the mixture. These sequences are then assigned figures of merit according to how well each of them accounts for the fragment ions in the tandem mass spectrum of the peptide.[135]

Carbohydrates.—The analysis of carbohydrates by mass spectrometry is receiving considerable attention, partly because of the introduction of the ES and MALDI techniques. Harvey has reviewed the role of MS in glycobiology,[136] with particular emphasis on oligosaccharides and glycoconjugates.

Chai *et al.*[137] identified 22 neutral O-linked oligosaccharides in bovine submaxilliary gland mucin glycoprotein using LSIMS. The sequences and linkages of subnanomolar amounts of thyroid gland oligosaccharides have been confirmed using FAB and MS/MS. The $[M + H]^+$ and $[M + Na]^+$ ions were compared, revealing several ions in the $[M + Na]^+$ spectra not present in the appropriate protonated form in the $[M + H]^+$ spectra. The utility of metal attachment for MS/MS analysis of oligosaccharides has been demonstrated by Leary and coworkers, who used dilithiated disaccharides labelled with ^{18}O and ^{2}H to determine the linkage positions of three isomeric disaccharides.[138] Centrifugal size-exclusion chromatography and membrane filtration have been used to remove salts from carbohydrates prior to analysis, resulting in an increase in the sensitivity and in the number of structurally significant peaks.[139]

Ammonia desorption chemical ionization MS of peracylated gentiobiose and

131 I. Papayannopoulos and K. Biemann, *Int. J. Mass Spectrom. Ion Processes*, 1992, **122**, 43.
132 J. P. Kiplinger, L. Contillo, W. L. Hendrick, and A. Grodski, *Rapid Commun. Mass Spectrom.*, 1992, **6**, 747.
133 A. N. Tyler, L. K. Romo, M. H. Frey, B. D. Musselman, J. Tamura, and R. B. Cody, *J. Am. Soc. Mass Spectrom.*, 1992, **3**, 637.
134 W. M. Hines, A. M. Falick, A. L. Burlingame, and B. W. Gibson, *J. Am. Soc. Mass Spectrom.*, 1992, **3**, 326.
135 R. S. Johnson and K. A. Walsh, *Protein Sci.*, 1992, **1**, 1083.
136 D. J. Harvey, *Glycoconjugate J.*, 1992, **9**, 1.
137 W. Chai, E. F. Hounsell, G. C. Cashmore, J. R. Rosankiewicz, C. J. Bauer, J. Feeney, T. Feizi, and A. M. Lawson, *Eur. J. Biochem.*, 1992, **203**, 257.
138 A. Staempfli, Z. Zhou, and J. A. Leary, *J. Org. Chem.*, 1992, **57**, 3590.
139 T. F. Chen, H. Yu, and D. F. Barofsky, *Anal. Chem.*, 1992, **64**, 2014.

isotopically labelled forms showed that the $[M + NH_4 - 42]^+$ ion is not formed directly from $[M + NH_4]^+$ by loss of ketene, but by way of a nucleophilic acyl substitution.[140] FT-ICR has been used to examine the origin of fragment ions created during desorption–ionization of disaccharides, avoiding the interference of matrix ions.[141] Three major product ions were observed, corresponding to loss of water, loss of monosaccharide, and a combination of both. Mass spectrometry of high-mannose oligosaccharides after trifluoroacetolysis and periodate oxidation was examined by FAB and electron impact.[142] The conditions used were designed to degrade terminal *N*-acetylglucosamine residues, and binding positions between monosaccharide residues were deduced. GC-MS has been used to study the carboxymethyl groups in *O*-(carboxymethyl)ated (1 → 3)-α-D-glucans and (1 → 3)-α-D-glucans.[143] The results showed a distribution characteristic of the glucan type, especially at position 2 on the D-glucosyl residues.

Three thyroid gland oligosaccharides have been characterized by FAB-MS/MS. The saccharides were derivatized with *p*-aminobenzoic acid ethyl ester, and detected in subnanomolar quantities. The $[M + Na]^+$ ions, in some cases, allowed assignments of the linkage positions between the sugar residues.[144] MS/MS has been used to analyse the B → C linkage position in neutral permethylated trisaccharides, of the formulae galactose(β-1 → 4)glucose(β-1 → X)glucose, where X = 3, 4, and 6. The relation of daughter ion to parent ratios was used to predict the linkage position in glucose(β-1 → X)glucose.[145] This method has been used for the $[M + Na]^+$ ions of permethylated oligosaccharides. The linkage positions in one compound may be detected by the presence or absence in a single spectrum of specific fragment ions arising from the cleavage of two ring bonds. The behaviour of the $[M + Na]^+$ ions of permethylated oligosaccharides at low energy leads to preferential cleavage of the glycosidic linkages, which become more efficient if argon replaces helium as the collision gas.[146] Distinctions between the linkage types 1 → 2, 1 → 3, and 1 → 4 of xylobioses can be achieved on the basis of the unimolecular decomposition spectra of the oxonium ions of the per-*O*-acylated methyl glycosides. This allows the detection of linkages between xylose residues.[147]

Oligosaccharides have been analysed using on-line liquid chromatography/thermospray mass spectrometry.[148] Molecular mass information on oligomers of up to a degree of polymerization of 10 was obtained using a mobile phase containing 0.1 mmol sodium acetate, which was compatible with thermospray ionization and gave sodiated and disodiated ions without fragmentation. Van der Greef and coworkers[149] have shown how thermospray and ES are useful for the detection of intact oligosaccharides using selected ion monitoring. The on-line coupling of high-performance anion-

[140] J. M. Peltier, R. W. Smith, D. B. McLean, and W. A. Szarek, *Org. Mass. Spectrom.*, 1992, **27**, 31.
[141] J. A. Carroll and C. B. Lebrilla, *Org. Mass. Spectrom.*, 1992, **27**, 639.
[142] P. Lipniunas, A. S. Angel, K. Erlansson, F. Lindh, and B. Nilsson, *Anal. Biochem.*, 1992, **200**, 58.
[143] S. Ukai, I. Yoshida, A. Honda, K. Nagai, and T. Kiho, *Carbohydr. Res.*, 1992, **224**, 201.
[144] D. A. Gage, E. Rathke, C. E. Costello, and M. Z. Jones, *Glycoconjugate J.*, 1992, **9**, 126.
[145] E. Yoon and R. A. Laine, *Biol. Mass Spectrom.*, 1992, **21**, 479.
[146] J. Lemoine, B. Fournet, D. Despeyroux, K. R. Jennings, R. Rosenberg, and E. de Hoffmann, *J. Am. Soc. Mass Spectrom.*, 1993, **4**, 197.
[147] L. Blok-Tip, A. van der Kerk-van Hoof, W. Heerma, J. Haverkamp, V. Kovacik, and J. Hirsch, *Biol. Mass Spectrom.*, 1993, **22**, 474.
[148] W. M. A. Niessen, R. A. M. Van der Hoeven, J. Van der Greef, H. A. Schols, and A. G. J. Voragen, *Rapid Commun. Mass Spectrom.*, 1992, **6**, 197.
[149] W. M. Niessen, R. A. M. Van der Hoeven, and J. Van der Greef, *Org. Mass. Spectrom.*, 1992, **27**, 341.

exchange chromatography and thermospray MS has been described for the analysis of neutral and acidic oligosaccharides obtained as degradation products of cell wall digestion.[150]

Lipids and Steroids.—Negative-ion mode FAB/MS/MS has been applied to 24 naturally occurring and synthetic diacylglycerylphosphocholines.[151] The method of selecting $[M - 15]^-$ as the parent ion and using the relative abundances of the carboxylate daughter ions to distinguish the esterification positions was found to be unreliable in the cases where the fatty acid group at S_N1 was much larger than that at S_N2. By use of the $[M - 86]^-$ ion as the parent, the relative intensities of the detected fragments resulting from the loss of the free fatty acids at S_N1 and S_N2 were found not to depend on the nature of the fatty acid groups present. Van Dorsselaer and coworkers have used negative FAB on phosphodiesters linked to oxysterols and nucleosides.[152]

The effect of structure on the electron-capture mass spectrometric response of steroids has been investigated.[153] Combinations of double bonds, carbonyl groups, epoxides, and halogens with a steroid nucleus were correlated with their electron-capture negative ionization responses. An epoxy keto steroid has a relative response 40 times greater than the corresponding enone.

A method to reduce degradative losses of steryl fatty acyl esters in high-temperature GC-MS employs selective deuteriation of the double bonds in the fatty acid moiety using a homogeneous catalyst.[154] GC behaviour is improved and structural information is preserved by labelling each double bond present in the original moiety. 16-Androsterones and their isomeric saturated analogues have been analysed by GC-MS, using selected ion monitoring.[155]

Desorption chemical ionization triple quadrupole mass spectrometry has been used to study the collision-induced dissociation of permethylated glycosphingolipids. The dissociation of positive ions produced mainly sugar sequence ions. A rearrangement reaction was suspected for unsaturated ceramides, as no fatty acid daughter ions were seen for compounds with unsaturated bonds at the C-4–C-5 position.[156] Lipids from *Entobacter agglomerans*, a Gram-negative bacterium found in cotton, have been identified using plasma desorption (PD)-MS.[157] FAB-MS/MS has been established as a useful technique for the identification of intact glycerophosphatidylethanolamine phospholipids, giving information about the polar head group and about fatty acid constituents. Scanning for a loss of 141 mass units was the strategy used to confirm the presence of a phosphoethanolamine polar head group, but this scan has now been found to discriminate against the common subgroup of phopholipids with a 1-*O*-alkyl-1′-enyl linkage (the plasmalogens) and against a 1-*O*-alkyl ether species.[158]

150 W. M. A. Niessen, R. A. M. Van der Hoeven, J. Van der Greef, H. A. Schols, G. Lucas-Lokhorst, A. G. J. Voragen, and C. Bruggink, *Rapid Commun. Mass Spectrom.*, 1992, **6**, 474.
151 Z. H. Huang, D. A. Gage, and C. C. Sweeley, *J. Am. Soc. Mass Spectrom.*, 1992, **3**, 71.
152 X. Pannecouke, A. Van Dorsselaer, and B. Luu, *Org. Mass. Spectrom.*, 1992, **27**, 140.
153 H. K. Mayer, W. Reusch, and J. T. Watson, *Org. Mass. Spectrom.*, 1992, **27**, 560.
154 R. P. Evershed, M. C. Prescott, and J. L. Goad, *J. Chromatogr.*, 1992, **590**, 305.
155 T. K. Kwan, D. B. Gower, D. J. H. Trafford, and H. L. J. Makin, *Biol. Mass Spectrom.*, 1992, **21**, 160.
156 J. S. Duh and G. R. Her, *Biol. Mass Spectrom.*, 1992, **21**, 391.
157 R. B. Cole, L. N. Domelsmith, C. M. David, R. A. Laine, and A. J. DeLucca, *Rapid Commun. Mass Spectrom.*, 1992, **6**, 616.
158 K. A. Kayganich and R. C. Murphy, *Anal. Chem.*, 1992, **64**, 2965.

DNA and Nucleotides.—McLuckey *et al.*[159] have examined multiply charged ions derived from ES ionization of the sodium salts of small oligonucleotides ($n = 4$ to 8) by MS/MS in a quadrupole ion trap. The ions were observed to dissociate with high efficiency. In every case the most facile reaction was the loss of the adenine anion, followed by the cleavage of the C-3–O bond of the phosphodiester linkage of the sugar from which the adenine had been lost. The collision-induced dissociation of protonated adenine and of monomethylated adenines has been studied by MS/MS spectrometry and isotopic labelling.[160] A DNA quadruplex has been detected by negative-ion ES at a low nozzle-skimmer bias (−150 V), but at higher bias values it is not seen. The complex is only seen in 10 mM sodium phosphate (pH 7.6) with 0.1 mM EDTA. Without the metal, only single-stranded DNA is seen.[161]

MALDI of oligodeoxythymidylic acids has been investigated by Huth-Fehre *et al.*;[162] using ferulic acid, sinapinic acid, or 2,5-dihydroxybenzoic acid as matrices for pure oligonucleotide $pd(T)_{12}$ and a mixture of $pd(T)_{12}$ through $pd(T)_{18}$. 2,5-dihydroxybenzoic acid gave the least fragmentation and adduct formation, and highest mass resolution. 3-hydroxypicolinic acid has been evaluated as a matrix for MALDI of oligonucleotides, and found to show good characteristics with regard to signal-to-noise ratio, mass range available, and ability to handle mixed-range oligomers. Spectra taken at 266 nm were found to give the best results in negative-ion mode.[163] RNA oligomers and transcripts up to 104 nucleotides long have been analysed successfully by MALDI. RNA is found to be more stable to MALDI than oligothymidylic acids, as the *N*-glycosidic bond seems to be broken during the desorption and bases are lost.[164] MALDI has been used to determine the molecular masses of synthetic methylphosphonate oligodeoxyribonucleotides of up to 18 nucleotides in length.[165] TOF mass spectra of DNA fragments as massive as 18 000 Da have been produced using laser ablation.[166]

Low picomole levels of nucleoside–carcinogen adducts were detected using FAB and MS/MS.[167] Chemical derivatization by trimethylsilyl, neutral loss scans, and daughter ion scans were all used to probe the structural specificity of the adducts.

FAB and MS/MS have been used to examine the interactions between the anticancer drug cisplatin and DNA.[168] FAB gives useful structural information, and high energy collisional-induced dissociation shows the location of the platinum binding sites.

For aglycon-halogenated nucleosides, an inverse linear relationship exists between the extent of dehalogenation during FAB and the electron affinity of a given nucleoside.[169]

[159] S. A. McLuckey, G. J. Van Berkel, and G. L. Glish, *J. Am. Soc. Mass Spectrom.*, 1992, **3**, 60.
[160] C. C. Nelson and J. A. McCloskey, *J. Am. Chem. Soc.*, 1992, **114**, 3661.
[161] D. R. Goodlett, D. G. Camp, II, C. C. Hardin, M. Corregan, and R. D. Smith, *Biol. Mass Spectrom.*, 1993, **22**, 181.
[162] T. Huth-Fehre, J. N. Gosine, K. J. Wu, and C. H. Becker, *Rapid Commun. Mass Spectrom.*, 1992, **6**, 209.
[163] K. J. Wu, A. Steding, and C. H. Becker, *Rapid Commun. Mass Spectrom.*, 1993, **7**, 142.
[164] E. Nordhoff, R. Cramer, M. Karas, F. Hillenkamp, F. Kirpekar, K. Kristainsen, and R. Roerpstorff, *Nucleic Acids Res.*, 1993, **21**, 3347.
[165] T. Keough, T. R. Baker, R. L. M. Dobson, M. P. Lacey, T. A. Riley, J. A. Hasselfield, and E. P. Hesselberth, *Rapid Commun. Mass Spectrom.*, 1993, **7**, 195.
[166] P. Williams, D. Schieltz, R. W. Nelson, C. W. Chou, C. W. Luo, and R. Thomas, in 'Proceedings of Advances in DNA Sequencing Technology', 1993, p. 86.
[167] M. S. Bryant, J. O. Lay, Jr., and M. P. Chiarelli, *J. Am. Soc. Mass Spectrom.*, 1992, **3**, 360.
[168] C. E. Costello, K. M. Comess, A. S. Plaziak, D. P. Bancroft, and S. J. Lippard, *Int. J. Mass Spectrom. Ion Processes*, 1992, **122**, 255.
[169] S. M. Musser and J. A. Kelley, *Org. Mass. Spectrom.*, 1993, **28**, 672.

Narcotics and Drugs of Abuse.—Mass spectrometry is recognized as being a formidable tool for the analysis of controlled substances, particularly in conjunction with GC, and a wealth of information on techniques for the analysis of narcotic compounds, in particular, has been published. A review of opiate MS has been given by Cone and Darwin.[170]

Using GC-MS in electron impact (EI) mode, the metabolites of danazol, an anabolic steroid banned by the International Olympic Committee, were detected in a urine sample. Ethisterone was identified as a nonconjugated metabolite, and the two stereoisomers of 2-hydroxymethylethisterone were observed, one in a conjugated form and the other primarily unconjugated.[171] Derivatives of seven commonly used sympathomimetic amines and two designer amines were isolated from urine, separated from amphetamine and methamphetamine by GC, and detected by MS using selected ion monitoring. False positive were only seen for methamphetamine, and only when $>5\,\mu g$ of ephedrine was present in the samples.[172] Several methods for introduction, derivitization, and ionization of lysergic acid diethylamide (LSD), iso-LSD, and *N*-dimethyl-LSD in body fluids have also been evaluated for MS/MS. Fragmentation pathways derived from low-energy collision-induced dissociation of LSD and derivatives have been proposed. Positive ion chemical ionization of these compounds provide a high degree of specificity for their identification in blood and urine at low $pg\,ml^{-1}$ levels, whilst negative ion chemical ionization and GC-MS/MS has been found to be of use in determining metabolites.[173]

7 Electrospray

Electrospray, atmospheric pressure ionization, and ion-spray are all variations on a common theme. Comparison between ES and MALDI is inevitable, but not straightforward. Green[174] has criticized a paper by Allen *et al.* that sought to compare ES and MALDI for analysis of protein mixtures. In Green's opinion, the paper did not compare the techniques impartially, but detracted from the utility of the ES technique.

ES was used to show the difference in the conformational states of a zinc-finger DNA-binding protein domain. Evidence for the induction and stabilization of a different conformational state with bound copper as opposed to zinc was revealed by a characteristic shift in the electrospray charge envelope.[175] A similar zinc-finger protein that binds to oestrogen was demonstrated to have a high affinity for copper, which inhibits its function, by the use of ES.[176] It was also found that non-covalently bound polypeptide and protein dimer ions are relatively common as low level contributions to ES mass spectra. The detection of these compounds requires gentle electrospray conditions, yielding low levels of internal excitation, but these conditions can lead to artifacts in MS/MS, due to incomplete solvation of ions.[177] The haem–apomyoglobin complex, which is non-covalently bound and has a mass of 17 568 Da, has been detected by ES at femtomolar quantities. The complex does not fragment under

[170] E. J. Cone and W. D. Darwin, *J. Chromatogr.*, 1992, **580**, 43.
[171] D. De Boer, E. G. Jing, and R. A. A. Maes, *J. Anal. Toxicol.*, 1992, **16**, 14.
[172] E. M. Thurman, M. J. Pedersen, R. L. Stout, and T. Martin, *J. Anal. Toxicol.*, 1992, **16**, 19.
[173] C. C. Nelson and R. L. Foltz, *Anal. Chem.*, 1992, **64**, 1578.
[174] B. N. Green, *Biochem. J.*, 1992, **284**, 603.
[175] W. T. Hutchens and M. H. Allen, *Rapid Commun. Mass Spectrom.*, 1992, **6**, 496.
[176] W. T. Hutchens, M. H. Allen, C. M. Li, and T. T. Yip, *FEBS Lett.*, 1992, **309**, 170.
[177] R. D. Smith, K. J. Light-Wahl, B. E. Winger, and J. A. Loo, *Org. Mass. Spectrom.*, 1992, **27**, 811.

kilovolt energy conditions.[178]

Allen and Vestal developed an ES system using heat as a means of desolvation, and no counterflow of heated gas.[179] The ability to obtain fragmentation data by collisional-induced dissociation was demonstrated clearly. ES mass spectra of human haemoglobin chains and transferrin were acquired on a magnetic sector mass spectrometer.[180] The observed isotopic distributions for the haemoglobin ion were in good agreement with the theoretical distribution within a resolution of 2000. Accurate mass measurements have been obtained for masses in the 500–16 000 range using a double-focusing magnetic-sector instrument and ES ionization.[181] Magnetic-field scans and accelerating-potential scans were used and gave mass accuracies in the low ppm range in both positive and negative-ion modes.

ES ionization has been used in a study of the reactions of multiply protonated melittin molecule ions. Flowing the ions entrained in gas through a heated metal capillary inlet served as a reaction vessel for measurements of ion reaction rates.[182] Differences in activation energies found for different charge states were attributed to Coulombic repulsion effects.

ES has been used to detect a ternary complex between the dimeric enzyme HIV-1 protease and a substrate-based inhibitor.[183] Under denaturing conditions, only the monomeric HIV-1 protease could be observed by mass spectrometry (see also ref. 184). Using the tightly binding competitive inhibitor JG365 the ternary complex was detected, and this was dissociated by raising the declustering potential at the ion sampling orifice. Electrospray has been used to detect and differentiate between the free human H-ras protein and its non-covalently bound GDP complex form.[185] Analysis of recombinant proteins has been performed using high-performance capillary electrophoresis interfaced with an electrospray ionization mass spectrometer.[186] The detection limit was 100 fmol, the standard deviation of peak migration times was less than 1%, and mass assignment was within 1 unit in 22 000. The results showed that individually neither CZE nor ES ionization could detect all impurities, but coupling enhanced the abilities of both.

Oligosaccharides digested from asparagine side chains of glycoproteins have been analysed by ES and MS/MS.[187] Oligosaccharides containing sialic acids were detected best by negative-ion mode, whilst others were best in the positive-ion mode. In positive-ion mode, the addition of sodium or ammonium acetate greatly enhanced the ion abundances. MS/MS was used to detect primary structural features of the oligosaccharides.

Boyle and Whitehouse coupled an ES source to a reflectron time-of-flight mass spectrometer, achieving a resolving power of over 1000 by orientation of the ion source

[178] J. A. Loo, A. B. Giordani, and H. Muenster, *Rapid Commun. Mass Spectrom.*, 1992, **7**, 186.
[179] M. H. Allen and M. L. Vestal, *J. Am. Soc. Mass Spectrom.*, 1992, **3**, 18.
[180] Y. Wada, J. Tamura, B. D. Musselman, D. B. Kassel, T. Sakurai, and T. Matsuo, *Rapid Commun. Mass Spectrom.*, 1992, **6**, 9.
[181] R. B. Cody, J. Tamura, and B. D. Musselman, *Anal. Chem.*, 1992, **64**, 1561.
[182] M. Busman, A. L. Rockwood, and R. D. Smith, *J. Phys. Chem.*, 1992, **96**, 2397.
[183] M. Baca and S. B. H. Kent, *J. Am. Chem. Soc.*, 1992, **114**, 3992.
[184] U. A. Mirza, S. L. Cohen, and B. T. Chait, *Anal. Chem.*, 1993, **65**, 1.
[185] A. K. Ganguly, B. N. Pramanik, A. Tsarbopoulos, T. R. Covey, E. Huang, and S. A. Fuhrman, *J. Am. Chem. Soc.*, 1992, **114**, 6559.
[186] K. Tsuji, L. Baczynskj, and G. E. Bronson, *Anal. Chem.*, 1992, **64**, 1864.
[187] K. L. Duffin, J. K. Welpy, E. Huang, and J. D. Henion, *Anal. Chem.*, 1992, **64**, 1440.

perpendicular to the field-free drift region.[188] The instrument was also sufficiently sensitive to allow the routine detection of low picomole quantities of multiply charged cytochromes.

8 Matrix-assisted Laser Desorption/Ionization

Matrix-assisted laser desorption/ionization (MALDI) remains one of the fastest-moving branches of mass spectrometry. MALDI in combination with time-of-flight (TOF) has been used for an extremely wide variety of applications, normally involving high molecular mass compounds, but some work involving smaller molecules has also been performed. The coupling of MALDI and FT-ICR has produced remarkable results; using laser fluences close to the MALDI threshold, resolutions of 30 000 have been presented for bovine insulin, rising to 100 000 for lower mass proteins. Detection of a doubly-charged carbonic anhydrase trimer ($m = 87\,000$) has also been achieved. Optimum deceleration times for a singly-charged MALDI-generated protein ion are a linear function of $\sqrt{m}$, when a 9.5 V deceleration potential is used.[189]

Nelson and Hutchens used MALDI-TOF to examine the affinity of a 26-residue synthetic peptide for copper(II) ions.[190] The chemistry was performed on the probe tip, and results showed the binding of up to one copper ion per histidine residue. MALDI was also used to identify metal-binding proteins in proteolytic digests.[191] Histidine-rich glycoprotein, a plasma and milk metal-transport protein, was purified and digested with trypsin; the fragments were analysed by MALDI-TOF. Copper(II) ions were added to the digest mixture and at least one metal-binding peptide was seen. This is the first direct observation by MS of differential peptide-metal digests in protein digest maps.

A TOF mass spectrometer with an electrostatic mirror has been used to measure the kinetic energy distributions of molecular ions ejected into an external field by MALDI. An energy deficit was observed by this experiment (*i.e.* the ions had less kinetic energy than imparted by the field), of around 24 eV in the typical case of insulin. This deficit increases with molecular ion signal intensity and with molecular weight of the protein. It is also larger for negative ions than for positive ions.[192] Pan and Cotter investigated the initial kinetic energy distributions of arginine–vasopressin molecule–ions generated by UV-MALDI.[193] The measured molecule ion kinetic energy distribution from vasopressin was found to be much broader than that from gramicidin S and was characterized by a high-energy tail possibly resulting from entrainment of analyte ions in the matrix ions as they leave the surface.

Ehring, Karas, and Hillenkamp have studied a large number of small polar organic molecules which have a strong resonant absorption at the laser wavelength.[194] In most cases, radical cations and ion species formed by single and multiple hydrogen cleavages and additions appeared as strong signals. A model was proposed which explains these features as a product of initial photoionization.

[188] J. G. Boyle and C. M. Whitehouse, *Anal. Chem.*, 1992, **64**, 2084.
[189] J. A. Castoro and C. L. Wilkins, *Anal. Chem.*, 1993, **65**, 3015.
[190] R. W. Nelson and W. T. Hutchens, *Rapid Commun. Mass Spectrom.*, 1992, **6**, 4.
[191] R. W. Nelson, W. T. Hutchens, and T. T. Yip, *FEBS Lett.*, 1992, **296**, 99.
[192] J. Zhou, W. Ens, K. G. Standing, and A. Verentchikov, *Rapid Commun. Mass Spectrom.*, 1992, **6**, 671.
[193] Y. Pan and R. J. Cotter, *Org. Mass. Spectrom.*, 1992, **27**, 3.
[194] H. Ehring, M. Karas, and F. Hillenkamp, *Org. Mass. Spectrom.*, 1992, **27**, 472.

Chan *et al.* demonstrated that cluster ions extending to very high masses (> 500 000 Da) can be formed and detected from proteins using the liquid 3-NBA matrix for UV-MALDI.[195] Excellent shot-to-shot reproducibility was reported for this liquid matrix. An organic dye added to the matrix was found to increase the efficiency of visible laser systems. The addition of Rhodamine 6G to 3-nitrobenzyl alcohol allowed the production of protein ions of m_R up to 250 000 using a 532 nm laser.[196,197]

It has been found that the UV (248 nm) desorption of neutral peptide molecules is greatly enhanced by the application of a thin layer of the sample on top of a UV-absorbing organic substrate (sinapinic acid).[198] Using this technique, gramicidin S is desorbed as an intact neutral, with approximately 1 eV less internal energy than if desorbed from a metal surface. α-Cyano-4-hydroxycinnaminic acid, when used as a matrix,[199] has been found to enhance the tendency of proteins to protonate multiply. Ionic and neutral products of matrix-assisted laser desorption have been detected simultaneously in a TOF mass spectrometer, the neutrals being photoionized by coherent UV radiation. The ratio of neutral : ions has been calculated at around 10 000 : 1 with desorption irradiances close to the threshold for ion formation. This ratio decreases with increasing irradiance from the laser beam. This dependence is consistent with a collisional mechanism for analyte-ion formation during the desorption, initiated by photoionization. Matrices giving large signal intensities of directly formed analyte ions gave large intensities of neutral ions as well.[200]

Metastable decay of laser-desorbed ions taking place in the first field-free drift region of a reflectron TOF mass spectrometer has been successfully monitored to obtain structural information from the larger peptides.[201] Fragments from such metastable decay are mass analysed by adjusting the potentials of the ion reflectron according to the kinetic energies of the ions. It has been discovered that post-source decay in MALDI is only 10% with nitrogen as the residual gas compared to helium.[202]

An excimer-pumped dye laser generating ultra-short pulses (560 fs full width at half maximum) in the UV (248 nm) and visible (496 nm) has been used to desorb protein ions from different matrices.[203] The existence of a threshold energy for the MALDI of insulin ions by the ultra-short pulses was established, and found to be similar to that observed at much longer pulse lengths. The amount of laser energy deposited into the sample, not the laser power, thus appeared to be important in MALDI. Results obtained from MALDI work involving bovine insulin in 3-nitrobenzyl alcohol or sinapinic acid matrices suggest that the threshold fluences for the formation of positive and negative ions are the same for a given matrix.[204] Laser generated plumes, such as those observed in MALDI, have been imaged by fluorescence measurements, giving concentration maps at time-intervals of 100 μs.[205]

[195] D. T. W. Chan, A. W. Colburn, and P. J. Derrick, *Org. Mass. Spectrom.*, 1992, **27**, 53.
[196] D. S. Cornett, M. A. Duncan, and J. I. Amster, *Org. Mass. Spectrom.*, 1992, **27**, 831.
[197] D. S. Cornett, M. A. Duncan, and J. I. Amster, *Anal. Chem.*, 1993, **65**, 2608.
[198] J. P. Speir and J. I. Amster, *Anal. Chem.*, 1992, **64**, 1041.
[199] R. C. Beavis, T. Chaudhary, and B. T. Chait, *Org. Mass. Spectrom.*, 1992, **27**, 156.
[200] C. D. Mowrey and M. V. Johnston, *Rapid Commun. Mass Spectrom.*, 1993, **7**, 569.
[201] B. Spengler, D. Kirsch, R. Kaufmann, and E. Jaeger, *Rapid Commun. Mass Spectrom.*, 1992, **6**, 105.
[202] B. Spengler, D. Kirsch, and R. Kaufmann, *J. Phys. Chem.*, 1992, **96**, 9678.
[203] P. Demirev, A. Westman, C. T. Reimann, P. Hakansson, D. Barofsky, B. U. R. Sundqvist, Y. D. Cheng, W. Seibt, and K. Siegbahn, *Rapid Commun. Mass Spectrom.*, 1992, **6**, 187.
[204] P. Y. Yau, D. T. W. Chan, P. G. Collis, A. W. Colburn, and P. J. Derrick, *Chem. Phys. Lett.*, 1993, **202**, 93.
[205] T. W. Heise and E. S. Yeung, *Anal. Chem.*, 1992, **64**, 2175.

Biemann and coworkers have studied MALDI using a double-focusing magnetic-sector mass spectrometer. Pulsed ion signals were generated by irradiation of a mixture of sample and matrix (2,5-dihydroxybenzoic acid) with either an XeF laser (353 nm) or a nitrogen laser (337 nm), and were detected by a focal-plane detector. A resolution (full width at half maximum) of 4500 was achieved for substance P ($m/z = 1347.7$), 2500 for CsI cluster ions at $m/z = 10\,005.7$, and 1250 for the isotope cluster of horse cytochrome c ($m/z = 12\,360$); 11 fmol of substance P was detected.[206]

A simple method for the off-line coupling of CZE and laser desorption has been described, with a detection limit of 100 fmol for bovine lactalbumin.[207,208] The challenge of interfacing liquid chromatography with MALDI has been examined.[209] Liquid sample has been introduced directly into the ion source of the mass spectrometer through a capillary.[210] Evaporative cooling was used to produce ice at the end of the capillary, which could be extruded using resistive heating. Immobilization procedures were used to improve the MALDI of hydrophobic cytochrome P-450 proteins containing high concentrations of buffer. A mass accuracy of 0.075% was achieved using bovine serum albumen as an internal standard.[211]

Small fragments of DNA, up to a 34-mer, have been analysed with various matrices.[212] Both positive and negative ions were seen with little fragmentation. A similar series of experiments have been performed on oligodeoxyribonucleotides, as various matrices were examined for their efficiency in analysis.[213] Mixtures of low picomole amounts of oligodeoxyribonucleotides with sizes from 6 to 30 units have been examined.[214] Again molecule–ions were observed, again with little or no fragmentation. MALDI has been used to analyse crude synthetic peptides.[215] A highly hydrophobic six-helix bundle template-assembled synthetic protein, a 19-mer peptide with an amphiphilic α-helix, and a 20-mer thrombin inhibitor were all successfully analysed. MALDI was used to verify the insertion of glycine into a C-terminal α-helix, thus solving a crystallographic problem in the inability to pinpoint the nature or location of an anomaly in a protein sequence.[216] MALDI was used to identify rapidly, sensitively, and sequence-specifically phosphopeptides in unfractionated digests of phosphoproteins.[217] Beavis produced a review of MALDI covering topics such as sample preparation, the nature of the spectra, nomenclature, and matrices.[218]

9 Fourier-transform Ion Cyclotron Resonance

Electrospray has been combined with Fourier-transform ion cyclotron resonance

[206] R. S. Annan, H. J. Kochling, J. A. Hill, and K. Biemann, *Rapid Commun. Mass Spectrom.*, 1992, **6**, 298.
[207] T. Keough, R. Takigiku, M. P. Lacey, and M. Purdon, *Anal. Chem.*, 1992, **64**, 1594.
[208] J. A. Castoro, R. W. Chiu, C. A. Monnig, and C. L. Wilkins, *J. Am. Chem. Soc.*, 1992, **114**, 7571.
[209] L. Li, A. P. L. Wang, and L. D. Coulson, *Anal. Chem.*, 1993, **65**, 493.
[210] E. R. Williams, G. C. Jones, L. L. Fang, T. Nagata, R. N. Zare, *Proc. SPIE-Int. Soc. Opt. Eng.*, 1992, 172.
[211] S. Lewis, K. K. Korsmeyer, and M. A. Correia, *Rapid Commun. Mass Spectrom.*, 1992, **7**, 16.
[212] K. Tang, S. L. Allman, and C. H. Chen, *Rapid Commun. Mass Spectrom.*, 1992, **6**, 365.
[213] E. Nordhoff, A. Ingedoh, R. Cramer, A. Overberg, B. Stahl, M. Karas, F. Hillenkamp, and P. F. Crain, *Rapid Commun. Mass Spectrom.*, 1992, **6**, 771.
[214] G. R. Parr, M. C. Fitzgerald, and L. M. Smith, *Rapid Commun. Mass Spectrom.*, 1992, **6**, 369.
[215] V. Steiner, K. O. Boernsen, M. Schaer, E. Gassmann, S. Hoffstetter-Kuhn, H. Rink, and M. Mutter, *Peptide Res.*, 1992, **5**, 25.
[216] L. J. Keefe, E. E. Lattman, C. Wolkow, A. Woods, M. Chevrier, and R. J. Cotter, *J. Appl. Crystallogr.*, 1992, **25**, 205.
[217] T. T. Yip and W. T. Hutchens, *FEBS Lett.*, 1992, **308**, 149.
[218] R. C. Beavis, *Org. Mass Spectrom.*, 1992, **27**, 653.

(FT-ICR), with the ion source positioned in the bore of the superconducting magnet adjacent to the trapped ion cell.[219] An electrospray current of 375 pA could be detected from gramicidin S at the cell, which is two orders of magnitude higher than an external ion source. Five stages of differential pumping within a series of concentric tubes of increasing radius were used to reduce pressure to 10^{-7} torr in the trapped cell.

Wilkins and coworkers explored the utility of 'screened' electrostatic ion trapping techniques. The observed rate of change of ion cyclotron resonance (ICR) frequency with trapping potential is independent of mass and charge (in the range of 1185–17 000 Da), and using a magnetic field of ~ 7 T, the electric potential within a screen cubic trap is reduced, decreasing frequency shifts as a function of trapping potential by a factor of 25. The average resolution for a bovine insulin spectrum was 10 000.[220] Hofstadler and Laude have interfaced an electrospray ionization system with FT-ICR. In their system, the electrospray occurs at atmospheric pressure in a 1.5 T field, and the FT-ICR detection takes place 25 cm away at 3.0 T in one or two cells, separated by a conductance limit and maintained at a pressure differential of 5×10^{-5} and 2×10^{-7} torr. The trapped ion cell was filled up to the ion space charge limit within a few hundred milliseconds. The rate of ion accumulation is independent of background pressure; hence the primary trapping mechanism is not solely collision dependent.[221]

Trapped ion cells which substitute open rectangular trapping electrodes for trapping plates perpendicular to the magnetic field have been evaluated.[222] Radical trapping electric fields are somewhat larger in the centre of the open cells, but axial excitation fields are a factor of two smaller compared with the closed cell. Axial ejection is significantly reduced in the open cell compared to the closed cell. The question of poor high-mass (high m/z) performance in FT-ICR has been addressed.[223] Theoretical calculations and experimental data showed the existence of resonances involving the z-motion and the radial motion of the ion, which result in energy transfer between modes, expansion of the ion cloud, and a signal loss. Polypeptide ions formed by MALDI have been found not to be in thermal equilibrium with their surroundings, but to exhibit a non-Boltzmann initial ion velocity distribution about a most probable ion velocity, which is independent of polypeptide mass. Radical and axial m/z upper limits and optimum trapping potentials were derived and discussed.[224]

Analytical expressions for the probability of radial or axial loss of thermal ions from a hyperbolic ion trap were developed, and gave rise to the observation that the probability of radial ion loss increases with m/z ratio and ion thermal energy, whilst axial ion loss is inversely proportional to trapping potential, and directly proportional to thermal energy. This model is based on the assumptions that non-interacting ions are formed initially along the trap y axis with a Maxwell–Boltzmann velocity distribution and trapped by a quadrupolar electrostatic potential.[225] The absolute number of trapped ions that produce a time-domain signal in FT-ICR has been estimated by comparing the experimentally observed signal voltage to that calculated

219 S. A. Hofstadler and D. A. Laude, Jr., *Anal. Chem.*, 1992, **64**, 569.
220 J. A. Castoro, C. Koster, and C. L. Wilkins, *Anal. Chem.*, 1993, **65**, 784.
221 S. A. Hofstadler and D. A. Laude, Jr., *J. Am. Soc. Mass Spectrom.*, 1992, **3**, 615.
222 S. C. Beu and D. A. Laude, Jr., *Int. J. Mass Spectrom. Ion Processes*, 1992, **112**, 2.
223 C. L. Holliman, D. L. Rempel, and M. L. Gross, *J. Am. Soc. Mass Spectrom.*, 1992, **3**, 460.
224 T. D. Wood, L. Schweikhard, and A. G. Marshall, *Anal. Chem.*, 1992, **64**, 1461.
225 M. A. May, P. B. Grosshans, and A. G. Marshall, *Int. J. Mass Spectrom. Ion Processes*, 1992, **120**, 193.

for a single ion orbiting at the ICR orbital radius of the ion packet. This gives an estimate of ~177 singly charged ions in a cubic trap at 3.0 T.[226] Detection limits have been reduced by subjecting a single population of multiply charged protein ions formed by ES to multiple FT-ICR without promoting ion loss from the trapped ion cell. This process is non-destructive, and allows at least 250 consecutive remeasurements of a single population of protein ions. A 25:1 signal:noise ratio has been observed for 30 fmol of bovine albumin dimer.[227]

A two-dimensional charged-disk model was used to explain Coulomb-induced frequency shifting in FT-ICR MS.[228] The model consists of a uniformly charged disk, whose excited cyclotron motion is perturbed by another uniformly charged disk with excited cyclotron motion. This model gave a better agreement with the experimental data than the charged-point model. Control of ion kinetic energy in ICR spectrometry, with respect to very low-energy collision-induced dissociation, has been examined by Boering *et al.*[229] Ions were continually accelerated and decelerated over a period of hundreds of milliseconds by repeatedly shifting the phase of the ratio frequency driving signal by 180°, using an external electronic circuit. At low kinetic energies, a succession of collisions which transferred small amounts of energy favoured the lowest energy fragmentation pathway and allowed ions with large activation energies to fragment at energies very close to threshold. Solutions have been reported for the frequency shift due to the induced image charges for both cylindrical and tetragonal ion traps, as a function of ion axial and radial position. For a cubic trap, the shift for an ICR orbital radius of one half the trap radius is ~10 μHz for one ion. Thus a frequency shift in the ppm range can be expected for typical ICR experiments involving 1000–100 000 ions at low *m/z* ratio (60–600). This has important implications for calibration.[230]

A new 20-dimensional FT-ICR experiment, stored-waveform ion modulation, or SWIM, 2-D-FT-ICR-MS/MS, has been developed. Each row of the two-dimensional data array is obtained by the use of a single stored excitation waveform whose frequency-domain magnitude spectrum is a sinusoid, the frequency of which changes from one row to the next. The conventional FT-ICR spectrum appears along the diagonal of the two-dimensional mass spectrum, with off-diagonal peaks corresponding to ion–neutral reactions whose ionic components may be identified by horizontal and vertical projections to the diagonal spectrum. Fragmentation due, for example, to collision-induced dissociation results in a peak on only one side of the diagonal; bidirectional processes give peaks on both sides. All ion–molecule reactions can therefore be seen without any prior knowledge of the system.[231]

Self-assembly alkanethiol monolayers on gold films have been studied by laser desorption at 193 nm radiation.[232] Adding sugars as comatrix elements permitted higher resolution than without sugar in MALDI-FT-ICR. Mass resolution of 6–10 000 was obtained for melittin, bovine insulin, and bovine insulin B chain, which rose if ribose or fructose was added to the matrix.[233] Capillary electrophoresis has been

[226] P. A. Limbach, P. B. Grosshans, and A. G. Marshall, *Anal. Chem.*, 1993, **65**, 135.
[227] Z. Guan, S. A. Hofstadler, and D. A. Laude, Jr, *Anal. Chem.*, 1993, **65**, 1588.
[228] S. P. Chen and M. B. Comisarow, *Rapid Commun. Mass Spectrom.*, 1992, **6**, 1.
[229] K. A. Boering, J. Rolfe, and J. I. Brauman, *Rapid Commun. Mass Spectrom.*, 1992, **6**, 303.
[230] X. Xhang, P. B. Grosshans, and A. G. Marshall, *Int. J. Mass Spectrom. Ion Processes*, 1993, **125**, 33.
[231] C. W. Ross III, S. Guan, P. B. Grosshans, T. L. Ricca, and A. G. Marshall, *J. Am. Chem. Soc.*, 1993, **115**, 7854.
[232] Y. Li, J. Huang, R. T. McIver, Jr, and J. C. Hemminger, *J. Am. Chem. Soc.*, 1992, **114**, 2428.
[233] C. Koster, J. A. Castoro, and C. L. Wilkins, *J. Am. Chem. Soc.*, 1992, **114**, 7572.

successfully interfaced with FT-ICR, using an electrospray interface and a high speed cryopumping arrangement. Initial results show that spectra with high sensitivity and excellent resolution (30 000–50 000) can be obtained every six seconds.[234]

FT-ICR has been used as a post-ionization technique for separation of electrosprayed protein mixtures.[235] The relative order of trapping of the ions is determined by the application of potential to the skimmer cone and the ion velocity. The optimum trapping potential has been found to depend primarily on the *m/z* ratio within the electrospray envelope, but also on protein-specific features, including velocity slip in the expansion.

10 Tandem Mass Spectrometry

The combination of electrospray and FT-ICR provides possibilities for tandem mass spectrometry (MS/MS) and hence elucidation of molecular structure, through collision-induced dissociation in the actual ion source (so called 'nozzle dissociation') in the case of very large ions (10^4 Da).[236] The difficulties inherent in the analysis of large, multiply-charged proline-containing proteins have been addressed by Smith and coworkers.[237] Low resolution tandem mass spectrometric data can be ambiguous due to the tendency of proline-containing proteins to form multiply-charged product ions. Collisional dissociation of proteins of up to 22 000 Da ionized by ES was considered. Fragmentation due to cleavage of the peptide bond near proline residues was usually dominant, and this process was found to be the only dissociation pathway observed for many large proteins. Multiply-charged ions of an IgG class monoclonal antibody, anti(human α_1-acid glycoprotein), which has a weight of 149 600 Da, were produced by pneumatically assisted ES, and then subjected to collisionally activated dissociation with argon at collision energies of 5–12 keV in a triple-quadrupole mass spectrometer. The structurally significant light-chain fragment (23 720 $\pm$ 30 Da) was collisionally released. There is evidence that this took place near an energetic threshold, and the energy required may well have come from the many collisions encountered by the antibody ions.[238]

Peptide and caesium iodide cluster ions decomposing to certain fragment ions following collision-induced decomposition have been shown to exhibit energy losses independent of transmission, whilst parent ions of certain other fragments show strong dependence of energy loss upon transmission. These results have been interpreted in terms of energy redistributions taking place in single collisions by direct momentum transfer.[239]

MS/MS results for high mass ions (*m/z* 1000–2500) comparable to those obtained by four-sector instruments were obtained using a magnetic sector/quadrupole instrument.[240]

Kenny and Orlando have reported MS/MS analysis of peptides at the femtomole

[234] S. A. Hofstadler, J. H. Wahl, J. E. Bruce, and R. D. Smith, *J. Am. Chem. Soc.*, 1993, **115**, 6983.
[235] S. A. Hofstadler, S. C. Beu, and D. A. Laude, Jr, *Anal. Chem.*, 1993, **65**, 312.
[236] J. A. Loo, J. P. Quinn, S. I. Ryu, K. D. Kent, M. W. Senko, and F. W. McLafferty, *Proc. Natl. Acad. Sci., U.S.A.*, 1992, **89**, 286.
[237] J. A. Loo, C. G. Edmonds, and R. D. Smith, *Anal. Chem.*, 1993, **665**, 425.
[238] R. Feng and Y. Konishi, *Anal. Chem.*, 1993, **65**, 645.
[239] C. D. Bradley and P. J. Derrick, *Org. Mass. Spectrom.*, 1993, **28**, 390.
[240] C. D. Bradley, J. M. Curtis, P. J. Derrick, and B. Wright, *Anal. Chem.*, 1992, **64**, 2628.

level using a four-sector instrument and array detection.[241] Employing continuous-flow FAB, the structures of several peptides (900–2000 Da) at the 900 fmol to 5.8 pmol level were elucidated without interference from the background. Metabolites in a crude plasma extract from spontaneously hypertensive rats were identified without chromatographic separation by FAB MS/MS.[242] The calcium-entry blocker AJ-2615 was detected and the structure elucidated. Compound-specific sequence ions were identified in the collision-induced dissociation spectra of the immonium ions of two isomeric hexaglycosylceramides.[243] Using standard EI, the spectra had been too complex to distinguish the isomers from one another. Structure-specific collision-induced fragmentations of ceramides has been achieved by complexation of the analytes with alkali metal ions. Structures of individual components of complex mixtures could be detected, with detection limits as low as 8 pmol.[244]

A number of different devices that can be used for surface-induced dissociation (SID) in tandem quadrupole instruments have been evaluated on the basis of fragmentation of $(C_6H_6)^{+\bullet}$, $[W(CO)_6]^{+\bullet}$ and $(Me_4N)^+$. Spectra obtained by in-line and 90° instruments were found to be similar.[245] The overall efficiency for even-electron ions was higher than for odd-electron ions of similar structure. High average internal energies and narrow distributions of energies were said to be imparted to the parent ions in SID.

SID of angiotensin I, renin tetradecapeptide, porcine pancreastatin fragment 33–49, and ACTH fragment 1–17, has been observed in a four-sector MS/MS spectrometer.[246] Abundant fragment ions were seen, facilitating determination of the amino acid sequences. SID has been applied to small protonated peptides generated by FAB.[247] The relative abundances of daughter ions were found to depend critically upon the energy of the ion-surface collision. At approximately 13 eV collision energy, the number of decomposition processes was maximized for the peptides studied. For the SID process, the kinetic energy distribution of the parent ions was said to be converted into a wider spread of centre-of-mass collision energies, leading to a larger variety of decomposition processes compared to collision-induced dissociation.

A comparison of the molecular and quasi-molecular ions of valinomycin formed using EI, FAB, field desorption (FD), and ES four-sector MS/MS has been made by Curtis *et al.*[248] Large differences were observed between the FD and FAB tandem mass spectra of $[M + Na]^{+\bullet}$ ions and between FD and EI spectra of $M^{+\bullet}$ ions. This was ascribed to formation of ions of different structures.

A number of small peptides (500–1800 amu) with various different masses were fragmented by SID in a tandem mass spectrometer as a comparison with collisional activation. The results allow the formulation of a model to explain the easy cleavage of

[241] P. T. M. Kenny and R. Orlando, *Anal. Chem.*, 1992, **64**, 957.
[242] M. Kurono, A. Itogawa, K. Yoshida, S. Narto, P. Rudewicz, and M. Kanai, *Biol. Mass Spectrom.*, 1992, **21**, 17.
[243] J. M. Curtis, P. J. Derrick, J. Holgersson, B. E. Samuelsson, and M. E. Breimer, *J. Am. Soc. Mass Spectrom.*, 1992, **3**, 353.
[244] Q. Ann and J. Adams, *Anal. Chem.*, 1993, **65**, 7.
[245] V. H. Wysocki, M. J. Ding, J. L. Jones, J. H. Callahan, and F. L. King, *J. Am. Soc. Mass Spectrom.*, 1992, **3**, 27.
[246] A. D. Wright, D. Despeyroux, K. R. Jennings, S. Evans, and A. Riddoch, *Org. Mass. Spectrom.*, 1992, **27**, 525.
[247] R. B. Cole, S. LeMeillour, and J. C. Tabet, *Anal. Chem.*, 1992, **64**, 365.
[248] J. M. Curtis, C. D. Bradley, P. J. Derrick, and M. M. Shiel, *Org. Mass. Spectrom.*, 1992, **27**, 502.

the amide bond, charge-remote backbone fragmentations, and the influence of intramolecular hydrogen bonding.[249] Collision-induced dissociation was again compared with surface-induced dissociation (at 100 eV) for peptides of mass 1300–1850. Only a small amount of the kinetic energy of the precursor ions is converted into internal energy for SID, as the spectra resemble low energy collision-induced dissociation rather than the high energy spectra.[250]

A quadrupole ion trap storage/reflectron TOF mass spectrometer has been developed as a detector for ES ionization. A continuous ES beam is converted into a pulsed beam for TOF measurement. Sensitivity in the low femtomole region was demonstrated with small peptides. The spectrum can be acquired in less than one second.[251,252] A reflectron TOF mass spectrometer for laser photodissociation has been described by Cornett *et al.*[253] Mass selection was performed by pulsed deflection plates at the end of the initial drift section. Laser photodissociation of ions was effected at the turning point of the ion trajectory in the reflectron, and the transit time through a second drift section defined the fragment masses (see also ref. 254). Scan laws havebeen derived for the detection of fragment ions produced by collisional activation of a multiply-charged precursor in a floated collision cell of a four-sector tandem mass spectrometer.[255]

A review of collisional activation in MS/MS has been given by McLuckey.[256] A review of trace analysis of organics in mixtures has been made by Biemann.[257]

11 Ion Trap

The ion-trapping technique holds great promise as an inexpensive, relatively high-performance mass spectrometer.

The selectivity of methylene substitution in a low-pressure quadrupole ion trap was elucidated by collisionally-activated dissociation.[258] Dimethyl ether and ethylene, the covalently bound adducts found at $[M + 45]^+$ and $[M + 41]^+$ respectively, are direct precursors to the methylene substitution product ions at $[M + 13]^+$. It was discovered that at low helium pressures, collisional-activated dissociation shows that the average internal energy deposition is higher and sensitivity is not lost if a collisional cooling delay is added.[259] Heavy target gases improved the trapping efficiency of caesium iodide clusters when added in small amounts to the helium buffer gas.[260] Trapped ions were activated by applying a short DC voltage pulse to one end of a quadrupole ion-trap mass spectrometer. High internal energy deposition was obtained, as evidenced by peak ratios in excess of 20 for the $m^+ = 91 : m^+ = 92$ ions for the dissociation of the n-butylbenzene ion.[261]

249 A. L. McCormack, A. Somogyi, A. R. Dongre, and V. H. Wysocki, *Anal. Chem.*, 1993, **65**, 2859.
250 D. Despeyroux, A. D. Wright, and K. R. Jennings, *Int. J. Mass Spectrom. Ion Processes*, 1993, **126**, 95.
251 S. M. Michael, B. M. Chien, and D. M. Lubman, *Anal. Chem.*, 1993, **65**, 2614.
252 S. M. Michael, B. M. Chien, and D. M. Lubman, *Rev. Sci. Instrum.*, 1992, **63**, 4277.
253 D. S. Cornett, M. Peschke, K. LaiHing, P. Y. Cheng, and K. F. Willey, *Rev. Sci. Instrum.*, 1992, **63**, 2177.
254 D. R. Jardine, J. Morgan, D. S. Alderdice, and P. J. Derrick, *Org. Mass. Spectrom.*, 1992, **27**, 1077.
255 R. Orlando and R. K. Boyd, *Org. Mass. Spectrom.*, 1992, **27**, 151.
256 S. A. McLuckey, *J. Am. Soc. Mass Spectrom.*, 1992, **3**, 599.
257 K. Biemann, *Pure Appl. Chem.*, 1993, **65**, 1021.
258 T. Donovan, C. C. Liou, and J. Brodbelt, *J. Am. Soc. Mass Spectrom.*, 1992, **3**, 39.
259 C. Paradisi, J. F. J. Todd, P. Traldi, and U. Vettori, *Org. Mass. Spectrom.*, 1992, **27**, 251.
260 K. L. Morand, K. A. Cox, and R. G. Cooks, *Rapid Commun. Mass Spectrom.*, 1992, **6**, 520.
261 S. A. Lammert and R. G. Cooks, *Rapid Commun. Mass Spectrom.*, 1992, **6**, 528.

The effect of the radial and axial dimensions on the Mathieu stability diagram (in terms of DC and RF voltages) has been discussed in relation to the stretched geometry employed in a commercial quadrupole ion-trap mass spectrometer.[262] The effects of storing ions of different values of the stability parameters a_z and q_z were studied in a quadrupole ion trap using helium or argon buffer.[263] A region was localized near the boundaries of the stability diagram where the ions experience an increase in their kinetic energy, reflected by the occurrence of fragmentation due to collisional activation, and a certain extent of ion loss due to unstable trajectories. These effects, known as boundary effects, depend on the buffer gas and on q_z of the ions, it is reported.[264] Simultaneous data obtained from each of the two steps of a MS/MS experiment show that the collisional history of the ions before the isolation process can greatly influence the degree to which ions can survive isolation. The facility with which ions absorb energy from the field within the trap, or whether the field is derived from the radiofrequency potential or a supplementary potential, can determine the extent to which ions are retained in the field.[265]

The fullerenes C_{60} and C_{70} were found to be ionized by desorption from a liquid matrix upon bombardment by caesium ions of 7 KeV kinetic energy.[266] The resulting radical cations, when activated in the ion trap by collisions with Xe target gas, undergo extensive dissociation by loss of multiple C_2 units.

Expressions for the ion-resonance ejection line shape have been derived, incorporating terms for ion–molecule collisions and variation, with time, of the resonance excitation frequency.[267] The results show that for any non-zero scan rate, there is an optimum collision damping value corresponding to a resolution maximum. Schwartz and Jardine discovered that the same techniques used for generation of high-resolution mass spectra on a quadrupole ion trap can be used for selection or storage of a very narrow mass window with nearly 100% efficiency.[268] Resonant ion ejection in a quadrupole ion trap was used to obtain mass measurement accuracy of better than 0.015% on gramacidin S.[269] Traldi, Catinella, and Bortolini suggested that mass displacements seen in the ion-trap mass spectra of certain fluorinated benzene derivatives were due to an interaction with the permanent dipole moment of the ion within the radiofrequency field.[270] It was found that at specific q_z values higher-order field effects caused a decrease in the intensity of chemical ionization reagent ions, due to inefficient storage. These are found to occur continuously along iso-β_2 lines for $\beta_2 = 1/3$, and 2/3.[271] McLuckey *et al.* found that random noise applied to the end caps of quadrupole ion traps was effective in inducing collisional activation of trapped ions independent of m/z ratio and number of ions.[272] Efficiency was a factory of three less

[262] J. V. Johnson, R. E. Pedder, and R. A. Yost, *Rapid Commun. Mass Spectrom.*, 1992, **6**, 760.
[263] K. L. Morand, S. H. Hoke, II, M. N. Eberlin, G. Payne, and R. G. Cooks, *Org. Mass. Spectrom.*, 1992, **27**, 248.
[264] C. Paradisi and P. Traldi, *Rapid Commun. Mass Spectrom.*, 1993, **7**, 690.
[265] R. E. March, M. R. Wier, M. Tkaczyk, F. A. Londry, R. L. Alfred, A. M. Franklin, and J. F. J. Todd, *Org. Mass. Spectrom.*, 1993, **28**, 499.
[266] H. F. Wu and J. Brodbelt, *Int. J. Mass Spectrom. Ion Processes*, 1992, **115**, 67.
[267] D. E. Goeringer, W. B. Whitten, M. J. Ramsey, S. A. McLuckey, and G. L. Glish, *Anal. Chem.*, 1992, **64**, 1434.
[268] J. C. Schwartz and I. Jardine, *Rapid Commun. Mass Spectrom.*, 1992, **6**, 313.
[269] J. D. Williams and R. G. Cooks, *Rapid Commun. Mass Spectrom.*, 1992, **6**, 524.
[270] P. Traldi, S. Catinella, and O. Bortolini, *Org. Mass. Spectrom.*, 1992, **27**, 927.
[271] D. M. Eades and R. A. Yost, *Rapid Commun. Mass Spectrom.*, 1992, **6**, 573.
[272] S. A. McLuckey, D. E. Goeringer, and G. L. Glish, *Anal. Chem.*, 1992, **64**, 1455.

than for conventional methods, and there is evidence for discrimination against low *m*/*z* product ions for multiply-charged biomolecules, however.

A quadrupole ion-trap mass spectrometer was used to examine internal energy deposition and fragmentation pathways of ionized pyrene and anthracene.[273] Through multiple collisional-activation steps, large amounts of energy can be deposited and multiple-stage dissociation reactions are observed. High-energy pathways can also be accessed in single stage MS/MS experiments by introducing a small amount of xenon as a target gas with the helium buffer, which also imparts a high dissociation efficiency.

A capillary SFC instrument with direct on-column injection was connected to a quadrupole mass filter/ion trap by a custom built interface.[274] The ion source was a standard electron impact-chemical ionization and operated using conventional ion trap electronics. This enterprising system was characterized and optimized. A sample of 80 fmol of anthracene provided a molecular-ion peak with a signal-to-noise ratio of 20. The interfacing of particle beam analyses with ion traps has been performed. The particle-beam interface was directly coupled with a quadrupole ion-trap mass spectrometer without the use of an external ion source, and vaporization of the particle beam and ionization by electron impact are all accomplished in the ion trap. Lower limits of detection for pesticides were seen than for conventional particle-beam systems using quadrupole mass analysers, although non-linear calibration curves seem to be the rule.[275]

Quantitative analysis by ion-trapping was investigated by Strife and Simms, when they used one gram plasma samples, spiked with antiinflammatory drugs at levels of 1–100 ng (with a standard at 50 ng).[276] Crude extracts of the sample were analysed using scan voltages combining radiofrequency (r.f.) and d.c. voltages, and an ion-trap tandem mass spectrometer. Ion-trap chemical ionization performance has been improved by the application of a modified scan function for the rejection of undesired electron-ionization-like ions formed at the beginning of the ionization period.[277] Desorption–ionization has been accomplished in a quadrupole ion trap by using a fibreoptic interface, with Gramicidin C as the reagent.[278] Trapping was found to be most efficient using a high gas pressure (≥ 1 mtorr), a long storage delay (≥ 50 ms) prior to detection, and a low radio-frequency trapping potential during the desorption pulse. Ion–molecule reactions were also observed during the experiments. A MALDI instrument for use with ion trap mass spectrometry has been developed. The laser beam and the sample are brought through holes in the ring electrode, allowing the laser-desorbed ions to expand and fill the ion trap. Detection limits in the femtomole range have been achieved for many small polypeptides such as bradykinin, neuromedim U-8, and leucine-enkephalin. However, a loss of sensitivity was seen at *m*/*z* above 3000.[279]

[273] B. D. Nourse, K. A. Cox, K. L. Morand, and R. G. Cooks, *J. Am. Chem. Soc.*, 1992, **114**, 2010.
[274] J. D. Pinkston, T. E. Delaney, K. L. Morand, and R. G. Cooks, *Anal. Chem.*, 1992, **64**, 1571.
[275] B. L. Kleintop, D. M. Eades, and R. A. Yost, *Anal. Chem.*, 1993, **65**, 1295.
[276] R. J. Strife and J. R. Simms, *J. Am. Soc. Mass Spectrom.*, 1992, **3**, 372.
[277] T. Cairns, K. S. Chiu, E. Siegmund, and M. Weber, *Rapid Commun. Mass Spectrom.*, 1992, **6**, 449.
[278] A. McIntosh, T. Donovan, and J. Brodbelt, *Anal. Chem.*, 1992, **64**, 2079.
[279] D. M. Chambers, D. E. Goeringer, S. A. McLuckey, and G. L. Glish, *Anal. Chem.*, 1993, **65**, 14.

3 Theoretical Organic Chemistry

By C. A. REYNOLDS
Department of Chemistry and Biological Chemistry, University of Essex, Wivenhoe Park, Colchester, CO4 3SQ, UK

1 Introduction

Theoretical organic chemistry is a wide ranging topic; theoretical chemists approach this discipline from a number of diverse perspectives ranging from extremely accurate methods applied to small model systems through to methods applicable to large biological molecules where the aim is to seek insight into complex processes rather than to produce quantitative results. Fortunately the author of the review has been encouraged to be selective rather than exhaustive. Consequently, the topics covered reflect the author's interests and have been grouped into three sections: density functional theory (DFT), because the author believes this will have a tremendous impact on theoretical organic chemistry in the years to come (for the reasons outlined below); calculations involving solvation, since although solvation is still largely neglected, it is clearly important in organic chemistry; and transition structures, because they are the key to understanding mechanism.

To conclude, a short section is appended at the end on neural networks, since these represent just one of many new computational techniques which are beginning to be applied in interesting ways to chemistry. Indeed, one of the criteria for including articles for review is the author's perception of the future potential of the methods for use in application to organic chemistry.

2 Density Functional Theory

In the opinion of the author, density functional theory (DFT)[1] is poised to make a revolutionary contribution to theoretical organic chemistry. The reasons for this are that it is potentially exact,[2] it scales very favourably with size (usually N^3 where N is the number of basis functions, though there is currently much research on order N methods in quantum chemistry) and it has recently been shown to give very promising results (see Table 1). Since the method may be unfamiliar to many readers, the relevant features are discussed below, followed by an overview of a selected proportion of the interesting results published in 1993.

Background.—Within traditional molecular orbital (MO) methods, where the total

[1] R. G. Parr and W. Yang, 'Density Functional Theory of Atoms and Molecules', Oxford University Press, New York, 1989.
[2] P. Hohenberg and W. Kohn, *Phys. Rev. B*, 1964, **136**, 864.

wavefunction, $\boldsymbol{\Psi}$, is constructed as a product (actually determinant) of N one-electron molecular orbitals, ϕ,

$$\boldsymbol{\Psi} = |\phi_1\phi_2\phi_3\cdots\phi_N| \quad (1)$$

the total energy of a molecule is composed of one-electron terms (the kinetic energy and the electron–nuclear potential terms) and two-electron terms. The two-electron terms consist of the coulomb energy and the exchange energy. All of these terms involve integrals over basis functions and the integrals are evaluated analytically.

In addition, accurate calculations require evaluation of the correlation energy, which is not present in self-consistent field (SCF) MO calculations. Generally, the correlation energy is evaluated by including excited states and so effectively the total wavefunction is now a linear combination of a large number of determinants similar to that on the right hand side of Equation (1). Such multi-determinant wavefunctions are essential for the accurate description of many chemical processes, particularly dissociation, and so they are also essential for a reliable description of transition states. *It is the calculation of the correlation energy which render impossible accurate ab initio calculations for systems of real chemical interest* because the calculations scale according to N^5 for the simplest approach [second order Møller–Plesset theory (MP2)], which does not even describe dissociation correctly, and according to N^7 for the more reliable approaches such as higher order Møller–Plesset perturbation theory or coupled cluster calculations. *In the light of this unfavourable scaling, density functional theory possibly offers the only reliable alternative.* The essentials of the method are given below.

Density Functional Methods.—Within the local density approximation (LDA), the exchange energy is evaluated as follows

$$E_x^{LDA} = C\int dr\rho(r)^{4/3} \quad (2)$$

where C is a constant and ρ is the electron density. The integral is now usually evaluated numerically over a grid. A key question when assessing DFT calculations is whether this integral has been evaluated accurately; a reasonable check is also to integrate $\rho(r)$ over the grid, which should give the total number of electrons in the molecule.

Probably the most significant change came in 1988 when Becke proposed that the gradient of the density was also important in evaluating the exchange energy.[3] (Such methods involving $\nabla\rho(r)$ are termed non-local or gradient methods.)

$$E_x^B = E_x^{LDA} - \beta\int dr\rho^{4/3}\frac{x^2}{1 + 6\beta x \sin h^{-1} x} \quad (3)$$

where β is a constant and $x = |\nabla\rho(r)|/\rho^{4/3}$. The correlation energy can be evaluated in a similar manner using local or non-local functionals. The non-local methods are usually denoted by a combination of letters to denote the combination of functionals for exchange and correlation; thus B-LYP denotes the Becke '88 exchange functional[3] plus the Lee, Yang, and Parr correlational functional;[4] B-VWN denotes the Becke

[3] A. D. Becke, Phys. Rev. A, 1988, **38**, 3098.
[4] C. Lee, W. Yang, and R. G. Parr, *Phys. Rev. B*, 1988, **37**, 785; B. Miehlich, A. Savin, H. Stoll, and H. Preuss, *Chem. Phys. Lett.*, 1989, **157**, 200.

exchange functional plus the Vosko, Wilk, and Nussair correlational functional.[5] There is currently much research into new functionals and we can certainly expect the current functionals to be superseded.

A number of choices exist as to how the density is evaluated. The integrals in Equations (2) and (3) may be evaluated using any density, and results show that the answers may not differ significantly whether the density is evaluated in a manner which is self consistent with equation (2) or (3);[6] alternatively any other density such as an SCF density may be used. In chemical applications, the density is usually constructed from atom-centred basis functions as in traditional molecular orbital methods though the resultant orbitals thus produced by the Kohn–Sham method may not have the same physical meaning. In studies of molecules interacting with surfaces, it is common to construct the density from plane waves (see section on Carr–Parrinello methods).

Model Studies on Small Molecules.—Since DFT is a reasonably new technique for chemists (but not for physicists who generally look at much larger energy differences), it has been important to establish the reliability of density functional theory in chemical problems. It is this area where significant and exciting progress has been made in the last twelve months. To date there have been few genuine application studies that have significantly enhanced our understanding of chemistry. However, there have been many significant studies on model systems. Some studies have considered the performance of DFT in predicting molecular properties for a large number of molecules; understandably, the most systematic study, by Johnson *et al.*, came out of the Pople group.[7] The general findings are: for bond lengths, the B-LYP functional gives bond lengths which are too long by about 0.018 Å – the error is comparable to that of Hartree–Fock methods but the sign of the error is reversed; for dipole moments, the errors are similar to those of Hartree–Fock methods, except that DFT has some success for difficult molecules like CO and NO; atomization energies are extremely well predicted by both the B-LYP and B-VWN methods, with a mean error close to zero; and harmonic frequencies are generally well produced, with the errors generally less than those obtained using MP2 methods.

Studies on bond breaking processes have also found that non-local DFT is valid for multideterminant problems,[8] and this is a very significant observation (see also transition state references in Table 1). For energetics, the majority of researchers find that density functional theory generally performs as well as MP2 (in the cases where MP2 is valid). This appears to be at odds with Johnson *et al.*[7], who generally find it vastly superior. However, as Johnson shows, the errors in traditional approaches tend to be systematic whereas the errors in DFT methods tend to be random and this tends to account for why DFT performs similarly to MP2.

Not all of the studies in Table 1 are significant by themselves. However, the weight of the evidence presented in this selection of papers is extremely strong, particularly for those involving calculations of dissociation energies or transition structures. Moreover, given the volume of research into new functionals, the preference for density functional theory over traditional approaches will grow and we will surely see DFT having a very profound influence on theoretical organic chemistry. Partly as a warning,

[5] S. H. Vosko, L. Wilk, and M. Nusair, *Can. J. Phys.*, 1980, **58**, 1200.
[6] C. T. Lee, G. Fitzgerald, and W. T. Yang, *J. Chem. Phys.*, 1993, **98**, 2971.
[7] B. G. Johnson, P. M. W. Gill, and J. A. Pople, *J. Chem. Phys.*, 1993, **98**, 5612.
[8] C. W. Murray, N. C. Handy, and R. D. Amos, *J. Chem. Phys.*, 1993, **98**, 7145.

some of the articles in Table 1 are based on local density functional theory, which the table shows may give good results. However, a closer look will show that reliable results often require non-local methods. (Needless to say the table does not show the cases where density functional theory performs badly.)

Applications.—At this point it is helpful to review briefly two applications that illustrate the power of density functional methods. The first is a study of the isomers of N_2H_2 and illustrates how favourably the energy barriers connecting the various isomers of N_2H_2 calculated using non-local density functional theory compared to those calculated using high level *ab initio* methods such as G2 theory.[9,10] The potential energy surface is illustrated in Figure 1, in which the relative energies given (in kcal mol^{-1}) are for Hartree–Fock; non-local DFT; G2 theory, respectively.)

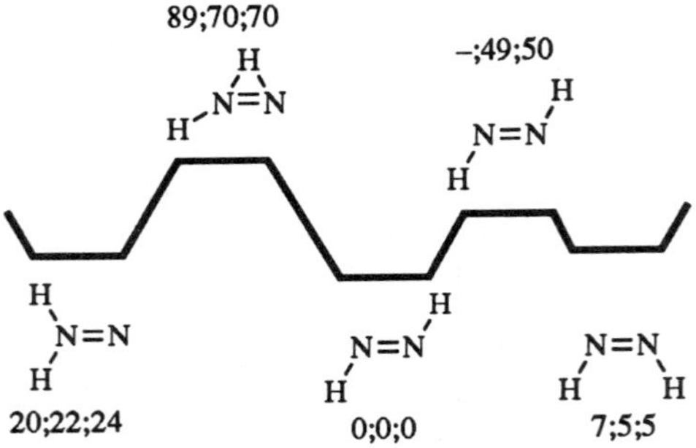

Figure 1

The use of DFT to study organometallics is aptly illustrated by studies on H–H and C–H bond activation by the σ-bond metathesis reaction (Scheme 1).[11] The bond dissociation energies of the complex have been calculated to within 20–30 kJ mol^{-1} (4.5–7.5 kcal mol^{-1}) of experiment. The transition structures for the above reaction have been located and the activation energies are consistent with the experimentally observed rate constants, thus the results are generally in line with experiment.

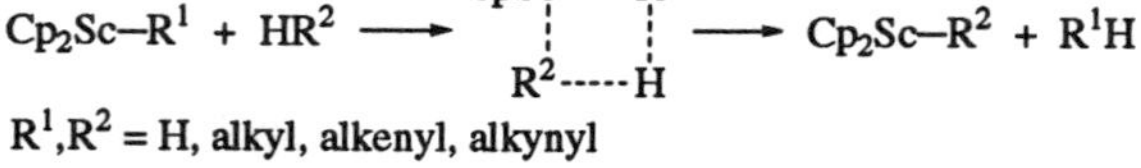

R^1,R^2 = H, alkyl, alkenyl, alkynyl

Scheme 1

[9] J. Andzelm, C. Sosa, and R. A. Eades, *J. Chem. Phys.*, 1993, **97**, 4664.
[10] B. J. Smith, *J. Phys. Chem.*, 1993, **97**, 10 513.
[11] T. Ziegler, E. Folga, and A. Berces, *J. Am. Chem. Soc.*, 1993, **115**, 636.

Table 1 *The performance of density functional methods**

Ref.	Geometry	Vib. frequencies	Ionization potential	Proton affinities	Atomic energies	Dissociation energy	Isomerization energy	Transition state	Other properties	Large no. of molec.	LDA only	Comp. to expt	Comp. to theory
[6]			•			•	•		•			•	•
[7]	•	•			•	•			•	•		•	•
[8]	•	•			•				•			•	•
[9]	•	•				•	•	•	•			•	•
[10]							•	•					
[17]	•											•	•
[18]			•	•	•	•				•		•	•
[19]			•	•	•	•			•	•		•	
[20]	•	•				•			•	•		•	
[21]	•								•	•	•	•	•
[22]	•	•	•			•						•	•
[23]									•				
[24]	•			•				•	•	•	•	•	•
[25]	•					•	•					•	
[26]	•	•				•			•			•	
[27]	•	•									•	•	•
[28]	•	•										•	•
[29]	•	•									•	•	
[30]		•					•		•			•	•
[31]	•	•										•	•
[32]									•			•	•
[33]	•	•											•
[34]			•						•		•	•	•
[35]			•							•		•	
[36]	•	•				•						•	•
[37]						•		•		•			
[38]		•				•							
[39]									•		•		•
[40]		•									•		
[41]									•	•		•	
[42]	•							•					•
[43]	•								•				•
[44]	•	•	•						•			•	•
[45]						•						•	•
[46]									•		•		
[47]	•	•							•		•	•	•
[48]	•											•	
[49]	•	•											
[50]	•					•		•				•	

*These studies generally include comparisons to experiment or traditional molecular orbital calculations; the latter usually include some treatment of electron correlation.

Finally, it is appropriate to note that DFT methods are now available in the popular GAUSSIAN[12] and CADPAC[13] traditional quantum chemical codes, as well as from other commercial sources.

Carr–Parrinello Methods.—Carr–Parrinello methods combine quantum mechanical methods with molecular dynamics. (The local density approximation is usually used, along with pseudopotentials to represent the core electrons and the basis set is typically constructed from plane-waves rather than atomic orbital-like basis functions.) The inherent advantage is that molecular dynamics can be performed without the need for crude force-fields containing parameters which may be far from ideal. In addition, bond breaking processes can be followed. The potential of this method for studying reactions is tremendous. However, to date there are a number of technical problems and applications have largely been applied to cluster chemistry and related problems. References 14 and 15 illustrate the method and show how it can be used in combination with simulated annealing to overcome the multiple minimum problem and determine the global minimum for molecular structures.

Excited States.—The Hohenberg–Kohn theorem[2], is, in principle, valid only for the ground state electron density, so there is as yet no universally held justification for applying DFT methods to excited states, despite much research in this area. However, a number of authors, for example Dual *et al.*,[16] have shown that DFT can yield meaningful results when applied to excited states.

3 Solvation

Background.—Although it is widely understood by organic chemists that solvent effects can have a dramatic effect on reactivity, it is rather surprising that most theoretical studies still do not include the effects of solvation as a matter of course. There are a number of reasons for this. One is the lack of suitable software, though this reason is rapidly becoming out of date. Another is that while solvation effects are indeed important, the errors due to the limitations of semiempirical treatments, the errors due to basis set size limitations, and the errors due to inadequate treatment of electron correlation (see above) are all still the largest potential errors in any calculation. A third reason is the lack of realistic models for treating solvation since all the models are defective in one way or another despite much research in recent years. A final reason is that some methods, particularly those involving computer simulations, require specialist skills different to those required for accurate gas-phase studies of potential energy surfaces.

One notable contribution towards providing readily available software for solvation

[12] 'GAUSSIAN 92/DFT', Gaussian Inc, 4415 Fifth Avenue, Pittsburgh, PA 15213, USA.
[13] R.D. Amos, 'CADPAC 5.1', University Chemical Laboratories, Lensfield Road, Cambridge, UK CB2 1EW.
[14] K. Laasonen, M. Parrinello, R. Car, C.Y. Lee, and D. Vanderbilt, *Chem. Phys. Lett.*, 1993, **207**, 208.
[15] R.O. Jones, *J. Chem. Phys.*, 1993, **99**, 1194.
[16] C. Dual, H.U. Gudel, and J. Weber, *J. Chem. Phys.*, 1993, **98**, 4023.

studies is the AMSOL program,[51] which is available through QCPE.[52] The authors of AMSOL, Cramer and Truhlar, have published an excellent review[53] of solvation models and despite the title ('Continuum solvation models: classical and quantum mechanical implementations') both empirical and explicit water simulation models are discussed, in addition to Poisson–Boltzmann, reaction-field methods, and generalized Born methods, which are discussed in some detail. The review also includes a comparison (with experiment) of the various methods for determining free energies of hydration. (Cramer and Truhlar's own generalized Born method[54] certainly does well in this comparison). An excellent review of Poisson–Boltzmann methods[55] has also been published in the last year. A particular advantage of this method is that it is applicable to large systems and can include ionic strength in the calculations. Typical uses include calculating the molecular electrostatic potential around macromolecules and calculating pK_a shifts in enzymes, but as yet there have been few applications to

[17] L. A. Eriksson, S. Lunell, and R. J. Boyd, *J. Am. Chem. Soc.*, 1993, **115**, 6896.
[18] A. D. Becke, *J. Chem. Phys.*, 1993, **98**, 5648.
[19] A. D. Becke, *J. Chem. Phys.*, 1993, **97**, 9173.
[20] H. Chen, M. Krasowski, and G. Fitzgerald, *J. Chem. Phys.*, 1993, **98**, 8710.
[21] R. M. Dickson and A. D. Becke, *J. Chem. Phys.*, 1993, **99**, 3898.
[22] L. Goodwin and D. R. Salahub, *Phys. Rev. A*, 1993, **47**, R774.
[23] D. P. Chong, D. Papousek, Y. T. Chen, and P. Jensen, *J. Chem. Phys.*, 1993, **98**, 1352.
[24] M. S. Stave and J. B. Nicholas, *J. Phys. Chem.*, 1993, **97**, 9630.
[25] E. Folga and T. Ziegler, *J. Am. Chem. Soc.*, 1993, **115**, 5169.
[26] I. Papai, J. Mink, R. Fournier, and D. R. Salahub, *J. Phys. Chem.*, 1993, **97**, 9986.
[27] K. O. Christe, W. W. Wilson, D. A. Dixon, S. I. Khan, R. Bau, T. Metzenthin, and R. Lu, *J. Am. Chem. Soc.*, 1993, **115**, 1836.
[28] K. O. Christe, D. A. Dixon, J. C. P. Sanders, G. J. Schrobilgen, and W. W. Wilson, *J. Am. Chem. Soc.*, 1993, **115**, 9461.
[29] K. O. Christe, D. A. Dixon, I. B. Goldberg, C. J. Schack, B. W. Walther, J. T. Wang, and F. Williams, *J. Am. Chem. Soc.*, 1993, **115**, 1129.
[30] A. Berces and T. Ziegler, *J. Chem. Phys.*, 1993, **98**, 4793.
[31] C. W. Murray, G. J. Laming, N. C. Hamdy, and R. D. Amos, *J. Phys. Chem.*, 1993, **97**, 1868.
[32] J. G. Guan, P. Duffy, J. T. Carter, D. P. Chong, K. C. Casida, M. E. Casida, and M. Wrinn, *J. Chem. Phys.*, 1993, **98**, 4753.
[33] M. M. Huhn, R. D. Amos, R. Kobayashi, and N. C. Handy, *J. Chem. Phys.*, 1993, **98**, 7107.
[34] D. P. Chong and C. Y. Ng, *J. Chem. Phys.*, 1993, **98**, 759.
[35] P. Duffy and D. P. Chong, *Org. Mass. Spec.*, 1993, **28**, 321.
[36] D. Heinemann and A. Rosen, *Theor. Chim. Acta.*, 1993, **85**, 249.
[37] P. Politzer, J. M. Seminario, M. C. Concha, and J. S. Murray, *Theor. Chim. Acta.*, 1993, **85**, 127.
[38] C. Sosa, C. Lee, G. Fitzgerald, and R. A. Eades, *Chem. Phys. Lett.*, 1993, **211**, 265.
[39] P. Fleukiger, J. Weber, R. Chiarelli, A. Rassat, and Y. Ellinger, *Int. J. Quant. Chem.*, 1993, **45**, 649.
[40] A. Berces and T. Ziegler, *Chem. Phys. Lett.*, 1993, **203**, 592.
[41] D. P. Chong and A. V. Bree, *Chem. Phys. Lett.*, 1993, **210**, 443.
[42] C. Mijoule, Z. Latajka, and D. Borgis, *Chem. Phys. Lett.*, 1993, **208**, 364.
[43] C. Lee and D. Vanderbilt, *Chem. Phys. Lett.*, 1993, **210**, 279.
[44] A. Ghosh and J. Almlof, *Chem. Phys. Lett.*, 1993, **213**, 519.
[45] J. M. Seminario, *Chem. Phys. Lett.*, 1993, **206**, 547.
[46] S. M. Colwell, C. W. Murray, N. C. Handy, and R. D. Amos, *Chem. Phys. Lett.*, 1993, **210**, 261.
[47] I. A. Topol and S. K. Burt, *Chem. Phys. Lett.*, 1993, **204**, 611.
[48] K. Waizumi, H. Masuda, H. Einaga, and N. Fukushima, *Chem. Lett.*, 1993, **7**, 1145.
[49] R. D. Amos, C. W. Murray, and N. C. Handy, *Chem. Phys. Lett.*, 1993, **202**, 489.
[50] T. Ziegler, E. Folga, and A. Berces, *J. Am. Chem. Soc.*, 1993, **115**, 636.
[51] C. J. Cramer, G. C. Lynch, G. D. Hawkins, D. G. Truhlar, and D. A. Liotard, AMSOL 4.0, QCPE 606, QCPE Bulletin, 1993, **13(4)**.
[52] Quantum Chemistry Program Exchange, Creative Arts Building 181, Indiana University, Bloomington, IN 47405, USA.
[53] C. J. Cramer and D. G. Truhlar, *Reviews in Computational Chemistry*, in the press.
[54] C. J. Cramer and D. G. Truhlar, *J. Comput.-Aided Mol. Des.*, 1992, **6**, 629.
[55] B. Honig, K. Sharp, and A. S. Yang, *J. Phys. Chem.*, 1993, **97**, 1101.

organic chemistry despite recent encouraging results referred to in the review. One recent application has been to non-natural DNA triplex-forming oligonucleotides.[56]

Computer Simulation Methods.—References to computer simulation reviews and texts are given in reference 19. Free energy methods[57] have yielded very high levels of agreement between calculated and experimental free energies, including free energies of hydration.[58] The real advantage of these methods is the treatment of specific solvent molecules. Jorgensen has shown how these methods, implemented within a Monte Carlo framework, can be useful for characterizing the hydration patterns of molecules. In particular, the solute–water hydrogen bonding patterns can be noted.[59–61] Such effects may be important in determining mechanistic effects and conformational preferences in solution.[62] An algorithm for calculating residence times at particular hydration sites has been presented and may be helpful for analysing such structural features, particularly for biomolecules.[63]

Combining Quantum Mechanical Calculations with Computer Simulations.—Computer simulations are of necessity based on molecular mechanics (but see above for references to Carr–Parrinello methods). They cannot in general therefore describe bond-breaking processes. However, a number of studies have described the combined use of quantum chemical studies with free energy perturbation (FEP) methods.[23] Such approaches have been used to study the mechanism of carbonic anhydrase (by studying the underlying elementary reactions)[64,65] and the Mentshutkin reaction.[66] The effect of solvation in reference 65 is quite marked: although the gas phase reaction is barrierless, the barrier in solution is over 19 kcal mol^{-1} and is mainly due to desolvation effects. Reference 64 actually describes the use of an empirical valence bond method, but the results are compared to *ab initio* calculations. The main value of this particular study is probably in illustrating the catalytic effect of the environment. The solvent was found to have a marked effect on the energetics, the structure of the transition state in the type II S_N2 Mentshutkin reaction, and the charge distribution in the transition structure, demonstrating the value of combined quantum mechanical–molecular mechanical calculations.

Combined calculations have also been applied to the torsional barrier in π-ethylimidazole.[67] Here it was found that the torsional barrier was raised in solution by nearly 2 kcal mol^{-1} and that the position of the barrier moved. Attempts were made to rationalize the effects in terms of solute–solvent hydrogen bonds, but the effects were considerably more complex than expected.

Tautomeric Equilibria. Tautomeric equilibria provide very popular test systems for the

56 V. Mohan, Y.K. Cheng, G.E. Marlow, and B.M. Pettitt, *Biopolymers*, 1993, **33**, 1317.
57 D.L. Beveridge and F.M. DiCapua, *Annu. Rev. Biophys. Biophys Chem.*, 1989, **18**, 431.
58 C.A. Reynolds, P.M. King, and W.G. Richards, *Mol. Phys.*, 1992, **76**, 251.
59 W.L. Jorgensen and T.B. Nguyen, *J. Comput. Chem.*, 1993, **14**, 195.
60 H.A. Carlson, T.B. Nguyen, M. Orozco, and W.L. Jorgensen, *J. Comput. Chem.*, 1993, **14**, 1240.
61 P.I. Nagy, G.J. Durant, and D.A. Smith, *J. Am. Chem. Soc.*, 1993, **115**, 2912.
62 P.I. Nagy, W.J. Dunn, G. Alagona, and C. Ghio, *J. Phys. Chem.*, 1993, **97**, 4628.
63 A.E. Garcia and L. Stiller, *J. Comput. Chem.*, 1993, **14**, 1396.
64 J. Aqvist, M. Fothergill, and A. Warshel, *J. Am. Chem. Soc.*, 1993, **115**, 631.
65 Z. Peng and K.M. Merz, *J. Am. Chem. Soc.*, 1993, **115**, 9640.
66 J.L. Gao and X.F. Xia, *J. Am. Chem. Soc.*, 1993, **115**, 9667.
67 D.R. Lowis, J.W. Essex, and W.G. Richards, *Molec. Sim.*, 1993, **9**, 369.

reliability of solvation free energy calculations, because of the chemical interest of the underlying problem; however the results are frequently ambiguous as they require solvation calculations to be combined with high-level quantum chemical calculations. Thus Burton *et al.* studied the tautomeric equilibria between the dihydroxy (1), monohydroxy-monoketo (2), and the diketo (3) forms of maleic hydrazide; they showed that both FEP methods and self-consistent-reaction field (SCRF) methods (see below) predict the monohydroxy-monoketo to be the most stable, and that solvation stabilizes the diketo form but that the dihydroxyform is unlikely to be observed, in agreement with experiment.[68] In contrast, however, the same group showed that neither the FEP method, nor the polarized continuum method (PCM),[69] nor the SCRF method yield results that are totally adequate for modelling the equilibria involving 3-hydroxypyrazole (4) and its tautomers (5)–(7) even though the errors were not large.[70]

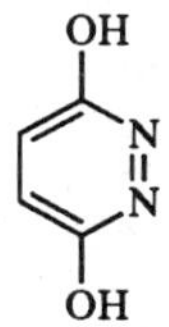
(1)

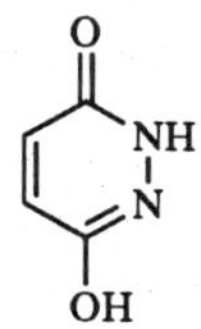
(2)

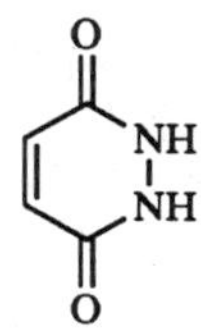
(3)

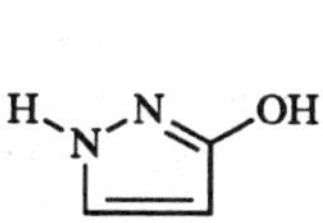
(4)

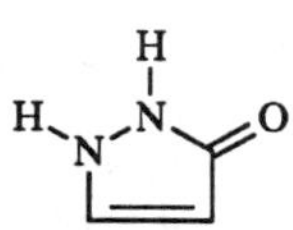
(5)

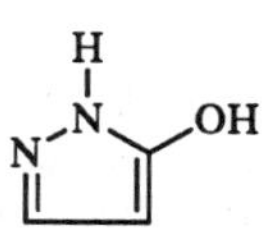
(6)

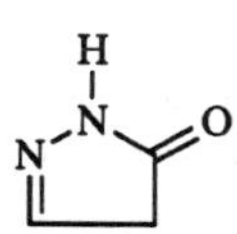
(7)

The calculations do show that polarization of the solute is important (which is not generally included in FEP methods). These two examples clearly show that great care is still required in applying these methods despite much recent progress.

Computational Cost. In contrast to many implementations of the SCRF methods, FEP methods remain computationally very expensive, though a particularly imaginative method for reducing the cost of FEP calculations is given in reference 71. However, many drawbacks of the method remain and some are listed in reference 72. If CPU time is a real issue, then the lower-level implementations of continuum methods remain more promising for small molecules; a real advantage of continuum methods is that the calculations are frequently much more easy to set up.

Continuum Methods.—The Kirkwood–Onsager result for the free energy of solvation in a medium of relative dielectric constant ε of a dipole μ centred in a sphere of radius α is given by Equation (4).

[68] N. A. Burton, D. V. S. Green, I. H. Hillier, P. J. Taylor, M. A. Vincent, and S. Woodcock, *J. Chem. Soc., Perkin Trans. 2*, 1993, 331.
[69] S. Miertus, E. Scrocco, and J. Tomasi, *Chem. Phys.*, 1981, **55**, 117.
[70] O. G. Parchment, D. V. S. Green, P. J. Taylor, and I. H. Hillier, *J. Am. Chem. Soc.*, 1993, **115**, 2352.
[71] G. King and R. A. Barford, *J. Phys. Chem.*, 1993, **97**, 8798.
[72] C. A. Reynolds, J. W. Essex, and W. G. Richards, *Chem. Phys. Lett.*, 1992, **199**, 257.

$$\Delta G = -\frac{(\varepsilon - 1)\mu^2}{(2\varepsilon + 1)\alpha^3} \quad (4)$$

The result includes polarization of the solvent by the solute. Quantum mechanical reaction field methods are frequently based on this result through modification of the self-consistent field (SCF) equations and are termed self-consistent reaction-field (SCRF) methods. Limitations of this approach frequently include: truncation of the multipole series at the dipole level; the use of a spherical (or ellipsoidal) cavity, which is frequently unrealistic; the wavefunction, Ψ, is approximate and does not usually include electron correlation (it may even be semiempirical); and the results depend markedly on the choice of cavity radius, α.

In recent work the quantum mechanical dipole approximation has been extended to higher multipoles within arbitrarily shaped cavities.[73] However, the convergence of such methods with regard to the multipole series and size/shape of cavity has recently been questioned;[74] in this work a distributed multipole series has been used – this approach certainly does need a molecular-shaped cavity.[39]

A number of studies have also extended the SCRF method beyond the SCF level to the Møller–Plesset second order level (MP2)[75] and to the configuration interaction (CI) level;[76] while this latter approach, like all continuum methods, is generally considered to be unable to treat strong specific solvation effects such as hydrogen bonds, it can treat excited states. An implementation of the polarizible continuum method of Miertus *et al.*, including electron correlation up to third order, has also been proposed.[77] However, given the rather crude description of the molecular shape usually used (but not in the PCM approach) and noting that the choice of radius has a marked effect on the predicted ΔG value, it is questionable whether the more accurate treatment of Ψ yields more accurate free energies of solvation, particularly given the other known errors in continuum methods.

In defence of traditional SCRF methods, Young *et al.* noted that the method gave results of comparable quality to those from molecular dynamics simulations when applied to tautomeric equilibria, partition coefficients, and amine basicity. However, the continuum methods systematically overestimated the free energies of hydration, though the errors tended to cancel upon taking differences. Moreover, they also found that going beyond the dipole level[78] in the SCRF method did not increase the accuracy. Good agreement with experiment was also found by the same group in studies on the conformational preferences of the alanine dipeptide.[79] Clearly the performance of these continuum methods, like any other method, depends on the problem to which it is applied.

A good illustration of the use of the PCM method is given by Tunon *et al.*, who have combined the PCM method with high level *ab initio* studies (up to MP4) to look at the difference in acid/base behaviour of a series of alcohols in both the gas phase and solution.[80]

[73] V. Dillet, D. Rinaldi, J. G. Anguyan, and J. L. Rivail, *Chem. Phys. Lett.*, 1993, **202**, 18.
[74] R. R. Pappalardo, E. S. Marcos, M. F. Ruiz-Lopez, D. Rinaldi, and J. L. Rivail, *J. Am. Chem. Soc.*, 1993, **115**, 3722.
[75] J. G. Anguyan, *Int. J. Quant. Chem.*, 1993, **47**, 469.
[76] M. V. Basilevsky, G. E. Chudinov, D. V. Napolov, and L. M. Timofeeva, *Chem. Phys.*, 1993, **173**, 345.
[77] F. J. O. Delvalle and M. A. Aguilar, *J. Mol. Struct. (THEOCHEM)*, 1993, **99**, 25.
[78] P. Young, D. V. S. Green, I. H. Hillier, and N. A. Burton, *Mol. Phys.*, 1993, **80**, 503.
[79] I. R. Gould and I. H. Hillier, *J. Chem. Soc., Chem. Commun.*, 1993, 951.
[80] I. Tunon, E. Silla, and J. L. Pascualahuir, *J. Am. Chem. Soc.*, 1993, **115**, 2226.

Semiempirical and Classical Implementations of Continuum Methods.—In terms of agreement with experiment in real applications, semiempirical continuum methods have probably given results as reliable as their *ab initio* quantum mechanical counterparts. The use of INDO/S-CI for studying excited states and spectra is quite widespread; in combination with a continuum method it has given quite good predictions of solvatochromic shifts of organic molecules.[81] Another semiempirical method, formulated within the NDDO framework, has given very encouraging results on excited states and again can predict solvatochromic shifts quite well. The particular advantages of this method[82] are the inclusion of the dispersion energy (which is not included in most continuum methods) as well as the cavity energy and very fast computation of the molecular electrostatic potential on the surface;[83] the dispersion energy calculations are relatively time-consuming but are essential to model correctly certain solvatochromic shifts. This latter method is again based on the polarizable continuum method of Miertus *et al.*,[69] which does use realistic molecular cavity shapes. Rather than trying to increase the level of treatment of Ψ in the method, the Tomasi group have shown that a classical implementation of the method using point charges (provided that they are potential-derived charges) can give quite good calculated solvation energies. This observation could be quite significant for studying macromolecules where a full molecular dynamics or quantum mechanical treatment could be very expensive.[84] (Pappalardo *et al.*[85] have also modelled the solvatochromic shift for the $n \rightarrow \pi^*$ transition of acetone using an MCSCF implementation of a continuum model with an ellipsoidal cavity).

Due to its implementation within MOPAC93,[86] the new COSMO method[87] is likely to become very popular. However, apart from showing that the glycine zwitterion is more stable in aqueous solution than the neutral form, and a very small number of comparisons with experimental values, as yet there is little indication of its general reliability.

Applications.—*Effect of Hydration on Cycloadditions.* The (2 + 2) cycloaddition of t-butylcyanoketene (8) to phenylethene (9) (Scheme 2) was studied using a four-orbital, four-electron CAS—MCSCF method, using earlier studies as a guide.

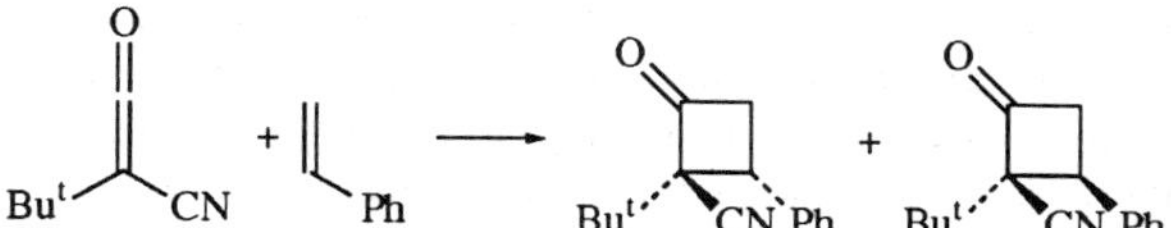

(8) (9)

Scheme 2

The most favourable route was found to proceed *via* a biradical intermediate in which steric interactions between the three substituents are minimized despite the observation that this mechanism does not produce the most stable product. The effect

[81] T. Fox and N. Rosch, *J. Mol. Struct.* (*THEOCHEM*), 1992, **95**, 279.
[82] G. Rauhut, T. Clark, and T. Steinke, *J. Am. Chem. Soc.*, 1993, **115**, 9174.
[83] G. Rauhut and T. Clark, *J. Comput. Chem.*, 1993, **14**, 503.
[84] R. Montagnani and J. Tomasi, *J. Mol. Struct.* (*THEOCHEM*), 1993, **98**, 131.
[85] R. R. Pappalardo, M. Reguero, M. A. Robb, and M. Frish, *Chem. Phys. Lett.*, 1993, **212**, 12.
[86] J. J. P. Stewart, MOPAC 93, *QCPE Bull.*, 1993, **13**.
[87] A. Klamt and G. Schuurmann, *J. Chem. Soc., Perkin Trans. 2*, 1993, 799.

of modelling hydration by a multipolar expansion (to order six) in an ellipsoidal cavity was found to enhance the gas phase product preference,[88] – though convergence problems in the multipolar expansion were noted. Significantly, the differential solvation energy corrections for such asynchronous pericyclic reactions are sufficiently large to have a marked effect on selectivity.

Nucleophilic Additions and the Supermolecule Approach to Solvation. The nucleophilic addition of ammonia to small molecules with activated double bonds, such as acrolein has been calculated at the MP2 level, with stationary points optimized at the HF/6-31G* level. The purpose of these calculations was to gain some understanding of the possible biological effects of such reactions involving DNA. Hydration energies were calculated using a polarizable dielectric continuum. However, the major reduction in the barrier height on solvation was found only when an explicit catalytic water molecule was included in the quantum mechanical treatment.[89] Such a supermolecule approach is currently the only viable method of studying solvent-assisted mechanisms. However, the calculation of hydration energies by the continuum method is probably not feasible, except in the AMSOL method, which is designed to treat explicit water molecules consistently.

Potential of Mean Force Calculations.—The free energy profile for the rotational isomerization of *N,N*-dimethylformamide has been determined using a combined quantum mechanical–molecular mechanical approach. The hybrid AM1/TIP3P model was used, with the AM1 energies scaled to reproduce high level *ab initio* results. The potential of mean field was determined by statistical perturbation theory implemented within a Monte Carlo framework.[90] The difference in polarity between the ground state and the transition state emphasizes the desirability of including polarization effects in such calculations.

4 Transition Structures

The theme of the majority of the articles reviewed in this section is inclusion of reports of calculated transition structures. Calculated transition structures and their associated barrier heights offer from theory a unique contribution to the study of organic reactivity. A useful general review on the interplay between experiment and calculated transition structures is given by Williams,[91] who remarks that the transition structure is frequently calculated more accurately than the energy barrier – as judged against very high level *ab initio* calculations. For this reason, the reader may wish to review the work reported below in the light of the earlier comments on *ab initio* methods and density functional theory; these ideas on the difficulty of obtaining definitive energy barriers are superbly illustrated in the high-quality articles described in the next section.

Diels–Alder Reactions.—The long running controversy as to whether Diels–Alder reactions occur *via* concerted or other mechanisms continues. Li and Houk[92] have

88 M. Reguero, R. R. Pappalardo, M. A. Robb, and H. S. Rzepa, *J. Chem. Soc. Perkin Trans. 2*, 1993, 1499.
89 L. Pardo, R. Osman, H. Weinstein, and J. R. Rabinowitz, *J. Am. Chem. Soc.*, 1993, **115**, 8263.
90 J. L. Gao, *J. Am. Chem. Soc.*, 1993, **115**, 2930.
91 I. H. Williams, *Chem. Soc. Rev.*, 1993, **22**, 277.
92 Y. Li and K. N. Houk, *J. Am. Chem. Soc.*, 1993, **115**, 7478.

applied state-of-the-art calculations to this problem and come down very firmly along the lines of the generally (but still not universally held) view that the process is indeed concerted for the reaction between butadiene (10) and ethylene (11) (Scheme 3). Their calculations on the reactants, concerted transition state (TS), biradical TS, and biradical intermediate for the reaction between butadiene and ethylene are summarized in Table 2.

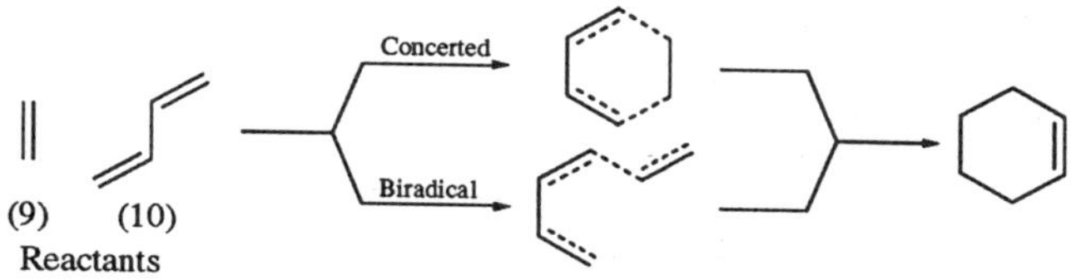

Scheme 3

The difference in energy of 5.8 kcal mol^{-1} between the concerted and stepwise transition structures calculated using the 3-21G basis set falls to 3.4 kcal mol^{-1} when zero-point energies are taken into account. For the 6-31G* basis set, the difference falls even further to 0.5 kcal mol^{-1}; thus when entropy is taken into account, the biradical process is predicted to be more favourable.

Since the CASSCF calculations include six electrons in six active orbitals, a good representation was indeed given. However, the MCSCF method overestimates the stability of biradical species. The more reliable quadratic CI results favour concerted pathways and the overall conclusion, based on several pieces of evidence, is given in the final column of the above table. As discussed elsewhere in this article, it should be noted that solvation may have a marked effect on these relative energies, and also that substituents may affect the pathway for analogues of either reactant in this reaction.

For butadiene dimerization a concerted mechanism is probably preferred, but the calculations, including eight electrons in eight active orbitals, are not sufficiently accurate to make such definitive statements. This particular controversy probably has a few more years to run!

There have been fewer studies of hetero-Diels–Alder reactions. The same group has studied the reaction of ten dienophiles *via* seventeen transition structures by assuming the reaction is concerted but not necessarily synchronous.[93] The reactivity differences and stereochemical preferences are discussed on the basis of the geometries and relative energies of the transition structures. It was shown that the lone pair orbitals on the dienophiles can exert a large destabilizing influence on the filled orbitals of the butadiene and that these lone-pair–π-electron interactions can have a large influence on stereoselectivities. In some cases it is found that solvation by dichloromethane, modelled by the Onsager reaction field, can actually stabilize the reactants more than the transition state, despite the large dipole moment in the transition state.

Likewise, there have been relatively few studies of inverse Diels–Alder reactions. An interesting claim to emerge from the reaction shown in Scheme 4 is that the regioselectivity of the reaction is controlled primarily by the (estimated) attractive

[93] M.A. McCarrick, Y.D. Wu, and K.N. Houk, *J. Org. Chem.*, 1993, **58**, 3330.

Table 2 *Relative energies (kcal mol^{-1}) of reactants, transition structures, and intermediates calculated and determined experimentally for the reaction of ethylene with butadiene (Scheme 3)*

Molecules	CASSCF/ 3-21G	CASSCF/ 6-31G*	UQCISD(T)	RQCISD(T)	Expt.	Conclusions
Reactants	0.0	0.0	0.0	0.0	0.0	
Concerted TS	37.3	43.8	29.4	25.5	25.1 ± 2	0
Biradical TS	43.1	45.7	39.2	35.7		5
Biradical intermediate	41.1	40.7	30.3	29.8	27.3	3

dispersion interactions between the phenyl rings of the reactants.[94] Semiempirical and Hartree–Fock methods are not able to take these factors into account (though molecular mechanics based transition state studies could – see below), and hence the treatment at the MP2 level was necessary to explain the experimental findings.

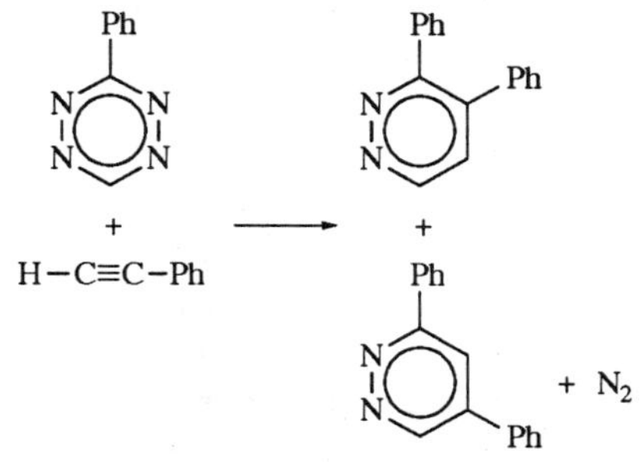

Scheme 4

An interesting AM1 semiempirical study (with limited configuration interaction) of the reaction between acrylonitrile (12) and allene (13) (Scheme 5) has been undertaken to help to explain the origin of the kinetic isotope effect observed with *gem*-dideuterioallene.[95] It is concluded that the kinetic isotope effect is due to the slower rotation of the heavier CD_2 group as it approaches the second transition structure in the stepwise biradical reaction.

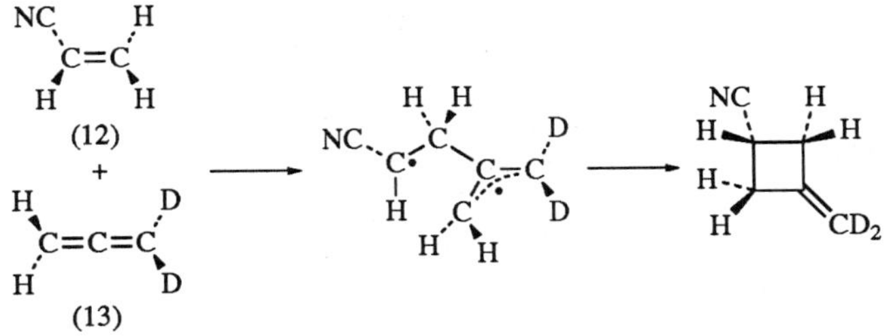

Scheme 5

Other interesting cycloaddition studies are reported in references 96–100.

Competition Between S_N2 and $E2$ Mechanisms.—In a high level (MP2/6-31G*) study of the effect of methyl substitution on the energetics of S_N2 substitution *versus* $E2$

[94] J. Cioslowski, J. Sauer, J. Hetzenegger, T. Karcher, and T. Hierstetter, *J. Am. Chem. Soc.*, 1993, **115**, 1353.
[95] E. A. Halevi and M. Wolfsberg, *J. Chem. Soc. Perkin Trans. 2*, 1993, 1493.
[96] D. A. Smith and C. W. Ulmer, *J. Org. Chem.*, 1993, **58**, 4118.
[97] R. Sustmann, W. Sicking, and R. Huisgen, *J. Org. Chem.*, 1993, **58**, 82.
[98] S. Yamabe, S. Kawajiri, T. Minato, and T. Machiguchi, *J. Org. Chem.*, 1993, **58**, 1122.
[99] F. P. Cossio, J. M. Ugalde, X. Lopez, B. Lecea, and C. Palomo, *J. Am. Chem. Soc.*, 1993, **115**, 995.
[100] B. E. Thomas, J. D. Evanseck, and K. N. Houk, *J. Am. Chem. Soc.*, 1993, **115**, 4165.

elimination in the reaction of F^- with $(CH_3)_2CHCl$ or $CH_3CH_2CH_2Cl$, Gronert has shown[101] that the *syn* transition state for *E*2 elimination adopts a *syn* clinal rather than a *syn* periplanar geometry (see also reference 102). The conditions under which *syn* periplanar transition states would be expected are postulated to be unsubstituted systems or systems with endothermic or marginally exothermic eliminations and hence late transition structures. The effect of methyl substitution upon the above energetics is discussed in terms of the balance between the steric effect of the methyl group and its ability to spread charge in both the transition state and the initial ion–dipole complexes. Aspects of this problem have also been addressed above; see reference 103.

Hynes and coworkers[102] have presented an interesting observation: for the reaction

$$(CH_3)_3CX \rightarrow (CH_3)_3C^+X^- \qquad (X = Cl, Br, I)$$

their calculations predict that solvent stabilization for the ionization actually decreases with increasing solvent polarity and yet the actual free energy also decreases (*i.e.* transition state ionic character and separation decrease), in complete contrast to conventional explanations. The calculations are based on a two-state valence-bond model and the aprotic solvents are modelled using a continuum method; the calculated activation free energies in all solvents agree to within ± 1.5 kcal mol^{-1} of experiment.

Lee and coworkers[105] have studied the effects of substituents on the S_N1 reactivities of cationic benzyl and mono and disubstituted benzhydryl with neutral leaving groups. It is interesting to note that a number of excellent linear relationships have been found between the enthalpy of activation and the reaction enthalpy, the length in the ground state of the bond being broken and change in the length of this bond as it goes to the transition state.

Hybrid Molecular Mechanical–Quantum Mechanical Models of the Transition State.—Full quantum mechanical determinations of transition state structures are now fairly commonplace, but they are not always routine and can therefore be time consuming. One way to increase the applicability of transition-state modelling to real problems in organic synthesis is to model the transition state using molecular mechanics, usually using the MM2 force-field. This approach has been very successful in rationalizing stereoselectivity. Recent examples of this approach include crotylborane addition to aldehydes,[106] the aldol reaction of enol borates,[107] and the synthesis of 1α,2β,25-trihydroxyvitamin D_3.[108]

Photochemistry.—In a study of the photochemistry of butadiene, Olivucci *et al.*[109] have shown that the traditional view that excitation from the ground $^1A_g'$ state to the first excited state, $^2A_g'$, is followed by decay from the avoided crossing region to the ground state, and that the efficiency of the decay is determined by the gap between the ground state; the excited state potential needs to be replaced by a mechanism where the

101 S. Gronert, *J. Am. Chem. Soc.*, 1993, **115**, 652.
102 F. M. Bickelhaupt, L. J. Dekoning, and N. M. M. Nibbering, *J. Org. Chem.*, 1993, **58**, 2436.
103 F. M. Bickelhaupt, E. J. Baerends, N. M. M. Nibbering, and T. Ziegler, *J. Am. Chem. Soc.*, 1993, **115**, 9160.
104 J. R. Mathis, H. J. Kim, and J. T. Hynes, *J. Am. Chem. Soc.*, 1993, **115**, 8248.
105 D. S. Chung, C. K. Kim, B. S. Lee, and I. Lee, *Tetrahedron*, 1993, **49**, 8359.
106 A. Vulpetti, M. Gardner, C. Gennari, A. Bernardi, J. M. Goodman, and I. Paterson, *J. Org. Chem.*, 1993, **58**, 1711.
107 A. Vulpetti, A. Bernardi, C. Gennari, J. M. Goodman, and I. Paterson, *Tetrahedron*, 1993, **49**, 685.
108 T. Takahashi, M. Nakazawa, Y. Sakamoto, and K. N. Houk, *Tetrahedron Lett.*, 1993, **34**, 4075.
109 M. Olivucci, I. N. Ragazos, F. Bernardi, and M. A. Robb, *J. Am. Chem. Soc.*, 1993, **115**, 3710.

funnel from the excited state to the ground state is actually a conical intersection. Since at the conical intersection the energy of the excited state and ground state are equal, the processes are fully efficient. The criteria for locating conical intersections are discussed; the process involves a constrained optimization on the excited state surface. Rather than several avoided crossings giving different products, a single conical intersection connecting the excited state to the ground state governs the formation of all products; the decay from the conical intersection to the ground state occurs within a single vibration. The work was based on hybrid molecular mechanics–valence bond calculations, which allowed for the rapid calculation of energies over a grid of points. The hybrid calculations are however fully supported by MCSCF calculations with a 4-31G basis set.

In a similar study, using the MCSCF/4-31G method throughout, Palmer *et al.* have studied the minima, transition structures, and conical intersections on the S_0, S_1, and S_2 surface of benzene, which includes the Kekulé forms of benzene, along with Dewar benzene, benzvalene, prismane, and the bicyclopropenyl structure (Figure 2).[110] A wonderfully complex set of stationary points and conical intersections are presented. For example, on the S_1 surface the only minimum is a D_{6h} structure, anti-Kekulé valence-bond isomer (g) connected to a prefulvene-like transition state (h); this is linked to a conical intersection which connects to the ground state.

The mechanism for the photosensitized cycloaddition reaction of penta-1,4-diene has been studied using the MP2/6-31G* method.[111] The mechanism proceeds *via* an inter-system crossing followed by rearrangement on S_0. Minimum energy crossing points between T_1 and S_0 were located and it was found that the preferred route involved the triplet cyclic intermediate crossing to S_0 followed by internal rotation of the terminal methylene group; the latter change involved relatively minor changes in the skeletal structure.

Tunnelling.—Calculation of the transition state geometry is not always sufficient for inferring the rate of reaction; dynamics calculations and tunnelling corrections may also be required. In an application on the 1,5-sigmatropic rearrangement in *cis*-penta-1,3-diene, Liu *et al.* have shown that the potential energy surface for dynamics can be calculated 'on-the fly' (*i.e.* by direct dynamics) using the semiempirical molecular orbital program MOPAC.[112] The advantage of direct dynamics is that it avoids the analytical representation of the potential energy function; it is this fitting problem that has prevented the application of modern dynamics theory to interesting organic and biochemical reactions. The calculations were based on variational transition-state theory (because this reduces the dynamics to contributions from reasonably localized regions of configurational space), including multidimensional semiclassical tunneling corrections. They show that tunnelling only occurred in a small region of the saddle point and that tunnelling can indeed account for the strong temperature dependence observed in the experimental kinetic isotope effect, as well as accounting for the magnitude of the kinetic isotope effect. Repeating the calculations without tunnelling leads to significant errors of about 6 eu in the entropy of activation and of about 4 kcal mol^{-1} in the enthalpy of activation. Although there were

110 I. J. Palmer, I. N. Ragazos, F. Bernardi, M. Olivucci, and M. A. Robb, *J. Am. Chem. Soc.*, 1993, **115**, 673.
111 M. Ohsaku, N. Koga, and K. Morokuma, *J. Chem. Soc., Perkin Trans. 2*, 1993, 71.
112 Y. P. Liu, G. C. Lynch, T. N. Truong, D. H. Lu, D. G. Truhlar, and B. C. Garrett, *J. Am. Chem. Soc.*, 1993, **115**, 2408.

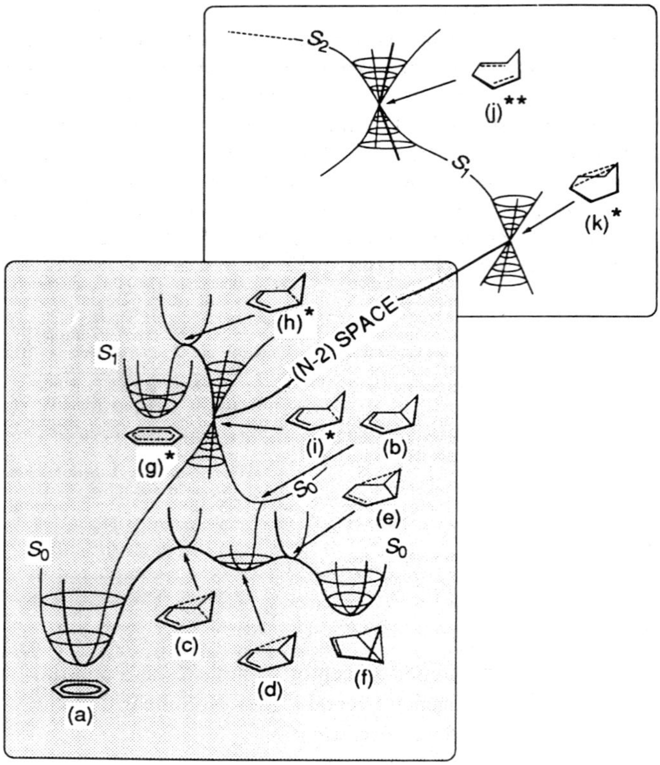

Figure 2
(Reproduced with permission from *J. Am. Chem. Soc.*, 1993, **115**, 673. © 1993 American Chemical Society)

differences in the results obtained with the AM1, PM3, and MINDO/3 Hamiltonians, all three gave essentially the same results.

The proton shift from CD_3H to CF_3 has been followed in a similar manner,[113] – except that here it was necessary to obtain AM1 parameters specific to this reaction because the predicted enthalpy of reaction was too large and this would have resulted in the calculated barrier being erroneously too asymmetric. (The parameters were fitted to both kinetic and thermodynamic data subject to the restriction that parameters did not change by more than 10%.)

5 Neural Networks

The use of theory to predict chemical reactivity with great reliability is a clear yet distant goal for many theoreticians. Intrinsic errors in many of the models still in use ensure that much work remains to be done. In some cases an expert chemist could

[113] Y. P. Liu, D. H. Lu, A. Gonzalezlafont, D. G. Truhlar, and B. C. Garrett, *J. Am. Chem. Soc.*, 1993, **115**, 7806.

make more reliable predictions than theory. With this in mind, artificial neural networks are being applied with increasing frequency to chemical problems. The article by Simon *et al.*[114] on the use of neutral networks to predict which bonds in aliphatic molecules will break preferentially is a good example that shows where quantum chemical calculations may not be the most powerful predictive tool. However, the neural network is able to learn the relationship between a number of empirical physical chemical descriptors (inputs) and reactivity (output). Schematically the neural network usually has the following form show in Figure 3 (this particular network was used to determine one-electron electrode potentials at pH 7 from semiempirical heats of formation and free energies of hydration).[115]

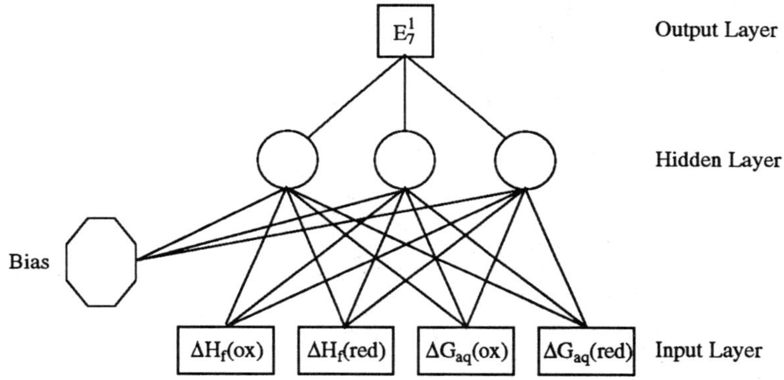

Figure 3

The inputs are initially multiplied by a (random) weight. The sum of the weighted inputs is then passed through a transfer function at the nodes in the hidden layer, the output of which is summed to give the output. If this total output is different to that desired, the weights are adjusted. When a given set of weights reproduces the desired output for a set of data the network may be considered to have learned the underlying relationship – assuming one exists. A full introduction (specific to chemistry) is given in the excellent review[116] and textbook[117] by Zupan and Gasteiger, and in the review by Rouvray.[118] The range of chemical problems that can be tackled using neural networks is probably limited only by the imagination and the availability of data suitable for training. Recent applications include the classification of inductive and resonance effects,[119] the prediction of ^{13}C NMR chemical shifts,[120,121] QSAR,[122] IR spectra

[114] V. Simon, J. Gasteiger, and J. Zupan, *J. Am. Chem. Soc.*, 1993, **115**, 9148.
[115] J.J. Wolfe, J.D. Wright, C.A. Reynolds, and A.C.G. Saunders, *Anti-Cancer Drug Design*, 1994, **9**, 85.
[116] J. Gasteiger, J. Zupan, *Angew. Chem., Int. Ed. Engl.*, 1993, **32**, 503.
[117] J. Zupan and J. Gasteiger, 'Neural Networks for Chemists', VCH, Weinheim, 1993.
[118] D.H. Rouvray, *Chem. Br.*, 1993, **29**, 495.
[119] V. Kvasnicka, S. Sklenak, and J. Pospichal, *J. Am. Chem. Soc.*, 1993, **115**, 1495.
[120] J.P. Doucet, A. Panaye, E. Feuilleaubous, and P. Ladd, *J. Chem. Inf. Comput. Sci.*, 1993, **33**, 320.
[121] G.M.J. West, *J. Chem. Inf. Comp. Sci.*, 1993, **33**, 577.
[122] A.C.G. Good, S.-S. So, and W.G. Richards, *J. Med. Chem.*, 1993, **36**, 433.

interpretation,[123] protein structure prediction,[124,125] and HPLC retention times. The neural network inputs may include descriptors calculated quantum mechanically.

6 Conclusions

It is appropriate to apologize to the authors of the many excellent works that have not been included in this review due to oversight, restrictions on space, or otherwise, but which show that theoretical organic chemistry remains a very healthy branch of chemistry.

[123] Q. C. Vanest, P. J. Schoenmakers, J. R. M. Smits, and W. P. M. Nijssen, *Vib. Spect.*, 1993, **4**, 263.
[124] B. Rost and C. Sander, *J. Mol. Biol.*, **232**, 584.
[125] G. Schneider and P. Wrede, *Angew. Chem., Int. Ed. Engl.*, 1993, **32**, 1141.

4 Reaction Mechanisms
Part (i) Pericyclic Reactions

By N. G. RAMSDEN

Glaxo Research and Development Limited, Greenford Road, Greenford, Middlesex, UB6 0HE, UK

1 Cycloadditions

The similarity index proposed recently as a quantitative measure of the extent of electron reorganization during chemical reactions allows discrimination between allowed and forbidden concerted processes. This is due to the difference in the extent of electron correlation between them. It also allows a simple evaluation of the relative energies of concerted and non-concerted reaction paths.[1] Several pericyclic reactions have been investigated for the variation in electron correlation during their course. The role of electron correlation is found to be most critical for high energy species along the reaction coordinate.[2] The similarity approach to chemical reactivity has also been used to study substituent effects in pericyclic chemistry.[3]

The use of transition state modelling with empirical force fields for reactions including the Diels–Alder reaction and Cope and Claisen rearrangements has been reviewed.[4]

Transition states for the concerted and stepwise Diels–Alder reaction and [2 + 2] dimerization of buta-1,3-diene have been located. On the basis of these, the diradical transition state for the Diels–Alder reaction of ethylene and butadiene appears to lie some 5 $kcal\,mol^{-1}$ above the concerted, synchronous transition state.[5] *Ab initio* MO calculations have been used to study the reaction of cyclopentadiene with a variety of dienophiles. *Endo cis* transition states are favoured and the lack of selectivity with acrylonitrile has been explained. Good agreement is found between enthalpies of activation and reactivities derived both experimentally and at the MB3 level.[6] Kinetic measurements on the Diels–Alder reaction of (1) show that a good correlation between $\log k_2$ and $(E_{HOMO} - E_{LUMO})^{-1}$ exists.[7] The Diels–Alder reactions of (2) with phenylacetylenes have been studied by MO calculations. At the HF/6-31G* level head-to-tail cycloadditions are favoured over head-to-head cycloadditions. This finding is in agreement with experimental data and provides the first evidence for

¹ R. Ponec, *J. Chem. Inf., Comput. Sci.*, 1993, **33**, 505.
² R. Ponec and M. Strnad, *Collect. Czech. Chem. Commun.*, 1993, **58**, 1751.
³ R. Ponec, *Collect. Czech. Chem. Commun.*, 1993, **58**, 11.
⁴ J. E. Eksterowicz and K. N. Houk, *Chem. Rev.*, 1993, **93**, 2439.
⁵ Y. Li and K. N. Houk, *J. Am. Chem. Soc.*, 1993, **115**, 7478.
⁶ W. L. Jorgensen, D. Lim, and J. K. Blake, *J. Am. Chem. Soc.*, 1993, **115**, 2936.
⁷ R. Sustmann and I. Siangouri-Feulner, *Chem. Ber.*, 1993, **126**. 1241.

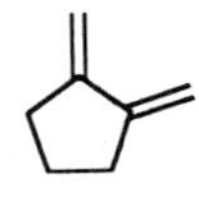

(1)

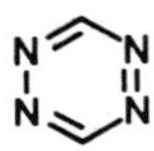

(2)

regioselectivity being controlled by Van de Waal's forces.[8] Frontier orbitals along the intrinsic reaction co-ordinate of [4 + 2] and [8 + 2] cycloadditions for ethylene and thiabutadiene have been studied by MO calculations. Even if the orbital phase of the reactant does not give favourable FMO interaction it may be converted into the correct phase on route.[9] The thermal, transannular Diels–Alder reaction of (3) has

(3)

been studied. Experimental results are in agreement with those predicted by theory.[10] In the aminium salt catalysed reaction a diene cation–radical reacts with a neutral ketene in a [4 + 2] cycloaddition (Scheme 1).[11] The influence of phenyl π electrons

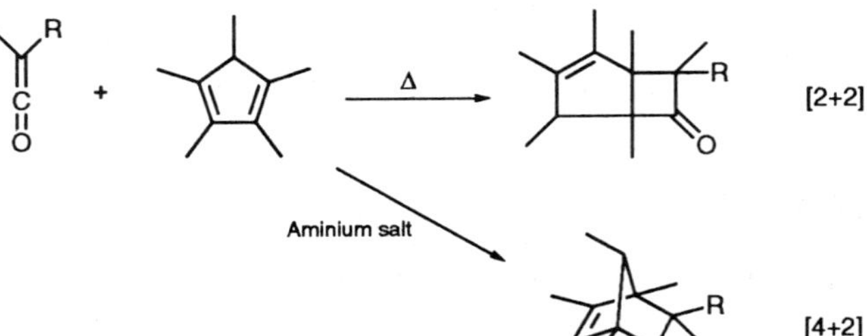

Scheme 1

during cycloaddition reactions of *endo*-3,3-diphenyltricyclo[3.2.1.0^{2,4}]oct-6-ene has been investigated.[12]

Diels–Alder reactions that are accelerated by increasing the electrophilic nature of the medium are accelerated by metal perchlorates in order of charge : radius ratio, *i.e.* $Mg^{2+} > Ba^{2+} > Li^{+} > Na^{+}$. This is presumably due to the effect of the cation on the electrophilic nature of the medium.[13]

[8] J. Cioslowski, J. Sauer, J. Hertzenegger, T. Karcher, and T. Hierstetter, *J. Am. Chem. Soc.*, 1993, **115**, 1353.
[9] S. Yamabe, S. Kawajiri, T. Minato, and T. Machiguchi, *J. Org. Chem.*, 1993, **58**, 1122.
[10] A. Ndibwami, S. Lamothe, P. Souay, S. Goldstein, and P. Deslongchamps, *Can. J. Chem.*, 1993, **71**, 714.
[11] R. M. Giuliano, A. D. Jordan, A. Gauthier, and K. Hoogsteen, *J. Org. Chem.*, 1993, **58**, 4979.
[12] M. Peeran, J. W. Wilt, R. Subramanian, and D. S. Crumrine, *J. Org. Chem.*, 1993, **58**, 202.
[13] A. Cassoschi, G. Desimoni, G. Faita, A. G. Invernizzi, S. Lanati, and P. Righetti, *J. Am. Chem. Soc.*, 1993, **115**, 8002.

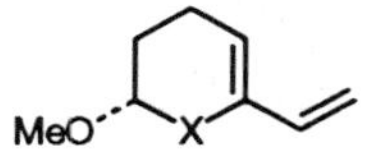

(4) X = CH_2 or O

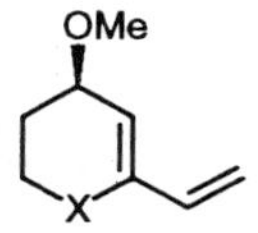

(5) X = CH_2 or O

A dienophile will add to the face opposite the methoxy group in both (4) and (5) due to both the diene conformation and secondary orbital interactions in the transition state.[14] A procedure for the conversion of *endo* (6) to *exo* (7) adducts in the Diels–Alder reaction has been determined (Scheme 2). The high *exo* selectivity was investigated by

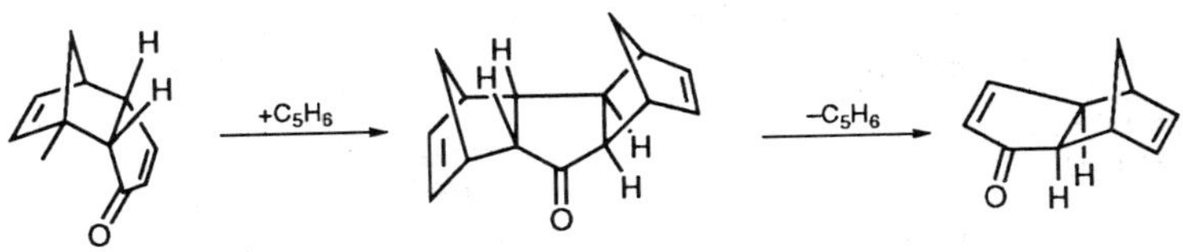

(6) (7)

Scheme 2

AM1 transition state calculations.[15] A mechanistically interesting and useful reversal in the facial selectivity of the Diels–Alder reaction has been achieved (Scheme 3).[16] The

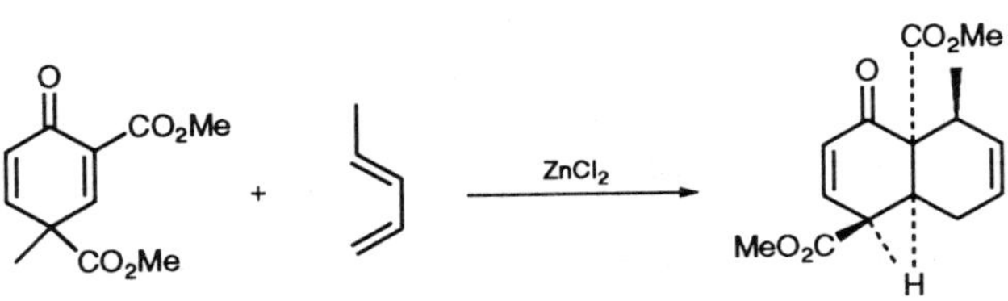

Scheme 3

β-sulfinyl dienophile (8) has been shown to react with cyclopentadiene with a very high degree of diastereoselectivity.[17]

The LUMO diene controlled [4 + 2] cycloaddition of azaazulenes with tetrazines has been described.[18] Fischer carbenes (9) undergo Diels–Alder reactions with comparable stereoselectivity to the Lewis acid catalysed reaction of the corresponding methyl ester. This makes them useful synthons in cases where the diene is intolerant of Lewis acids.[19] *N*-arylcyclohepta-2,4,6-triene-1-imines react with benzonitrile oximes

14 M. Schmittel and H. Van Seggern, *J. Am. Chem. Soc.*, 1993, **115**, 2165.
15 P. P. M. Dol, A. J. H. Klunder, and B. Zwannenberg, *Tetrahedron Lett.*, 1993, **34**, 3181.
16 H. J. Liu and Y. Hai, *Tetrahedron Lett.*, 1993, **34**, 423.
17 T. K. Yang, S. M. Hung, C. Z. Chen, Y. Jiang, and A. Q. Mi, *Youji Huaxue*, 1993, **13**, 183.
18 G. Frenzen, W. Maisa, U. Reimes, and G. Seitz, *Chem. Ber.*, 1993, **126**, 441.
19 W. D. Wulff and T. S. Powers, *J. Org. Chem.*, 1993, **58**, 2381.

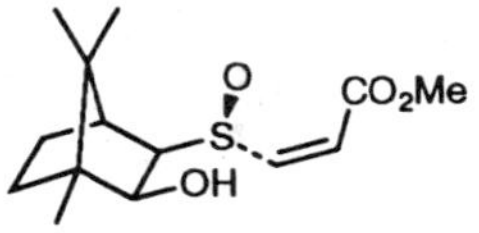
(8)

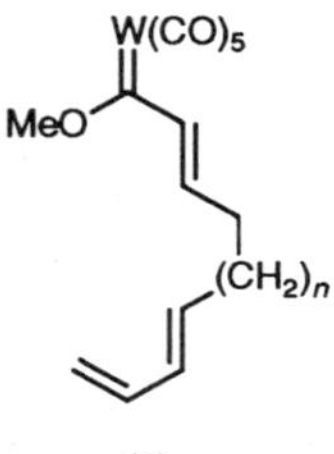
(9)

via [4 + 2] cycloadditions. A Hammett ρ-value of 0.99 suggests nucleophilic attack of the imines to the nitrile oxides.[20]

(η^6-Cyclohepta-1,3,5-triene)tricarbonylchromium(0) undergoes a thermal [6 + 4] cycloaddition with both electron-rich and electron-poor butadienes to give the *endo* product.[21]

The reaction between the chirally enriched allene (10) and the perhaloalkene (11) has been studied both experimentally and by molecular modelling. Results suggest that (12) is formed by one continuous minimum energy pathway, whilst (13) is formed from

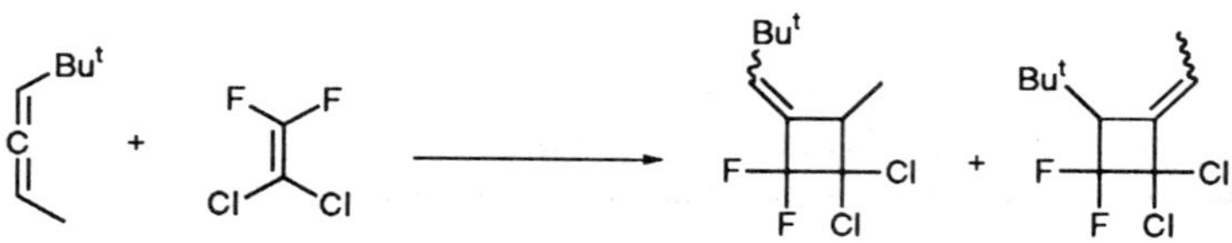
(10) (11) (12) (13)

two competing pathways. The two stereoisomers of (13) have opposite absolute configurations and thus no chirality transfer is seen from (10) to (13).[22] The reaction of fluorine and ethylene has been studied at the MP2/6-31*G level.[23] Fluorine approaches the ethylene double bond vertically to form a perpendicular complex as the intermediate. This then rearranges to a rhombic-type complex as a transition state to give the final *syn* product.

Cycloadditions between ketenes and cyclopentadiene have been studied by a semiempirical AM1 method. Ketene and methyl ketene react *via* a concerted but non-synchronous twisted transition state, with a small degree of charge transfer from the cyclopentadiene to the ketene. Chloromethyl ketene, however, reacts in a stepwise manner.[24] Acetylenic links can act as antarafacial components in [2 + 2] processes.[25] *N*-Methylhex-3-yn-1-amine will react in a nominal $[2_s + 2_a + 2_s + 2_s]$ Möbius process. Evidence for the concertedness of this reaction has been discussed.

Silenes undergo [2 + 4] and [2 + 2] cycloadditions with imines.[26] The [2 + 4]

[20] K. Ito, K. Saito, and K. Takahashi, *Heterocycles*, 1993, **36**, 21.
[21] J. H. Rigby, H. S. Ateeq, N. R. Charles, S. V. Cuisiat, M. D. Ferguson, J. A. Henshilwood, A. C. Krueger, C. O. Ogbu, K. M. Short, and M. J. J. Heeg, *J. Am. Chem. Soc.*, 1993, **115**, 1382.
[22] D. J. Pasto and J. Brophy, *J. Phys. Org. Chem.*, 1993, **6**, 95.
[23] T. Iwaoka, C. Kaneko, A. Shigiwara, and H. Ichikawa, *J. Phys. Org. Chem.*, 1993, **6**, 195.
[24] J. Jiang, D. Fang, and X. Fu, *Gaodeng Xuexiao Huaxue Xuebao*, 1993, **14**, 696.
[25] A. Viola, J. J. Collins, N. Filipp, and J. S. Locke, *J. Org. Chem.*, 1993, **58**, 5067.
[26] A. G. Brook, W. J. Chatterton, and R. Kimarathasan, *Organometallics*, 1993, **12**, 3666.

adducts generally rearrange to [2 + 2] cycloadducts and these silazetidines decompose by a variety of mechanisms. Computations suggest a multistep, diradical mechanism for the head-to-tail cycloaddition of two silaethylenes.[27] Polarization of the double bond in silenes does not lead to a concerted mechanism. The energy surface is similar to that of the ethylene–ethylene reaction, but both the activation energy and the energy of the diradical are much lower. The cycloaddition of tetrafluoroethylene and dicyclopentadiene is first order in dicyclopentadiene and second order in tetrafluoroethylene.[28] The activation energy is 118.56 kcal mol^{-1}.

Oxazolium-5-oxide (14) cycloadds to 1,2-dicyanocyclobutene (15) to give the imino acid (16) and the dihydroazepine (17).[29] AM1 calculations have been used to rationalize the observed product stereochemistry and evaluate possible reaction mechanisms.

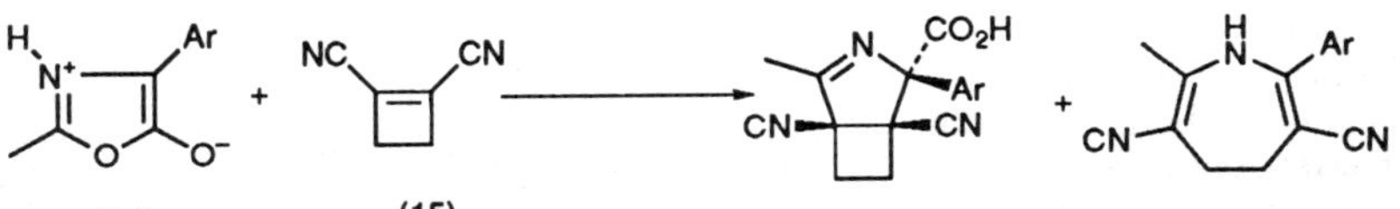

(14) (15) (16) (17)

The effect of solvent and substituents on the [2 + 2] cycloaddition of t-butylcyanoketene and styrene has been studied using a CAS-MCSCF method. The reaction proceeds *via* a diradical in which steric interactions are minimized. However this minimization leads to a product with *cis* stereochemistry.[30]

New models for the [2 + 2] Staudinger reaction have been discussed.[31] Simple feed-forward, three layer neural networks have been used to classify and predict the regioselectivity of 1,3-dipolar cycloadditions.[32]

A quantitative analysis of the transition states and energy barriers of the *syn* and *anti* cycloadditions of norbornene and *cis*-3,4-dichlorocyclobutene with formonitrile oxime has investigated the role of many factors involved in facial selectivity.[33] Further studies have allowed the development of a conceptual framework involving repulsion, delocalization, and deformation.[34] This has provided a detailed account of the geometry and energy differences between the *syn* and *anti* transition states. The cyclization of imine oxides has been studied. In the reaction of (19) with the furanone (18) the major product is derived from an *exo* transition state.[35] *Exo* adducts are also the major products from reactions of (20).[36] With furanones the stereochemistry of the products is *cis* due to an interaction between the nitrogen lone pair and the carbonyl group. When this interaction is not present, for example in the dihydrofurans, *trans* products begin to predominate.

[27] F. Bernardi, A. Bottoni, M. Olivucci, M. A. Robb, and A. Venturni, *J. Am. Chem. Soc.*, 1993, **115**, 3322.
[28] N. B. Christova, S. D. Pavlova, and G. K. Kostov, *Kinet. Catal. Lett.*, 1993, **49**, 393.
[29] C. A. Maryanoff and I. J. Turchi, *Heterocycles*, 1993, **35**, 649.
[30] M. Reguero, R. R. Pappalardo, M. A. Robb, and H. S. Rzepa, *J. Chem. Soc., Perkin Trans. 2*, 1993, 1499.
[31] F. P. Cossio, B. Lecea, C. Cuevas, A. Mieglo, and C. Palomo, *An. Quim.*, 1993, **89**, 119.
[32] S. Sklenak, V. Kvasnicka, and J. Pospical, *Acta Chem. Hung.*, 1993, **130**, 103.
[33] A. Rastelli, M. Bagatti, A. Ori, R. Gandolfi, and M. Burdisso, *J. Chem. Soc., Faraday Trans.*, 1993, **89**, 29.
[34] A. Rastelli, M. Bagatti, and R. Gandolfi, *J. Chem. Soc., Faraday Trans.*, 1993, **89**, 3913.
[35] D. Alonso-Peramau, P. de March, M. Figuerdo, J. Font, and A. Seria, *Tetrahedron*, 1993, **49**, 4267.
[36] P. Cid, P. de March, M. Figuerdo, J. Font, S. Milan, A. Soria, and A. Virgili, *Tetrahedron*, 1993, **49**, 3857.

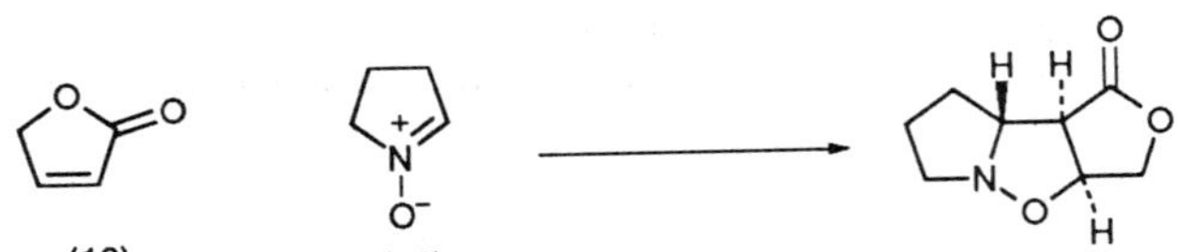

(18) (19)

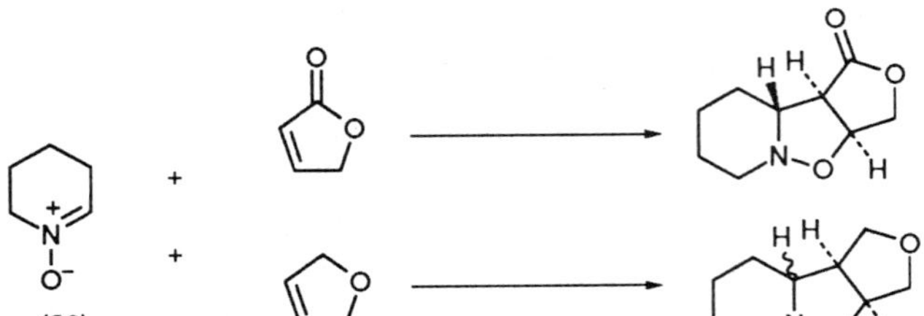

(20)

Diastereoselectivity in nitrile oxide additions to Oppolzer's chiral sultams cannot be explained solely by bulky groups shielding faces of the alkene.[37] Transition states were investigated by PM3 methods and the origin of the selectivity appears to be a repulsion between the dipolar oxygen and the sultam oxygens. This is supported by the fact that selectivity is lost in the corresponding isothiazolidines. The work supports Hehre's claim that electrostatic effects are a general determinant of facial selectivity. 2-Ethoxy and 2-ethylthio-1-azetidines (21) undergo 1,3-dipolar cycloadditions with nitrile oxides and nitrile ylids to give stable products. With nitrilimines, 1,2,4-triazoles (22)

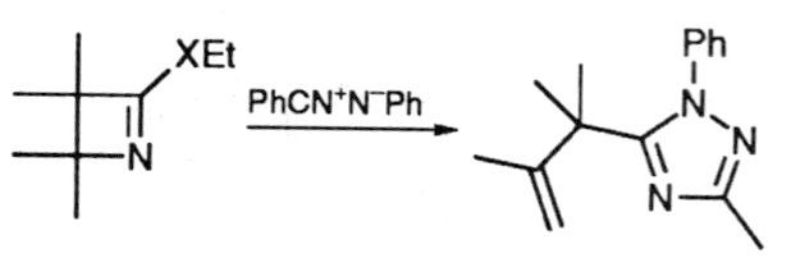

(21) X = O or S (22)

are unexpectedly produced.[38] Benzonitrile oxides undergo 1,3-cycloaddition reactions to give 1,2,4-oxadiazoles as the major products. Isoxazoles and 4,5-dihydroxazoles are also produced. The kinetics of all these reactions have been determined and substituent and solvent effects on product distribution are discussed.[39]

The regioselectivity of the intramolecular reaction of *N*-4-methylpent-4-enyl nitrones has been investigated by AM1 MO methods and transition state theory. The difference in selectivity between the *E* and *Z* isomers is enhanced by the effect on the relative activation entropies caused by introduction of the methyl group.[40] The reaction of mesoionic 1,3-dithiolium-4-oate (23) with phenyl isocyanate is not a

[37] K. S. Kim, B. H. Kim, W. M. Park, S. J. Cho, and B. J. 'Mihn, *J. Am. Chem. Soc.*, 1993, **115**, 7472.
[38] K. Hemming, A. B. N. Luheshi, A. D. Redhouse, R. K. Smalley, J. R. Thompson, P. D. Kennewell, and R. Westwood, *Tetrahedron.*, 1993, **49**, 4383.
[39] P. Beltrane, E. Cadoni, C. Floris, and G. Gelli, *Tetrahedron.*, 1993, **49**, 7001.
[40] S. Ma and X. Fu, *Hauxue Xuebao.*, 1993, **51**, 496.

1,3-dipolar cycloaddition as the product is an azetidinone. The phenyl isocyanate reacts with the open ketene isomer (24) of the mesoionic compound.[41] The existence of

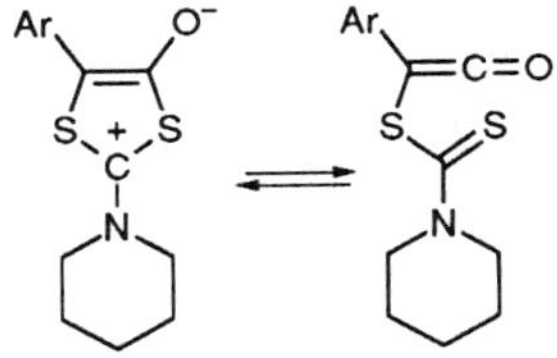

(23) (24)

a free-resonant, unstabilized azomethine ylid in a 1,3-dipolar cycloaddition has been demonstrated by ^{13}C labelling.[42] The cycloaddition of phenylsulfonyl allenes with diazomethane has been studied.[43] The product pyrazolines undergo a variety of mechanistically interesting transformations.

2 Sigmatropic Rearrangements

Chiral Lewis acids derived from binaphthol and diisopropyltitanium dichloride catalyse the ene reaction of glyoxylate esters with silyl enol ethers rather than a Mukaiyama aldol reaction.[44] *Syn* diastereoselectivity is very high with either isomer of the silyl enol ether and this can be explained by consideration of transition state geometries. A mechanistic pathway to explain the observed asymmetric induction during the Lewis acid catalysed ene reaction of (25) has been presented.[45] The reaction

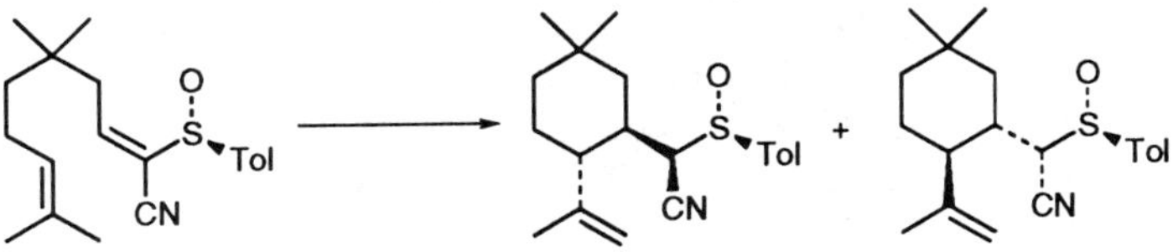

(25)

of enamines with ethyl propiolate has been explained as an ene-addition reaction (Scheme 4).[46] Alloxan (26) and 1,3-dimethylalloxan (27) undergo ready ene reactions at the central CO.[47] The stereoselectivity of both reactions is determined by steric factors. The transition states for the ene reaction between propene and formaldehyde imine have been located by *ab initio* MO calculations.[48] The *endo* transition state is lower in energy when C–C bond formation occurs and the *exo* transition state is lower in energy when C–N bond formation occurs. This is due to interactions between the nitrogen lone pair and the central carbon atom of the ene.

[41] M. Bssaibis, A. Robert, and A. Souizi, *J. Chem. Soc., Chem. Commun.*, 1993, 998.
[42] J. M. Bentley, D. M. Smith, H. J. Wadsworth, and C. L. Willis, *J. Chem. Res.*, (*S*), 1993, 240.
[43] A. Padwa, M. A. Filipkowski, D. N. Kline, S. S. Murphree, and P. E. Yeske, *J. Org. Chem.*, 1993, **58**, 2061.
[44] K. Mikami and S. Matsukawa, *J. Am. Chem. Soc.*, 1993, **115**, 7039.
[45] K. Hiroi and M. Umemura, *Tetrahedron.*, 1993, **49**, 1831.
[46] Z. Huang and M. Wang, *J. Chem. Soc., Perkin Trans. 1*, 1993, 1085.
[47] G. B. Gill and M. S. Hidris, *Tetrahedron.*, 1993, **49**, 219.
[48] B. E. Thomas, IV and K. N. Houk, *J. Am. Chem. Soc.*, 1993, **115**, 790.

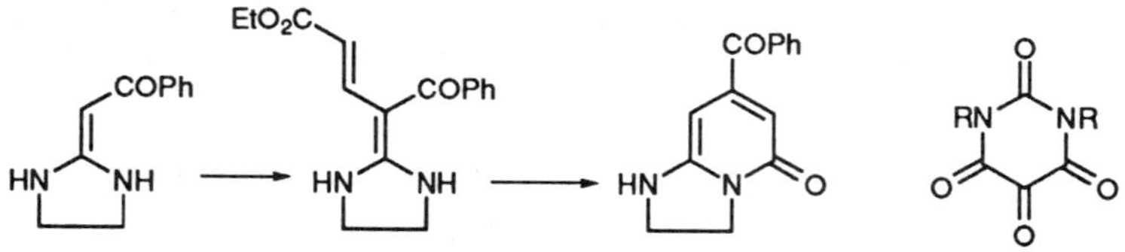

(26) R = H
(27) R = Me

Scheme 4

Terminal fluorine atoms exert a significant inhibitory steric effect on both disrotatory electrocyclic closures of hexa-1,3,5-trienes and Cope rearrangements.[49] Cope rearrangements of perfluorodienes proceed *via* a 1,4-cyclohexyl diradical, in contrast to the concerted rearrangement of non-fluorinated dienes.[50] Mass spectra of 2,5-diphenylhexa-1,5-dienes have provided evidence for a cation–radical Cope rearrangement in the gas phase. The reaction directionality is different to that for the thermal process.[51] Activation entropies and energies have been calculated for the competing C–C cleavage and Cope rearrangement of (28).[52]

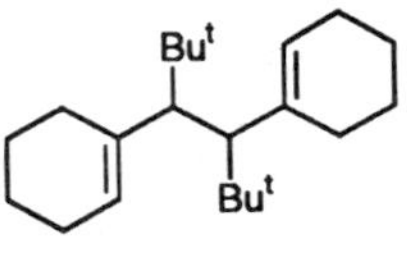

(28)

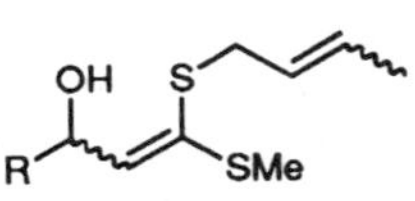

(29)

The stereoselectivity of the thio-Claisen rearrangement of *S*-crotyl-α-hydroxy ketene dithioacetals (29) is determined by the classical internal control obtained with a [3,3] sigmatropic shift.[53] External control also exists based on the stereochemistry of the *S*-crotyl double bond.

The solvent effect on the retro-Claisen reaction has been studied and the rate constants found to be a linear function of Reichardt's $E_T(30)$ solvent polarity parameter.[54] This suggests that, like the forward reaction, the retro-Claisen is concerted with partial charge separation in the transiton state.

An investigation of the migratory tendencies of several groups when undergoing a [1,5] sigmatropic shift suggests that both good conjugative electron-withdrawing ability and the availability of a low-energy vacant orbital lead to good migratory ability.[55] However, steric, conformational, and secondary orbital interaction effects can mask these. Further evidence is provided for the concerted nature of the reaction. Activation for a [1,5] hydride transfer and cycloaddition have been determined.[56] A diradical with little ionic character is generated on thermolysis of bicyclo[2.1.0]pen-

[49] W. R. Dolbier, Jr and K. W. Palmer, *J. Am. Chem. Soc.*, 1993, **115**, 9349.
[50] N. Jing and D. M. Lemal, *J. Am. Chem. Soc.*, 1993, **115**, 8481.
[51] H. Ikeda, T. Takasaki, Y. Takahashi, and T. Miyashi, *J. Chem. Soc., Chem. Commun.*, 1993, 367.
[52] C. Herberg, H. D. Beckhaus, T. Koertvelyesi, and C. Ruechardt, *Chem. Ber.*, 1993, **126**, 117.
[53] P. Baslim and S. Perrio, *Tetrahedron*, 1993, **49**, 3131.
[54] G. Desimoni, G. Faita, S. Gratti Comini, P. P. Righetti, and G. Tacconi, *Tetrahedron*, 1993, **49**, 2093.
[55] D. W. Jones and R. J. Morman, *J. Chem. Soc., Perkin Trans. 1*, 1993, 681.
[56] T. H. Petersen and B. K. Carpenter, *J. Am. Chem. Soc.*, 1993, **115**, 5466.

tane derivatives. The energetics of [1,5] sigmatropic hydrogen shifts in cyclopentadiene, pyrrole, and phosphole and have been determined from *ab initio* calculations.[57] Activation enthalpies have been used to explain the reactions of these heterocycles. The kinetic isotope effect for the [1,5] sigmatropic rearrangement of *cis*-penta-1,3-dienes has been studied by a direct dynamics method.[58] The contribution to the shifts by and the nature of tunnelling effects have been discussed. Kinetic studies on the rearrangement of tetraphenylcyclopentadienes (30) suggest that they occur by [1,3] and [1,5] sigmatropic shifts.[59] *Ab initio* MO calculations show that the

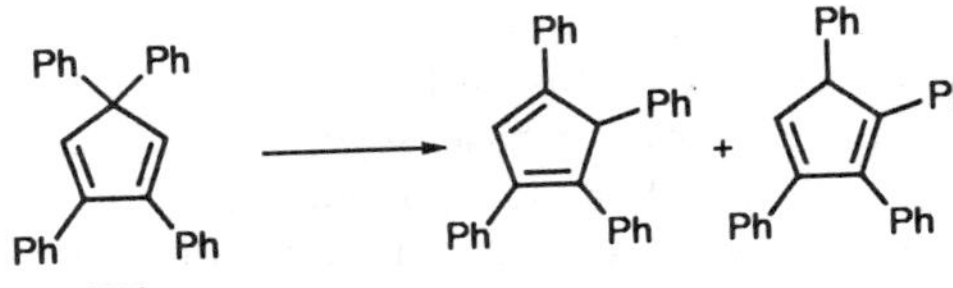

(30)

suprafacial [1,3] hydrogen shift of $H_3CHC{=}PH$ to $H_2C{=}CHPH_2$ and the degenerate $H_2PCH{=}PH$ to $HP{=}CHPH_2$ shift exist and compete favourably with the antarafacial process.[60] A [5,5] sigmatropic shift has been observed in the thermal

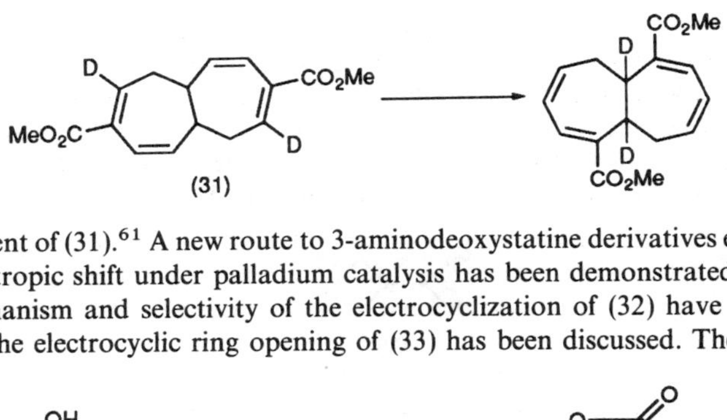

(31)

rearrangement of (31).[61] A new route to 3-aminodeoxystatine derivatives employing a [3,3] sigmatropic shift under palladium catalysis has been demonstrated.[62]

The mechanism and selectivity of the electrocyclization of (32) have been investigated.[63] The electrocyclic ring opening of (33) has been discussed. The calculated

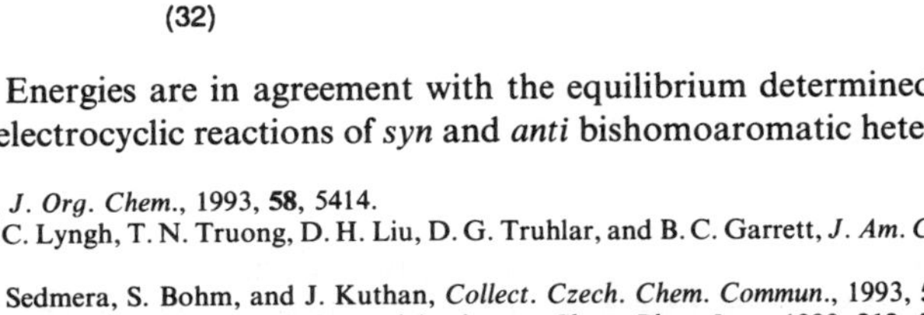

(32)

Gibb's Free Energies are in agreement with the equilibrium determined experimentally.[64] The electrocyclic reactions of *syn* and *anti* bishomoaromatic heterocycles (34)

57 S. Bachrach, *J. Org. Chem.*, 1993, **58**, 5414.
58 Y. P. Liu, G. C. Lyngh, T. N. Truong, D. H. Liu, D. G. Truhlar, and B. C. Garrett, *J. Am. Chem. Soc.*, 1993, **115**, 2408.
59 P. Sebek, P. Sedmera, S. Bohm, and J. Kuthan, *Collect. Czech. Chem. Commun.*, 1993, **58**, 882.
60 M. T. Nguyen, L. Landuyt, and L. G. Vanquickenborne, *Chem. Phys. Lett.*, 1993, **212**, 543.
61 K. Hafner, H. J. Linder, W. Luo, K. P. Meinhardt, and T. Zink, *Pure Appl. Chem.*, 1993, **65**, 17.
62 A. M. Doherty, B. E. Kornberg, and M. D. Rielly, *J. Org. Chem.*, 1993, **58**, 795.
63 M. Leclaire, S. Bathnagar, and J. Y. Lallemand, *Bull. Soc. Chim. Fr.*, 1993, **130**, 310.
64 I. J. Turchi, J. B. Press, J. J. McNally, M. P. Bonner, and K. L. Sorgi, *J. Org. Chem.*, 1993, **58**, 4629.

(33)

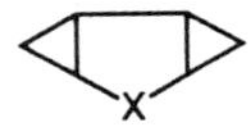

(34) X = O,S

have been studied experimentally.[65] The electrocyclic ring opening of singlet oxirane has been studied and it is suggested that the stereochemistry of oxirane cycloadditions is due to competition between a conrotatory, asynchronous process and a single methylene rotation process.[66] AM1 MO calculations of the reaction path for ring expansion of the cyclobutylidenyl carbene suggest that the process produces a lumomer of cyclopentyne rather than cyclopentyne itself.[67] Products obtained from subsequent cycloadditions are consistent with this. Further calculations on the reaction path of these [2 + 2] and [4 + 2] cycloadditions of the lumomer show concertedness for the former process and stepwise for the latter. Results are consistent with Dewar's theory of orbital isomerization.[68]

3 Other Reactions

A [2,3] Wittig rearrangement, *via* a concerted, five-ring transition state, has been accomplished overcoming a severe six-ring deformation barrier.[69] Interactions of buta-1,3-diene with oxygen on a silver surface have been studied by extended Hückel tight binding calculations.[70] Interaction of a terminal carbon of the diene with surface oxygen gives an intemediate which undergoes preferential 1,4 ring closure to a tetrahydrofuran. Phenylchlorocarbene reacts with [1.1.1]propellane to give (35).

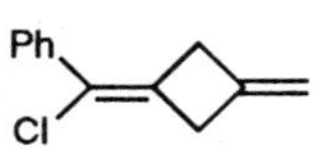

(35)

There is no evidence for a diradical intermediate as is found for the corresponding reaction of diphenylcarbene.[71]

[65] T. Golz, S. Hammes, and F. G. Klaerner, *Chem. Ber.*, 1993, **126**, 485.
[66] Y. Yamaguchi, H. F. Schaefer, III, and I. L. Alberts, *J. Am. Chem. Soc.*, 1993, **115**, 5790.
[67] J. C. Gilbert and S. Kirschner, *Tetrahedron Lett.*, 1993, **34**, 599.
[68] J. C. Gilbert and S. Kirschner, *Tetrahedron Lett.*, 1993, **34**, 603.
[69] Z. You and M. Koreeda, *Tetrahedron Lett.*, 1993, **34**, 2597.
[70] B. Schioett and K. A. Jorgensen, *J. Phys. Chem.*, 1993, **97**, 10738.
[71] J. C. Scaiano and P. F. McGarry, *Tetrahedron Lett.*, 1993, **34**, 1243.

4 Reaction Mechanisms
Part (ii) Free-radical Reactions

By S. CADDICK
School of Chemistry and Molecular Sciences, University of Sussex, Falmer, Brighton, BN1 9QJ, UK

1 Introduction

Research into the development of free-radical processes continues to attract significant interest and many notable advances have been published in the 1993 literature. This year's highlights will focus on four general themes, but there will inevitably be some overlap between the areas: initiators, promoters, and mediators; intramolecular reactions; intermolecular reactions; applications of radical processes to synthetic and biological chemistry.

2 Initiators, Promoters, and Mediators

Fraser-Reid[1] and coworkers have demonstrated the utility of di-t-butylhyponitrite (DBH) as a convenient reagent for the generation of carbon-centred radicals. Many synthetically useful applications have been illustrated; the hydroxymethylation of enone (1) is a particularly good example.

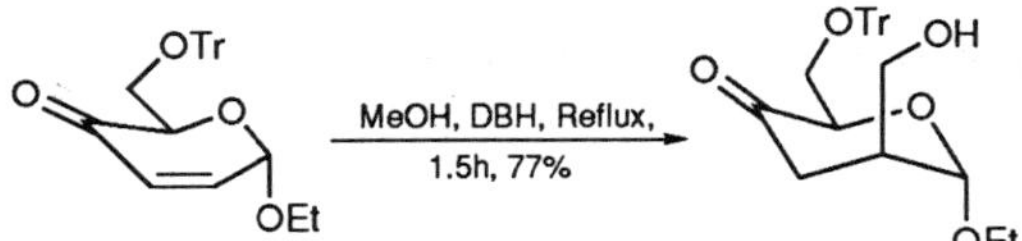

(1) (2)

Given the popularity of the $SnBu_3H$/AIBN system, new methods for the generation of tributyltin radicals are of considerable interest. Previous work had examined the utility of bis(trimethylstannyl)benzopinacolate for the generation of trimethylstannyl radicals. Two recent reports[2,3] demonstrate the ability of bis(tri-n-butylstannyl)benzopinacolate (3) to generate tributyltin radicals and benzophenone under thermal conditions.

Silicon reagents continue to attract interest as promoters; Chatgilialoglu's group

[1] B. Venkateswara Rao, J. B. Chan, N. Moskowitz, and B. Fraser-Reid, *Bull. Soc. Chim. Fr.*, 1993, **130**, 428.
[2] D. J. Hart, R. Krishnamurthy, L. M. Pook, and F. L. Seely, *Tetrahedron Lett.*, 1993, **34**, 7819.
[3] M. J. Tomaszewski and J. Warkentin, *J. Chem. Soc., Chem. Commun.*, 1993, 1407.

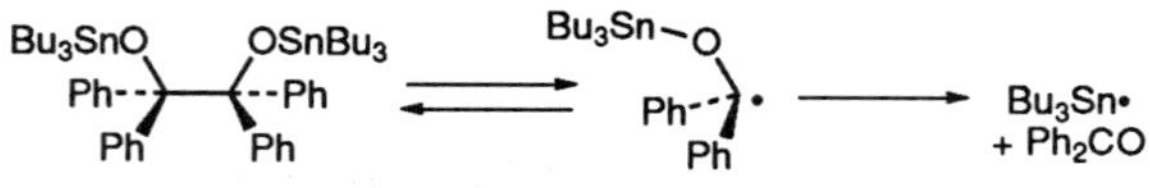

(3) (4)

has, in an interesting report,[4] demonstrated that care must be taken when choosing initiators for these reagents. Thus reduction of (5) with tris(trimethylsilyl)silane (TTMSS) gave markedly different ratios of products (6) depending on the initiator

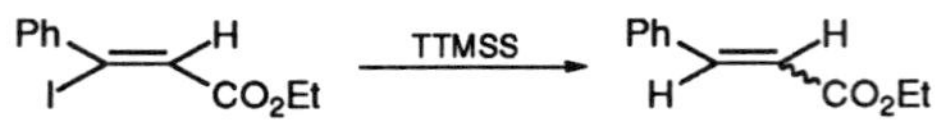

(5) (6)

	Z	*E*
AIBN	17	83
$(PhCO_2)_2$	71	29

used. When AIBN is used the *E* isomer predominates; initiation using benzoyl peroxide leads to the *Z* isomer. Further experiments suggest that AIBN is more efficient as an initiator than benzoyl peroxide when utilizing TTMSS as a radical promoter. The tris(trimethylsilyl) radicals can undergo *reversible* addition to the alkene to furnish the *E* alkene.

Recently Murphy and coworkers have developed a particularly interesting radical process utilizing tetrathiofulvene (TTF) as a promoter.[5] Thus treatment of diazonium

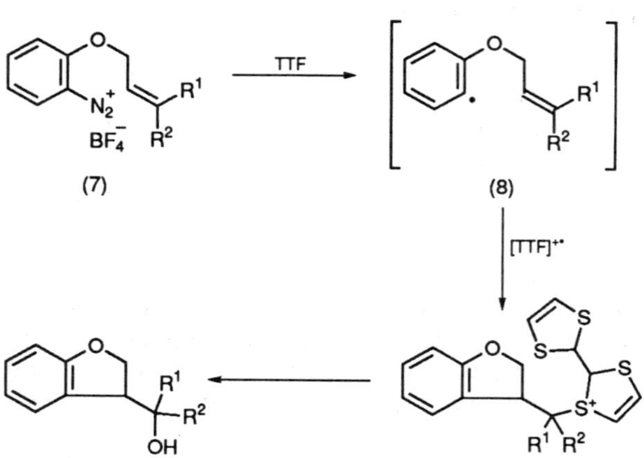

(7) (8) (9) (10)

[4] C. Ferreri, M. Ballestri, and C. Chatgilialoglu, *Tetrahedron Lett.*, 1993, **34**, 5147.
[5] C. Lampard, J. A. Murphy, and N. Lewis, *J. Chem. Soc., Chem. Commun.*, 1993, 295.

ion (7) with TTF in moist solvent leads to the adducts (10). The proposed mechanism of these reactions is electron transfer from the electron donor, TTF, to the diazonium species to generate an aryl radical (8). This then undergoes cyclization and capture with TTF radical–cation to generate (9), which reacts with water to provide product (10). Intermediate (9) has been characterized and shown to react with methanol and acetonitrile to give products (11) and (12).

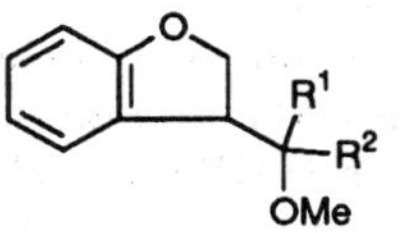

(11) $R^1 = R^2$ = H or alkyl

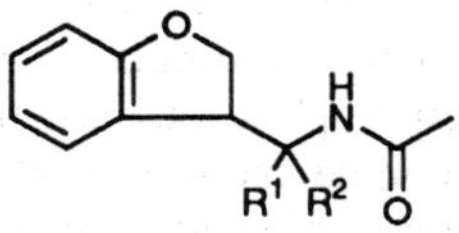

(12) $R^1 = R^2$ = H or alkyl

Rhodium and ruthenium are also beginning to gain an accepted place in the armoury of reagents used to promote free-radical reactions in synthesis. In a full account,[6] Ishibashi *et al.* have reported the synthesis of the naturally occurring alkaloid skeleton (14) using an interesting chlorine atom transfer cyclization.

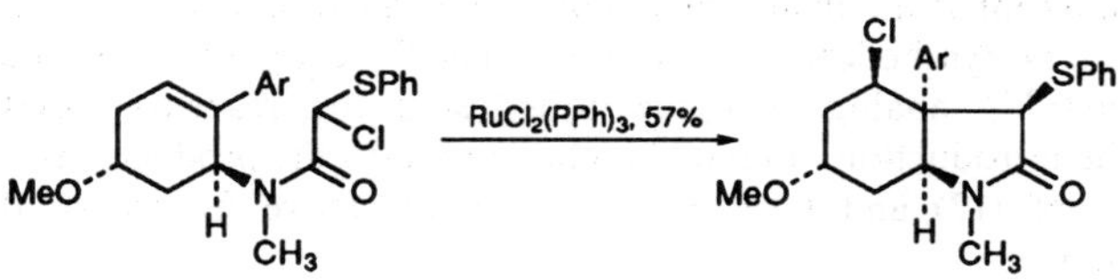

(13) (14)

Giese, Hartung, *et al.* have been examining rhodium complexes as alkyl radical precursors.[7] Although alkylrhodoximes (15) are much less photoreactive than the

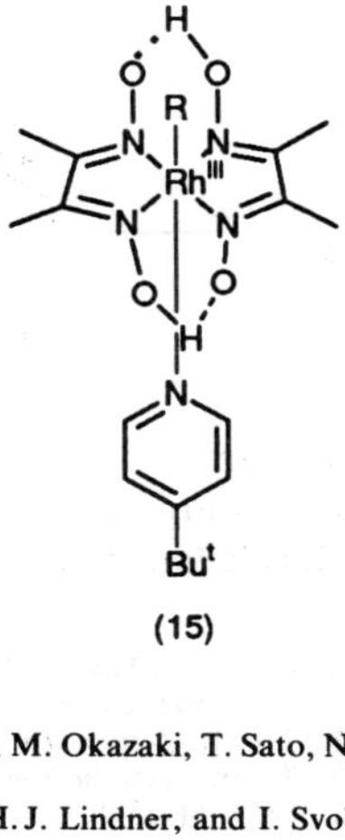

(15)

[6] H. Ishibashi, N. Uemura, H. Nakatani, M. Okazaki, T. Sato, N. Nakamura, and M. Ikeda, *J. Org. Chem.*, 1993, **58**, 2360.
[7] B. Giese, J. Hartung, C. Kesselheim, H. J. Lindner, and I. Svoboda, *Chem. Ber.*, 1993, **126**, 1193.

corresponding cobaloxime reagents, they have been shown to participate in a number of reactions. One of the advantages these stable reagents may have over their cobalt counterparts is the ability to act as precursors to tertiary radicals.

Hill and co-workers have been investigating the utility of polyoxotungstate photocatalysed radical–addition reactions.[8] They have found that secondary and tertiary radicals can be generated from the corresponding alkanes *via* C–H activation. These radicals have been utilized in addition reactions with alkenes and alkynes; the reported yields are high at low conversion rates (< 10%).

Samarium diiodide continues to attract attention as a useful promoter and numerous synthetic applications have been reported. Many transformations promoted by samarium diiodide require additives such as HMPA or DMPU to proceed efficiently. A highly significant study by Hasegawa and Curran[9] details investigations of alkyl/aryl halide and carbonyl group reductions with SmI_2. They have shown that water is a safe and convenient alternative to the traditional additives and, that in halide reductions, water accelerates the rate of the reaction. The origin of the enhancement is unclear, but it appears that the water does not act solely as a proton source.

3 Intramolecular Reactions

General.—In recent years tin and silicon promoted diene and enyne cyclizations have been shown to be synthetically useful procedures. Recently Simpkins and coworkers have reported[10,11] analogous cyclizations using the reagents $TolSO_2SePh$ or $(Ph)_2PH$. The introduction of useful functionality into the products, as illustrated in the isolation of (17) and (19), is always an attractive feature of this type of transformation.

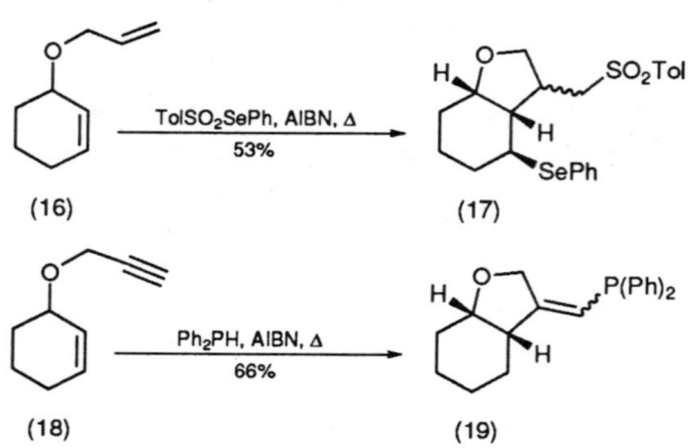

Acyl tellurides have been shown to act as acyl radical precursors;[12] cyclization of telluride (20) can be promoted by photolysis in very good chemical yields. The retention of the useful vinyl aryl telluride functionality is particularly attractive as the

[8] B.S. Jaynes and C.L. Hill, *J. Am. Chem. Soc.*, 1993, **115**, 12212.
[9] E. Hasegawa and D.P. Curran, *J. Org. Chem.*, 1993, **58**, 5008.
[10] J.E. Brumwell, N.S. Simpkins, and N.K. Terrett, *Tetrahedron Lett.*, 1993, **34**, 1219.
[11] J.E. Brumwell, N.S. Simpkins, and N.K. Terrett, *Tetrahedron Lett.*, 1993, **34**, 1215.
[12] C. Chen and D. Crich, *Tetrahedron Lett.*, 1993, **34**, 1545.

products can undergo further synthetic elaboration as exemplified in the production of the phenyl selenide product (21) shown.

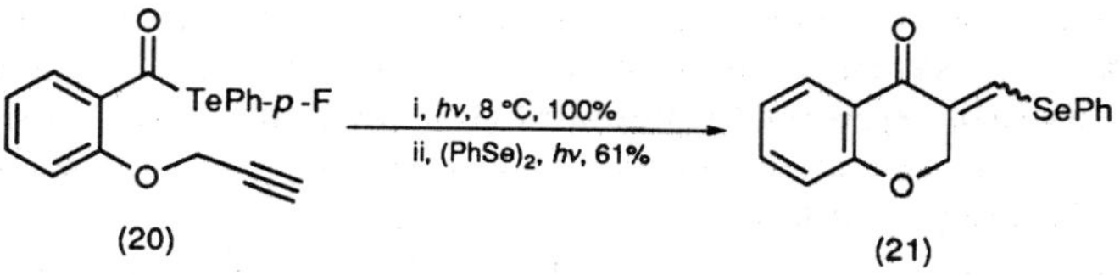

Uneyama and coworkers have also utilized telluride precursors for useful radical-based transformations; in this case however, vinyl radicals rather than acyl radicals are generated. Precursors such as (22)[13] have been utilized in an appealing new indole synthesis; photolysis of (22) leads to good yields of indole (23).

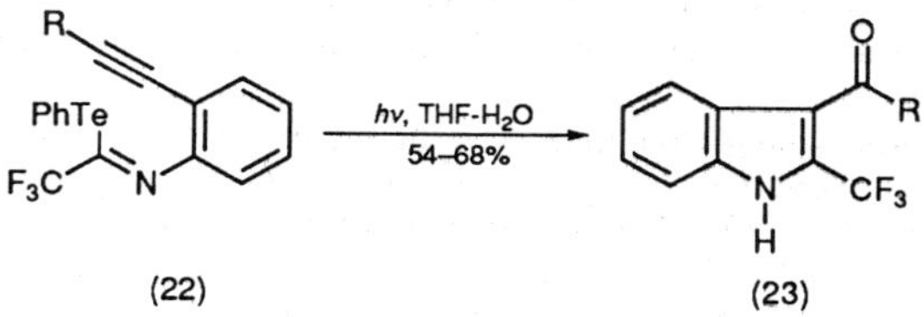

Intramolecular radical addition to aromatic systems has also been of recent interest. Ziegler and co-workers have continued their work in the indole area by demonstrating that both dioxolanyl[14] and oxiranyl[15] radicals derived from thiohydroxamate ester precursors (24) and (26) undergo intramolecular addition to provide entry into enantiomerically pure dihydroindoles (25) and (27). The initial photoadduct is a dimer

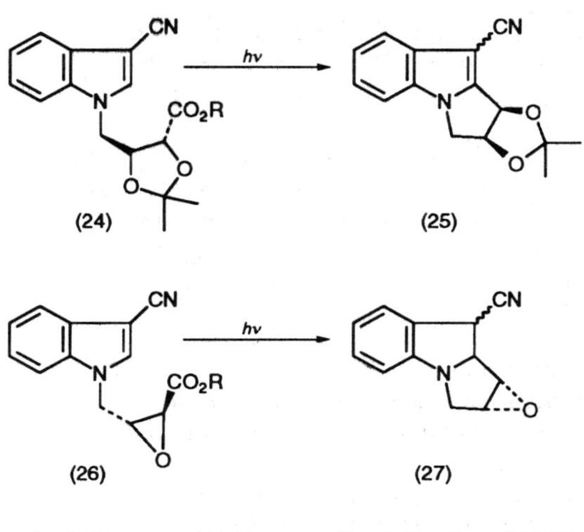

[13] Y. Ueda, H. Watanabe, J. Uemura, and K. Uneyama, *Tetrahedron Lett.*, 1993, **34**, 7933.
[14] F. E. Ziegler and P. G. Harran, *J. Org. Chem.*, 1993, **58**, 2768.
[15] F. E. Ziegler and P. G. Harran, *Tetrahedron Lett.*, 1993, **34**, 4505.

which can then undergo dissociation and disproportionation to give the products shown in moderate yields.

Ipso **Substitution.**—Some recent approaches to the formation of carbon–carbon bonds in aromatic systems employ *ipso*-substitution processes. Treatment of aryl ether (28) with Bu_3SnH/AIBN gives the *C*-substituted aromatic system (29),[16] possibly *via* an

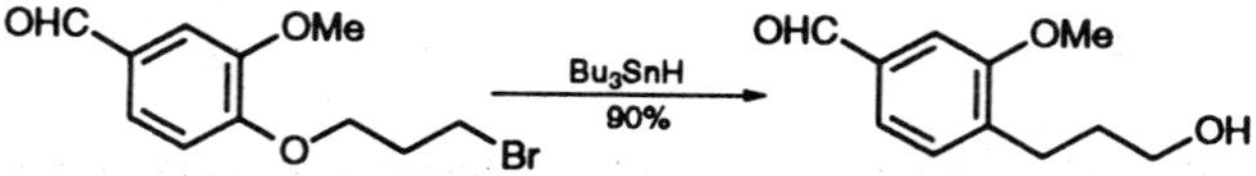

(28) (29)

addition–elimination process. These reactions can proceed efficiently to give rearranged products in good yield; however the competing reduction process can dominate if the aromatic ring is not suitably activated with electron withdrawing or donating substituents.

An alternative *ipso* substitution using sulfone substituted aromatic systems has also been reported. Treatment of sulfones (30) with Bu_3SnH/AIBN leads to the fused indole systems (31) in moderate to good yields.[17]

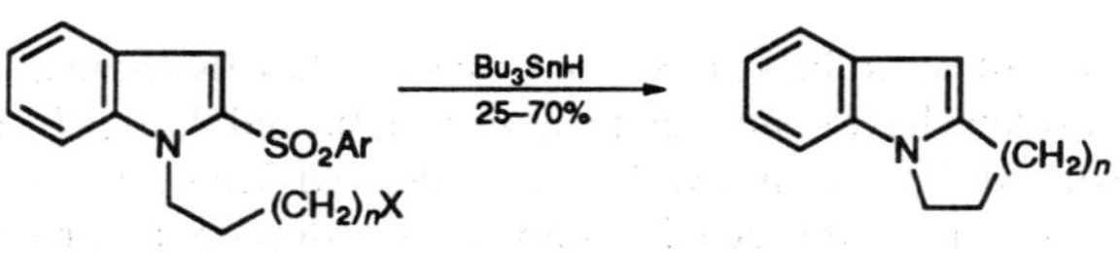

(30) $n = 1, 2, 3$ (31) $n = 1, 2, 3$

In an interesting recent report Cadogan, McNab, and coworkers[18] have demonstrated the ability of carboxylic esters to act as radical leaving groups. Flash vacuum pyrolysis (FVP) of aryl allyl ether (32) leads to the benzofuran derivatives (33) in moderate to good yields (39–95%).

Ipso-substitution is of course not limited to aromatic systems; it is now known that acyl germanes can also participate as exemplified in the transformation of (34) to (35).

However, the related acyl silane (36) undergoes addition and then formally a 'radical Brook' rearrangement to give a cyclopentanol, isolated as the benzoate (37).

This process has recently been put to excellent synthetic utility; treatment of (38) with allyl tributylstannane and AIBN, followed by desilylative lactonization gives (40) in reasonable yield.[19]

Rearrangements.—*Hydrogen-atom transfer*. 1,5-Hydrogen abstractions continue to interest synthetic chemists. A notable example is reported by Kraus and Andersh;[20]

[16] E. Lee, C. Lee, J. S. Tae, H. S. Whang, and K. S. Li, *Tetrahedron Lett.*, 1993, **34**, 2343.
[17] S. Caddick, K. Aboutayab, and R. West, *SYNLETT*, 1993, 231.
[18] M. Black, J. I. G. Cadogan, G. A. Cartwright, H. McNab, and A. D. MacPherson, *J. Chem. Soc., Chem. Commun.*, 1993, 959.
[19] D. P. Curran, W.-T. Jiaang, M. Palovich, and Y.-M. Tsai, *SYNLETT*, 1993, 403.
[20] G. A. Kraus and B. Andersh, *J. Chem. Soc., Chem. Commun.*, 1993, 646.

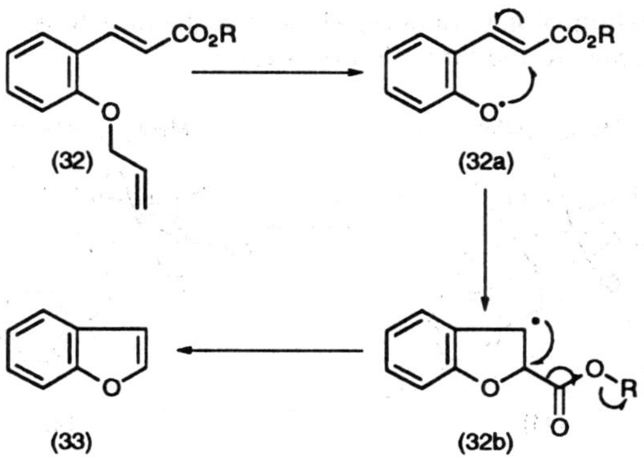

(32) (32a) (33) (32b)

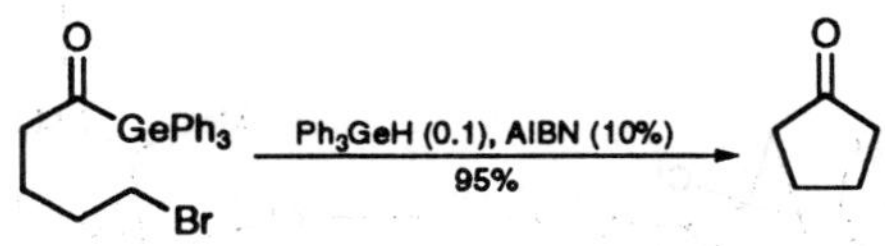

(34) (35)

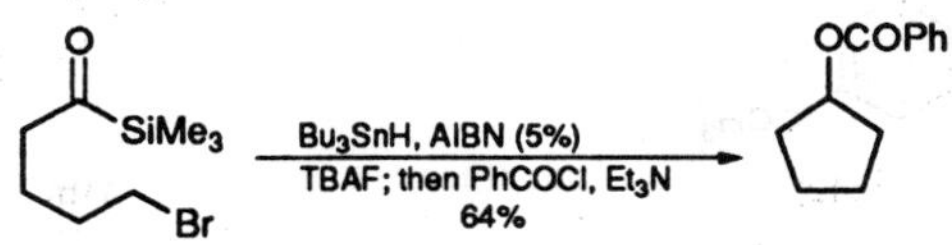

(36) (37)

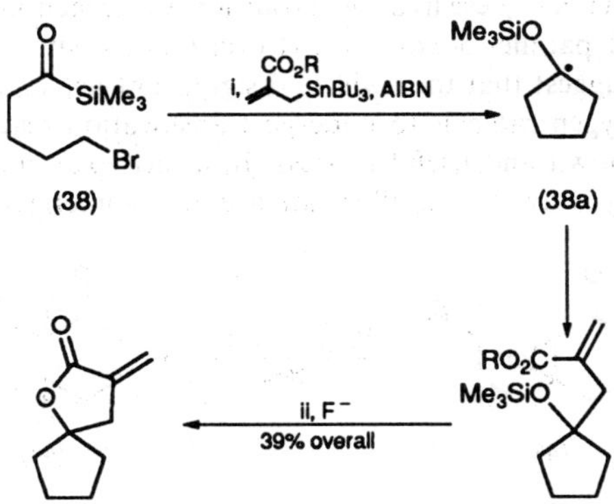

(38) (38a) (40) (39)

here 1,5-translocation of aryl radical (41) is used to initiate an interesting radical sequence, which leads to the formation of (42).

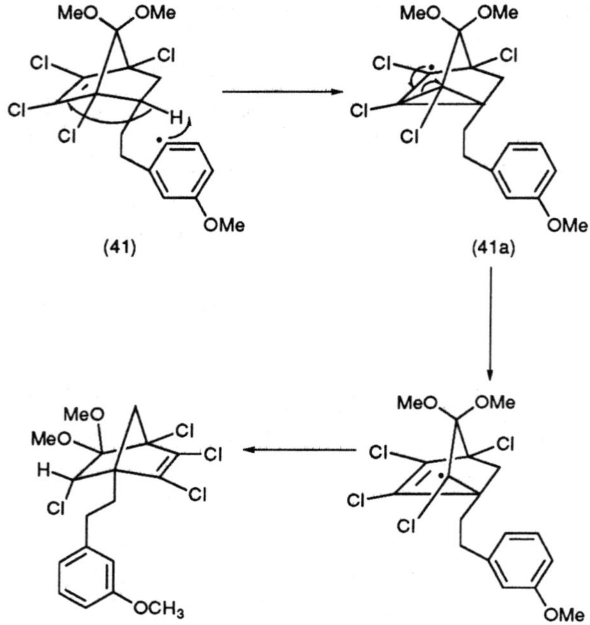

(41) (41a) (41b) (42)

The well-established hydrogen-atom abstraction capability of vinyl radicals is the subject of an illuminating full account by Curran and Shen.[21] They investigate the effect of substituents on these processes and calculate rate constants (typically $10^6\ s^{-1}$). The authors conclude that these hydrogen translocation reactions proceed *via* an early transition state and parallel 5-*exo* hexenyl cyclizations with respect to substituent effects. They also suggest that they will find significant use in synthesis.

The ability of oxygen radicals to undergo translocation reactions is, of course, a particularly well known and useful process. In a nice extension of their previously developed chemistry, Rawal *et al.* illustrate a useful isomerization of (43) to (44).[22]

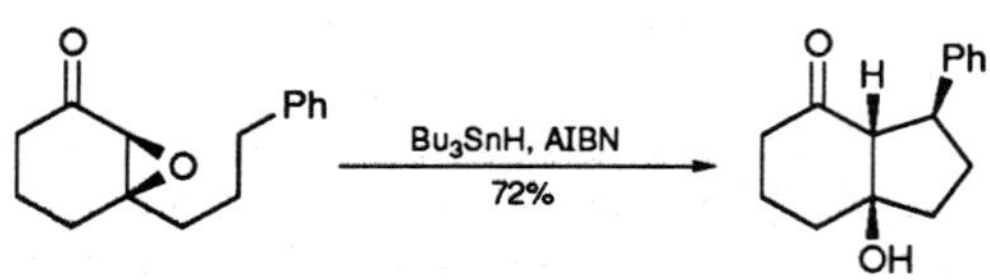

(43) (44)

Although catalytic in principle and practice, the authors found that the highest yields were obtained when mediated full equivalent each of Bu_3SnH and AIBN was used.

[21] D. P. Curran and W. Shen, *J. Am. Chem. Soc.*, 1993, **115**, 6051.
[22] V. H. Rawal, V. Krishnamurthy, and A. Fabre, *Tetrahedron Lett.*, 1993, **34**, 2899.

Group Transfer. Exciting possibilities exist for radical transformations which involve the transfer of groups as opposed to atoms. Kim and Lim have been investigating the homolytic 1,5- and 1,6-transfer of the tributyltin group; their recent work has demonstrated that these transfers [(45) to (46) and (47); (48) to (49) and (50)

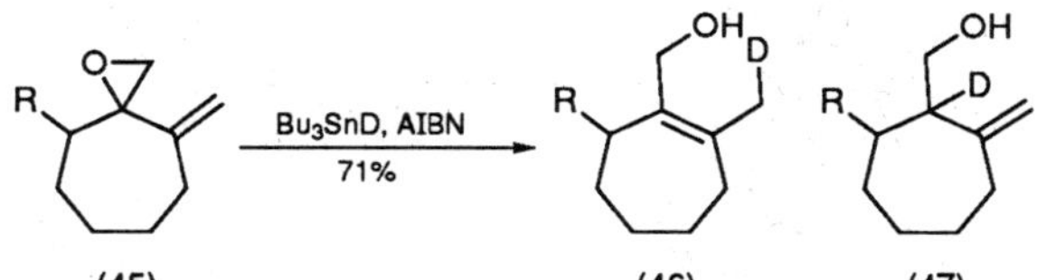

(45) (46) (47)

(46):(47) 66:34

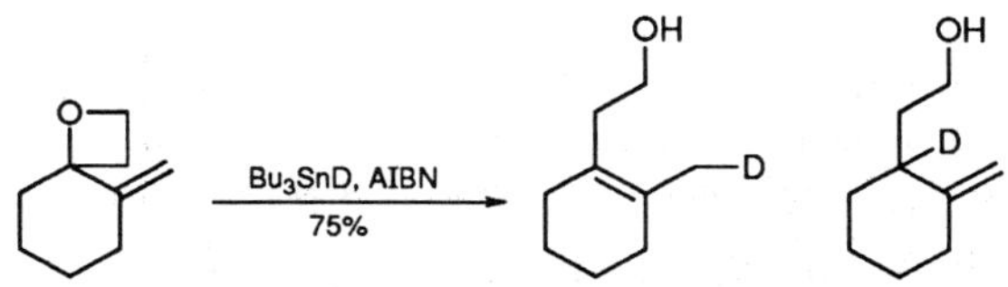

(48) (49) (50)

(49):(50) 88:12

respectively] are more favourable than the analogous hydrogen-atom transfers.[23,24] The authors estimate the rate constants for 1,5- and 1,6-Bu_3Sn transfer at *ca.* $10^9\ s^{-1}$; the reported rate constant for 1,5-hydrogen transfer is *ca.* $10^8\ s^{-1}$.

The 1,2-radical migration reaction, generalized below as (51) to (52), has attracted

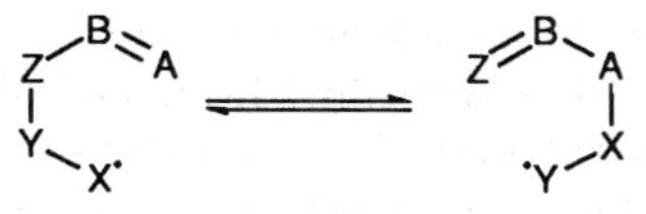

(51) (52)

significant interest because of its application to a range of synthetically useful systems. The reaction mechanism is complex and recent studies on *β*-acyloxytetrahydropyranyl radicals by Beckwith and Duggan[25] have suggested three mechanistic possibilities: concerted *via* a five-membered transition structure (53a); dissociation into an intermediate radical-cation–anion pair (53b); or 1,2-oxygen shift *via* a three-membered transition structure (53c).

The development of related migration reactions and associated mechanistic investigations are ongoing.[26] One particularly nice example is reported by Crich and Yao; treatment of vinyl ether (54) with Bu_3SnH leads to the formation of ketone (55) *via* a carbon–carbon bond-forming migration process.[27]

[23] S. Kim and K. M. Lim, *Tetrahedron Lett.*, 1993, **34**, 4851.
[24] S. Kim and K. M. Lim, *J. Chem. Soc., Chem. Commun.*, 1993, 1152.
[25] A. L. J. Beckwith and P. J. Duggan, *J. Chem. Soc., Perkin Trans. 2*, 1993, 1673.
[26] D. Crich and Q. Yao, *Tetrahedron Lett.*, 1993, **34**, 5677.
[27] D. Crich and Q. Yao, *J. Chem. Soc., Chem. Commun.*, 1993, 1265.

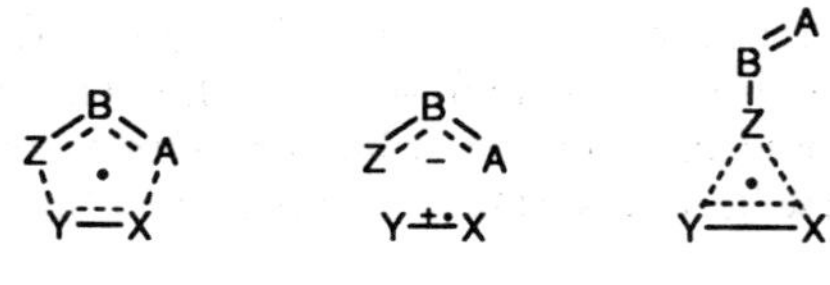

(53a) (53b) (53c)

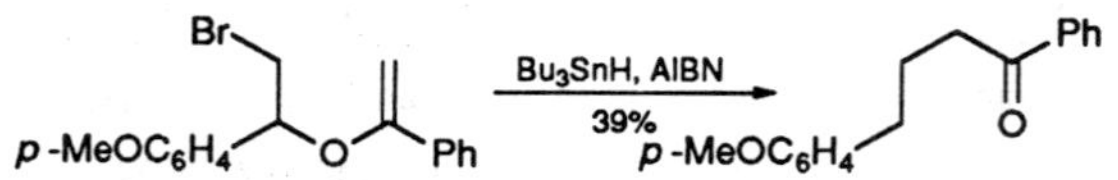

(54) (55)

These and other related group-transfer reactions have found synthetic utility in the carbohydrate area. Jung and Tinachoe have recently utilized a stereospecific formyl group transfer for the transformation of (56) to (56a).[28]

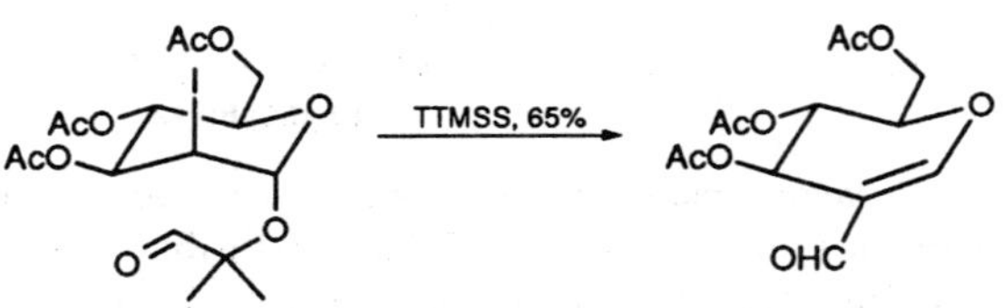

(56) (56a)

Ring Expansions. Zhang and Dowd continue to demonstrate the generality and synthetic applications of ring expansions which are initiated by addition of carbon-centred radicals to carbonyl groups.[29] In a nice recent development Kim and coworkers have shown[30] that aminyl radicals generated from azide precursor (57) will undergo intramolecular addition to a ketone to generate an oxygen radical (57b). This then undergoes β-scission to give the ring-expanded product (57c) in good to excellent yields.

The β-scission of alkoxy radicals is now a well-established path for the initiation of ring-expansion reactions. Recently Pattenden and Schulz have reported[31] an impressive cascade process in which the primary cyclization step involves addition of a carbon radical to an imine. Thus, treatment of cyclobutanone oxime (58) with TTMSS leads to (59) in good yield by the sequence shown below.

Methylene cyclopropanes also serve as useful precursors for ring-expansion reactions *via* addition–fragmentation processes as demonstrated in the transformation of (60) to (61) by Destabel, Kilburn, and Knight.[32]

[28] M. E. Jung and S. W. Tinachoe, *Tetrahedron Lett.*, 1993, **34**, 6247.
[29] W. Zhang and P. Dowd, *Tetrahedron.*, 1993, **49**, 1965.
[30] S. Kim, G. H. Joe, and J. Y. Do, *J. Am. Chem. Soc.*, 1993, **115**, 3328.
[31] G. Pattenden and D. J. Schulz, *Tetrahedron Lett.*, 1993, **34**, 6787.
[32] C. Destabel, J. D. Kilburn, and J. Knight, *Tetrahedron Lett.*, 1993, **34**, 3151.

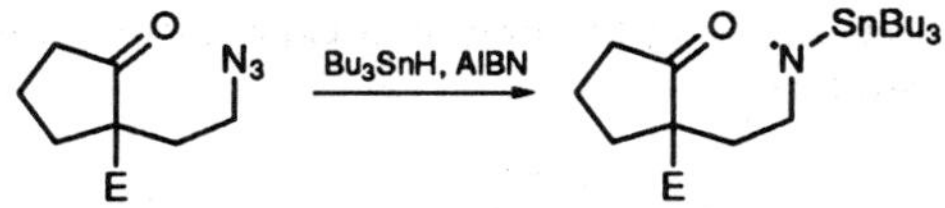

(57) (57a)

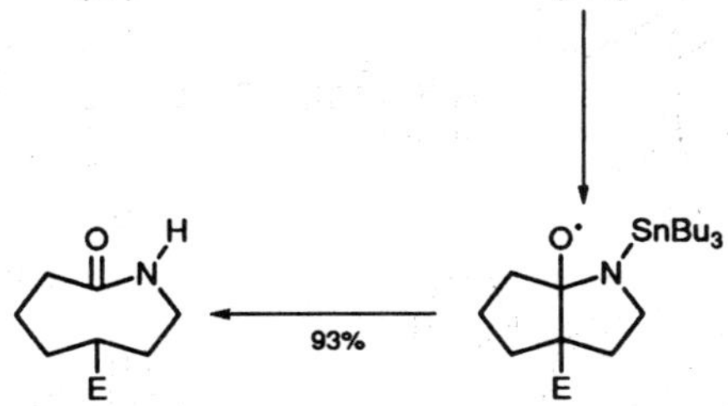

(57c) (57b)

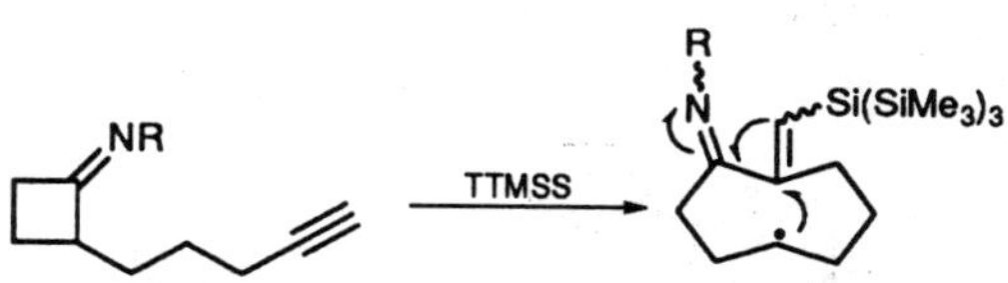

(58) (58a)

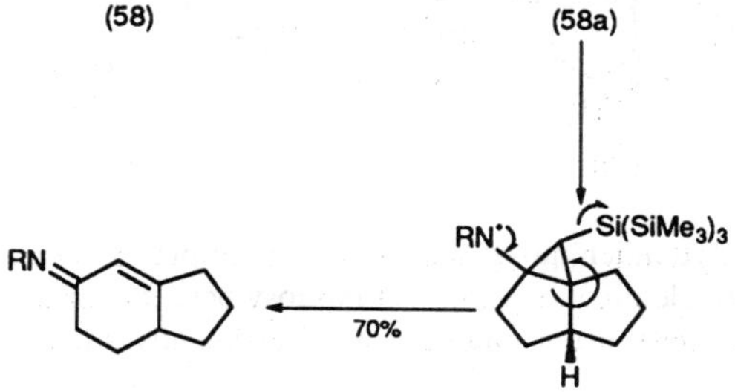

(59) (59a)

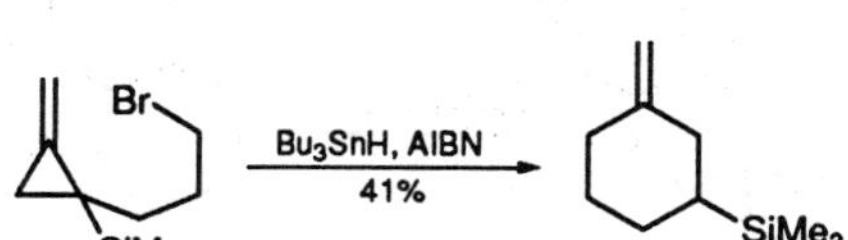

(60) (61)

Many radical-mediated ring-expansion methods are based upon the propensity of oxygen radicals to undergo β-scission. There are, of course, a range of methods which can be used for the formation of such radicals and these have been used to promote ring-expansion sequences. Rawal and Zhong have demonstrated.[33] that radical-induced ring opening of epoxides can lead to ring-expanded products; this process can be incorporated into a useful tandem sequence as shown in the transformations below.

[33] V. H. Rawal and H. M. Zhong, *Tetrahedron Lett.*, 1993, **34**, 5197.

Care is required when choosing the optimum experimental conditions in order to isolate the desired product.

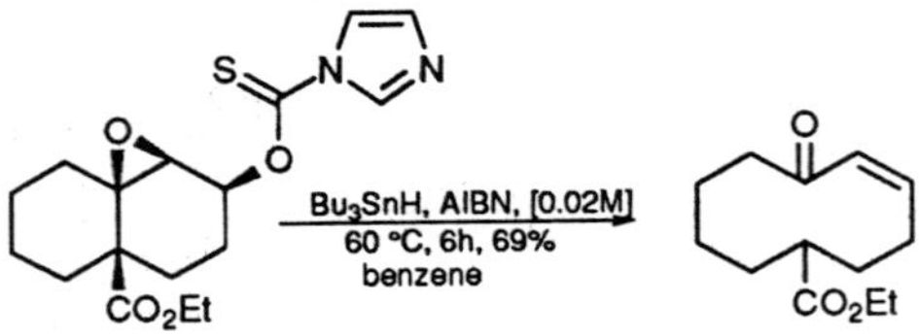

(62) (63)

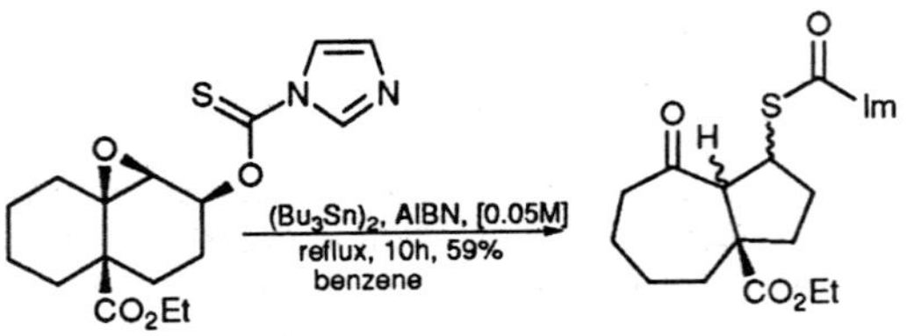

(64) (65)

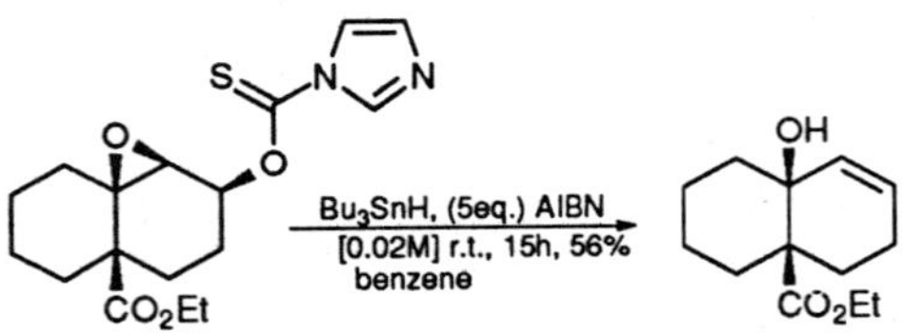

(66) (67)

Mowbray and Pattenden have used alcohols in impressive sequences;[34] photolysis of the readily available precursor (68) with iodosylbenzene diacetate and iodine leads to the isolation of (69) in reasonable yield. Further investigations also highlight a

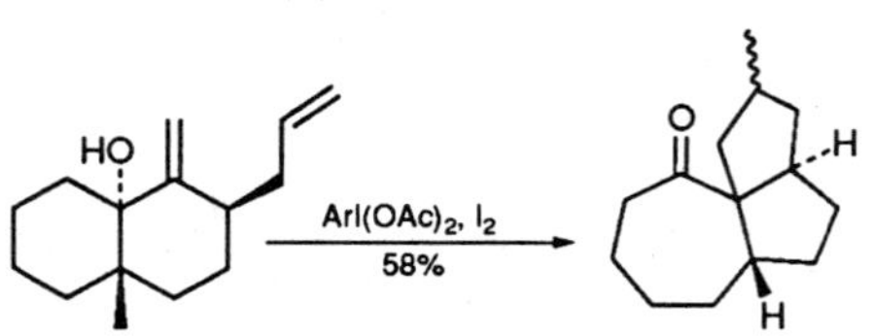

(68) (69)

potential limitation of the methodology; if a competing hydrogen-atom abstraction is geometrically possible, it can lead to significantly reduced yields of desired products. This is, however, a potential drawback with most methodologies incorporating oxygen-radical intermediates.

Booker-Milburn and Thompson have utilized ferric chloride to promote a range of diastereoselective ring-expansion–cyclization reactions;[35] more recently this group

[34] C.E. Mowbray and G. Pattenden, *Tetrahedron Lett.*, 1993, **34**, 127.
[35] K.I. Booker-Milburn and D.F. Thompson, *SYNLETT*, 1993, 592.

has demonstrated the potential of alternative iron(III) salts. Treatment of (70) with iron(III) nitrate leads to the isolation of (71) in reasonable yield (49%).

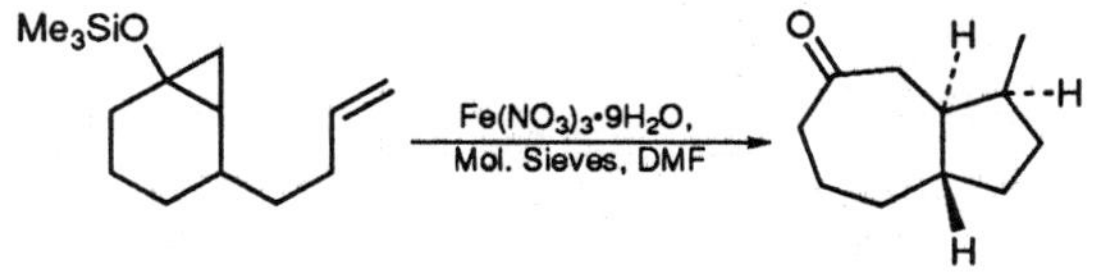

(70) (71)

Tandem Cyclizations.—Tandem cyclizations continue to attract attention from synthetic chemists as methods for forming complex polycyclic systems. Applications in natural product synthesis are now very common; recent examples include Curran and Shen's approach[36] to Modhephene utilizing sequential 5-*exo*-trig cyclizations, (72) to (73), and Parson and coworkers' tandem approach to lysergic acid derivatives, (74) to (75).[37]

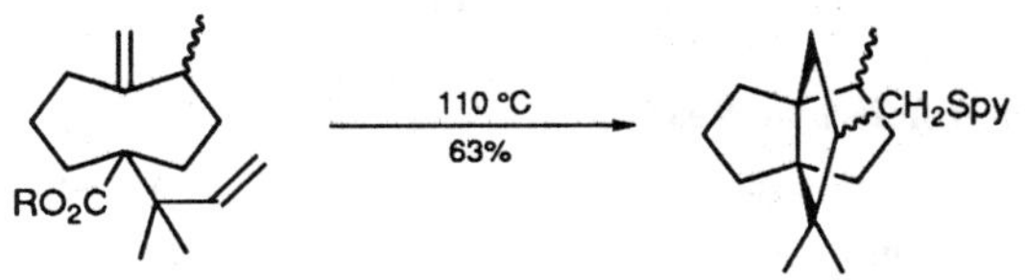

(72) (73)

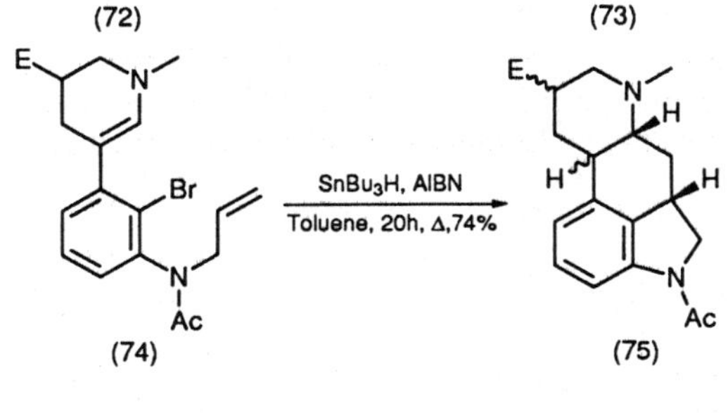

(74) (75)

The enediynes have attracted a huge amount of synthetic interest because of their ability to generate diradicals which can be utilized to good effect in biological systems. Grissom *et al.* have utilized similar materials in ingenious tandem cyclization reactions;[38] simple thermolysis of the trienediyne (76) leads to (77), *via* bis-cyclization of the aromatic diradical (76a), itself produced by a Bergman cyclization.

A spectacular example of tandem cyclization has been used in a very elegant synthesis of a tetracyclic steroidal skeleton using enyne-allenes.[39] Thermolysis of

[36] D. P. Curran and W. Shen, *Tetrahedron.*, 1993, **49**, 755.
[37] Y. Ozlu, D. E. Cladingboel, and P. J. Parsons, *SYNLETT*, 1993, 357.
[38] J. W. Grissom, T. L. Calkins, and H. A. McMillen, *J. Org. Chem.*, 1993, **58**, 6556.
[39] Y. W. Andemichael, Y. Huang, and K. K. Wang, *J. Org. Chem.*, 1993, **58**, 1651.

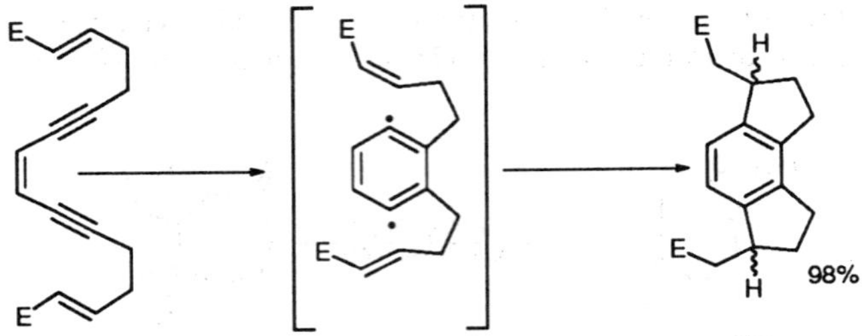

(76) (76a) (77)

precursor (78) leads to the tetracyclic product (79) *via* the radical cyclization–cycloaddition sequence shown. This useful transformation is particularly noteworthy because of the absence of any reagents, although the yields are moderate in the cases reported; the authors propose this to be a consequence of the cycloaddition component of the sequence. The utilization of radical reactions in sequence with efficient cycloaddition reactions is potentially very powerful and likely to find considerable synthetic application in future years.

Another unusual and noteworthy tandem cyclization sequence is reported by Harrowven.[40] Treatment of ketenethioacetal (80) with $SnBu_3H$/AIBN leads to thiophene (81); the proposed sequence for this novel transformation is shown below.

4 Intermolecular Reactions

General.—A synthetically useful intermolecular radical reaction is reported by Barton and Zhu.[41] White-light irradiation of PTOC esters (82) in the presence of two equivalents of white phosphorus, followed by oxidation leads to good yields (70–86%) of alkyl phosphonic acids (83). Competition experiments demonstrated that white phosphorus is a highly efficient trap for alkyl radicals.

Some interesting and synthetically useful carbonylation reactions have recently been published demonstrating that the use of 'slow' hydride donors, such as triphenylgermanium hydride, leads to improved conversion into carbonylation products.[42] Thus Gupta and Kahne incorporate a carbonylation into their hydroxymethylation approach using *in situ* triphenylgermanium hydride. Reductive carbonylation of iodide (84) led to (85) in moderate yield.

Ryu and coworkers are also active in this area and have recently demonstrated the benefits of TTMSS in mediating free-radical carbonylation. Alkyl and vinyl halides undergo carbonylation using Bu_3SnH or TTMSS as promoter; however TTMSS permits the use of lower CO pressures.[43]

A useful alternative vinylation–allylation sequence has been recently developed by Huval and Singleton;[44] in contrast to the use of tin reagents described by Keck and others, this work utilizes vinyl and allyl halides, including (86) and (88), to give good yields of addition products, (87) and (89).

[40] D.C. Harrowven, *Tetrahedron Lett.*, 1993, **34**, 5653.
[41] D.H.R. Barton and J. Zhu, *J. Am. Chem. Soc.*, 1993, **115**, 2071.
[42] V. Gupta and D. Kahne, *Tetrahedron Lett.*, 1993, **34**, 591.
[43] I. Ryu, M. Hasegawa, A. Kurihara, A. Ogawa, S. Tsunoi, and N. Sonoda, *SYNLETT*, 1993, 143.
[44] C.C. Huval and D.A. Singleton, *Tetrahedron Lett.*, 1993, **34**, 3041.

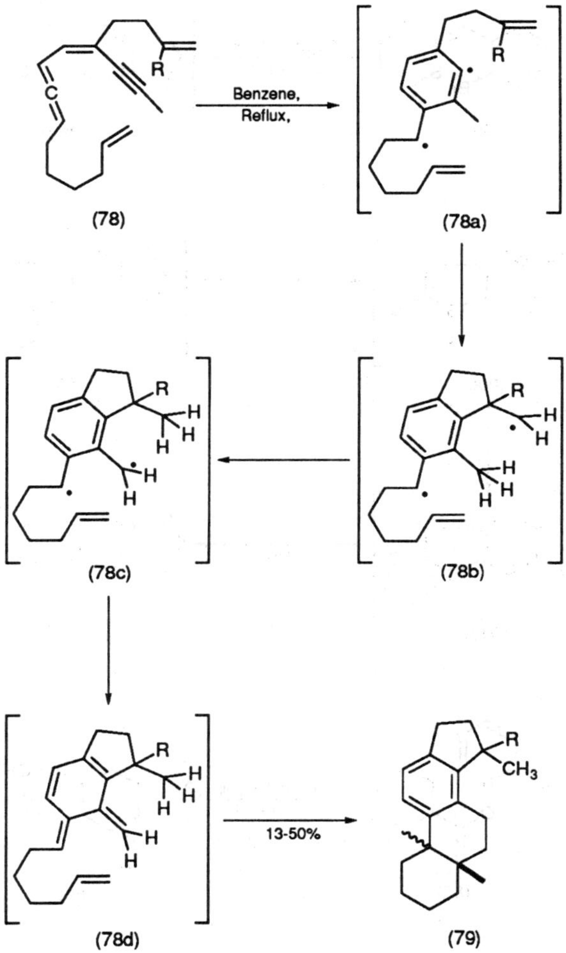

Intermolecular addition to aromatic systems continues to be a subject of intensive investigation.[45–47] The Bacchiochi group has demonstrated that a range of systems (Fe^{II}/H_2O_2; Et_3B/O_2; Mn^{III}) can promote addition of various alkyl radicals to aromatic systems, exemplified in the transformation of (90) to (91).[48–51]

Other work has demonstrated the ability of organostannyl radicals to undergo

[45] Q.-Y. Chen and Z.-T. Li, *J. Chem. Soc., Perkin Trans. 1*, 1993, 1705.
[46] M. Yoshida, R. Imai, Y. Komatsu, Y. Morinaga, N. Kamigata, and M. Iyoda, *J. Chem. Soc., Perkin Trans. 1*, 1993, 501.
[47] Q.-Y. Chen and Z.-T. Li, *J. Chem. Soc., Perkin Trans. 1*, 1993, 645.
[48] E. Baciocchi and E. Muraglia, *Tetrahedron Lett.*, 1993, **34**, 5015.
[49] E. Baciocchi and E. Muraglia, *Tetrahedron Lett.*, 1993, **34**, 3799.
[50] E. Baciocchi and E. Muraglia, *J. Org. Chem.*, 1993, **58**, 7610.
[51] E. Baciocchi, B. Floris, and E. Muraglia, *J. Org. Chem.*, 1993, **58**, 2013.

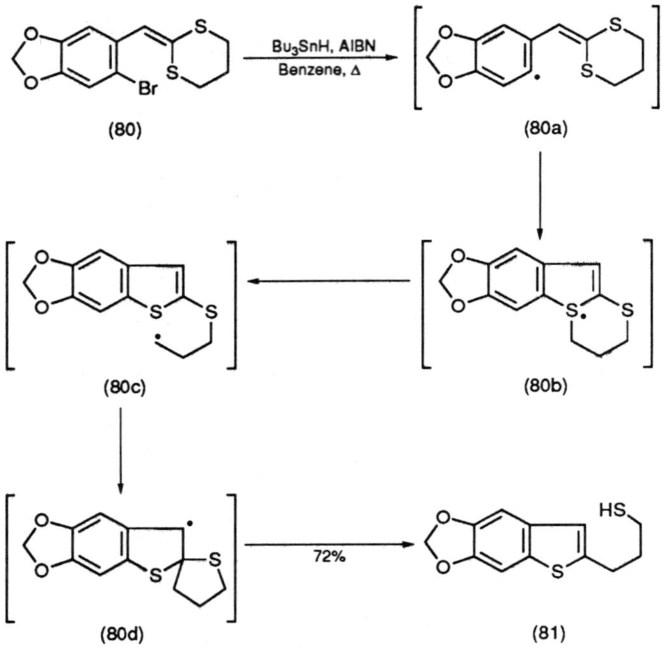
Bu3SnH, AIBN
Benzene, Δ
(80)
(80a)
(80b)
(80c)
(80d)
72%
(81)

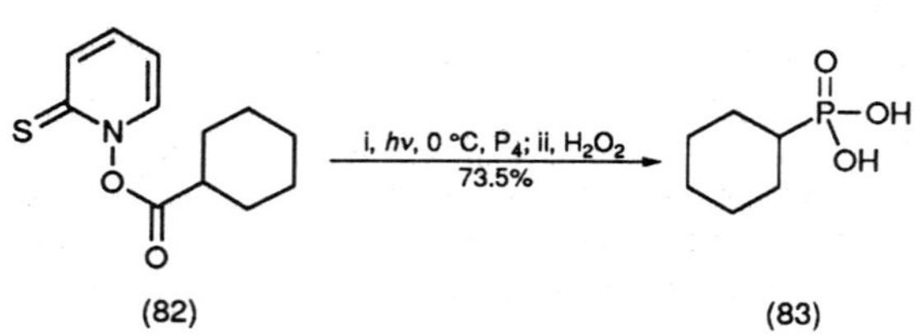
i, hν, 0 °C, P4; ii, H2O2
73.5%
(82)
(83)

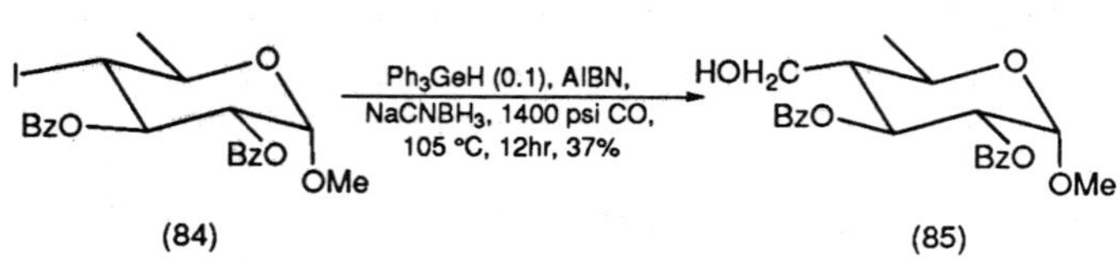
Ph3GeH (0.1), AIBN,
NaCNBH3, 1400 psi CO,
105 °C, 12hr, 37%
(84)
(85)

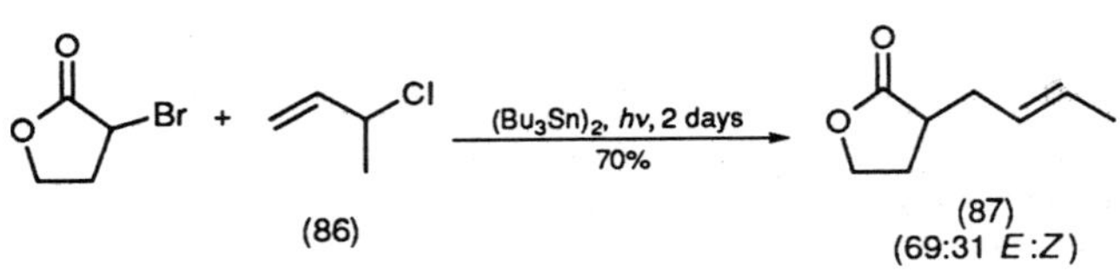
(Bu3Sn)2, hν, 2 days
70%
(86)
(87)
(69:31 E:Z)

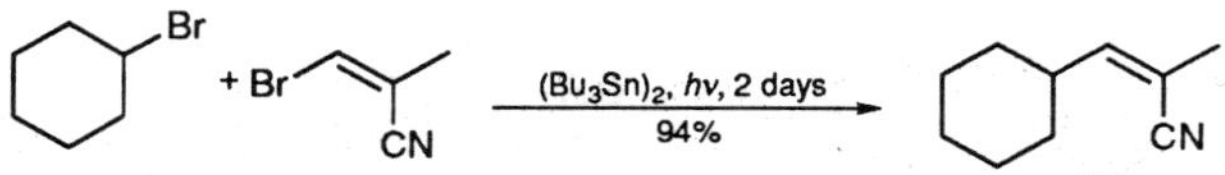

(88) (89)
(23:77 *E*:*Z*)

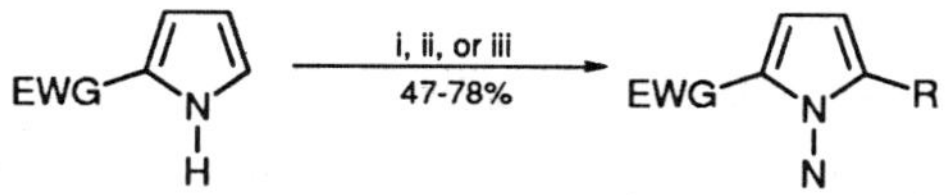

(90) (91)

Reagents: i, Et_3B, O_2, benzene or DMSO, RI; ii, $FeSO_4$, H_2O_2, DMSO, RI; iii, $Mn(OAc)_3$, NaOAc, RH

addition to sulfone-substituted aromatic systems as shown in the transformation of (92) to (93).[52]

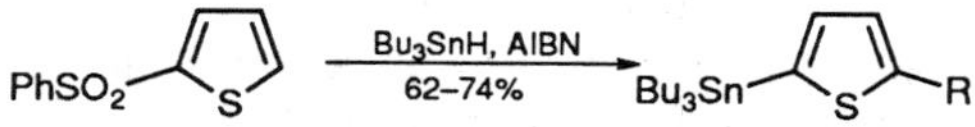

(92) (93)

Acyclic Stereocontrol.—*π-Substituted Radicals.* One of the most important recent synthetic developments in free-radical chemistry is the stereocontrolled formation of C–C bonds in acyclic systems. Most investigations in this area have concentrated on substrate-controlled reactions and further important contributions have been reported in the last year. Giese *et al.* have recently demonstrated[53] that polar groups can enhance the stereoselectivity of 'enolate radical' reductions. In earlier work it had been demonstrated that substituted enolate radicals, *e.g.* from (94), give products which can be explained by invoking a conformation which minimizes A-strain effects. By incorporating polar groups it was found that the degree of stereocontrol was significantly enhanced [from 66 : 33 (X = Me) to >95 : 5 (X = F, OMe)] because of the additional unfavourable dipole–dipole interactions between X and CO_2R in conformer (94b).

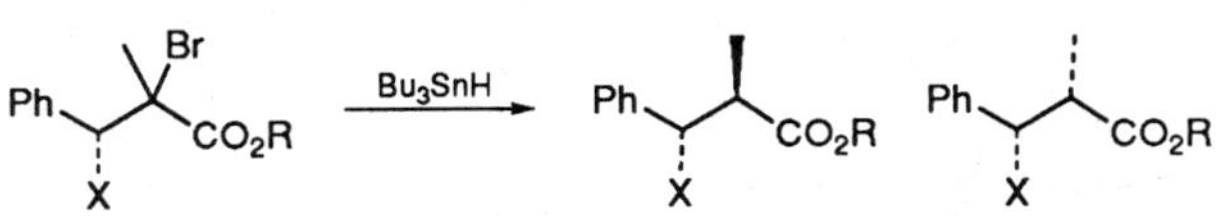

(94) (95) (96)

X	(95)	(96)
Me	66	3
F	95	5
OMe	97	3

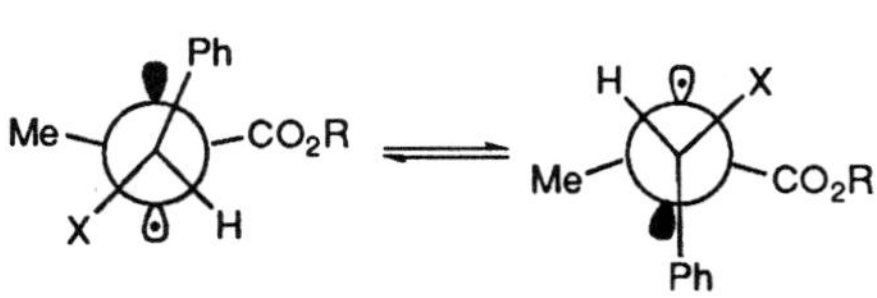

(94a) (94b)

[52] S. Caddick and S. Khan, *Tetrahedron Lett.*, 1993, **34**, 7469.

[53] B. Giese, W. Damm, F. Wetterich, H.-G. Zeitz, J. Rancourt, and Y. Guindon, *Tetrahedron Lett.*, 1993, **34**, 5885.

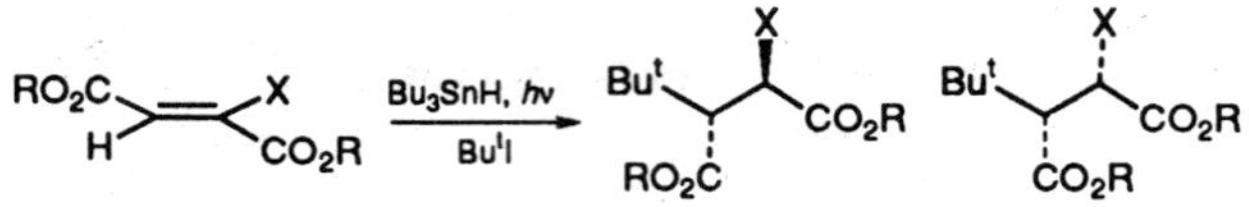

(97) (98) (99)

X	(98)	(99)
Me	96	4
Cl	88	12
OMe	84	16

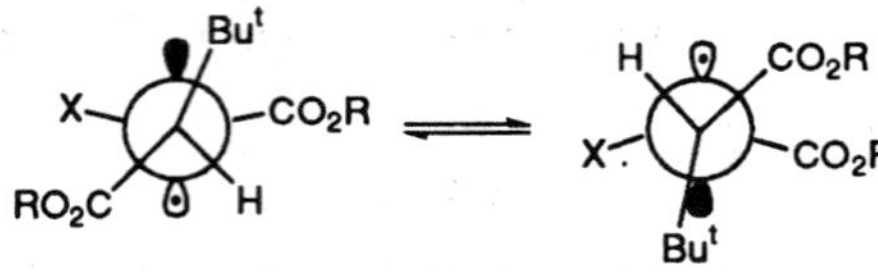

(97a) (97b)

In keeping with this analysis it is found that introduction of a polar substituent at the *radical centre*, *e.g.* from (97), reduces the energy difference between the two conformers. In these cases selectivity decreases when X = Cl, OMe and is optimal when X = CH_3.

Although the use of the A-strain model is useful for predicting stereoselectivity in cases of this type, recently Curran and Ramamoorthy have discussed at length[54] the fact that such models are not particularly tolerant to substituent effects. In addition to being substrate dependant, the stereochemical outcome of many reactions of this type may be additionally dependant on the nature of the incoming reagent.

Heteroatom-substituted Radicals. Recent work has suggested that, in general, α-alkoxy radicals show selectivity which can be explained using the Felkin–Anh rule. Thus, Giese and coworkers have optimized the transition states of radical (100), which undergoes reduction with hydride donor TTMSS.[55] There are similarities between this type of reduction, proposed by Giese to proceed *via* conformer (100a), and the analogous hydride-mediated reduction. Eksterowicz and Houk's calculations[56] support this proposal and go on to show that the radical transition structure is more pyramidalized than the transition state for the corresponding ionic reduction.

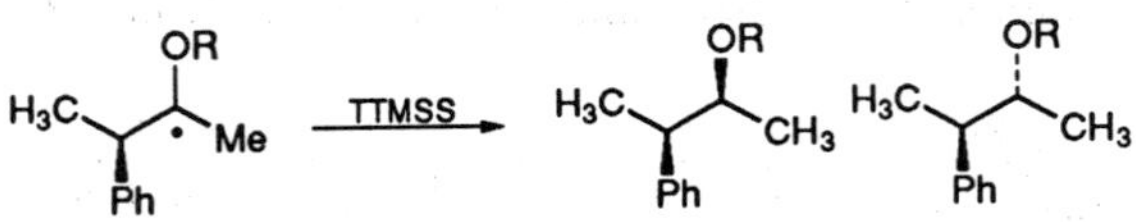

(100) R = H, $Si(SiMe_3)_3$ *syn* : *anti* 75:25

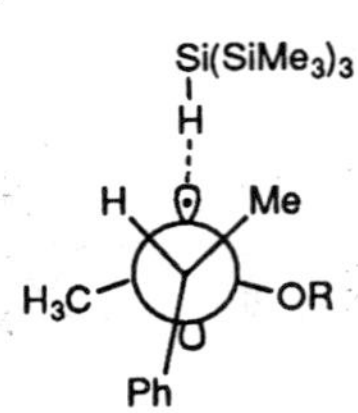

(100a)

[54] D. P. Curran and P. S. Ramamoorthy, *Tetrahedron*, 1993, **49**, 4841.
[55] W. Damm, J. Dickhaut, F. Wetterich, and B. Giese, *Tetrahedron Lett.*, 1993, **34**, 431.
[56] J. E. Eksterowicz and K. N. Houk, *Tetrahedron Lett.*, 1993, **34**, 427.

Curran and Sun have examined related systems; early work had demonstrated that nitrogen substituted radicals give *syn* products, which can be explained using the A-strain model. However, with α-H–N substituted radicals *anti* products are obtained; treatment of (101) with TTMSS provides the products (102) and (103). The high degree of *anti* selectivity can be rationalized by invoking a 'steric' Felkin–Anh conformation (101a) or a Cram chelate conformation (101b).[57]

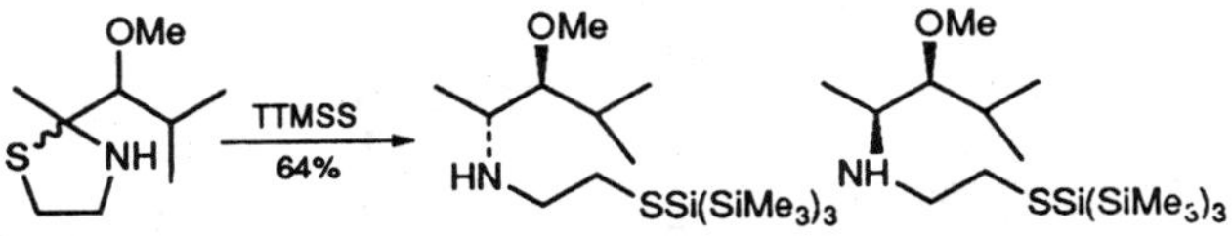

(101) (102) (103)

(102):(103) 86:14

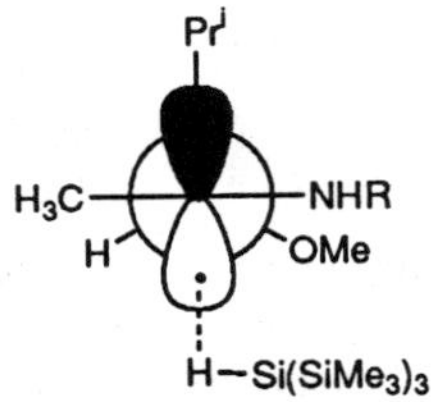

(101a)

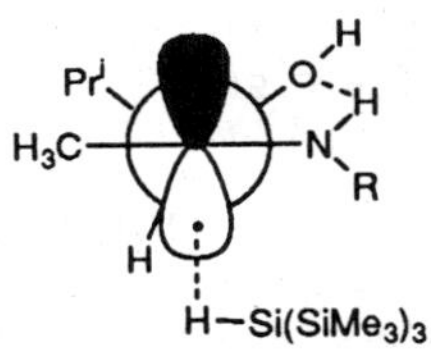

(101b)

Previous studies had demonstrated that simple alkyl radicals can also participate in highly stereoselective reactions. In a recent investigation Curran, Giese, and co-workers examined the iodine atom transfer reaction of (104) and (105) which gives in a highly *syn*-selective manner (106) as the major product (98 : 2 *syn* : *anti*). These authors propose that the iodine donor approaches conformer (106a) between the medium and large groups and *anti* to the smallest group.[58] This transition state minimizes torsional strain as the radical centre pyramidalizes in the transfer process.

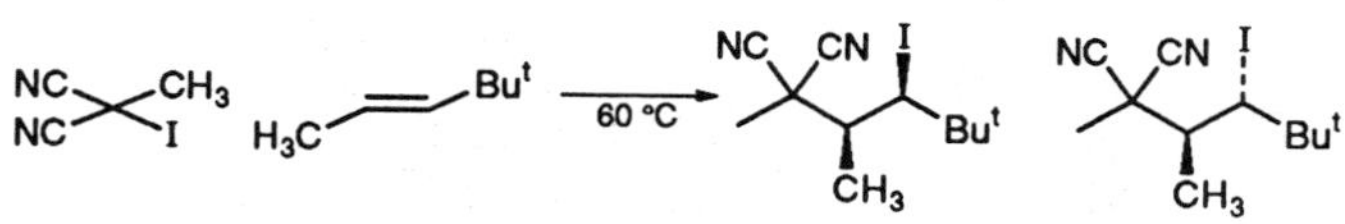

(104) (105) (106) (107)

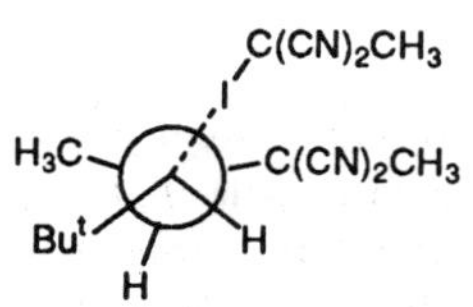

(106a)

[57] D. P. Curran and S. Sun, *Tetrahedron Lett.*, 1993, **34**, 6181.

[58] G. Thoma, D. P. Curran, S. V. Geib, B. Giese, W. Damm, and F. Wetterich, *J. Am. Chem. Soc.*, 1993, **115**, 8585.

Reagent-controlled Stereoselectivity. Some highly diastereoselective reactions have been recently reported by Toru and coworkers.[59] In this very exciting development α-sulfinylcyclopentenones (108) were found to undergo β-addition with t-butyl radicals to give adducts (109) and (110). In the presence of Lewis acid, product (109) could be isolated in good yield; the high degree of diastereoselectivity can be attributed to chelation of the sulfoxide and the carbonyl with the Lewis acid. In the absence of Lewis acid the other diastereoisomer predominated.

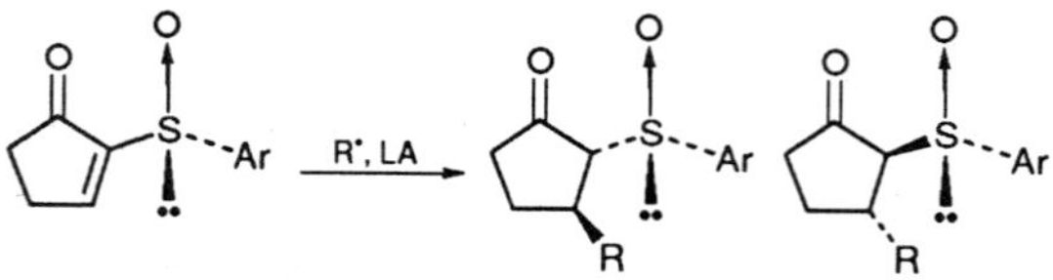

(108) (109) (110)

Conditions	Ratio (109)	Ratio (110)
$(Bu^t)_3B$, $Ti(Cl)_2(OPr^i)_2$, CH_2Cl_2, 0 °C, 94%	98	2
$(Bu^t)_3B$, CH_2Cl_2, 0 °C, 91%	38	62

Radical Sequences.—The use of inter- and intramolecular radical reactions in sequence provides opportunity for the rapid assembly of complex organic materials. Lee and coworkers have developed an approach to cyclohexenecarboxylates using this type of approach as shown in the transformation of (111), (112), and (113) to (114).[60]

Bachi recently disclosed an elegant approach to β-lactams as illustrated in the transformation of (115) to (116).[61]

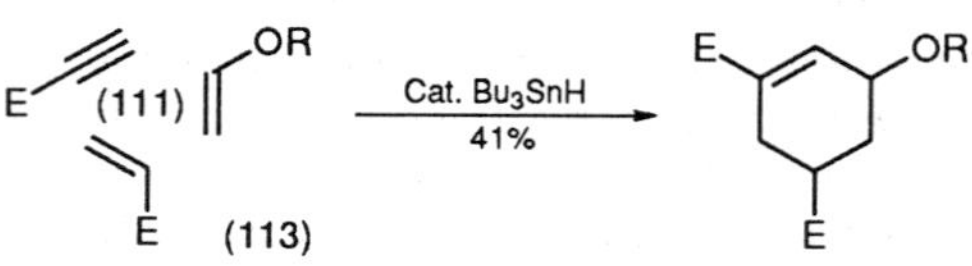

(112) (114) $E = CO_2R$

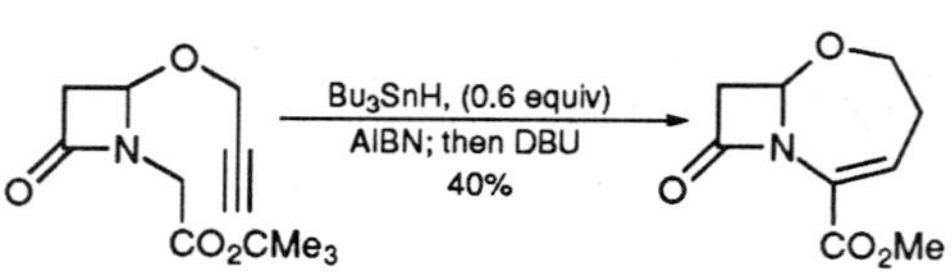

(115) (116)

This impressive transformation involves a radical sequence incorporating an intermolecular radical addition to give (115a), an intramolecular hydrogen atom transfer to provide (115b), and an intramolecular cyclization–fragmentation to give (115d). Treatment of (115d) with DBU gives the observed product (116) in reasonable yield. It is interesting to note that this sequence incorporates a relatively unusual

[59] T. Toru, Y. Watanabe, M. Tsusaka, and Y. Ueno, *J. Am. Chem. Soc.*, 1993, **115**, 10464.
[60] E. Lee, C. U. Hur, Y. H. Rhee, Y. C. Park, and S. Y. Kim, *J. Chem. Soc., Chem. Commun.*, 1993, 1466.
[61] E. Bosch and M. D. Bachi, *J. Org. Chem.*, 1993, **58**, 5581.

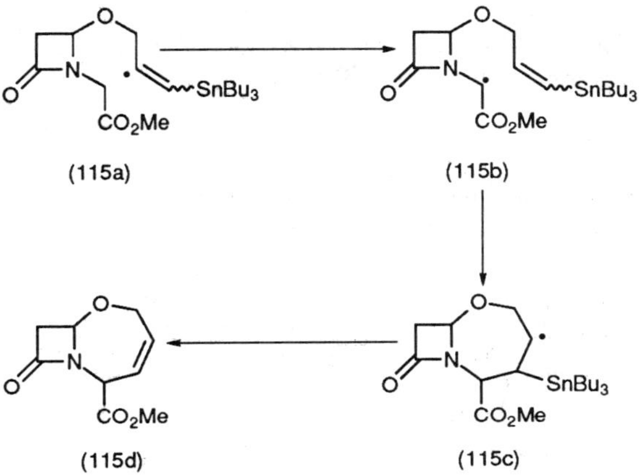

(115a) (115b) (115c) (115d)

1,6-hydrogen atom transfer step. A similar transformation incorporating a 1,5-transfer was also attempted; treatment of (117) with $SnBu_3H$ led to the isolation of (119) as the major product; the desired product (118) was isolated in a disappointing 19% yield.

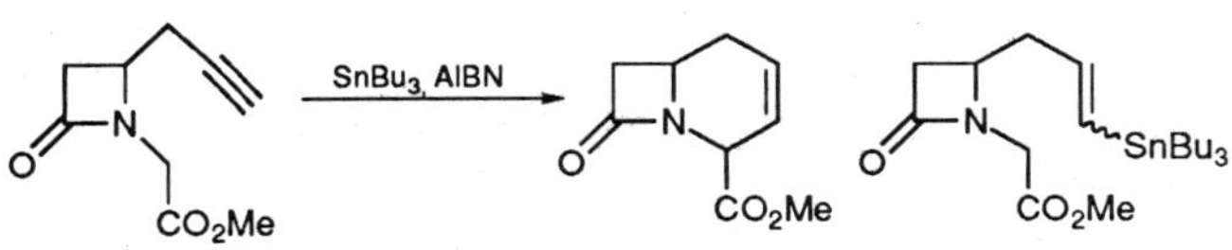

(117) (118) 19% (119) 56%

A carbonylation component can also be incorporated into synthetically efficient radical sequences as exemplified by the transformation of (120) to (121) in 72% yield.[62]

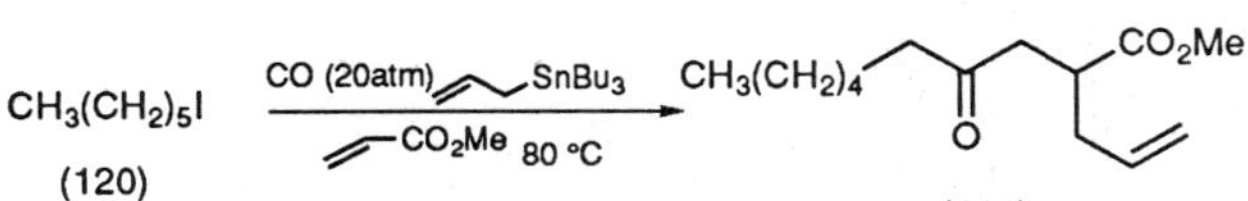

(120) (121)

5 Applications of Radical Process to Synthetic and Biological Chemistry

Synthetic Chemistry.—Corey and Helal have recently developed a convenient enantioselective synthesis of monosubstituted oxiranes;[63] the utilization of a selective radical dechlorination of chloroalkanes nicely illustrates the potential for radical reactions in sensitive systems. Selective bis-dechlorination of an enantiomerically pure trichloromethyl carbinol (122) gives (123), which under basic work-up gives the oxirane (124) in good yields (>90% ee).

The iodine atom transfer method developed by Curran and coworkers is finding significant utility in synthesis; in a full account[64] Ziegler and Fields describe the application of this powerful methodology to the synthesis of functionalized penicillins

[62] I. Ryu, H. Yamazaki, A. Ogawa, N. Kambe, and N. Sonoda, *J. Am. Chem. Soc.*, 1993, **115**, 1187.
[63] E. J. Corey and C. J. Helal, *Tetrahedron Lett.*, 1993, **34**, 5227.
[64] C. B. Ziegler, Jr and T. L. Fields, *Tetrahedron.*, 1993, **49**, 3919.

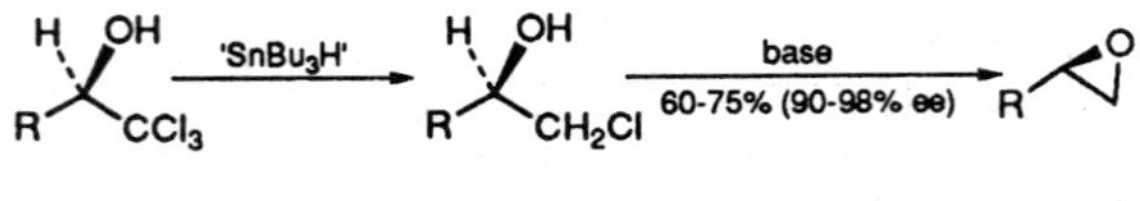

(122) (123) (124)

[(125 to (126)]. Kawecki and Welch have exploited the iodine atom transfer methodology in the synthesis of functionalized fluoro-β-lactams (128).[65]

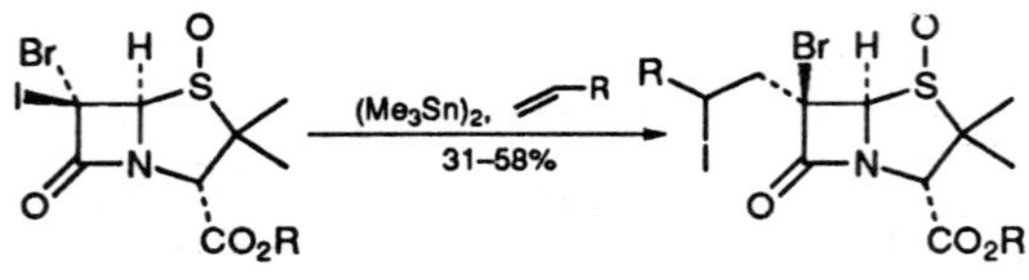

(125) (126)

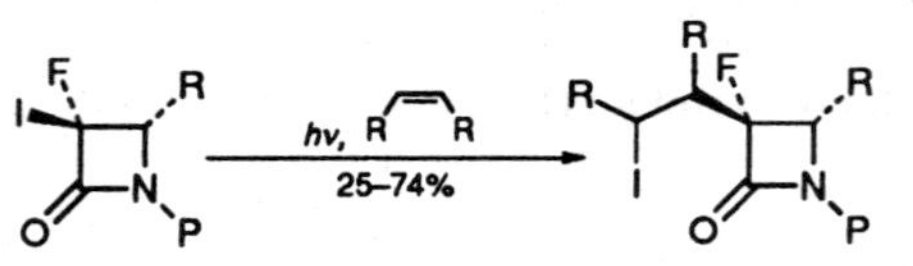

(127) (128)

Various alkaloid syntheses have incorporated free-radical transformations; Schultz and coworkers have utilized a simple intramolecular Bu_3SnH mediated cyclization of (129) to give (130) in their synthesis of (+)-1-deoxylycorine.[66]

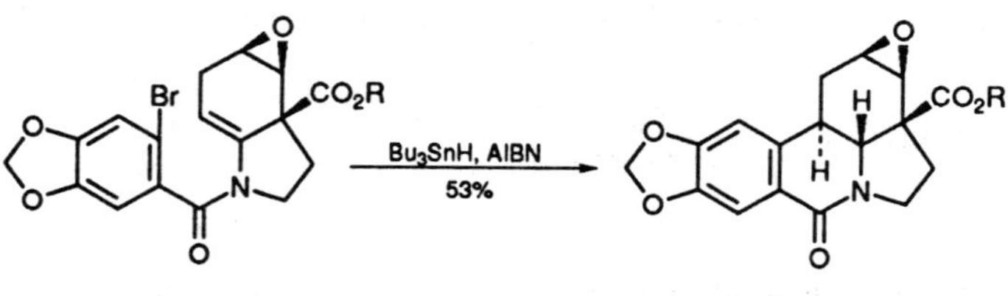

(129) (130)

The application of Bn_3SnH in alkaloid synthesis is nicely exploited by Bonjoch, Solé, and Bosch; in an impressive one-pot transformation (131) undergoes desulfurization and reductive cyclization to give tubifolidine (132) in 50% yield.[67]

Biological Chemistry.—The enediynes continue to attract a great deal of attention; C-1027 chromophore is a particularly interesting addition to this fascinating family of natural products. Yoshida and coworkers have recently shown that this molecule (133), which exhibits DNA cleaving capabilities, can undergo a *direct* rapid cyc-

[65] R. Kawecki and J.T. Welch, *Tetrahedron Lett.*, 1993, **34**, 3087.
[66] A.G. Schult, M.A. Holoboski, and M.S. Smyth, *J. Am. Chem. Soc.*, 1993, **115**, 7904.
[67] J. Bonjoch, D. Solé, and J. Bosch, *J. Am. Chem. Soc.*, 1993, **115**, 2064.

(131) (132)

loaromatization to give the potentially DNA-damaging benzenoid diradical (134).[68]

The enediynes have stimulated an increased interest in the interaction of carbon-centred radicals with DNA, because of the prospect of developing therapeutic agents. Breen and Murphy have shown[69] that certain simply vinyl epoxides react with thiol radicals to generate DNA-damaging, oxygen-centred radicals; they

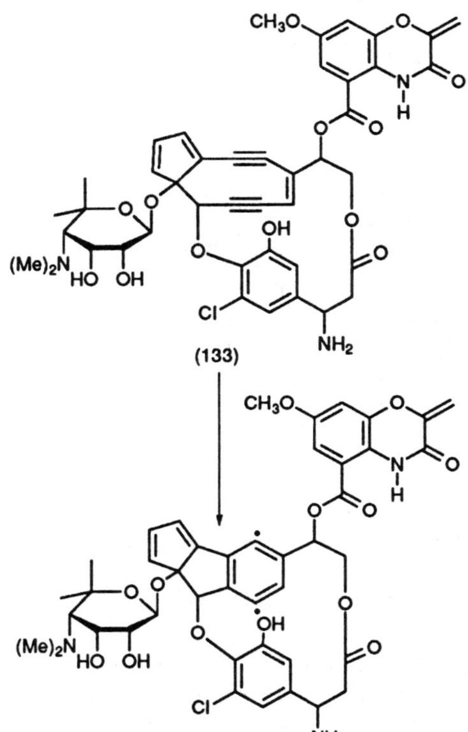

(133)

(134)

[68] K. Yoshida, Y. Minami, R. Azuma, M. Saeki, and T. Otani, *Tetrahedron Lett.*, 1993, **34**, 2637.
[69] A. P. Breen and J. A. Murphy, *J. Chem. Soc., Chem. Commun.*, 1993, 191.

then report[70] the DNA-cleaving capabilities of substrate (135), which is readily synthesized and incorporates a phenanthrolinium unit (a known DNA intercalator). Treatment of commercially available supercoiled DNA with (135), glutathione, and horseradish peroxidase leads to DNA strand cleavage; control experiments show that the cleavage is dependant upon the presence of the glutathionyl radicals.

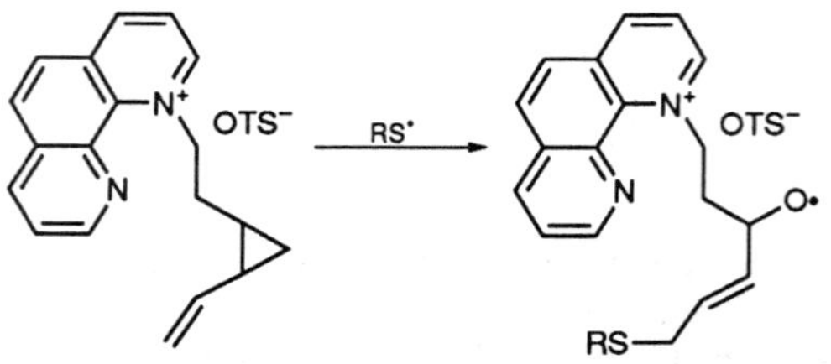

(135) (136)

[70] A. P. Breen and J. A. Murphy, *J. Chem. Soc., Perkin Trans. 1*, 1993, 2979.

5 Aliphatic and Alicyclic Chemistry

By P. QUAYLE
Department of Chemistry, The Victoria University of Manchester, Manchester, M13 9PL, UK

1 Introduction

This has been a good year for synthetic organic chemistry, as shown by no less than three[1] total syntheses of the immunosuppressive agent rapamycin (1). These embody a range of synthetic methodology which was only recently deemed esoteric, including Stille cross-coupling reactions, samarium-promoted Tischenko reactions, and titanium-mediated aldol reactions. Application of the traditional skills of the synthetic organic chemist to increasingly more complex biological problems[2] is well demonstrated in Schreiber's studies of the synthesis of FK-506 derived ligands capable of binding proteins.[3] In a similar vein, the use of biochemical techniques to solve chemical problems is ever more apparent, as witnessed by the number of reviews in this area: particularly noteworthy are appraisals of the use of catalytic antibodies[4] in organic synthesis. On more familiar ground, extensive reports on the use of enzymes in organic synthesis[5] and a review on chemoenzymatic approaches[6] to azasugars provide much useful information and stimulus. Other major research themes in recent years – supramolecular chemistry[7] and molecular recognition[8] phenomena – have again enjoyed extensive coverage.

In new synthetic methodology, organometallic chemistry continues to provide

[1] K.C. Nicolaou, T.J. Chakraborty, A.D. Piscopio, N. Minowa, and P. Bertinato, *J. Am. Chem. Soc.*, 1993, **115**, 4419; D. Romo, S.D. Meyer, D.D. Johnson, and S.L. Schreiber, *J. Am. Chem. Soc.*, 1993, **115**, 7906; C.M. Hayward, D. Yohannes, and S.J. Danishefsky, *J. Am. Chem. Soc.*, 1993, **115**, 9345.

[2] A.T. Sneden, *SYNLETT*, 1993, 313; R.S. Varma, *SYNLETT*, 1993, 621; G.G. Cross and T.M. Fyles, *SYNLETT*, 1993, 449; R.S. Varma, *SYNLETT*, 1993, 621; M Kahn, *SYNLETT*, 1993, 821; A. Giannis and T. Kolter, *Angew. Chem., Int. Ed. Engl.*, 1993, **32**, 1244; C. Unverzagt, *Angew. Chem., Int. Ed. Engl.*, 1993, **32**, 1691; C.A.A. van Boeckel and M. Petitou, *Angew. Chem., Int. Ed. Engl.*, 1993, **32**, 1671.

[3] D.M. Spencer, T.J. Wandless, S.L. Schreiber, and G.R. Crabtree, *Science*, 1993, **262**, 1019; M.B. Andrus and S.L. Schreiber, *J. Am. Chem. Soc.*, 1993, **115**, 10420.

[4] C. Laumann, *Angew. Chem., Int. Ed. Engl.*, 1993, **32**, 1291; P.G. Schultz and R.A. Lerner, *Acc. Chem. Res.*, 1993, **26**, 391; D. Hilvert, *Acc. Chem. Res.*, 1993, **26**, 552.

[5] D.H.G. Crout, S.M. Roberts, and J.B. Jones, *Tetrahedron: Asymmetry*, 1993, **4**, 757; 1081.

[6] G.C. Look, C.H. Fotsch, and C.-H. Wong, *Acc. Chem. Res.*, 1993, **26**, 182.

[7] S. Shinkai, *Tetrahedron*, 1993, **49**, 8933; S. Anderson, H.L. Anderson, and J.K.M. Saunders, *Acc. Chem. Res.*, 1993, **26**, 469; *Top. Curr. Chem.*, 1993, 165; R.M. Izatt and J.S. Bradshaw, *Pure Appl. Chem.*, 1993, **65**, 355; D.N. Reinhoudt, *Pure Appl. Chem.*, 1993, **65**, 2313; A.F. Danil De Namor, *Pure Appl. Chem.*, 1993, **65**, 1457.

[8] T.H. Webb and C.S. Wilcox, *Chem. Soc. Rev.*, 1993, **22**, 383; K.E. Krakowiak, J.S. Bradshaw, and R.M. Izatt, *SYNLETT*, 1993, 611; Z. Asfari, J. Weiss, and J. Vicens, *SYNLETT*, 1993, 719; N.T. Thuong and C. Helene, *Angew. Chem., Int. Ed. Engl.*, 1993, **32**, 666.

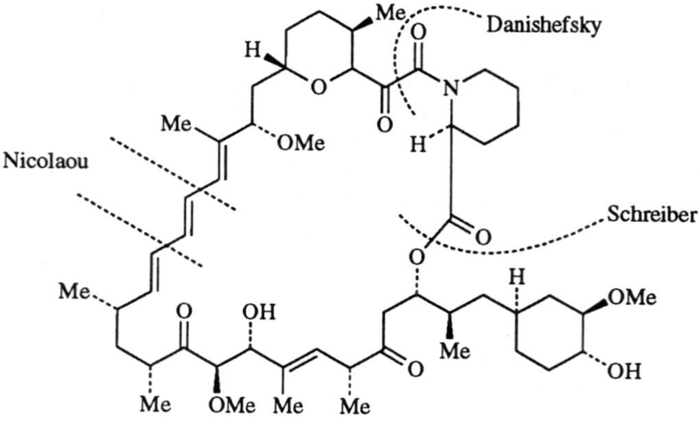

(1)

fruitful avenues for investigation,[9] especially in the development of new asymmetric reactions.[10] Asymmetric methodology in general[11] continues to be a recurrent theme, as is the use of free-radical reactions in organic synthesis.[12] Environmental considerations have prompted investigation of reactions carried out in the solid state[13] and in aqueous solution[14] rather than in organic solvents. There has been renewed interest also in the application of photochemical[15] and electrochemical[16] techniques to organic synthesis.

The search for complex medicinal agents[17] and the isolation of novel structural types,[18] most notably from marine sources,[19] continues to pose enormous methodological challenges,[20] and few molecules are more challenging than the enediyne family

9 P. Knochel and R. D. Singer, *Chem. Rev.*, 1993, **93**, 2117; Y. Yamamoto and N. Asao, *Chem. Rev.*, 1993, **93**, 2207; B. Bennetau and J. Dunogues, *SYNLETT*, 1993, 171; S. G. Davies and T. J. Donohoe, *SYNLETT*, 1993, 323; U. Schuchardt, W. A. Carvalho, and E. V. Spinace, *SYNLETT*, 1993, 713; P. Wipf, *Synthesis*, 1993, 537; K. Ritter, *Synthesis*, 1993, 735; G. G. Melikyan, *Synthesis*, 1993, 833; A. Furstner, *Angew. Chem., Int. Ed. Engl.*, 1993, **32**, 164; J. Sundermeyer, *Angew. Chem., Int. Ed. Engl.*, 1993, **32**, 1144; Y. Nishigaichi, A. Takuwa, A. Tauwa, Y. Naruta, and K. Murayama, *Tetrahedron*, 1993, **49**, 7395; M. Schlosser, O. Desponds, R. Lehmann, E. Moret, and G. Rauchschwalbe, *Tetrahedron*, 1993, **49**, 10 175; A. J. Pearson, *Tetrahedron*, 1993, **49**, 5415; J. H. Rigby, *Acc. Chem. Res.*, 1993, **26**, 579.

10 J. M. Brown, *Chem. Soc. Rev.*, 1993, **22**, 25; C. Bolm, *Angew. Chem., Int. Ed. Engl.*, 1993, **32**, 232; W. Herrmann and C. W. Kohlpaintner, *Angew. Chem., Int. Ed. Engl.*, 1993, **32**, 547; O. Reiser, *Angew. Chem., Int. Ed. Engl.*, 1993, **32**, 547; A. Pfaltz, *Acc. Chem. Res.*, 1993, **26**, 39.

11 A. H. Hoveyda, D. A. Evans, and G. C. Fu, *Chem. Rev.*, 1993, **93**, 1307; K. Fuji, *Chem. Rev.*, 1993, **93**, 2037; M. Lautens, *SYNLETT*, 1993, 177; S. W. McCombie, 1993, 807; L. Banfi, *Synthesis*, 1993, 1029; H. Kunz and K. Rück, *Angew. Chem., Int. Ed. Engl.*, 1993, **32**, 336; B. H. Kim and D. P. Curran, *Tetrahedron*, 1993, **49**, 293; D. J. Ager and M. B. East, *Tetrahedron*, 1993, **49**, 5683; K. Koja and T. Shioiri, *Tetrahedron*, 1993, **49**, 1711; M. T. Reetz, *Acc. Chem. Res.*, 1993, **26**, 462; K. Sakai and H. Suemune, *Tetrahedron: Asymmetry*, 1993, **4**, 2109.

12 A. L. J. Beckwith, *Chem. Soc. Rev.*, 1993, **22**, 143; P. Dowd and W. Zhang, *Chem. Rev.*, 1993, **93**, 2091.

13 F. Toda, *SYNLETT*, 1993, 303.

14 C.J. Li, *Chem. Rev.*, 1993, **93**, 2023.

15 M. T. Crimmins and T. L. Reinhold, *Org. React.*, 1993, **44**, 297; G. Pandey, *Top. Curr. Chem.*, 1993, **168**, 175.

16 M. E. Niyazymbetov and D. H. Evans, *Tetrahedron*, 1993, **49**, 9627.

17 P. W. Collins and S. W. Djuric, *Chem. Rev.*, 1993, **93**, 1533; H. Kessler, *Angew. Chem., Int. Ed. Engl.*, 1993, **32**, 543; S. F. Wunk, *Tetrahedron*, 1993, **49**, 9877; S. L. Beacage and R. P. Iyer, *Tetrahedron*, 1993, **49**, 10 441.

18 J. P. Michael and G. Pattenden, *Angew. Chem., Int. Ed. Engl.*, 1993, **32**, 1.

19 D. J. Faulkner, *Chem. Rev.*, 1993, **93**, 1671; D. J. Faulkner, *Top. Curr. Chem.*, 1993, **167**.

20 R. Gleiter and D. Kratz, *Angew. Chem., Int. Ed. Engl.*, 1993, **32**, 842.

of antitumour agents.[21] Recent advances in the development of more concise approaches to complex molecules are reviewed in two timely articles on double-Michael reactions[22] and 'sequential reactions'.[23] Ultimately, fundamental advances in synthetic design will depend upon a more quantitative[24] appreciation of stereochemical and electronic factors controlling organic reactivity. These considerations are doubtless responsible in part for the continuing intense interest in the chemistry of C_{60} and related compounds.[25] The possibility of synthesizing molecular arrays possessing some characteristics of living organisms has been addressed.[26] Advances in this area may have fundamental implications, but it is a sobering thought, however, that our understanding of complex biosynthetic problems (*e.g.* B_{12}) is still incomplete.[27]

2 Aliphatic Chemistry

General.—The synthesis of planar four-coordinate carbon compounds has been reviewed.[28] Olah[29] has presented a fascinating overview on his work on 'super electrophiles'. Rearrangement reactions are frequently used in organic synthesis and several timely accounts of current progress in this area have appeared.[30] The synthetic applications of α-amino acids,[31] alkoxyallenes,[32] amine oxides,[33] β-lactones,[34] and organofluorine[35] compounds provide much interesting reading. Organic synthesis still relies heavily upon protecting group strategies for functional group differentiation, and two excellent reviews, dealing with use of silicon-based protecting groups for alcohols[36] and with esters as a protecting group for carboxyl functionality,[37] have appeared. Heteroatom-based methodologies,[38] in particular those of phosphorus and selenium, have transformed contemporary organic synthesis. Reich's review[39] on selenoxide eliminations in particular provides many useful pointers for the practising synthetic chemist. An extensive review of the Baeyer–Villiger oxidation of aldehydes and ketones is also most welcome.[40]

There have been theoretical studies on a number of text-book reactions, including

21 M. M. Campbell, M. Sainsbury, and P. A. Searle, *Synthesis*, 1993, 179; R. E. Wolf, *Pure Appl. Chem.*, 1993, **65**, 1103; J.-C. Harmange and Figatère, *Tetrahedron: Asymmetry*, 1993, **4**, 1711.
22 M. Ihara and K. Fukumoto, *Angew. Chem., Int. Ed. Engl.*, 1993, **32**, 1010.
23 L. F. Tietze and U. Beifuss, *Angew. Chem., Int. Ed. Engl.*, 1993, **32**, 131.
24 J. E. Eksterowicz and K. N. Houk, *Chem. Rev.*, 1993, **93**, 2439; I. Ugi, J. Bley, A. Dengler, A. Dietz, E. Fontain, B. Gruber, R. Herges, M. Knauer, K. Reitsam, and N. Stein, *Angew. Chem., Int. Ed. Engl.*, 1993, **32**, 201; C. D. Johnson, *Acc. Chem. Res.*, 1993, **26**, 476.
25 A. Hirsch, *Angew. Chem., Int. Ed. Engl.*, 1993, **32**, 1138; H. Schwartz, *Angew. Chem., Int. Ed. Engl.*, 1993, **32**, 1412.
26 D. W. Urry, *Angew. Chem., Int. Ed. Engl.*, 1993, **32**, 819.
27 A. I, Scott, *Angew. Chem., Int. Ed. Engl.*, 1993, **32**, 1223.
28 M. Albrecht, G. Erker, and C. Krüger, *SYNLETT*, 1993, 26.
29 G. A. Olah, *Angew. Chem., Int. Ed. Engl.*, 1993, **32**, 767.
30 S. R. Wilson, *Org. React.*, 1993, **43**, 93; R. F. C. Brown and F. W. Eastwood, *SYNLETT*, 1993, 9; H. R. Sonawane, N. S. Bellur, D. G. Kulkarni, and J. R. Ahuja, *SYNLETT*, 1993, 875; J.-L. Ripoll and Y. Vallee, *Synthesis*, 1993, 659; H. M. L. Davies, *Tetrahedron*, 1993, **49**, 5203.
31 A. Golebiowski and J. Jurczak, *SYNLETT*, 1993, 241.
32 R. Zimmer, *Synthesis*, 1993, 165.
33 A. Albini, *Synthesis*, 1993, 263.
34 A. Pommier and J.-M. Pons, *Synthesis*, 1993, 441.
35 H. Uno and H. Suzuki, *SYNLETT*, 1993, 91; G. Resnati, *Tetrahedron*, 1993, **49**, 9385.
36 J. Muzart, *Synthesis*, 1993, 11; C. J. Salomon, E. G. Mata, and O. A. Mascaretti, *Tetrahedron*, 1993, **49**, 3691.
37 A. Fava, *Pure Appl. Chem.*, 1993, **65**, 595.
38 I. Ugi, *Synthesis*, 1993, 1.
39 H. J. Reich and S. Wollowitz, *Org. React.*, 1993, **44**, 4.
40 G. R. Krow, *Org. React.*, 1993, **43**, 251.

Michael additions,[41] the S_N2–$E2$ spectrum,[42] pinacol rearrangements,[43] and electrophilic additions to alkenes.[44] Many stereocontrolled reactions involve the interaction of a carbonyl group with one or more metal centres. Much experimental evidence concerning the mode of such interactions has appeared this year.[45–50] The use of strong bases (*e.g.* Bu^nLi/TMEDA) is now common place, but often with little understanding of the nature of the species reacting in the medium. Recent mechanistic studies[51] and an authoritative survey of organoalkali metal structures[52] should promote such understanding and, perhaps, encourage more efficient use. An evaluation of directing groups in *ortho*-lithiation reactions poses a number of interesting mechanistic questions.[53] A theoretical investigation into the generalized anomeric effect[54] and further reports on the transition state structures involved in cyclohexanone reductions have appeared.[55]

The ability to prepare probes containing chiral methyl groups has had a profound effect upon our understanding of a number of biosynthetic pathways. A personal account of work in this area is most instructive.[56]

Coupling Reactions.—Synthetic methods concerned with carbon–carbon bond formation continue to be a major preoccupation. Variants on the Heck and Stille cross-coupling reactions for the stereoselective formation of olefins are abundant (Schemes 1–11).[57–68] A growing body of empirical evidence suggests that the addition of copper(I) salts as a cocatalyst in such reactions may have a pronounced effect upon the rate-determining transmetallation step.[69] In several cases, reaction times are

41 L. Pardo, R. Osman, H. Weinstein, and J. Rabinowitz, *J. Am. Chem. Soc.*, 1993, **115**, 8263.
42 S. Gronert, *J. Am. Chem. Soc.*, 1993, **115**, 652.
43 K. Nakamura and Y. Osamura, *J. Am. Chem. Soc.*, 1993, **115**, 9112.
44 M. Fujita, M. Ishida, K. Manako, K. Sato, and K. Ogura, *Tetrahedron Lett.*, 1993, **34**, 645.
45 S. E. Denmark and N. G. Almstead, *J. Am. Chem. Soc.*, 1993, **115**, 3133.
46 S. Castellino and W. J. Dwight, *J. Am. Chem. Soc.*, 1993, **115**, 2986; P. G. Willard and Q.-Y. Liu, *J. Am. Chem. Soc.*, 1993, **115**, 3380.
47 H. Sasai, T. Suzuki, N. Itoh, K. Tanaka, T. Date, K. Okamura, and M. Shibasaki, *J. Am. Chem. Soc.*, 1993, **115**, 10 372.
48 G. Quinkert, H. Becker, M. Del Grosso, G. Dambacher, J. W. Bats, and G. Durner, *Tetrahedron Lett.*, 1993, **34**, 6885.
49 M. P. Bernstein and D. B. Collum, *J. Am. Chem. Soc.*, 1993, **115**, 789.
50 M. Simard, J. Vaugeois, and J. D. Wuest, *J. Am. Chem. Soc.*, 1993, **115**, 370.
51 F. E. Rosenberg, M. P. Bernstein, J. Gilchrist, A. T. Harrison, D. J. Fuller, and D. B. Collum, *J. Am. Chem. Soc.*, 1993, **115**, 3475; M. A. Nichols and P. G. Willard, *J. Am. Chem. Soc.*, 1993, **115**, 1568.
52 E. Weiss, *Angew. Chem., Int. Ed. Engl.*, 1993, **32**, 1501.
53 P. Beak, S. T. Kerrick, and D. J. Gallagher, *J. Am. Chem. Soc.*, 1993, **115**, 10 628.
54 U. Salzner and P. von R. Schleyer, *J. Am. Chem. Soc.*, 1993, **115**, 10 231.
55 J. M. Coxon and R. T. Luibrand, *Tetrahedron Lett.*, 1993, **34**, 7097; Y. Wu and K. N. Houk, *J. Am. Chem. Soc.*, 1993, **115**, 10 992.
56 H. G. Floss and S. Lee, *Acc. Chem. Res.*, 1993, 26, 166.
57 P. Melnyk, J. Gasche, and C. Thal, *Tetrahedron Lett.*, 1993, **34**, 5449.
58 A. R. Hunt, S. K. Stewart, and A. Whiting, *Tetrahedron Lett.*, 1993, **34**, 3599.
59 T. Sakamoto, Y. Kondo, N. Takazawa, and H. Yamanaka, *Tetrahedron Lett.*, 1993, **34**, 5955.
60 D. A. Evans and T. Bach, *Angew. Chem., Int. Ed. Engl.*, 1993, **32**, 1326.
61 H. Finch, N. A. Pegg, and B. Evans, *Tetrahedron Lett.*, 1993, **34**, 8353.
62 J. P. Michael, S.-F. Chang, and C. Wilson, *Tetrahedron Lett.*, 1993, **34**, 8365.
63 D. Badone and U. Guzzi, *Tetrahedron Lett.*, 1993, **34**, 3603.
64 Z. Xu and J. S. Moore, *Angew. Chem., Int. Ed. Engl.*, 1993, **32**, 1354.
65 S. W. Lee and P. L. Fuchs, *Tetrahedron Lett.*, 1993, **34**, 5209; Z. Jin and P. L. Fuchs, *Tetrahedron Lett.*, 1993, **34**, 5205.
66 T. Jeffery, *Tetrahedron Lett.*, 1993, **34**, 1133.
67 M. Amat, S. Hadida, and J. Bosch, *Tetrahedron Lett.*, 1993, **34**, 5005.
68 K. F. McClure and S. J. Danishefsky, *J. Am. Chem. Soc.*, 1993, **115**, 6094.
69 R. J. Hinkle, G. T. Poulter, and P. J. Stang, *J. Am. Chem. Soc.*, 1993, **115**, 11 626; see also J. M. Saa and G. Martorell, *J. Org. Chem.*, 1993, **58**, 1963 for related observations.

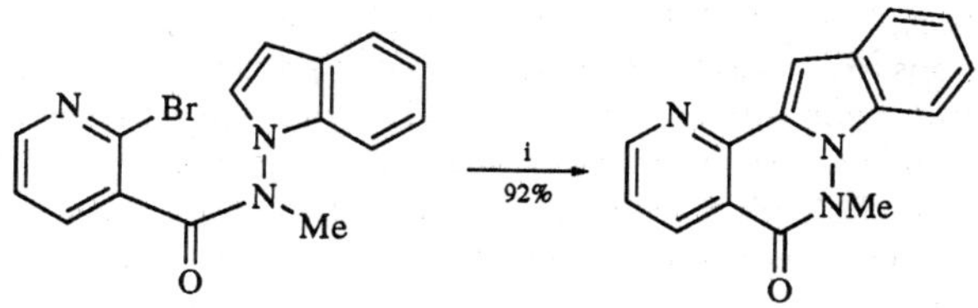

(ref 57)

Reagents: i, $Pd(OAc)_2$, Ph_3P, Bu_4NBr, K_2CO_3, DMF

Scheme 1

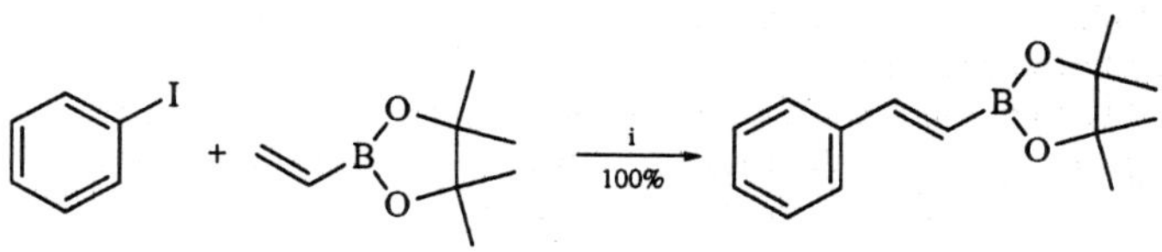

(ref 58)

Reagents: i, $Pd(OAc)_2$, NEt_3, CH_3CN, 80 °C

Scheme 2

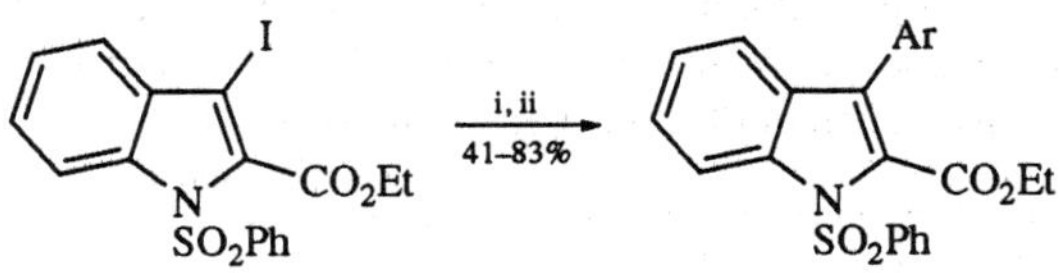

(ref 59)

Reagents: i, Zn*; ii, ArX, $Pd(PPh_3)_3$, THF, 80 °C

Scheme 3

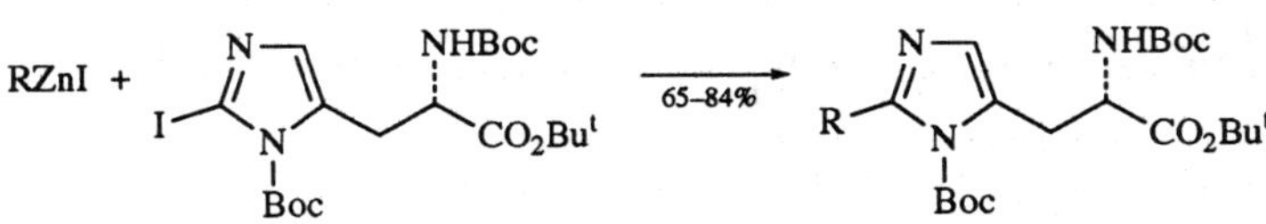

(ref 60)

Reagents: Pd(0)

Scheme 4

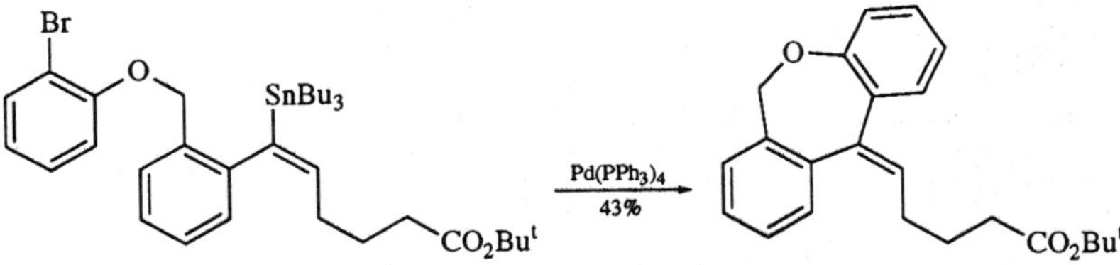

(ref 61)

Scheme 5

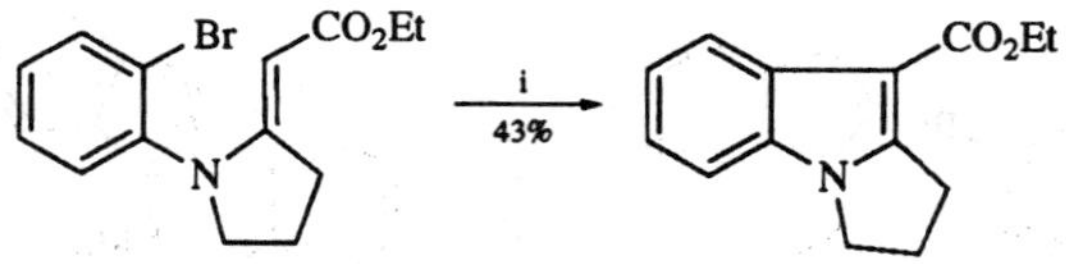

(ref 62)

Reagents: i, $Pd(OAc)_2$, NEt_3, CH_3CN, 80 °C

Scheme 6

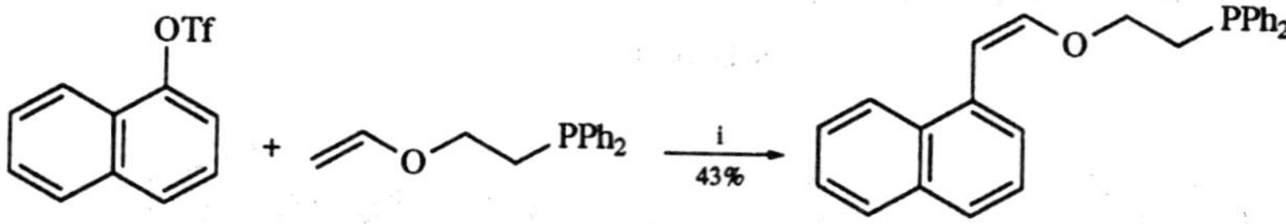

(ref 63)

Reagents: i, proton sponge, $Pd(OAc)_2$, DMF, 80 °C

Scheme 7

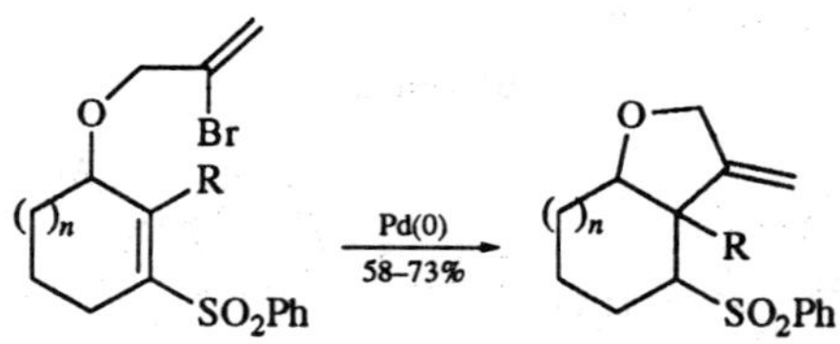

(ref 65)

Scheme 8

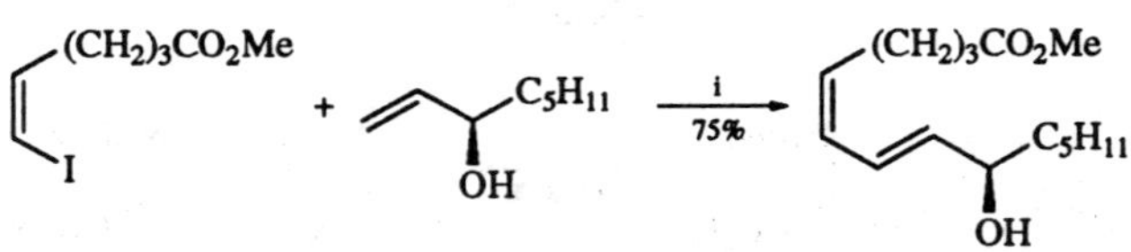

(ref 66)

Reagents: i, $Pd(OAc)_2$, AgOAc, DMF, 45 °C

Scheme 9

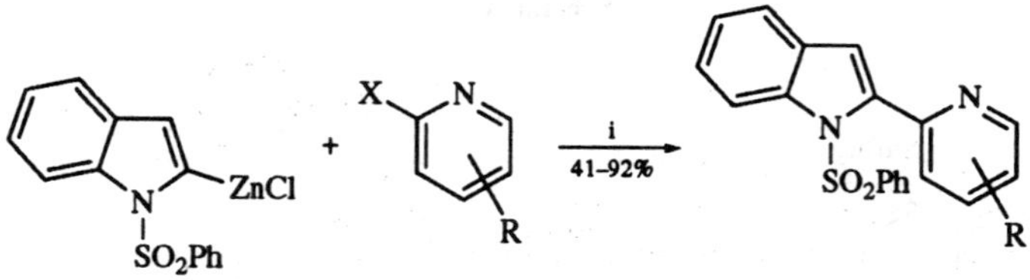

(ref 67)

Reagents: i, $PdCl_2(PPh_3)_2$, DIBAL, THF, 80 °C

Scheme 10

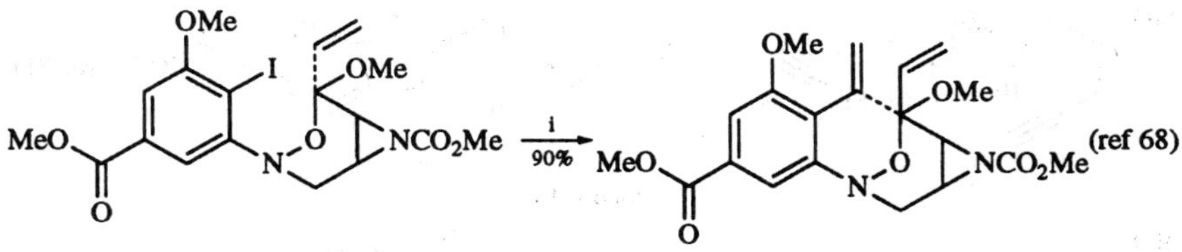

Reagents: i, Pd(0), NEt_3, CH_3CN, 80 °C

Scheme 11

reduced and yields are higher especially in those reactions which are sluggish when using Pd(0) alone, Scheme 12. Remarkably, Piers has shown that, in at least one system, the palladium catalyst can be dispensed with altogether if stoichiometric quantities of copper chloride are added, Scheme 13.[70] This result obviously warrants further investigation.

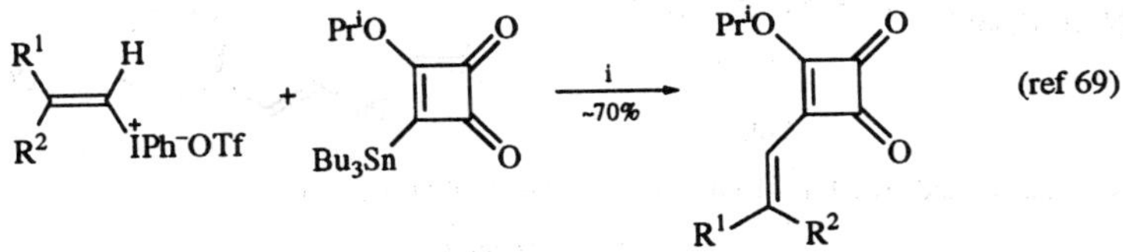

Reagents: i, $PdClBn(PPh_3)_2$, CuCl, DMF, rt

Scheme 12

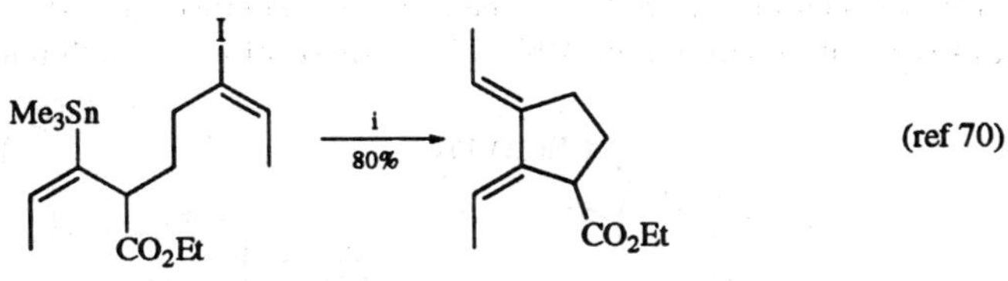

Reagents: i, $CuCl_2$ (2.2 eq), DMF, 62 °C, 3 min

Scheme 13

'Contrasteric' functionalization of 1,1-dihaloalkenes has been observed in Heck or Stille reactions by a number of workers.[71] The generation of chelated vinyl palladium species has been invoked in order to rationalize these observations (Scheme 14). The use of vinyl boronic acids[72] (Suzuki cross-coupling) and vinyl silanes[73] in Stille-type reactions continues to gain acceptance (Scheme 15). Cascade sequences which utilize

[70] E. Piers and T. Wong, *J. Org. Chem.*, 1993, **58**, 3609.

[71] J. Suffert, A. Eggers, S. W. Scheuplein, and R. Bruckner, *Tetrahedron Lett.*, 1993, **34**, 4177; S. Torii, H. Okumoto, T. Tadokoro, A. Nishimura, and M. A. Rashid, *Tetrahedron Lett.*, 1993, **34**, 2139; J. M. Nuss, R. A. Rennels, and B. H. Levine, *J. Am. Chem. Soc.*, 1993, **115**, 6991; for related examples in organolithium chemistry see D. Grandjean and P. Pale, *Tetrahedron Lett.*, 1993, **34**, 1155; M. Braun and K. Opdenbusch, *Angew. Chem., Int. Ed. Engl.*, 1993, **32**, 578.

[72] N. Yasuda, L. Xavier, D. L. Rieger, Y. Li, A. E. DeCamp, and U.-H. Dolling, *Tetrahedron Lett.*, 1993, **34**, 3211.

[73] K. Takahashi, T. Minami, Y. Ohara, and T. Hiyama, *Tetrahedron Lett.*, 1993, **34**, 8263.

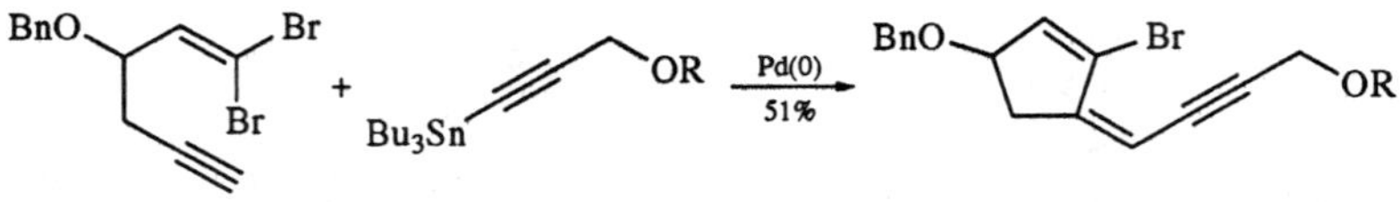

(ref 71)

Scheme 14

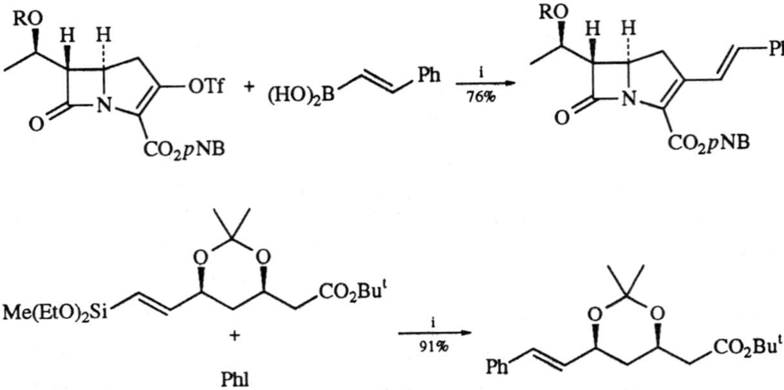

(ref 72)

(ref 73)

Reagents: i, $Pd(dbq)_2$, THF, KOH(aq), rt; ii, TBAF, $P(OEt)_3$, Pd(o)

Scheme 15

sequential Heck and carbonylation reactions have been reported by Grigg (Scheme 16).[74] The use of ketone enolates in palladium-mediated cross-coupling reactions has hitherto been neglected, but a fine example demonstrating the synthetic potential of this reaction is featured in Albizati's synthesis of (+)-hapallindole Q (Scheme 17).[75]

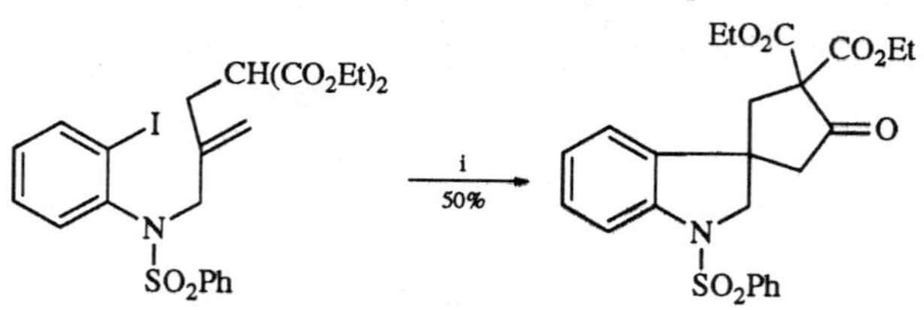

(ref 74)

Reagents: i, Pd(o), CO, CH_3CN, TlOAc

Scheme 16

The stereospecific cross coupling of a homochiral α-alkoxystannane with an acyl halide has been reported by Falck[76] in his synthesis of (+)-gonifuranone (Scheme 18). The stereochemical course in this reaction was opposite to that expected; further work in this area is required in order to ascertain the generality of this observation.

Modifications[77,78] of well-established copper-mediated carbon–carbon bond-form-

[74] R. Grigg and V. Sridharan, *Tetrahedron Lett.*, 1993, **34**, 7471.
[75] V. Vaillancourt and K. F. Albizati, *J. Am. Chem. Soc.*, 1993, **115**, 3499.
[76] J. Ye, R. K. Bhatt, and J. R. Falck, *Tetrahedron Lett.*, 1993, **34**, 8007.
[77] M. Bergdahl, M. Eriksson, M. Nilsson, and T. Olsson, *J. Org. Chem.*, 1993, **58**, 7238.
[78] J. Westermann and K. Nickisch, *Angew. Chem., Int. Ed. Engl.*, 1993, **32**, 1368.

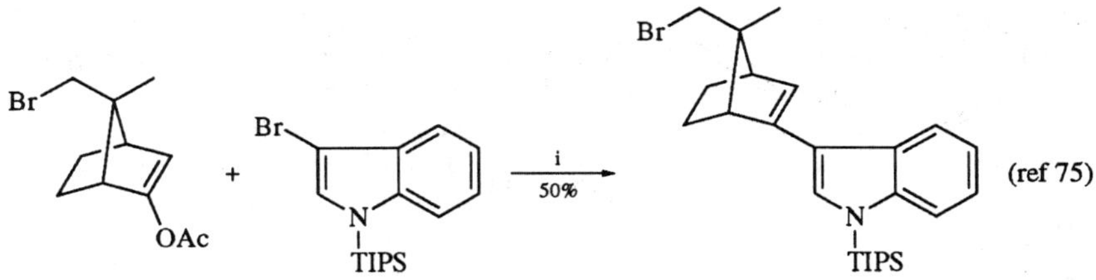

Reagents: i, Bu_3SnOMe, Pd(0), toluene, 100 °C

Scheme 17

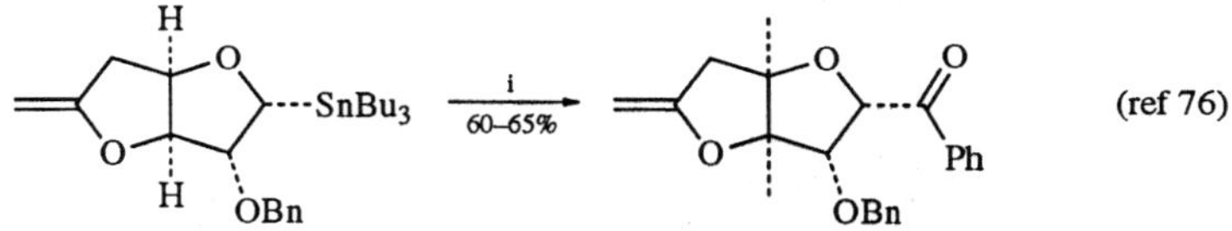

Reagents: i, PhCOCl, $PdCl_2(PPh_3)_2$, CuCN, toluene, 80 °C

Scheme 18

ing reactions continue to be the focus of attention (Scheme 19). A major advance this year includes the use of organolithium zincates as a 'soft' source of organolithium reagents in the conjugate addition reactions of enones. Knochel[79] has published full experimental details of the preparation of the synthetically versatile functionalized organometallics (2) (Scheme 20).

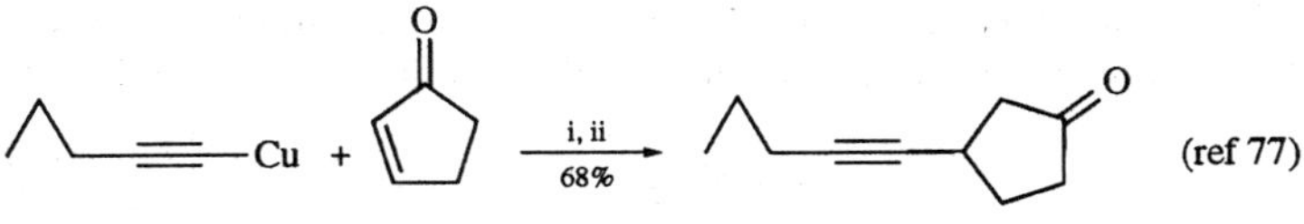

Reagents: i, LiI, TMSCl

Scheme 19

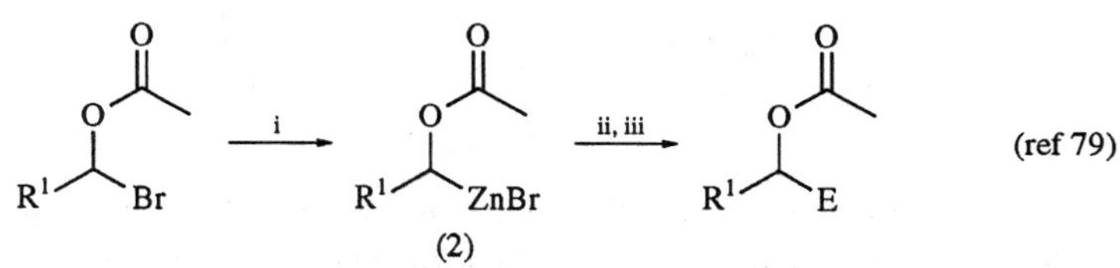

Reagents: i, Zn, THF, 12 °C; ii, CuCN·2LiCl, 30 °C, DMF; iii, E^+

Scheme 20

Asymmetric Catalytic Coupling Reactions.—Asymmetric Heck reactions, as developed by Shibasaki, have been utilized in the synthesis of indolizidine alkaloids[80] and in a

[79] P. Knochel, T. S. Chou, C. Jubert, and D. Rajapohal, *J. Org. Chem.*, 1993, **58**, 508; M. J. Rozema, C. Eisenberg, H. Lutjens, R. Ostwald, K. Belyk, and P. Knochel, *Tetrahedron Lett.*, 1993, **34**, 3115.

[80] S. Nukui, M. Sodeoka, and M. Shibisaki, *Tetrahedron Lett.*, 1993, **34**, 4965.

total synthesis of (−)-eptazocene (Scheme 21).[81] Ligand systems which enable enantioselective palladium-catalysed allylic substitution reactions have again been the subject of numerous reports.[10] Several authors have demonstrated that oxazoline-derived ligands can induce high levels of enantioselectivity in these reactions (Scheme 22).[82,83] Pfaltz[84] has also demonstrated that ligands such as (3) enable the enantioselective 1,4-addition of Grignard reagents to enones in poor to moderate levels of induction (16–83% e.e.) (Scheme 23).

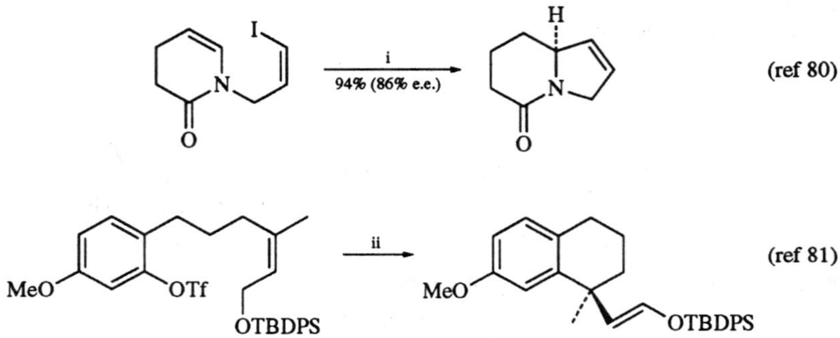

Reagents: i, Pd*L_n, Ag^+; ii, Pd*(0)

Scheme 21

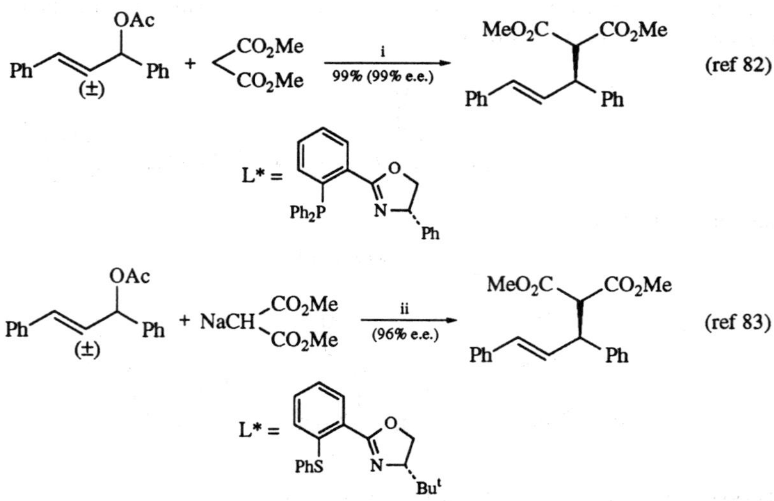

Reagents: i, 1 mol % [{Pd(C_3H_5)Cl}$_2$], 2.5 mol % L*, BSA, KOAc, CH_2Cl_2, 23 °C; ii, [{Pd(C_3H_5)Cl}$_2$]L*

Scheme 22

[81] S. Takemoto, M. Sodeoka, H. Sasai, and M. Shibasaki, *J. Am. Chem. Soc.*, 1993, **115**, 8477.
[82] P. von Matt and A. Pfaltz, *Angew. Chem., Int. Ed. Engl.*, 1993, **32**, 566.
[83] G. J. Dawson, C. G. Frost, C. J. Martin, J. M. J. Williams, and S. J. Coote, *Tetrahedron Lett.*, 1993, **34**, 7793.
[84] Q.-L. Zhou and A. Pfaltz, *Tetrahedron Lett.*, 1993, **34**, 7725.

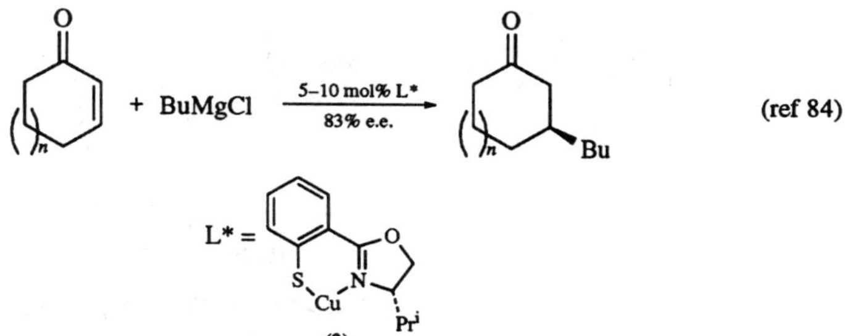

(3)

Scheme 23

Alkenes and Polyenes.—Schlosser has developed a general method for the synthesis of *Z* vinyl halides[85] using the modified Wittig reagents (4). *Erythro* phosphinyl alcohols (5) afford *E* olefins upon exposure to PCl_3 and NEt_3, in a reaction similar to a Ramberg–Bäcklung rearrangement (Scheme 24).[86]

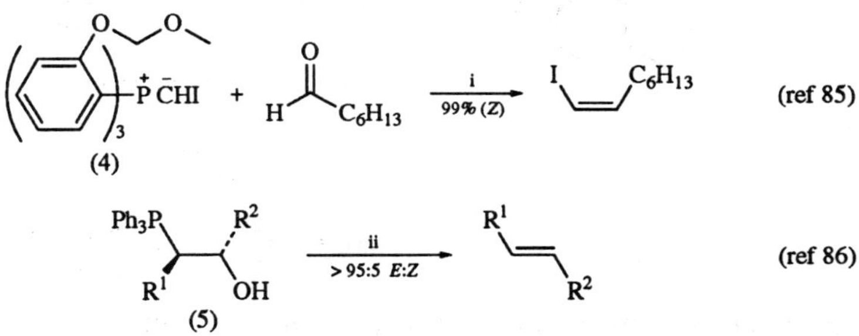

Reagents: i, THF, 50 °C; ii, PCl_3, NEt_3, CH_2Cl_2, 0 °C

Scheme 24

Denmark[87] has reported a highly stereoselective synthesis of trisubstituted olefins, a structural motif present in many natural products but difficult to introduce in a stereoselective fashion using existing Wittig methodology (Scheme 25). Two

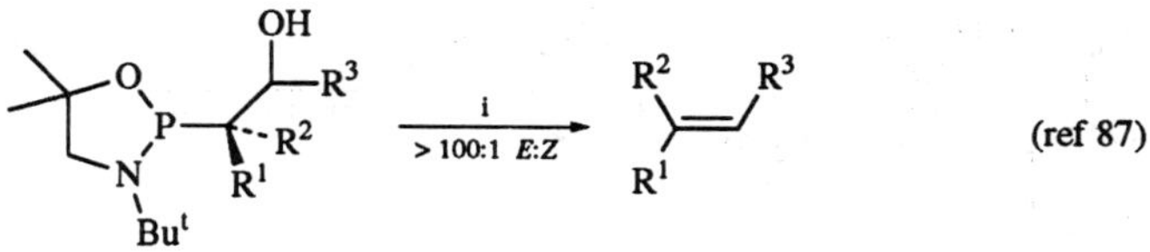

Reagents: i, THF, 105 °C

Scheme 25

[85] X.-p. Zhang and M. Schlosser, *Tetrahedron Lett.*, 1993, **34**, 1925.
[86] N. Lawrence and F. Muhammad, *J. Chem. Soc., Chem. Commun.*, 1993, 1187.
[87] S. Denmark and J. Amburgey, *J. Am. Chem. Soc.*, 1993, **115**, 10 386.

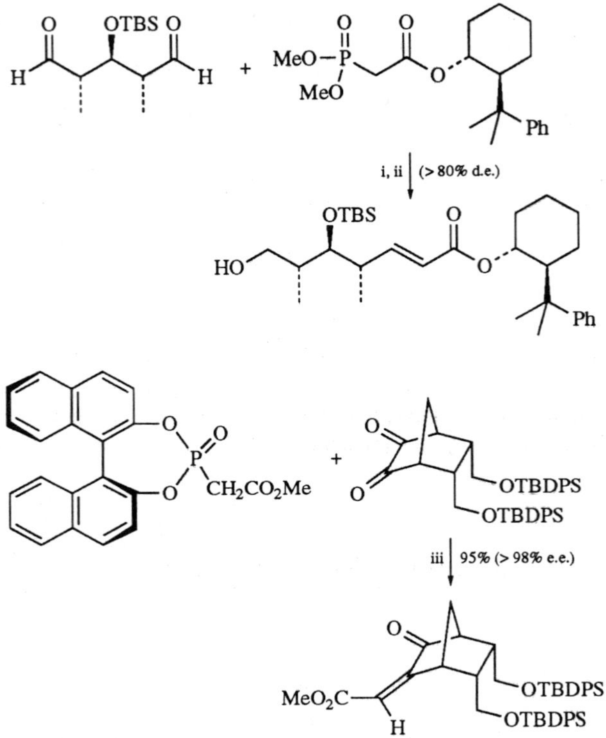

(ref 88)

(ref 89)

Reagents: i, KHMDS, 18-C-6, −100 °C; ii, $NaBH_4$; iii, NaH, −78 °C; THF

Scheme 26

groups[88,89] have described the use of homochiral Wadsworth–Emmons reagents for the desymmetrization of *meso* carbonyl compunds (Scheme 26). Similarly, the kinetic resolution of racemic ketones[90] using the phosphonate (6) afforded trisubstituted alkenes with useful levels of asymmetric induction (Scheme 27). The use of homochiral

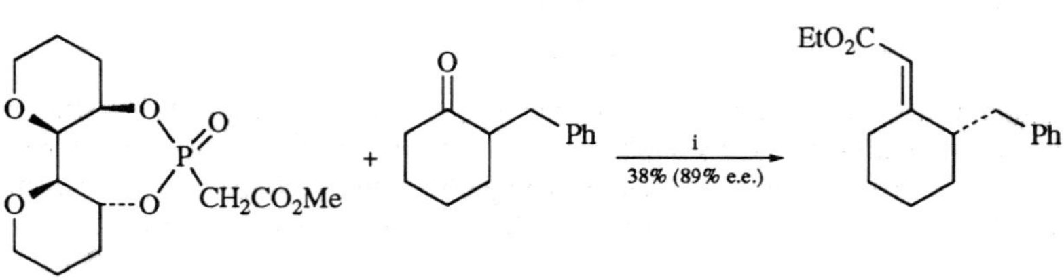

(ref 90)

(6)

Reagents: i, KHMDS, THF, −80 °C

Scheme 27

[88] N. Kann and T. Rein, *J. Org. Chem.*, 1993, **58**, 3802.
[89] K. Tanaka, Y. Ohta, K. Fuji, and T. Taga, *Tetrahedron Lett.*, 1993, **34**, 407.
[90] K. Narasaka, E. Hidai, Y. Hayashi, and J.-I. Gras, *J. Chem. Soc., Chem. Commun.*, 1993, 102.

bases in asymmetric elimination reactions[91] also appears to be an attractive route to scalemic exocyclic olefins (Scheme 28).

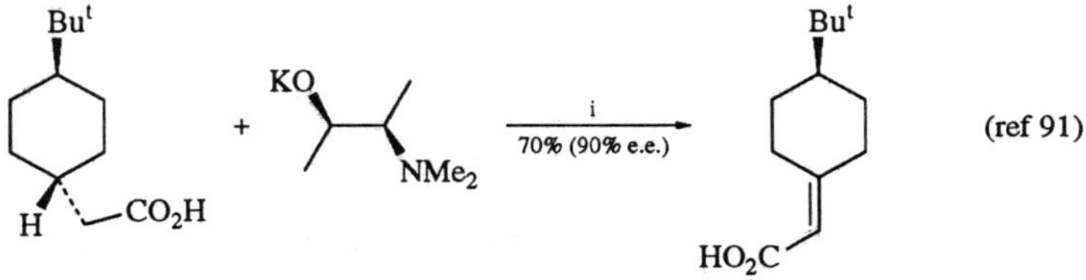

Reagents: i, THF 90 °C

Scheme 28

Organozirconium reagents have been used in several guises for the stereoselective synthesis of alkenes and polyenes (Scheme 29).[92–94] In a ground-breaking develop-

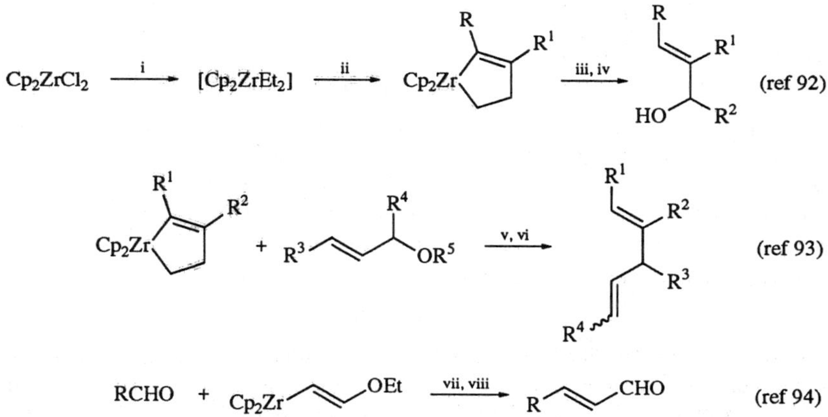

Reagents: i, 2 × EtMgBr; ii, RC≡CR′; iii, R_2CHO; iv, H_3O^+; v, 35 °C; vi, H_3O^+, rt; vii, cat. $AgClO_4$, rt, 30 min; viii, H_3O^+

Scheme 29

ment, two groups have reported the use of an olefin metathesis strategy for the synthesis of functionalized olefins (Scheme 30).[95,96] Applications of this methodology will doubtless arise in the near future. The related Tebbe-mediated synthesis of olefins from unactivated carbonyl compounds (esters, *etc.*) has been extensively reviewed.[97]

[91] J. Vadecord, J.-C. Plaquevent, L. Duhamel, and P. Duhamel, *J. Chem. Soc., Chem. Commun.*, 1993, 116.
[92] T. Takahashi, M. Kageyama, V. Denisov, R. Hara, and E. Negishi, *Tetrahedron Lett.*, 1993, **34**, 687.
[93] T. Takahashi, N. Suzuki, M. Kageyama, D. Y. Kondakov, and R. Hara, *Tetrahedron Lett.*, 1993, **34**, 4811.
[94] H. Maeta and K. Suzuki, *Tetrahedron Lett.*, 1993, **34**, 341.
[95] G. C. Fu and R. H. Grubbs, *J. Am. Chem. Soc.*, 1993, **115**, 3800; G. C. Fu, S. T. Nguyen, and R. H. Grubbs, *J. Am. Chem. Soc.*, 1993, **115**, 9856.
[96] H. Junga and S. Blechert, *Tetrahedron Lett.*, 1993, **34**, 3731.
[97] S. H. Pine, *Org. React.*, 1993, **43**, 1.

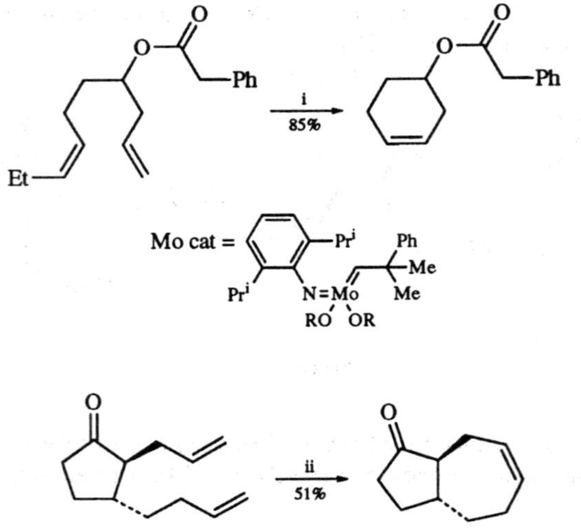

(ref 95)

(ref 96)

Reagents: i, Mo catalyst, benzene, 20 °C; ii, MTO/SiO_2, Al_2O_3, CH_2Cl_2, rt

Scheme 30

Transfer hydrogenation of acetylenes[98] has been shown to occur under mild reaction conditions to provide *Z* olefins with levels of high stereochemical purity (typically $Z:E > 97:3$). γ-Hydroxy-α,β-ynones are cleanly transformed into *E*,*E* dienones upon treatment with triphenyl phosphine.[99] Saladie[100] has reported a highly stereoselective synthesis of trienes *via* a 1,6 reductive–elimination sequence (Scheme 31).

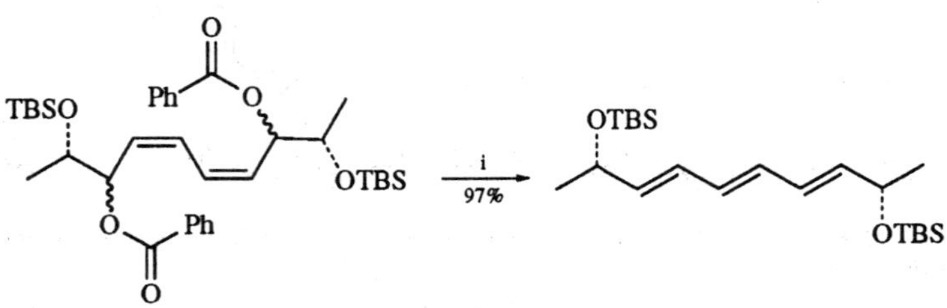

(ref 100)

Reagents: i, $TiCl_3$, $LiAlH_4$, THF, 65 °C

Scheme 31

The Ramberg–Bäcklund reaction has been applied with good effect in the synthesis of unsaturated α-amino acids without loss of stereochemical fidelity (Scheme 32).[101] 1-Nitroalkenes are transformed into ω-unsaturated nitriles in a one-pot sequence (Scheme 33).[102]

[98] K. Tani, N. Ono, S. Okamoto, and F. Sato, *J. Chem. Soc., Chem. Commun.*, 1993, 386.
[99] G. Guo and X. Lu, *J. Chem. Soc., Chem. Commun.*, 1993, 394.
[100] G. Solladie, G.B. Stone, and A. Rubio, *Tetrahedron Lett.*, 1993, **34**, 1803.
[101] Z.-X. Guo, M.J. Schaeffer, and R.J.K. Taylor, *J. Chem. Soc., Chem. Commun.*, 1993, 874.
[102] H.-W. Tso, B.A. Gilbert, and J.R. Hwu, *J. Chem. Soc., Chem. Commun.*, 1993, 669.

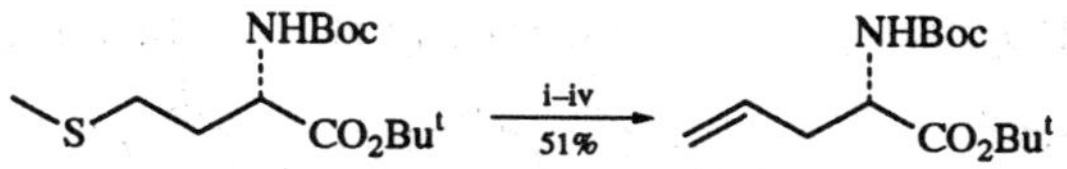

(ref 101)

Reagents: i, Oxone, MeOH; ii, CaO, SO_2Cl_2; iii, *m*CPBA; iv, $KOBu^t$

Scheme 32

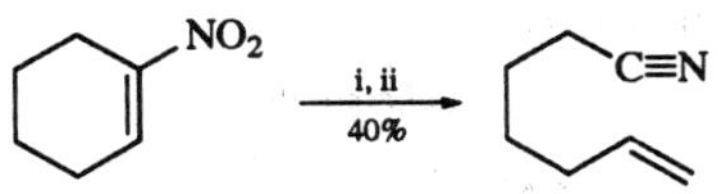

(ref 102)

Reagents: i, Me_3SiCH_2MgCl; ii, PCl_3, THF

Scheme 33

Stereoselective functionalization of olefins continues to be an area of much methodological interest.[103] Of note this year is the report of a palladium-catalysed asymmetric acetalization of acrylamides (Scheme 34),[104] the stereoselective cyclization

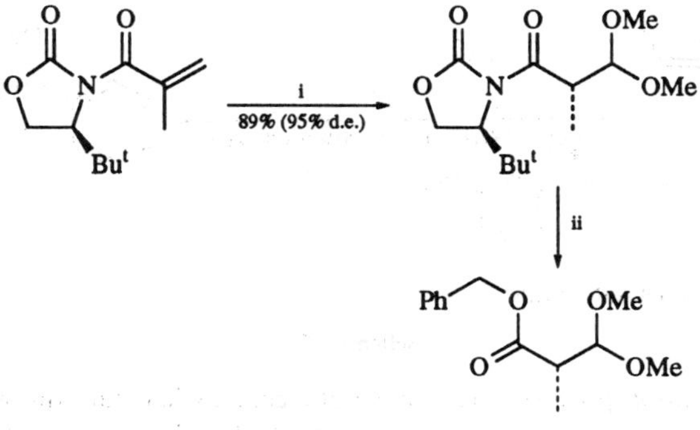

(ref 104)

Reagents: i, MeOH, $PdCl_2$, $CuCl_2$, O_2, rt; ii, $PhCH_2OLi$

Scheme 34

of 3-aminopent-1-en-5-ols to the corresponding tetrahydrofurans (Scheme 35),[105] and an asymmetric synthesis of α-amino acids *via* a stereoselective iodolactonization reaction (Scheme 36).[106] Details of an asymmetric selenoetherification reaction have also appeared (Scheme 37).[107]

Epoxides, Diols, and Related Compounds.—The Sharpless enantioselective dihydroxylation (ADHT) of achiral olefins has been the focus of mechanistic investigation; two distinct interpretations of kinetic data have been advanced.[108] Sharpless favours a non-concerted [2 + 2] cycloaddition–rearrangement reaction rather than a classical

103 J. Rodriguez and J.-P. Dulcere, *Synthesis*, 1993, 1177.
104 T. Hosokawa and S.-i. Murahashi, *J. Chem. Soc., Chem. Commun.*, 1993, 117.
105 Y. Tamura, H. Harayama, and T. Bando, *J. Chem. Soc., Chem. Commun.*, 1993, 1601.
106 O. Kitigawa, T. Hanano, N. Kikuchi and T. Taguchi, *Tetrahedron Lett.*, 1993, **34**, 2165.
107 R. Deziel, R. Goulet, L. Grenier, J. Bordeleau, and J. Bernier, *J. Org. Chem.*, 1993, **58**, 3619.
108 E. J. Corey, M. C. Noe, and S. Sarshar, *J. Am. Chem. Soc.*, 1993, **115**, 3828; H. C. Kolb, P. G. Andersson, Y. L. Bennani, G. A. Crispino, K.-S. Jeong, H.-L. Kwong, and K. B. Sharpless, *J. Am. Chem. Soc.*, 1993, **115**, 12226.

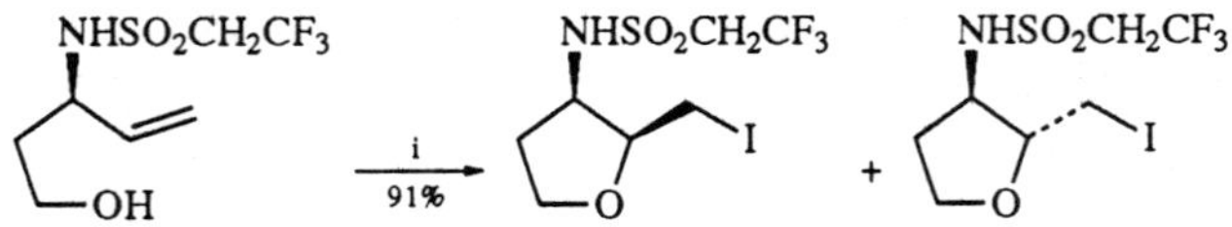

(*cis* : *trans* = 85 : 15)

(ref 105)

Reagents: i, I_2, $NaHCO_3$, 0 °C, H_2O, Et_2O

Scheme 35

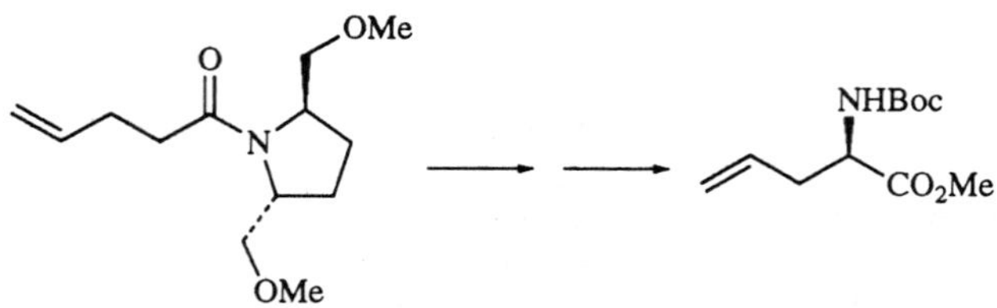

(ref 106)

Scheme 36

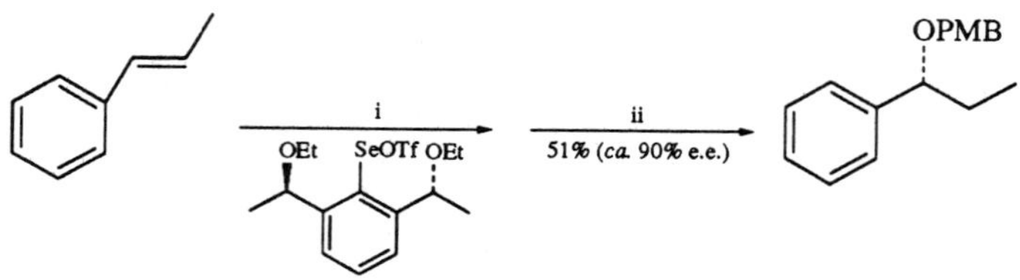

(ref 107)

Reagents: i, *p*MBA; ii, Ph_3SnH, AIBN

Scheme 37

[3 + 2] cycloaddition process. Irrespective of mechanistic detail, this experimentally simple reaction is now an indispensible synthetic tool as shown below (Scheme 38).[109]

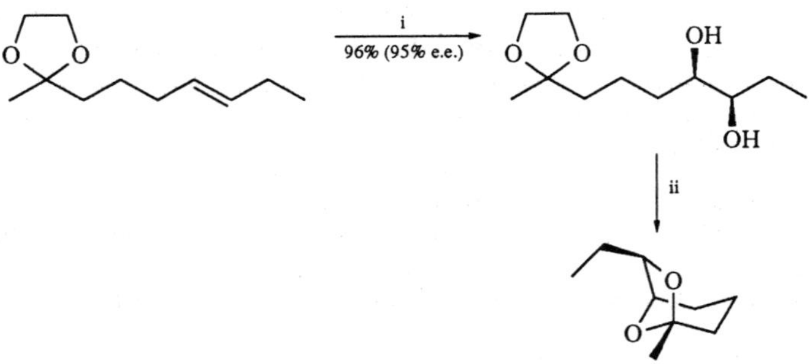

(ref 109)

(+)-*exo*-brevicomin

Reagents: i, OsO_4, $(DHDQ)_2$–PHAL; ii, TsOH

Scheme 38

[109] J. A. Soderquist and A. M. Rane, *Tetrahedron Lett.*, 1993, **34**, 5034.

Other workers have also developed both catalytic (Scheme 39)[110,111] and stoichiometric (Scheme 40)[112] variations, which in several instances are complementary to the Sharpless system.

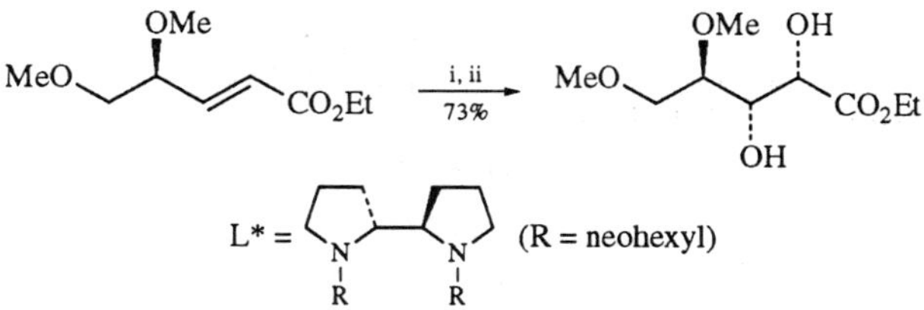

(ref 111)

Reagents: i, OsO_4, L*, CH_2Cl_2, −8 °C; ii, $Na_2S_2O_5$

Scheme 39

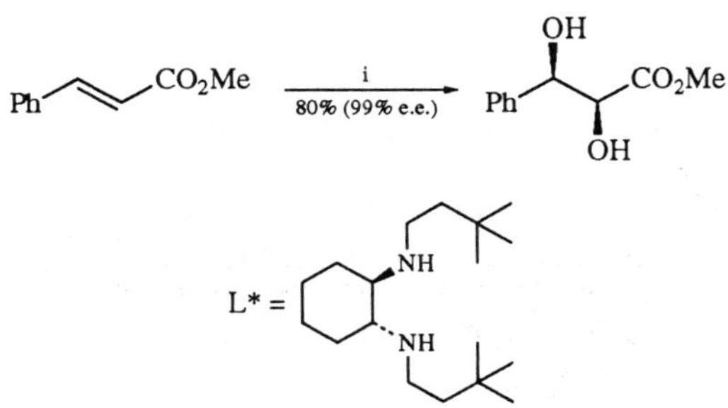

(ref 112)

Reagents: i, OsO_4, L*, toluene, −90 °C

Scheme 40

Diastereoselective[113] and enantioselective[114] epoxidation reactions continue to be exploited synthetically (Scheme 41). Jacobsen's enantioselective epoxidation provides particularly rapid access to a range of functionalized olefins which are useful in

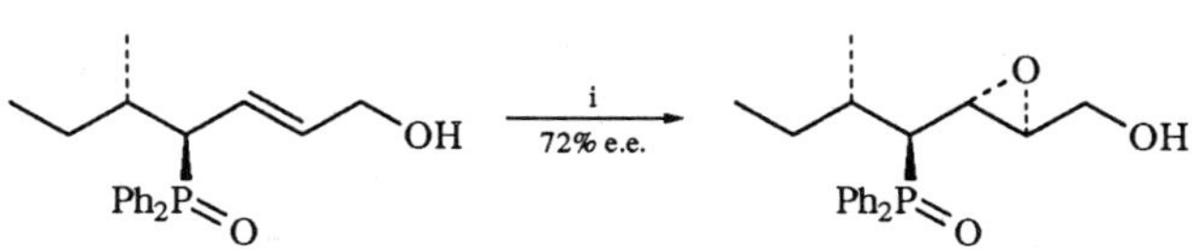

(ref 113)

Reagents: i, Bu^tOOH, 0.5 eq $Ti(OPr^i)_4$, L-(+)-DIPT, −16 °C

Scheme 41

110 E. J. Corey, M. C. Noe, and W.-C. Shieh, *Tetrahedron Lett.*, 1993, **34**, 5995.
111 T. Oishi, K. Iida, and M. Hirama, *Tetrahedron Lett.*, 1993, **34**, 3573.
112 S. Hanessian, P. Moffre, M. Girard, S. Beaudon, J.-Y. Sanceau, and Y. Bennani, *J. Org. Chem.*, 1993, **58**, 1991.
113 J. Clayden, E. W. Collington, R. B. Lamont, and S. Warren, *Tetrahedron Lett.*, 1993, **34**, 2203.
114 E. J. Allain, L. P. Hager, L. Deng, and E. N. Jacobsen, *J. Am. Chem. Soc.*, 1993, **115**, 4415.

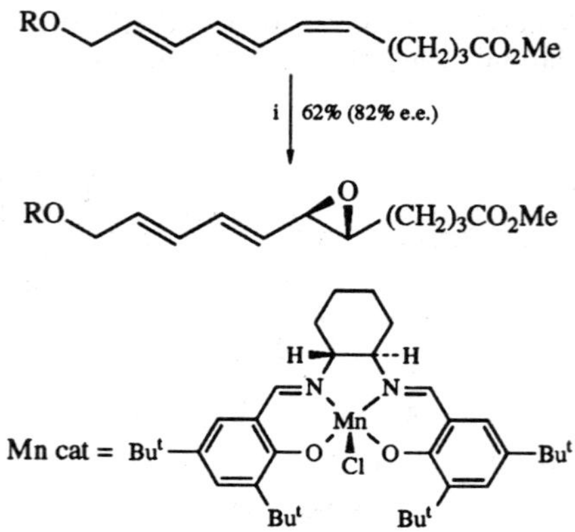

(ref 115)

Reagents: i, Mn cat, NaOCl, pH 11.5

Scheme 42

target-orientated synthesis, as demonstrated by his synthesis of leukotriene A_4 methyl ester (Scheme 42).[115]

Ito[116] has developed a potentially powerful method for the stereoselective synthesis of polyols which utilizes an intramolecular bis-silylation–oxidation sequence (Scheme 43). Adam[117] has adapted the photooxidation of allylic alcohols into a titanium-catalysed epoxy–hydroxylation sequence, in which the relative configuration of four contiguous chiral centres can be controlled in a single reaction sequence (Scheme 44).

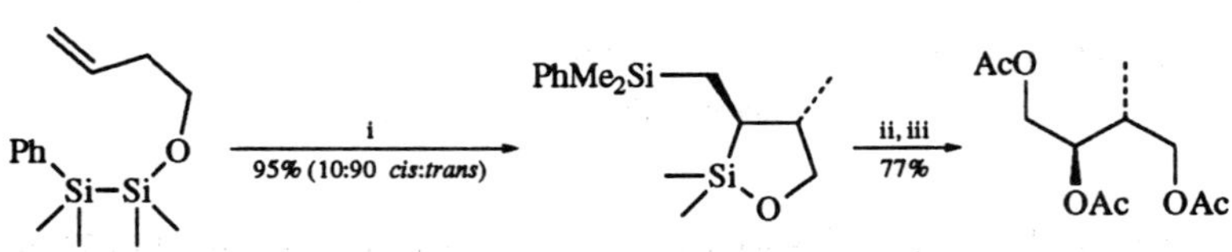

(ref 116)

Reagents: i, $Pd(OAc)_2$, RNC; ICl; ii, KF, KHF_2, $KHCO_3$; iii, Ac_2O

Scheme 43

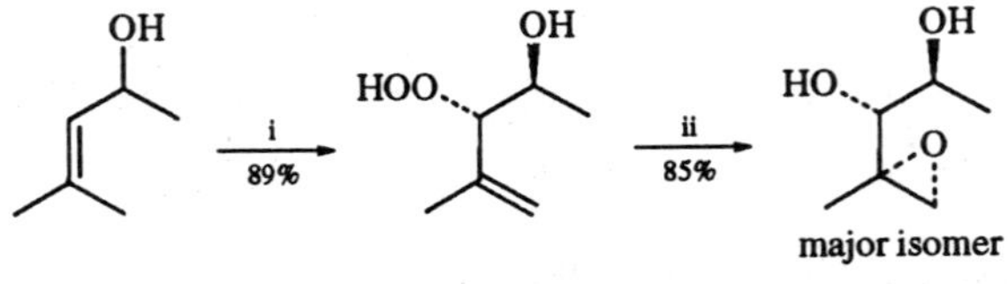

(ref 117)

Reagents: i, O_2, TPP, *hv*, 0 °C; ii, 0.05 eq $Ti(OPr^i)_4$, CH_2Cl_2, −25 °C

Scheme 44

[115] S. Chang, N. H. Lee, and E. N. Jacobsen, *J. Org. Chem.*, 1993, **58**, 6939.

[116] M. Murakami, M. Suginome, K. Fujimoto, H. Nakamura, P. G. Andersson, and Y. Ito, *J. Am. Chem. Soc.*, 1993, **115**, 6487. (*n.b.* For an improvement to the Fleming oxidative desilylation reaction see I. Fleming and S. B. D. Winter, *Tetrahedron Lett.*, 1993, **34**, 7287).

[117] W. Adam and B. Nestler, *Angew. Chem., Int. Ed. Engl.*, 1993, **32**, 733.

Enantioselective aziridination[118,119] of olefins has been achieved with high levels of asymmetric induction (up to 96% e.e.). Such compounds prove to be useful synthetic intermediates (Scheme 45).

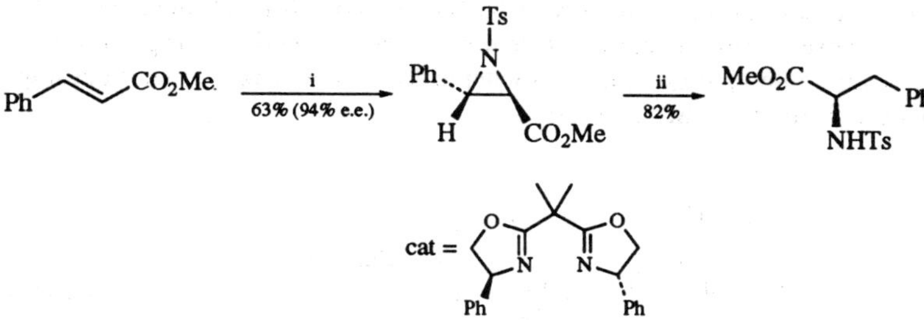

(ref 118)

Reagents: i, cat. CuOTf, PhI=NTs, rt; ii, HCO_2H, Pd(0), MeOH

Scheme 45

Alkynes.—Tetraethynylmethane (7), a high-carbon monomer whose polymerization could lead to a network homologous to diamond, has been readily prepared from a dibromide (Scheme 46) in an eight-step sequence.[120] Solid tetraethynylmethane

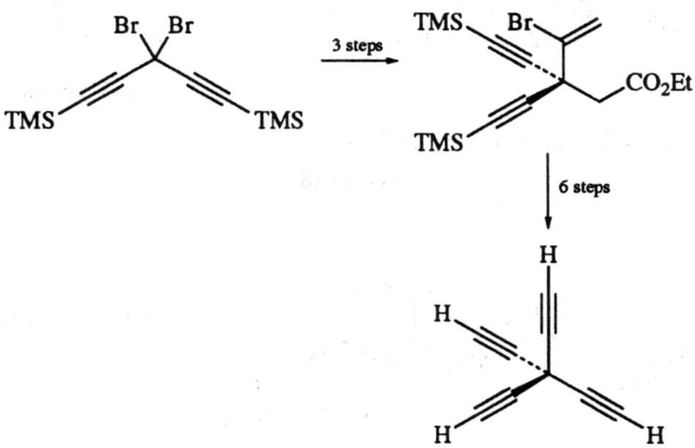

(7)

(ref 120)

Scheme 46

decomposes rapidly at room temperature, but a trimethylsilylated precursor could be handled under ambient conditions, and an *X*-ray crystal structure of this material confirms the structure and reveals that bonds from the central C(sp^3) to the C(sp)

[118] D. A. Evans, M. M. Faul, M. T. Bilodeau, B. A. Anderson, and D. M. Barnes, *J. Am. Chem. Soc.*, 1993, **115**, 5328.
[119] Z. Li, K. R. Conser, and E. N. Jacobsen, *J. Am. Chem. Soc.*, 1993, **115**, 5326.
[120] K. S. Feldman, C. M. Kraebel, and M. Parvez, *J. Am. Chem. Soc.*, 1993, **115**, 3846.

atoms of the attached alkynyl units are lengthened by *ca.* 0.04 Å over standard values, presumably reflecting steric compression at the central carbon. Medium ring and macrocyclic acetylenic lactones are readily available (65–90%) from tosylhydrazones of vinylogous lactones upon treatment with NBS in aqueous Bu^{t}OH (Scheme 47).[121] Acetylenes are versatile intermediates as demonstrated by synthetic applications of Livinghouse's amidometallation reaction (Scheme 48)[122] and Takai's metallocyclopropene functionalization[123] (Scheme 49).

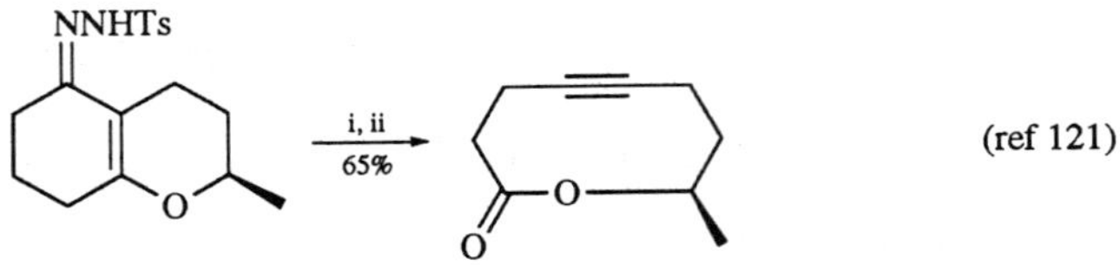

Reagents: i, Bu^{t}OH, NBS, −10 °C; ii, $NaHSO_3$, 55 °C

Scheme 47

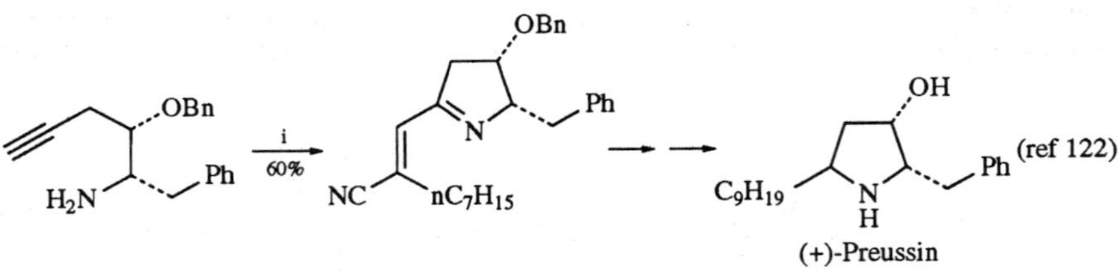

Reagents: i, $CpTiMeCl_2$, 25 °C; ii, $C_7H_{15}COCN$, 25 °C

Scheme 48

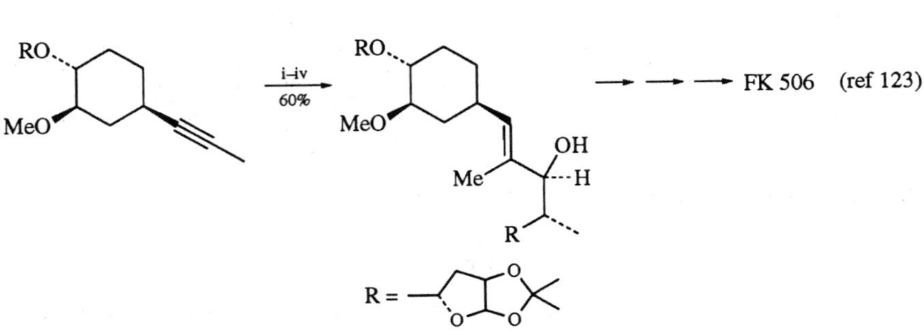

Reagents: i, $TaCl_5$, Zn; ii, py, THF; iii, RCHO; iv, NaOH, DME

Scheme 49

121 J. R. Mahajan and I. S. Resck, *J. Chem. Soc., Chem. Commun.*, 1993, 1748.
122 P. L. McGrane and T. Livinghouse, *J. Am. Chem. Soc.*, 1993, **115**, 11 485.
123 M. E. Maier, B.-U. Haller, R. Stumpf, and H. Fisher, *SYNLETT*, 1993, 863.

Enolate, Aldol, and Related Reaction.—Collum[124] has demonstrated that the readily available hindered amide base (8) enolizes ketones with very high levels of $E:Z$ selectivity (Scheme 50). Use of manganese[125] and zinc[126] enolates in selective alkylation reactions has been amply demonstrated; asymmetric deprotonations and alkylations have found synthetic applications;[127] asymmetric protonation of thiol ester enolates has been reported[128] to proceed with high levels of asymmetric induction: future developments in this area are eagerly awaited.

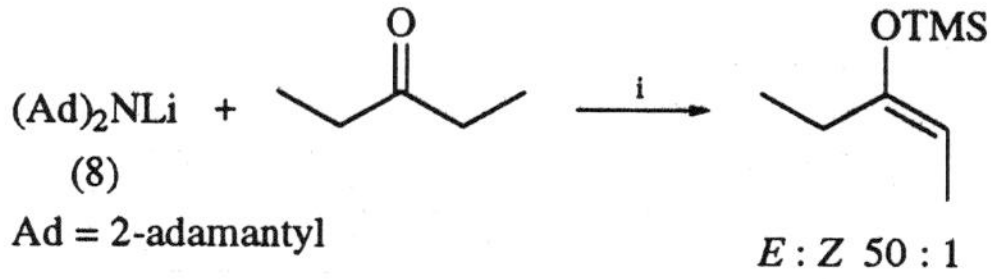

(ref 124)

Reagents: i, TMSCL, THF, −20° C

Scheme 50

Evans-type methodology is now commonplace for the synthesis of *syn* aldol products of high diastereoisomeric and enantiomeric purity.[129] Note however that there may be substrate control of the sense of induction, as in the case of α,α,α-trifluoroketones, which afford 'non-Evans' *syn* aldol products (Scheme 51).[130]

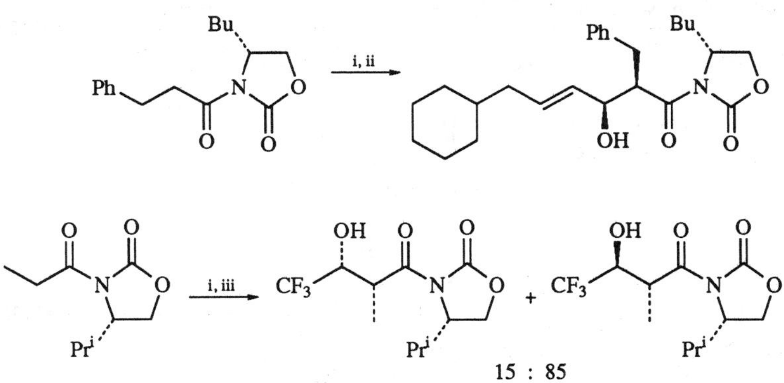

(ref 129)

(ref 130)

Reagents: i, Bu_2BOTf, Pr^i_2NEt; ii, 4-cyclohexylcrotonaldehyde; iii, CF_3CHO

Scheme 51

124 K. Sakuma, J. H. Gilchrist, F. E. Romesberg, C. E. Cajthaml, and D. B. Collum, *Tetrahedron Lett.*, 1993, **34**, 5213.
125 M. T. Reetz and H. Haning, *Tetrahedron Lett.*, 1993, **34**, 7395.
126 U. Groth, T. Huhn, and N. Richer, *Liebigs Ann. Chem.*, 1993, 49.
127 A. J. Edwards, S. Hockey, F. S. Mair, P. R. Raithby, R. Snaith, and N. S. Simpkins, *J. Org. Chem.*, 1993, **58**, 6942; M. Sobukawa and K. Koja, *Tetrahedron Lett.*, 1993, **34**, 5101; D. Milne and P. J. Murphy, *J. Chem. Soc., Chem. Commun.*, 1993, 884; Y. Hasegawa, H. Kawasaki, and K.Koga, *Tetrahedron Lett.*, 1993, **34**, 1963.
128 C. Fehr, I. Stempf, and J. Galindo, *Angew. Chem., Int. Ed. Engl.*, 1993, **32**, 1043.
129 B. D. Dorsey, K. J. Plzak, and R. G. Ball, *Tetrahedron Lett.*, 1993, **34**, 1851.
130 K. Iseki, S. Oishi, T. Taguchi, and Y. Kobayashi, *Tetrahedron Lett.*, 1993, **34**, 8147.

Subtle changes to the chiral auxiliary and of additives in the reaction medium[131] can also affect dramatically the stereochemical course of these reactions (Scheme 52).

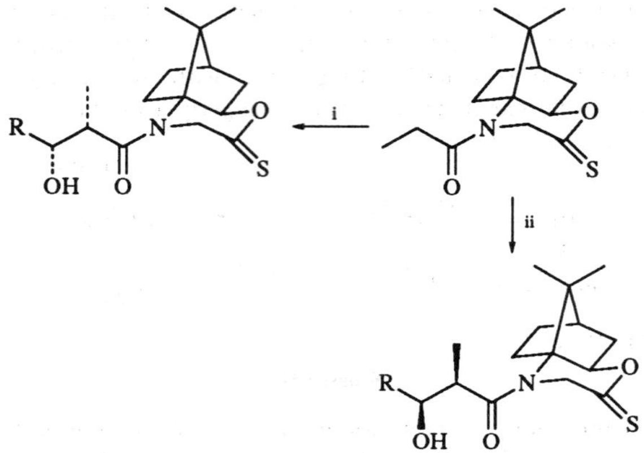

(ref 131)

Reagents: i, $TiCl_4$, RCHO; ii, Bu_2BOTf, RCHO

Scheme 52

Evans[132] has shown that Z enolates derived from *β*-hydroxy ketones undergo *anti* Felkin aldol reactions with high levels of double diastereodifferentiation, (Scheme 53).

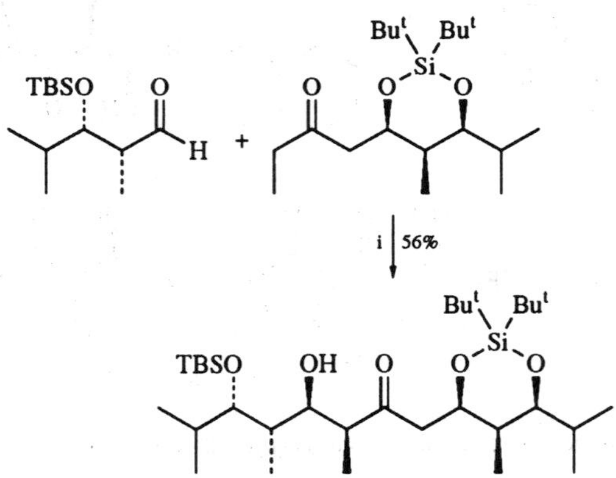

(ref 133)

> 99 : 1 diastereoselection

Reagents: i, $PhBCl_2$, Pr^i_2NEt, CH_2Cl_2, $-78\,°C$

Scheme 53

Titanium enolates derived from glycine are observed to undergo aldol reactions[133] (presumably *via* a boat-type transition state) with high *anti* selectivity (up to >99 : 1

[131] T.-H. Yan, C.-W. Tan, H.-C. Lee, H.-C. Lo, and T.-Y. Huang, *J. Am. Chem. Soc.*, 1993, **115**, 2613.
[132] D. A. Evans and M. A. Calter, *Tetrahedron Lett.*, 1993, **34**, 6871.
[133] S. Kanemasa, T. Mori, and A. Tatsukawa, *Tetrahedron Lett.*, 1993, **34**, 8293.

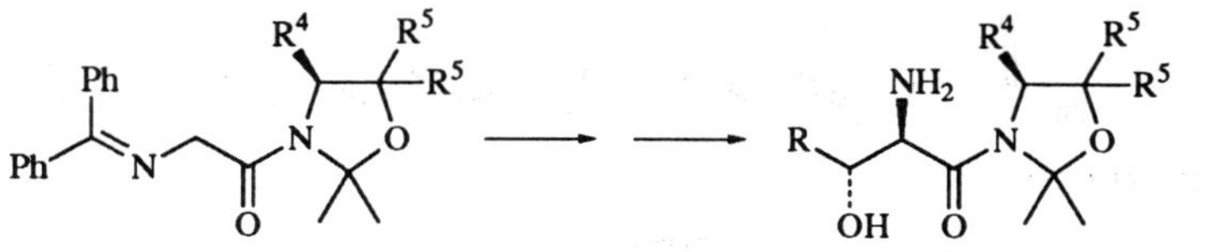

(ref 133)

($R^4 = CH_2Ph$; $R^5 = H$; > 99 : 1 d.e.)

Scheme 54

d.e.) (Scheme 54). In a related area, intense interest has focused upon the Mukaiyama reaction,[134–136] with particular emphasis on the development of new catalyst systems (Scheme 55) and on synthetic applications of allyl metal species[137] (Scheme 56).

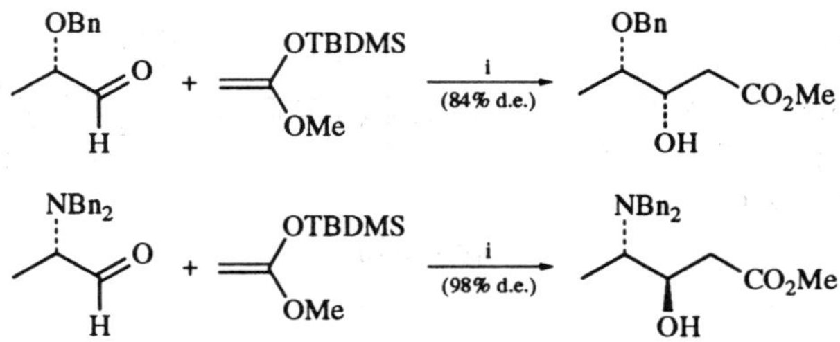

(ref 135)

Reagents: i, 3 mol % $LiClO_4$, CH_2Cl_2, −30 °C

Scheme 55

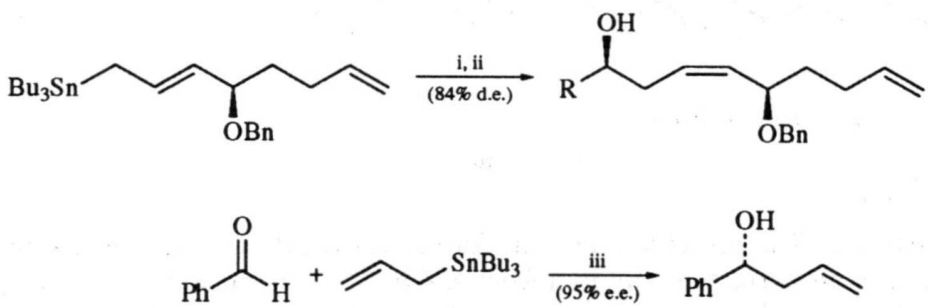

(ref 137)

(ref 138)

Reagents: i, $SnCl_4$, CH_2Cl_2; ii, RCHO; iii, 1 : 1 (*R*)-BINAP–Ti(OPr^i)$_4$

Scheme 56

Realization that readily available BINOL-derived catalysts impart high levels of asymmetric induction in these reactions represents a major advance.[138] The observation that allyl stannanes undergo highly chemoselective alkylations (aldehydes > ketones, acid chlorides) in the presence of protic acids (2 M HCl) is also of some practical significance (Scheme 57).[139] The ene reaction, especially when

[134] K. Mikami and S. Matsukawa, *J. Am. Chem. Soc.*, 1993, **115**, 7039.
[135] M. Reetz and D. N. A. Fox, *Tetrahedron Lett.*, 1993, **34**, 1119.
[136] K. Mikami, M. Terada, and T. Nakai, *J. Chem. Soc., Chem. Commun.*, 1993, 343.
[137] A. H. McNeill and E. J. Thomas, *Tetrahedron Lett.*, 1993, **34**, 1669.
[138] G. Keck, K. H. Tarbet, and L. S. Geraci, *J. Am. Chem. Soc.*, 1993, **115**, 8467.
[139] A. Yanagisawa, H. Inoue, M. Morodome, and H. Yamamoto, *J. Am. Chem. Soc.*, 1993, **115**, 10 356.

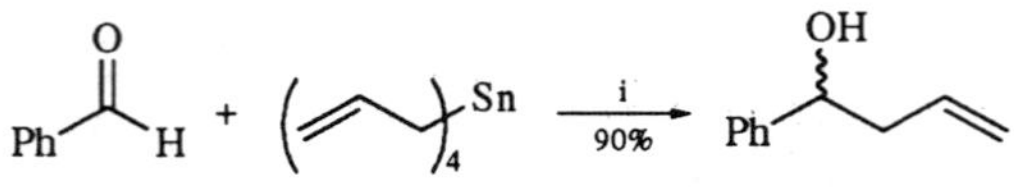

(ref 139)

Reagents: i, HCl, THF, 20 °C

Scheme 57

conducted in a chelation-controlled mode, has proven to be highly effective for the creation of up to three contiguous chiral centres in acyclic examples (Scheme 58).[140] Enantioselective Reformatsky[141] and nitro-aldol reactions[142] have been reported; zinc[143] and copper[144] homoenolates continue to be developed as useful synthetic reagents as do α- and β-heterosubstituted organolithium reagents.[145] 1,7-Asymmetric induction has been observed in the reduction of 4-ketoalkylboronates (Scheme 59).[146]

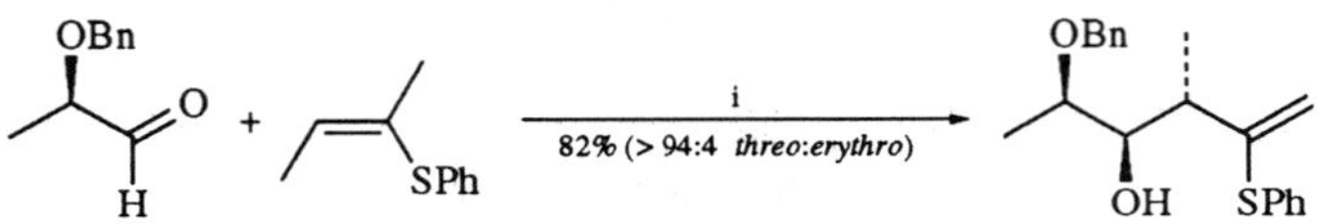

(ref 140)

Reagents: i, $SnCl_4$, −8 °C

Scheme 58

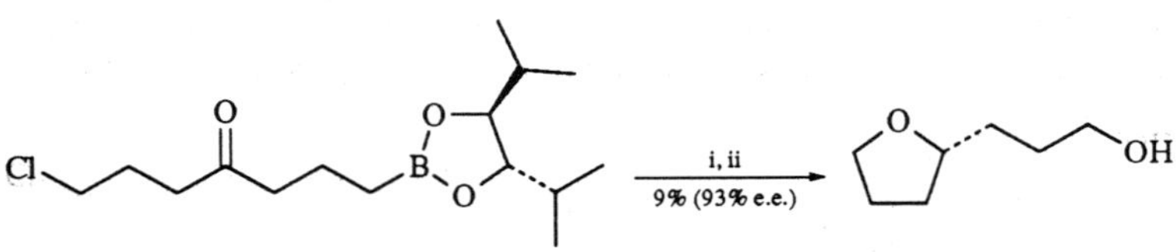

(ref 146)

Reagents: i, $BH_3{\cdot}SMe_2$, 0 °C; ii, NaOH(aq), H_2O_2

Scheme 59

Radical Reactions.—The use of free radicals for the construction of carbon–carbon bonds continues to attract attention (Scheme 60).[147] A particularly interesting observation this year is that scalemic α-sulfinylenones undergo conjugate addition with

[140] T. Nakamura, K. Tanino, and I. Kuwajima, *Tetrahedron Lett.*, 1993, **34**, 7591; T. Nakamura, K. Tanino, and I. Kuwajima, *Tetrahedron Lett.*, 1993, **34**, 477; K. Mikami and S.-i. Sakuda, *J. Chem. Soc., Chem. Commun.*, 1993, 710.

[141] K. Soai, A. Oshio, and T. Saito, *J. Chem. Soc., Chem. Commun.*, 1993, 121; Y. Yamamoto, M. Naito, Y. Uozumi, and T. Hayashi, *Tetrahedron Lett.*, 1993, **34**, 1468.

[142] H. Sasai, T. Suzuki, N. Itoh, S. Arai, and M. Shibasaki, *Tetrahedron Lett.*, 1993, **34**, 2657; P. Wipf and S. Lim, *J. Chem. Soc., Chem. Commun.*, 1993, 1654.

[143] R. F. W. Jackson and A. B. Rettie, *Tetrahedron Lett.*, 1993, **34**, 2985.

[144] I. Ryu, K. Matsumoto, Y. Kameyama, M. Ando, N. Kusumoto, A. Ogawa, N. Kambe, S. Murai, and Sonada, *J. Am. Chem. Soc.*, 1993, **115**, 12 330.

[145] J. Almena, F. Foubelo, and M. Yus, *Tetrahedron Lett.*, 1993, **34**, 1645; A. F. Burchat, J. M. Chong, and S. B. Park, *Tetrahedron Lett.*, 1993, **34**, 51; A. Guijarro and M. Yus, *Tetrahedron Lett.*, 1993, **34**, 3487; R. E. Gawley and Q. Zhang, *J. Am. Chem. Soc.*, 1993, **115**, 7515; H. J. Reich and R. R. Dykstra, *Angew. Chem., Int. Ed. Engl.*, 1993, **32**, 1469; W. H. Pearson, A. C. Lindbeck, and J. W. Kampf, *J. Am. Chem. Soc.*, 1993, **115**, 2622.

[146] G. A. Molander and K. L. Bobbitt, *J. Am. Chem. Soc.*, 1993, **115**, 7517.

[147] A. G. Myers, D. Y. Gin, and D. H. Rogers, *J. Am. Chem. Soc.*, 1993, **115**, 2036;

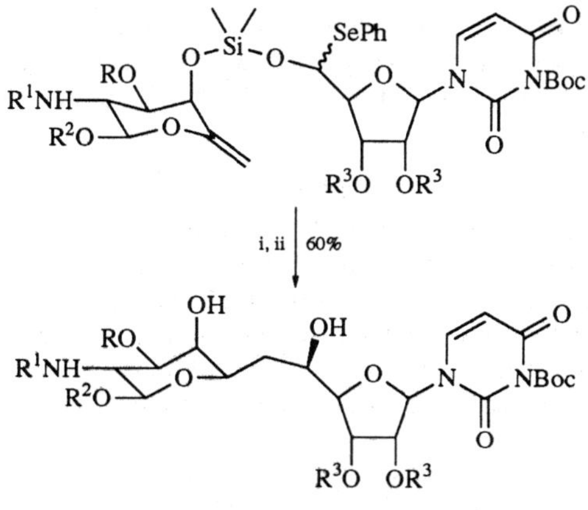

(ref 147)

Reagents: i, Bu_3SnH, BEt_3, 0 °C; ii, KF, MeOH

Scheme 60

carbon-centred radicals in a highly stereoselective fashion in the presence of bidentate Lewis acids (Scheme 61).[148] The stereoselective functionalization of α-functionalized radicals has also come under further investigation.[149,150] A potentially useful observation by Alper[151] suggests that carbon radicals may be trapped by carbon monoxide, albeit at high pressure, to afford carboxyl-containing products (Scheme 62).

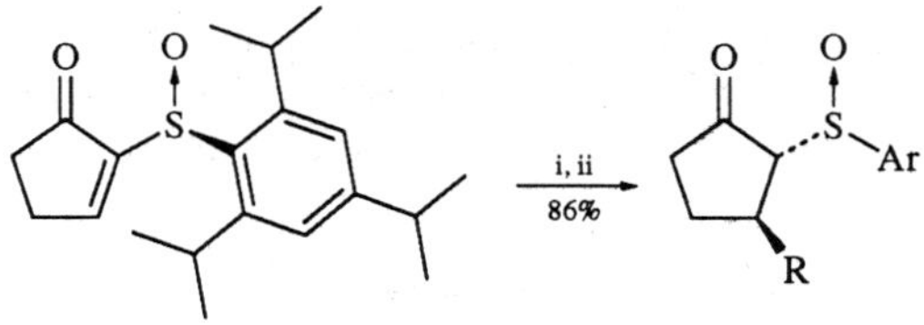

(ref 148)

Reagents: i, $TiCl_2(OPr^i)_2$, CH_2Cl_2, 0 °C; ii, BR_3

Scheme 61

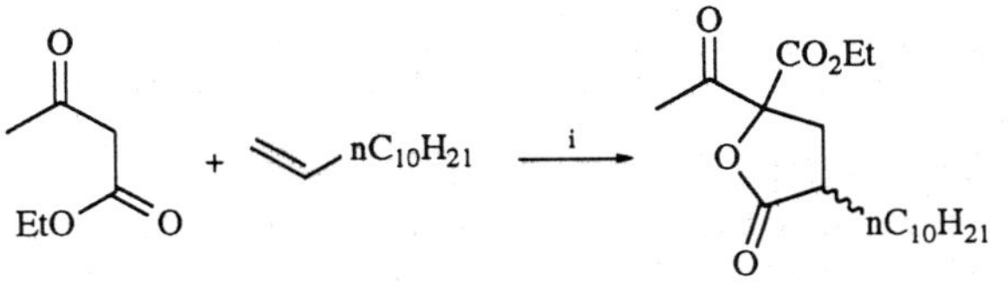

(ref 151)

Reagents: i, CO (600 psi), 0 °C, $Mn(OAc)_3{\cdot}2H_2O$

Scheme 62

[148] T. Toru, Y. Watanabe, M. Tasuaka, and Y. Ueno, *J. Am. Chem. Soc.*, 1993, **115**, 10464.

[149] W. Damm, J. Dickhaut, F. Wellerich, and B. Giese, *Tetrahedron Lett.*, 1993, **34**, 431; C. Ferreri, M. Ballestri, and C. Chatgilialoglu, *Tetrahedron Lett.*, 1993, **34**, 5147.

[150] *e.g.* D. P. Curran and S. Sun, *Tetrahedron Lett.*, 1993, **34**, 6181; B. Giese, W. Damm, F. Wetterich, H.-G. Zeltz, J. Rancourt, and Y. Guindon, *Tetrahedron Lett.*, 1993, **34**, 5885.

[151] I. Ryu and H. Alper, *J. Am. Chem. Soc.*, 1993, **115**, 7543.

Reductions.—The use of oxazaborilidine-derived reducing agents[152] for ketone reduction provides rapid access to a variety of functionalized alcohols with a predictable sense of induction (Scheme 63). Lithium aminoborohydrides[153] have

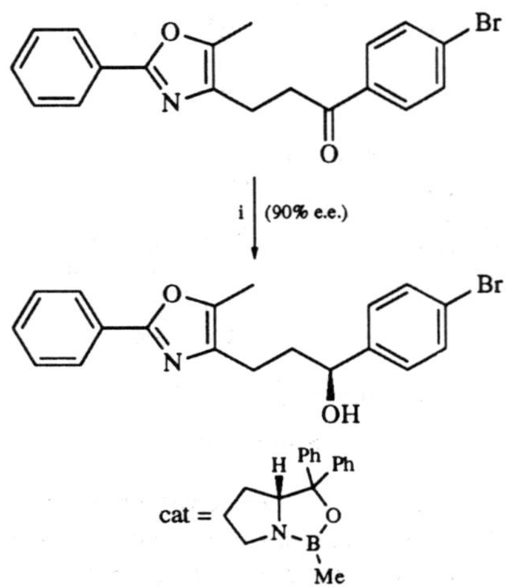

(ref 152)

Reagents: i, BH_3·THF, 25 °C, cat

Scheme 63

proved to be highly selective reducing agents. Williams[154] has developed a highly diastereoselective synthesis of *syn* 1,2-amino alcohols from the corresponding oximes (Scheme 64). Matsuki has demonstrated that *meso* anhydrides may be desymmetrized

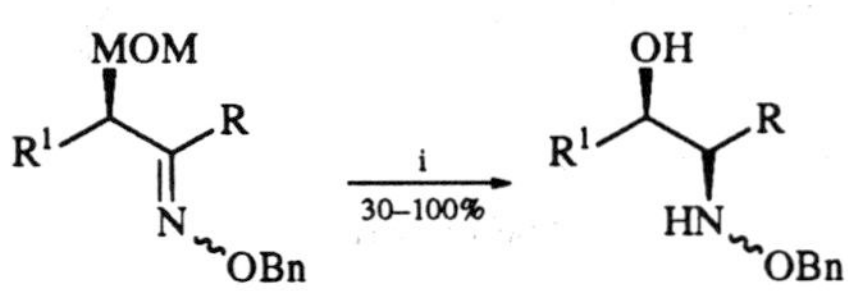

(ref 154)

Reagents: i, 5 eq TABH, CH_3CO_2H, CH_3CN, 35 °C

Scheme 64

by reduction with the LAH–EtOH–BINAL–H complex, to afford lactones with high optical purity[155] (84 to 95% e.e.) (Scheme 65). Evans[156] has developed a chiral samarium catalyst for the promotion of asymmetric Meerwein–Ponndorf–Verley reductions; high levels of induction (usually in excess of 97% e.e.) are obtained for reduction of aryl alkyl ketones (Scheme 66).

[152] C. Bolm and M. Felder, *Tetrahedron Lett.*, 1993, **34**, 6041; G. J. Quallich and T. M. Woodall, *Tetrahedron Lett.*, 1993, **34**, 4145.
[153] J. C. Fuller, E. L. Stageland, C. T. Goralski, and B. Singaram, *Tetrahedron Lett.*, 1993, **34**, 257.
[154] D. R. Williams, M. H. Osterhout, and J. P. Reddy, *Tetrahedron Lett.*, 1993, **34**, 3271.
[155] K. Matsuki, H. Inoue, and M. Takeda, *Tetrahedron Lett.*, 1993, **34**, 1167.
[156] D. A. Evans, S. G. Nelson, M. R. Gagne, and A. R. Muci, *J. Am. Chem. Soc.*, 1993, **115**, 9800.

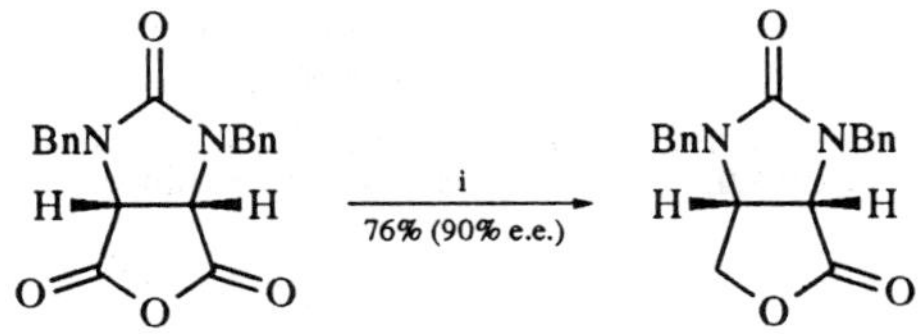

(ref 155)

Reagents: i, (*R*)-BINAL-H, −78 °C to rt

Scheme 65

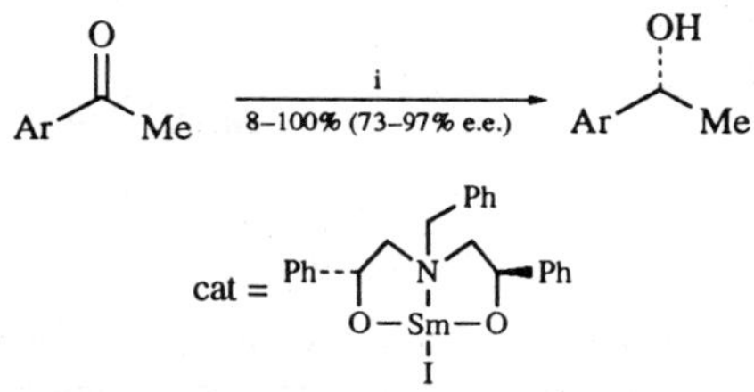

(ref 156)

Reagents: i, Me_2CHOH, 5% cat

Scheme 66

Asymmetric Hydrogenation and Related Reactions.—Significant advances this year include the use of the DUPHOS ligand system (9) in conjunction with rhodium(I) for the enantioselective reduction (typically >99% e.e.) of dehydroamino acids (Scheme 67).[157] Buchwald[158] has developed a titanium catalyst for the asymmetric reduction of

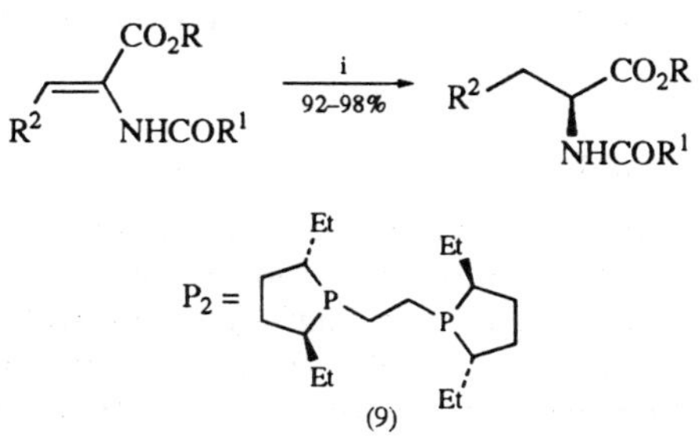

(9)

(ref 157)

Reagents: i, cat [(COD)Rh(P_2)]OTf, H_2

Scheme 67

trisubstituted olefins (>92% e.e.). The synthesis and exploitation of novel phosphine ligands continues to be fashionable.[159] Brown[160] has shown that the hybrid phosphine–isoquinoline ligand system (10) can be effectively employed in the asymmetric hydroboration[161] of simple olefins (Scheme 68).

[157] M. J. Burk, J. E. Feaster, W. A. Nugent, and R. L. Harlow, *J. Am. Chem. Soc.*, 1993, **115**, 10 125.
[158] R. D. Broene and S. L. Buchwald, *J. Am. Chem. Soc.*, 1993, **115**, 2569.
[159] H. Brisset, Y. Gourdel, P. Pellon, and M. Le Corre, *Tetrahedron Lett.*, 1993, **34**, 4523.
[160] J. M. Brown, D. I. Hulmes, and T. P. Layzell, *J. Chem. Soc., Chem. Commun.*, 1993, 1673.
[161] N. Sakai, S. Mano, K. Nozaki, and H. Takaya, *J. Am. Chem. Soc.*, 1993, **115**, 7033.

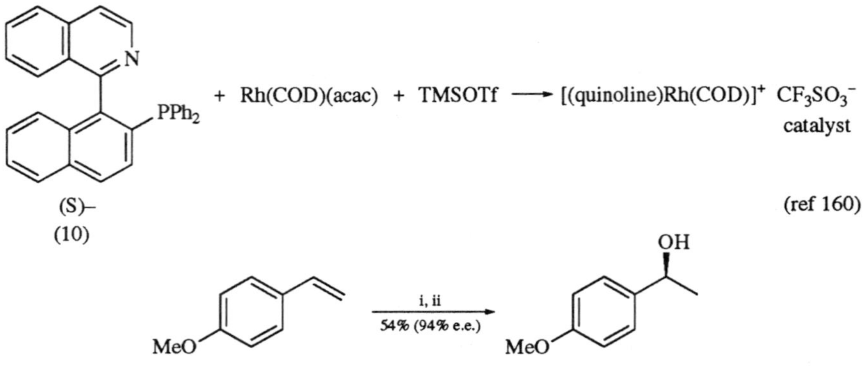

Reagents: i, Catalyst, catechol borane; ii, NaOH, H_2O_2

Scheme 68

In a rather novel departure – asymmetric catalyst poisoning – Faller[162] has demonstrated that hydrogenation of *racemic* cyclohexene-1-ol with *racemic* $RuCl_2$(BINAP)DMF_x in the presence of (1*R*,2*S*)-ephedrine affords the alcohol (11) with high levels of induction (93% e.e.) at 72% conversion (Scheme 69).

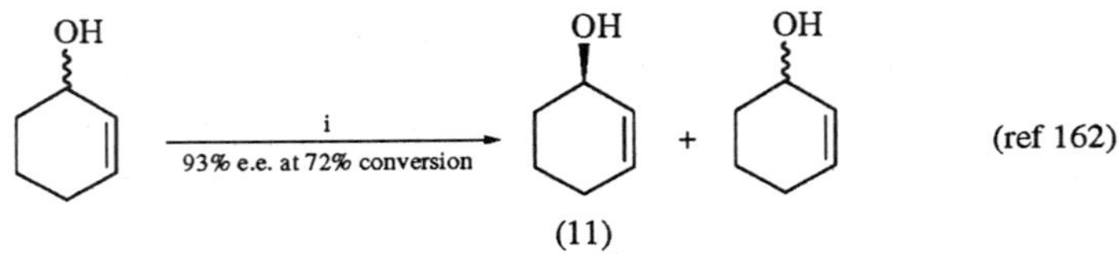

Reagents: i, H_2, (*R*)-Ru(BINAP), (−)-(1*R*,2*S*)-ephedrine$_2$

Scheme 69

Pot Pourri.—Readily available, scalemic α-epoxy alcohols,[163] (hydroxymethyl)aziridines,[164] and α,β-unsaturated epoxides[165] continue to find novel uses in synthesis (Scheme 70), as do the now-fashionable oxidizing agents dialkyl dioxiranes[166] and TPAP[167] (Scheme 71). The preparation of optically pure β-amino acids is presently an area of much methodological interest. A particularly ready entry[168] into this class of compounds utilizes an asymmetric conjugate addition of homochiral lithium amides to α,β-unsaturated carboxylic acid derivatives, as outlined in Scheme 72. Finally, a fascinating report by Yamamoto[169] details the preparation of stable Lewis acid complexes of reactive aldehydes using highly hindered aluminium alkoxide complexes.

162 J. W. Faller and M. Tokunaga, *Tetrahedron Lett.*, 1993, **34**, 7359.
163 M. Poch, M. Alcon, A. Moyano, M. A. Percas, and A. Riera, *Tetrahedron Lett.*, 1993, **34**, 7781.
164 T. Kawabata, Y. Kiryu, Y. Sugiura, and K. Fuji, *Tetrahedron Lett.*, 1993, **34**, 5127.
165 M. Miyashita, K. Yoshihara, K. Kawamine, M. Hoshino, and H. Irie, *Tetrahedron Lett.*, 1993, **34**, 6285.
166 R. Curci, L. D'Accolti, A. Detomaso, C. Fusco, K. Takeuchi, Y. Ohga, P. E. Eaton, and Y. C. Yip, *Tetrahedron Lett.*, 1993, **34**, 4559.
167 K. R. Guertin and A. S. Kende, *Tetrahedron Lett.*, 1993, **114**, 5639.
168 N. Asao, N. Tsuda, and Y. Yamamoto, *J. Chem. Soc., Chem. Commun.*, 1993, 1660.
169 K. Maruoke, A. B. Conçepcion, N. Murase, M. Oishi, N. Hirayama, and H. Yamamoto, *J. Am. Chem. Soc.*, 1993, **115**, 3943.

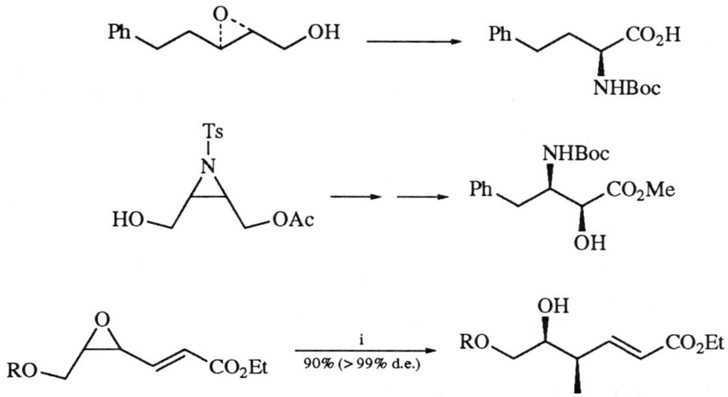

(ref 163)

(ref 164)

(ref 165)

Reagents: i, $AlMe_3$, H_2O; $-30\,°C$

Scheme 70

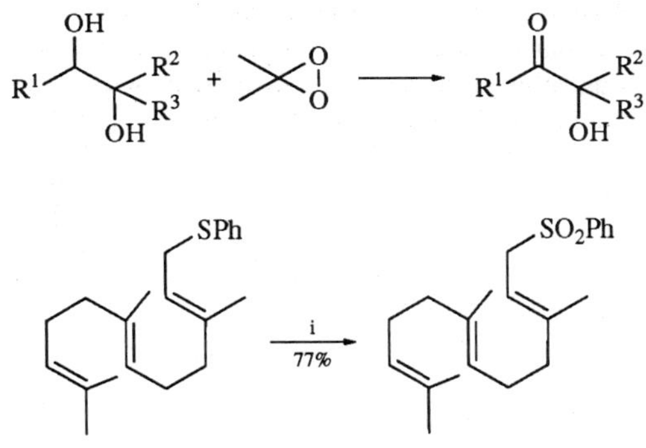

(ref 166)

(ref 167)

Reagents: i, cat. TPAP, NMO, CH_3CN

Scheme 71

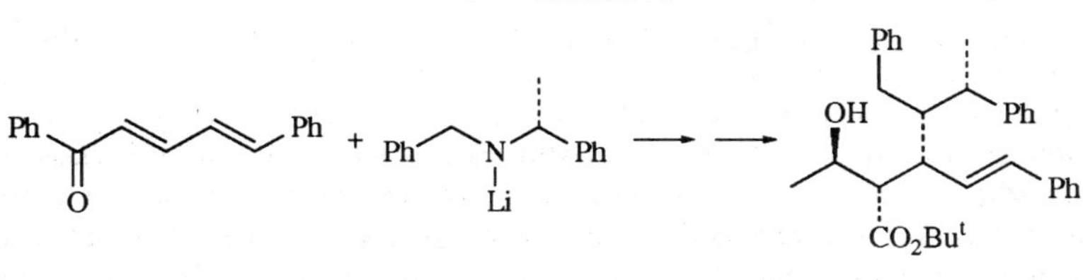

(ref 168)

Scheme 72

The reaction of such species with a variety of nucleophiles is also disclosed (Scheme 73).

3 Alicyclic Chemistry

Cyclopropanes.—Keck[170] has developed a facile synthesis of vinyl cyclopropanes (12), which themselves are useful synthetic intermediates (Scheme 74). Cohen[171] has shown

[170] G. E. Keck and S. D. Tonnies, *Tetrahedron Lett.*, 1993, **34**, 4607.
[171] C. A. Shook, M. L. Romberger, S.-H. Jung, M. Xiao, J. P. Sherbine, B. Zhang, F.-T. Lin, and T. Cohen, *J. Am. Chem. Soc.*, 1993, **115**, 10 754.

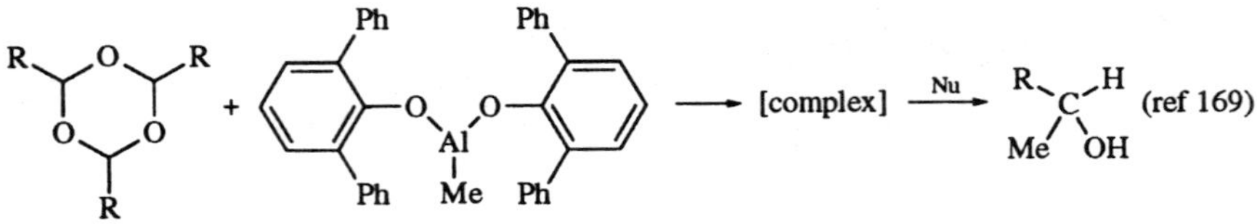

Scheme 73

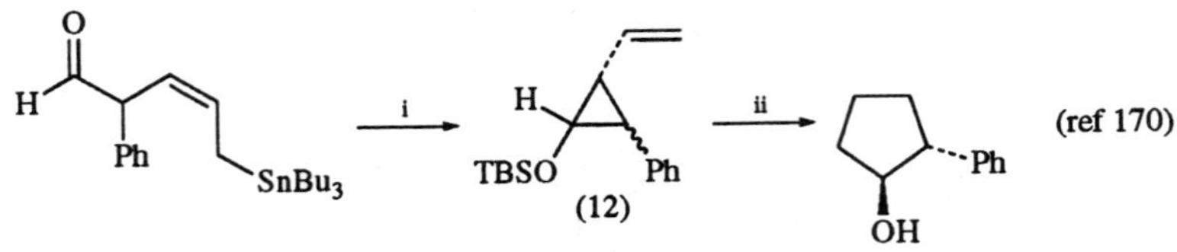

Reagents: i, TBSOTf, CH_2Cl_2, 0 °C; ii, KF, DMSO, 18-C-6

Scheme 74

that the cyclopropyl sulfides (13) suffer sequential single-electron reduction, alkylation, and Peterson olefination to generate the methylene cyclopropanes, rearrangement of which affords a general route to hydrazaazulenes (Scheme 75). Rather

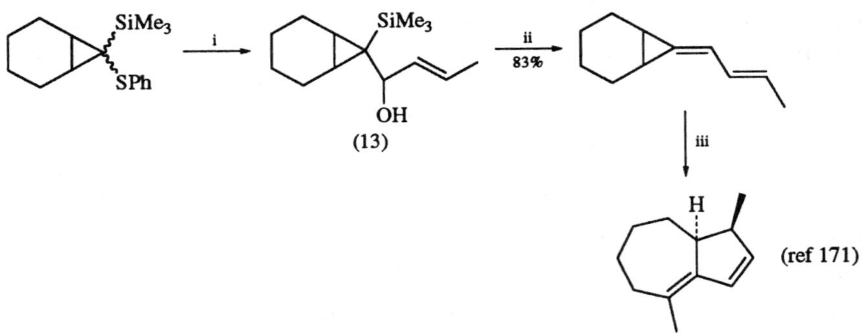

Reagents: i, e^-, $CH_3CH{=}CHCHO$; ii, $KOBu^t$; iii, 180 °C

Scheme 75

unexpectedly,[172] treatment of α,β-unsaturated aldehydes with $CrCl_2$–DMF complex provides a general synthesis of cyclopropanols (Scheme 76). Hoffmann[173] has described a rare example of nucleophilic attack at the internal position of a π-allylpalladium complex. Reaction of the complex (14) with a variety of ketone or ester enolates in the presence of TMEDA affords the cyclopropanes (15) in good isolated yields (Scheme 77). The generality of this reaction has yet to be established, but it adds a further dimension to the reactivity of this most investigated class of organometallic reagent.

Bäckvall[174] has demonstrated high site selectivity in the cyclopropanation of 2-phenylsulfonyl-1,3-dienes with nucleophilic and electrophilic cyclopropanating

[172] D. Mongomery, K. Reynolds, and P. Stevenson, *J. Chem. Soc., Chem. Commun.*, 1993, 363.
[173] A. Wilde, A. R. Otte, and H. M. R. Hoffmann, *J. Chem. Soc., Chem. Commun.*, 1993, 615.
[174] J.-E. Bäckvall, C. Löfström, S. K. Juntunen, and M. Mattson, *Tetrahedron Lett.*, 1993, **34**, 2007.

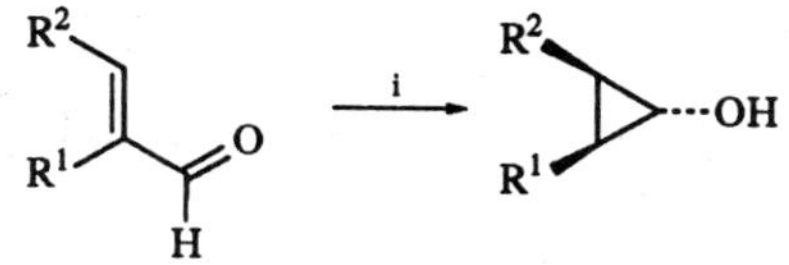

(ref 172)

Reagents: i, $CrCl_2$, DMF

Scheme 76

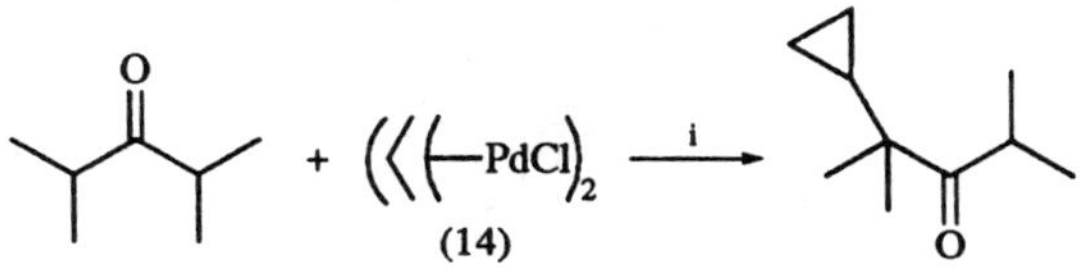

(14) (15)

(ref 173)

Reagents: i, LDA, TMEDA, 70 °C

Scheme 77

agents; intramolecular cyclopropanations of iodonium ylides[175] have been utilized in the synthesis of a vitamin D ring-A synthon. Enantioselective rhodium-mediated cyclopropanation reactions have again come under much scrutiny.[176] Lee[177] has developed a simple synthesis of functionalized vinylcyclopropanes which serve as precursors to cyclopentenones and furans (Scheme 78). Normant[178] has utilized a

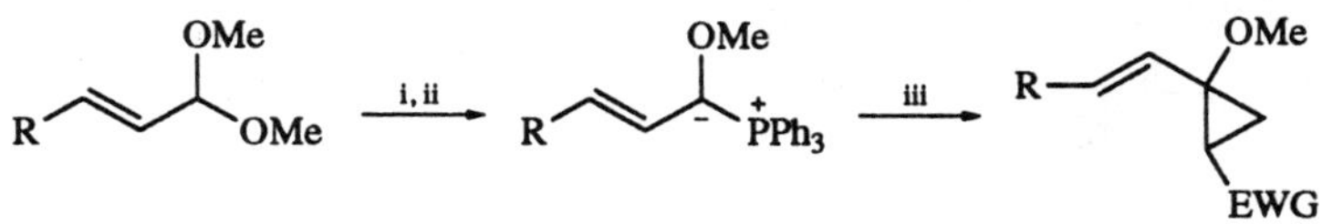

(ref 177)

Reagents: i, PPh_3, TMSOTf, THF, 40 °C; ii, Bu^nLi; iii, CH_2CH–EWG

Scheme 78

metalla-Claisen rearrangement in the stereoselective elaboration of cyclopropanes (Scheme 79). A biomimetic synthesis[179] of cyclopropane-containing eicosanoids

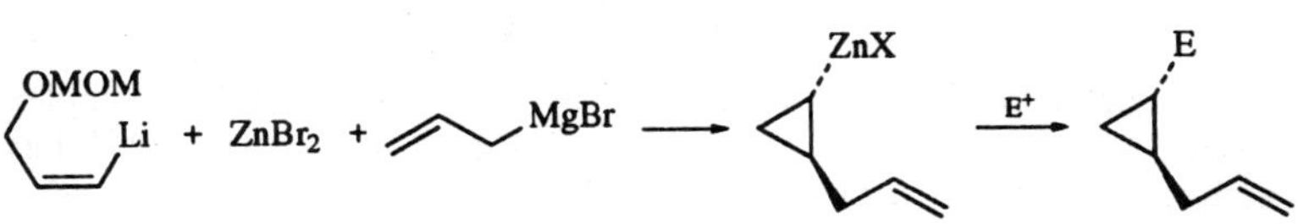

(ref 178)

Scheme 79

utilized an intramolecular S_N2' substitution reaction of an allylic epoxide (Scheme 80).

Cyclobutanes.—The synthesis of the tetraethynylcyclobutadiene complex (16) has been

[175] R. M. Moriarty, J. Kim, and L. Guto, *Tetrahedron Lett.*, 1993, **34**, 4129.
[176] M. P. Doyle, W. R. Winchester, J. A. Hoorn, U. J. Lynch, S. H. Simonsen, and R. Ghosh, *J. Am. Chem. Soc.*, 1994, **115**, 9968.
[177] P. H. Lee, J. S. Kim, Y. C. Kim, and S. Kim, *Tetrahedron Lett.*, 1993, **34**, 7583.
[178] D.Beruben, I. Marek, J.-F. Normant, and N. Platzer, *Tetrahedron Lett.*, 1993, **34**, 7575.
[179] J. D. White and M. S. Jensen, *J. Am. Chem. Soc.*, 1993, **115**, 2970.

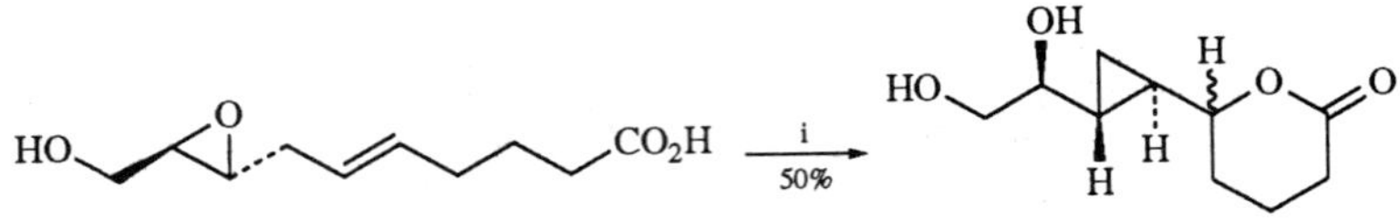

(ref 179)

Reagents: i, $SnCl_4$, $MeNO_2$, 0 °C

Scheme 80

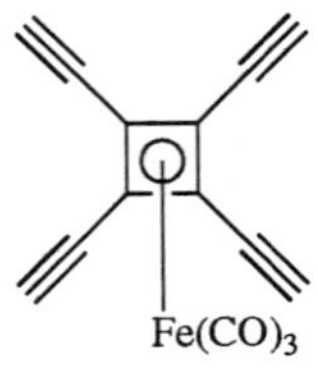

(16)

(ref 180)

described.[180] Liebskind[181] has developed a versatile route to the bisquaryl system (17) utilizing a Pd/Cu homocoupling procedure (Scheme 81). Organometallic approaches[182] to cyclobutane synthesis have featured heavily this year, as outlined in Scheme 82. Corey [183] has described a novel samarium-mediated aldol-type reaction in

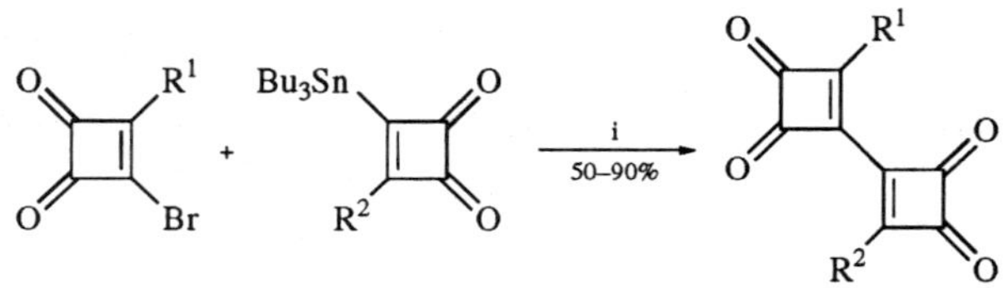

(17)

(ref 181)

Reagents: i, 5% $PdClBn(PPh_3)_2$, MeCN, 50 °C

Scheme 81

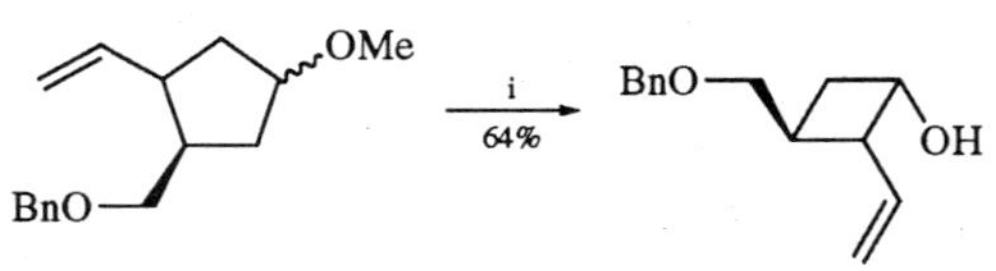

(ref 182)

Reagents: i, Cp_2ZrCl_2, BuLi, $BF_3{\cdot}OEt_2$, −78 °C

Scheme 82

the synthesis of (±)-paeoniflorigenin (Scheme 83); similarly Fukumoto[184] has

[180] U. H. F. Bunz and V. Enkelmann, *Angew. Chem., Int. Ed. Engl.*, 1993, **32**, 1653.
[181] L. S. Liebeskind, M. S. Yu, R. H. Yu, J. Wang, and K. S. Hagen, *J. Am. Chem. Soc.*, 1993, **115**, 9048.
[182] H. Ito, T. Taguchi, and Y. Hanzawa, *Tetrahedron Lett.*, 1993, **34**, 7639; M. Hojo, K. Tomita, Y. Hirohara, and A. Hosomi, *Tetrahedron Lett.*, 1993, **34**, 8123; B. M. Trost, K. Imi, and A. F. Indolese, *J. Am. Chem. Soc.*, 1993, **115**, 8831; B. M. Trost, M. Yanai, and K. Hoogsteen, *J. Am. Chem. Soc.*, 1993, **115**, 5294; S. Kobbing and G. Raabe, *Chem. Ber.*, 1993, **126**, 1849.
[183] E. J. Corey and Y.-J. Wu, *J. Am. Chem. Soc.*, 1993, **115**, 8871.
[184] M. Ihara, T. Taniguchi, K. Makita, M. Takano, M. Ohnishi, N. Taniguchi, K. Fukumoto, and C. Kabuto, *J. Am. Chem. Soc.*, 1993, **115**, 8107.

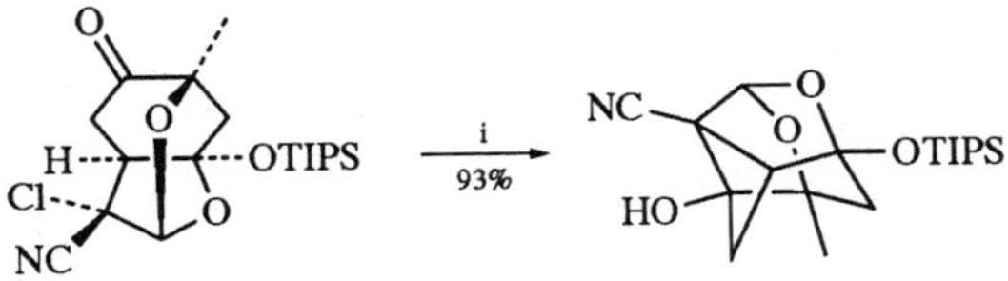

(ref 183)

Reagents: i, SmI_2, THF, $-45\,°C$

Scheme 83

disclosed details of a tandem intramolecular Michael–Aldol reaction for the synthesis of polycyclic cyclobutanes (Scheme 84).

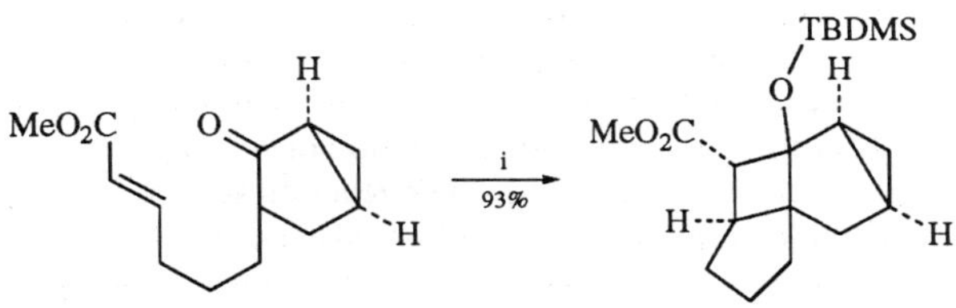

(ref 184)

Reagents: i, TBDMSOTf, NEt_3

Scheme 84

[2 + 2]-Cycloaddition strategies[185–187] have been employed in the synthesis of a variety of novel cyclobutane systems (Scheme 85). The synthetic utility of the

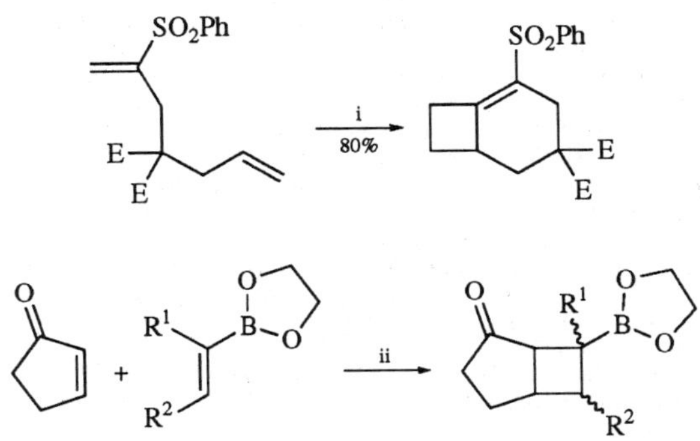

(ref 185)

(ref 187)

Reagents: i, 80 °C; ii, *hν*

Scheme 85

cyclobutane ring system in natural product synthesis has been nicely exemplified by Crimmins[188] in his synthesis of (±)-bilobalide (Scheme 86).

Cyclopentanes.—The synthesis of cyclopentanes (and hexanes) from sugar derivatives

[185] A. Padwa, M. A. Filipkowski, M. Meske, S. H. Watterson, and Z. Ni, *J. Am. Chem. Soc.*, 1993, **115**, 3776.
[186] F. Jin, Y. Xu, and W. Huang, *J. Chem. Soc., Chem. Commun.*, 1993, 814.
[187] W. G. Hollis, W. C. Lappenbusch, K. A. Everberg, and C. M. Woleben, *Tetrahedron Lett.*, 1993, **34**, 7517.
[188] M. T. Crimmins, D. K. Jung, and J. L. Gray, *J. Am. Chem. Soc.*, 1993, **115**, 3146.

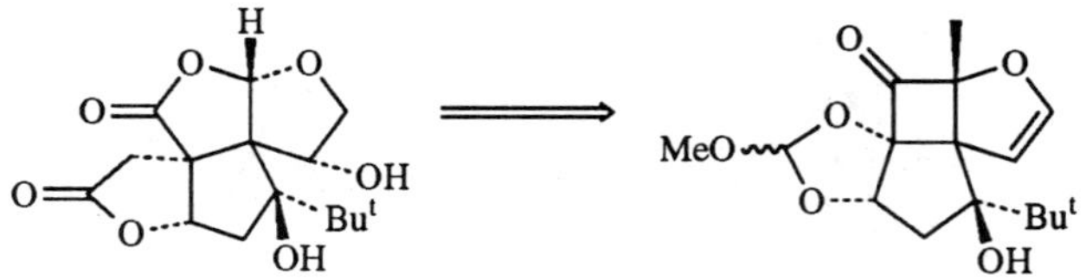

(ref 188)

Scheme 86

has been exhaustively reviewed by Ferrier.[189] In keeping with recent trends, radical[190–196] and radical-anion-based[197] approaches to functionalized cyclopentanes have again been the subject of intense methodological investigation (Scheme 87); the use of Mn^{3+} for the generation of malonate-type[198] radicals provides a useful alternative to more standard methods of radical initiation. In related studies, cyclization of radical cations[199] provides a direct entry to spirocyclic systems.

A novel two-fold addition–rearrangement reaction of vinyl anions to squarate esters generates tetracyclic cyclopentanoids in a single reaction sequence (Scheme 88).[200] A variety of organometallic-based strategies[201] have also emerged for the preparation of functionalized cyclopentanes (Scheme 89). Most notable amongst these is the Knochel[202]–Normant[203] cyclization of ω-olefinic organozinc reagents (Scheme 90), a zirconium-mediated ring contraction of carbohydrate derivatives (Scheme 91),[204] and a titanocene-catalysed conversion of eneynes into bicyclic cyclopentenones (Scheme 92).[205] The latter provides a useful alternative to the more common Pauson–Khand

[189] R. J. Ferrier and S. Middleton, *Chem. Rev.*, 1993, **93**, 2779.
[190] D. L. J. Clive and H. W. Manning, *J. Chem. Soc., Chem. Commun.*, 1993, 666.
[191] J. Fidalgo, L. Castedo, and D. Dominguez, *Tetrahedron Lett.*, 1993, **34**, 7317.
[192] Y.-J. Chen, C.-M. Chen, and W.-Y. Lin, *Tetrahedron Lett.*, 1993, **34**, 2961.
[193] V. H. Rawal, V. Krishnamurthy, and A. Fabre, *Tetrahedron Lett.*, 1993, **34**, 2899.
[194] C.-K. Shan, C.-Y. Shen, T.-S. Jean, R.-T. Chiu, and W.-H. Tseng, *Tetrahedron Lett.*, 1993, **34**, 7641; K. S. Feldman, H. M. Berven, and P. H. Weinreb, *J. Am. Chem. Soc.*, 1993, **115**, 11 364; J. W. Grissom, T. L. Calkins, and M. Egan, *J. Am. Chem. Soc.*, 1993, **115**, 11 744.
[195] F.-H. Wartenberg, H. Junga, and S. Blechert, *Tetrahedron Lett.*, 1993, **34**, 5251.
[196] J. E. Brumwell, N. S. Simpkins, and N. K. Terrett, *Tetrahedron Lett.*, 1993, **34**, 1215; D. F. Taber, Y. Wang, and S. J. Stacel, *Tetrahedron Lett.*, 1993, **34**, 6209.
[197] J. M. Aurrecoechea and A. Fernandez-Acebes, *Tetrahedron Lett.*, 1993, **34**, 549; T. Gillmann, *Tetrahedron Lett.*, 1993, **34**, 607; S. Arseniyadis, D. V. Yashunsky, M. Dorado, R. B. Alves, E. Toromanoff, L. Toupet, and P. Poitier, *Tetrahedron Lett.*, 1993, **34**, 4927.
[198] J. Cossy and A. Bouzide, *Tetrahedron Lett.*, 1993, **34**, 5583.
[199] J. Cossy and A. Bouzide, *J. Chem. Soc., Chem. Commun.*, 1993, 1218.
[200] J. T. Negri, T. Morwick, J. Doyon, P. D. Wilson, E. R. Hickey, and L. A. Paquette, *J. Am. Chem. Soc.*, 1993, **115**, 12 189.
[201] B. A. Anderson, J. Bao, T. A. Brandvold, C. A. Challener, W. D. Wulff, Y.-C. Wu Xu, and A. L. Rheingold, *J. Am. Chem. Soc.*, 1993, **115**, 10 671; M. Yamaguchi, M. Sehata, A. Hayashi, and M. Hirama, *J. Chem. Soc., Chem. Commun.*, 1993, 1708; B. M. Trost and Y. Shi, *J. Am. Chem. Soc.*, 1993, **115**, 9421; K. Takeda, M. Fujisawa, T. Makino, E. Yoshii, and K. Yamaguchi, *J. Am. Chem. Soc.*, 1993, **115**, 9351; T. J. Katz, A. M. Gilbert, M. E. Huttenloch, G. Min-Min, and H. H. Brintzinger, *Tetrahedron Lett.*, 1993, **34**, 3551; H. Pellissier, A. Tubul, and M. Santelli, *Tetrahedron Lett.*, 1993, **34**, 827; M. Toyota, Y. Nishikawa, K. Motoki, N. Yoshida, and K. Fukumoto *Tetrahedron Lett.*, 1993, **34**, 6099; X.-M. Wu, K. Funakoshi, and K. Sakai, *Tetrahedron Lett.*, 1993, **34**, 5927; H.-J. Knoller and R. Graf, *Tetrahedron Lett.*, 1993, **34**, 4765.
[202] H. Stadtmüller, R. Lentz, C. E. Tucker, T. Stüdemann, W. Dörner, and P. Knochel, *J. Am. Chem. Soc.*, 1993, **115**, 7027.
[203] C. Meyer, I. Marek, G. Courtemanche, and J.-F. Normant, *Tetrahedron Lett.*, 1993, **34**, 6053; H. Stadmüller, C. E. Tucker, A. Vaupel, and P. Knochel, *Tetrahedron Lett.*, 1993, **34**, 7911.
[204] H. Ito, Y. Motoki, T. Taguchi, and Y. Hanzawa, *J. Am. Chem. Soc.*, 1993, **115**, 8835.
[205] S. E. Berk, R. B. Grossman, and S. L. Buchwald, *J. Am. Chem. Soc.*, 1993, **115**, 4912.

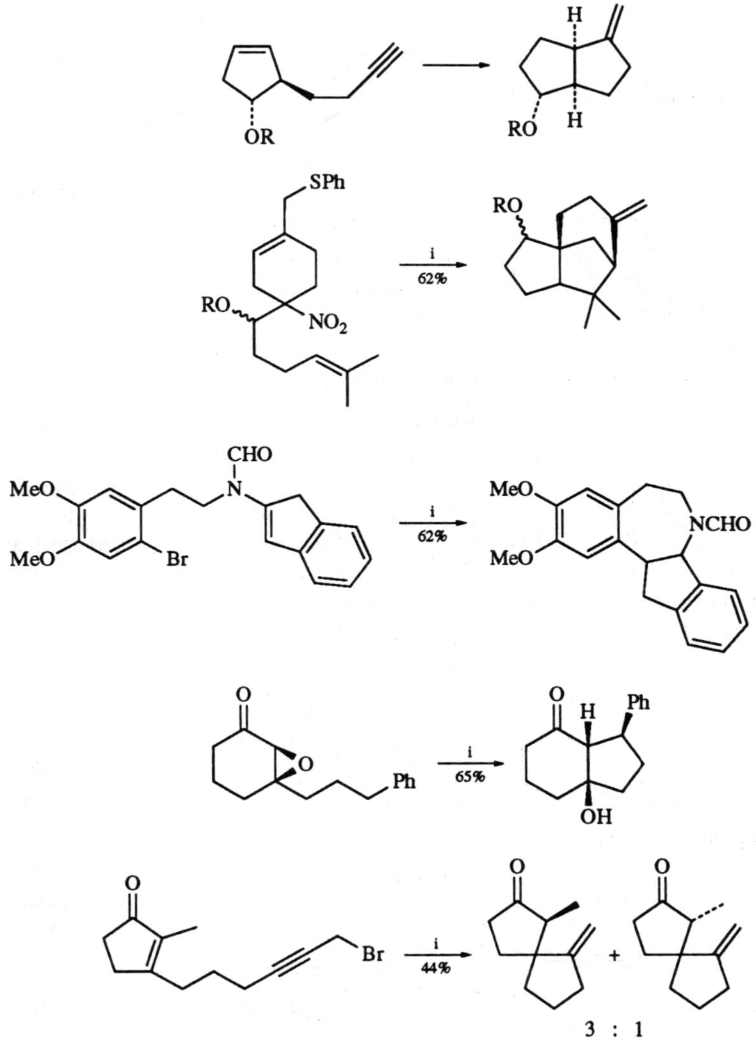

(ref 190)

(ref 192)

(ref 191)

(ref 193)

(ref 195)

Reagents: i, Bu_3SnH, AIBN

Scheme 87

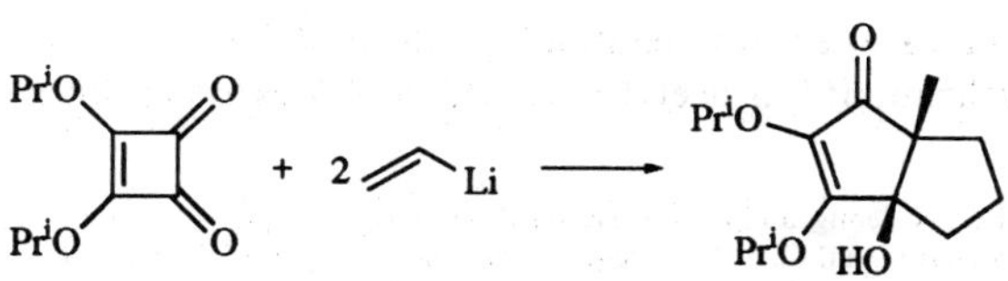

(ref 200)

Scheme 88

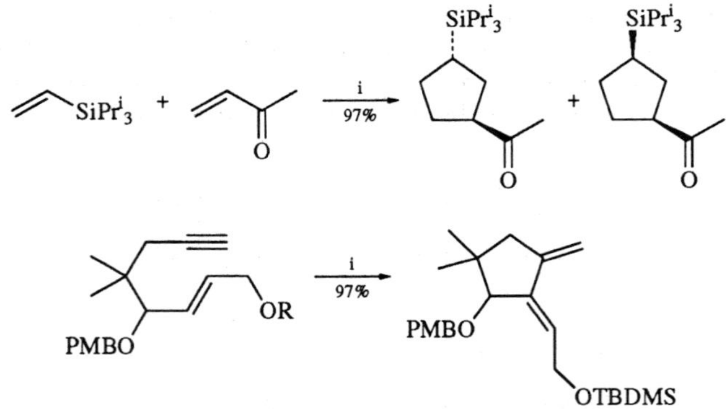

(ref 201)

Reagents: i, $TiCl_4$, $-78\,°C$; ii, $Pd_2(dba)_3$, $CHCl_3$, KOAc, BDEDA

Scheme 89

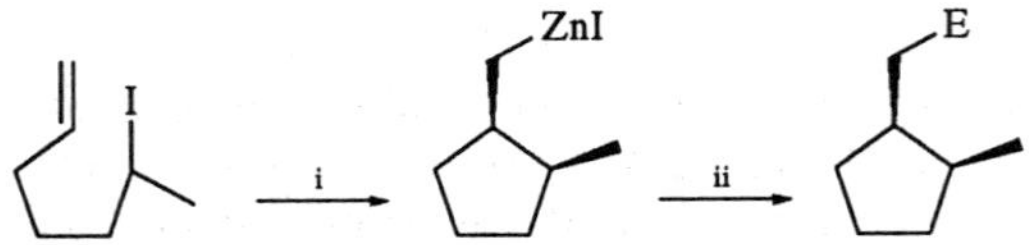

(ref 202 and 203)

Reagents: i, Zn, Et_2O; ii, E^+

Scheme 90

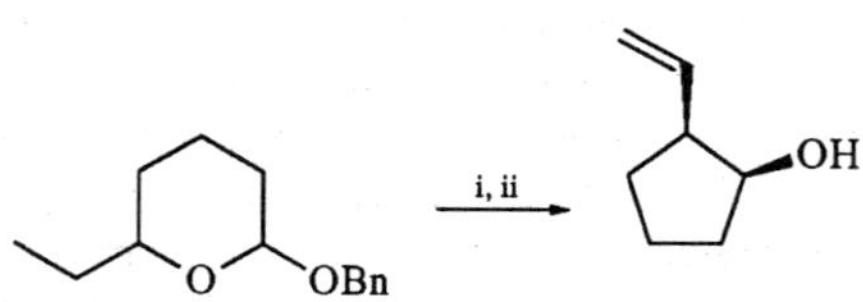

(ref 204)

Reagents: i, 'Cp_2Zr'; ii, $BF_3{\cdot}OEt_2$

Scheme 91

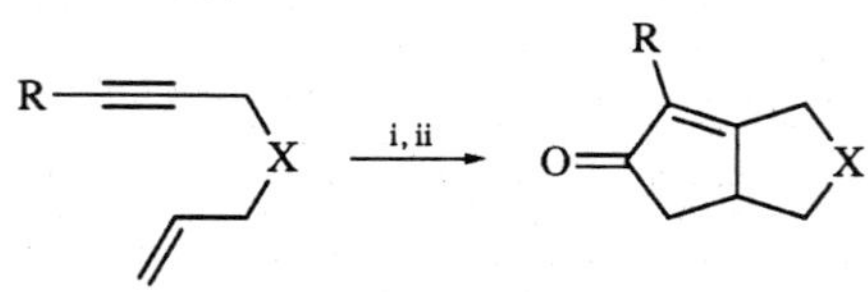

(ref 205)

Reagents: i, 'Cp_2Ti',R′NC; ii, H_3O^+

Scheme 92

reaction, which has also enjoyed continuing interest.[206]

5-*Endo*-dig-cyclization[207] of acetylenic ethers promises to provide a rapid entry into

[206] S.-e. Yoo, S.-H. Lee, N. Jeong, and I. Cho, *Tetrahedron Lett.*, 1993, **34**, 3435; M. E. Kraft, I. L. Scott, S. H. Romero, S. Feibelmann, and C. E. Van Pelt, *J. Am. Chem. Soc.*, 1993, **115**, 7199.

[207] R. L. Funk, G. L. Bolton, K. M. Brummond, K. E. Ellstead, and J. B. Stallman, *J. Am. Chem. Soc.*, 1993, **115**, 7023.

a variety of cyclopentane derivatives as does the Enders[208] diastereoselective intramolecular conjugate addition of SAMP–RAMP enolates to α,β-unsaturated esters. Finally,[209] thermal reaction of α,ω-bis-diazoketones affords a general synthetic route to both cyclohexenones and cyclopentenone derivatives (Scheme 93).

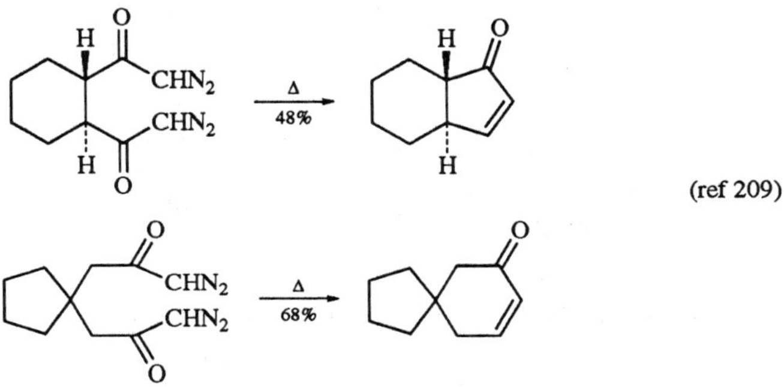

Scheme 93

Cyclohexanes.—The use of α-santonin in natural product synthesis has been reviewed.[210] The 65th birthday of the Diels–Alder reaction was celebrated in style with a review[211] on recent developments and the numerous examples of synthetic applications of existing and novel diene–dienophile partners (Figure 1).[212] The use of tandem intramolecular Diels–Alder reactions (TIMDA) is a potentially powerful strategy if the timing of sequential cycloadditions can be controlled. Synthetic applications of this process have appeared this year, which shows that this is achievable (Scheme 94)[213,214]

The development of effective catalysts for asymmetric intermolecular Diels–Alder reactions continues to be a major preoccupation[215,216] and some examples are illustrated in Scheme 95. Of particular interest are the reports of Brown[217] and Sanders;[218] the latter is concerned with the development of chemically engineered enzyme mimics which enhance both the rate of Diels–Alder reactions *and* promote

208 D. Enders, H. J. Scherer, and J. Runsink, *Chem. Ber.*, 1993, **126**, 1929.
209 S. Ohira, M. Moritani, T. Ida, and M. Yamato, *J. Chem. Soc., Chem. Commun.*, 1993, 1299; K. Nakatani, K. Takada, Y. Odagaki, and S. Isoe, *J. Chem. Soc., Chem. Commun.*, 1993, 556.
210 A. K. Banerjee, W. J. Vera, and N. C. Gonzalez, *Tetrahedron*, 1993, **49**, 4761.
211 K. Krohn, *Angew. Chem., Int. Ed. Engl.*, 1993, **32**, 1582.
212 (*a*) H. J. Reich, I. L. Reich, K. E. Yelm, J. E. Holladay, and D. Gschneider, *J. Am. Chem. Soc.*, 1993, **115**, 6625; (*b*) B. L. Williamson, P. J. Stang, and A. M. Arif, *J. Am. Chem. Soc.*, 1993, **115**, 2590; (*c*) M. E. Jung, C. N. Zimmerman, G. T. Lowen, and S. I. Khan, *Tetrahedron Lett.*, 1993, **34**, 4453; (*d*) M. Nakagawa, Y. Torisawa, T. Hosaka, K. Tanabe, T. Da-te, K. Okamura, and T. Hino, *Tetrahedron Lett.*, 1993, **34**, 4543; (*e*) V. H. Rawal, C. Michoud, and R. F. Monstel, *J. Am. Chem. Soc.*, 1993, **115**, 3030; (*f*) M. C. Clasby, D. Craig, and A. Marsh, *Angew. Chem., Int. Ed. Engl.*, 1993, **32**, 144; (*g*) Y.-F. Lu and A. G. Fallis, *Tetrahedron Lett.*, 1993, **34**, 3367; (*h*) D. A. Singleton, K. Kim, and J. P. Martinez, *Tetrahedron Lett.*, 1993, **34**, 3071; (*i*) J. Barluenga, F. Aznar, C. Valdes, A. Martin, S. Garcia-Granda, and E. Martin, *J. Am. Chem. Soc.*, 1993, **115**, 4403; (*j*) J.-P. Dulcère, V. Agati, and R. Faure, *J. Chem. Soc., Chem. Commun.*, 1993, 270; (*k*) J. E. Kerrigan, P. G. McDougal, and D. VanDerver, *Tetrahedron Lett.*, 1993, **34**, 8055; (*l*) J. Martynow, M. Dimitroff, and A. G. Fallis, *Tetrahedron Lett.*, 1993, **34**, 8201; (*m*) J. P. Konopelski and R. A. Kasar, *Tetrahedron Lett.*, 1993, **34**, 4587; (*n*) A. Kambuchi, N. Miyaura, and A. Suzuki, *Tetrahedron Lett.*, 1993, **34**, 4827.
213 I. E. Marko, P. Seres, G. R. Evans, and T. M. Swarbick, *Tetrahedron Lett.*, 1993, **34**, 7305.
214 D. R. Goldberg, J. A. Hansen, and R. J. Giguère, *Tetrahedron Lett.*, 1993, **34**, 8003.
215 E. J. Corey and Z. E. Wang, *Tetrahedron Lett.*, 1993, **34**, 4001.
216 K. Ishihara, O. Gao, and H. Yamamoto, *Tetrahedron Lett.*, 1993, **34**, 6917.
217 D. C. Braddock, J. M. Brown, and P. Guiry, *J. Chem. Soc., Chem. Commun.*, 1993, 1244.
218 C. J. Walter, H. L. Anderson, and J. K. Saunders, *J. Chem. Soc., Chem. Commun.*, 1993, 458.

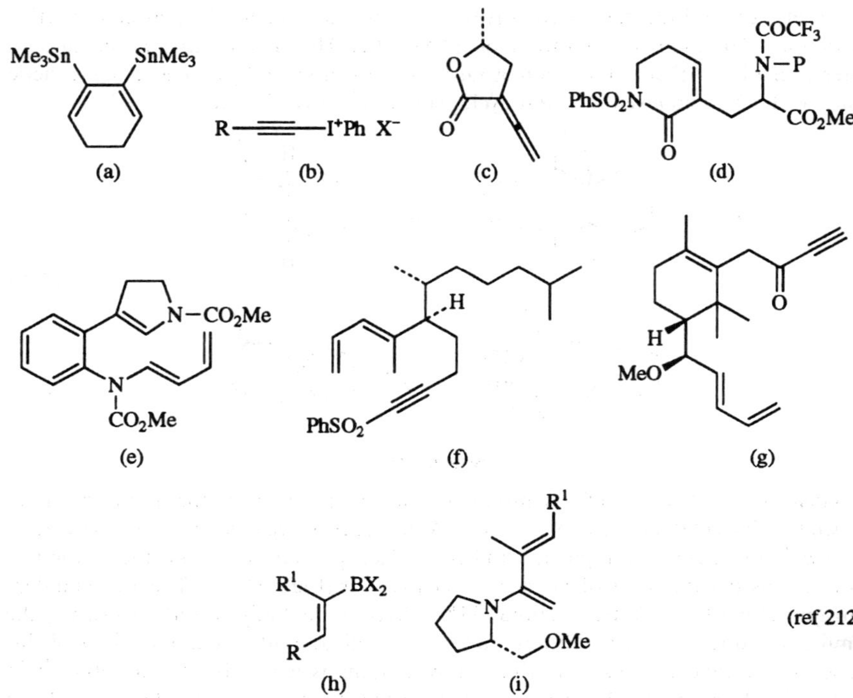

(ref 212)

Figure 1

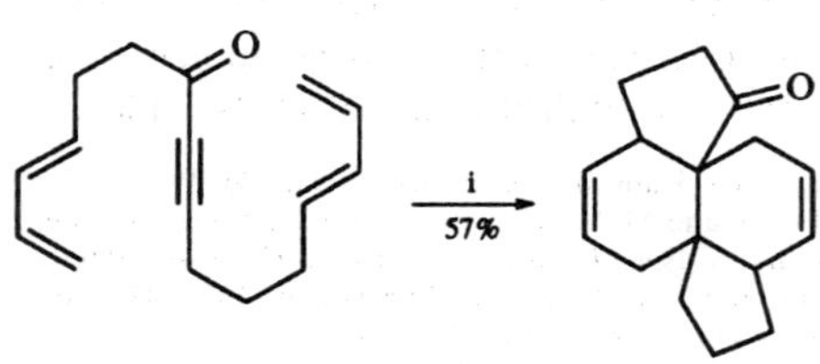

(ref 214)

Reagents: i, $BF_3{\cdot}OEt_2$, CH_2Cl_2, 0 °C

Scheme 94

exo–endo selectivity (Scheme 96). Photo-induced Diels–Alder reactions are also reported to be *exo* selective.[219]

Free radical reactions have been exploited to good effect in the synthesis of a range of cyclohexane-contining systems (Scheme 97),[220] as have a variety of tandem Michael-type reactions[221] (Scheme 98). Harman[222] has developed a novel dearomatization procedure *via* the intermediacy of osmium(II)(η^2-aryl) complexes (Scheme 99). This

[219] B. Pandey and P. V. Dalvi, *Angew. Chem., Int. Ed. Engl.*, 1993, **32**. 1612.
[220] D. L. J. Clive and M. H. D. Postema, *J. Chem. Soc., Chem. Commun.*, 1993, 429.
[221] C. Kuroda, Y. Ohnishi, and J. Y. Satoh, *Tetrahedron Lett.*, 1993, **34**, 2613.
[222] J. Gonzalez, M. Sabat, and W. D. Harman, *J. Am. Chem. Soc.*, 1993, **115**, 8857.

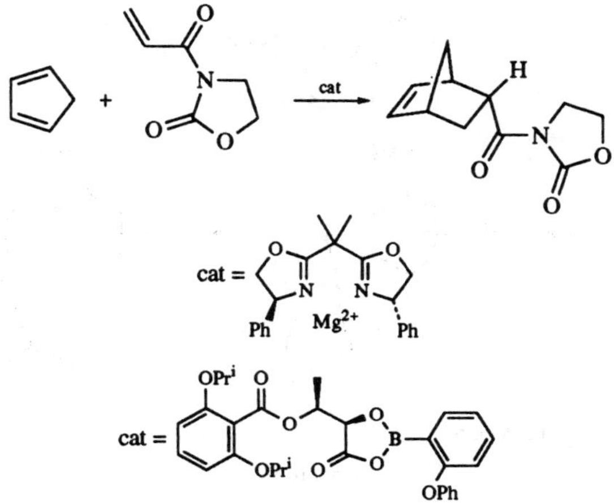

(ref 215)

(ref 216)

Scheme 95

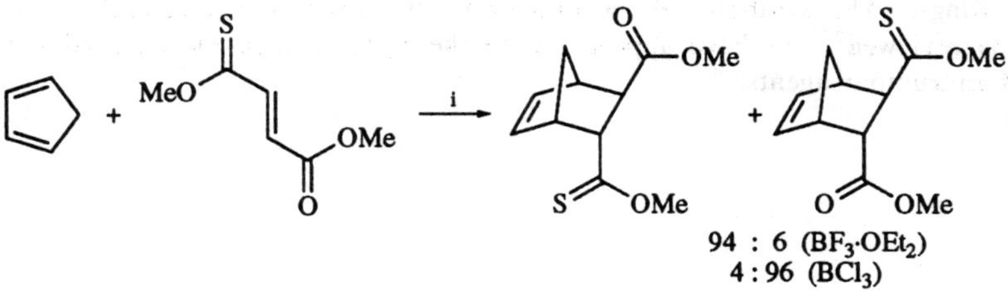

(ref 218)

Reagents: i, Lewis Acid, CH_2Cl_2

Scheme 96

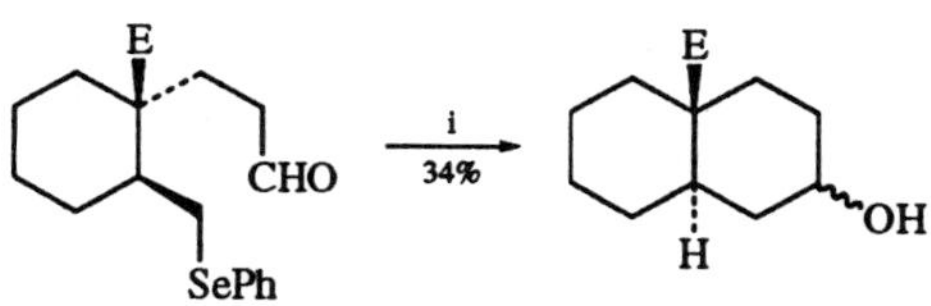

(ref 220)

Reagents: i, Ph_3SnH, BEt_3, 'air'

Scheme 97

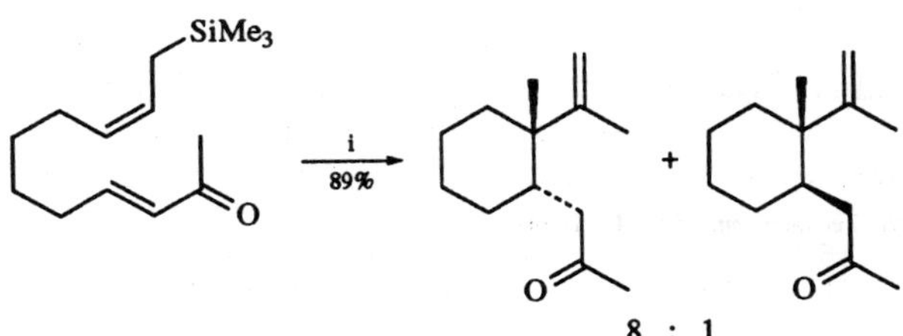

(ref 221)

Reagents: i, $TiCl_4$, CH_2Cl_2

Scheme 98

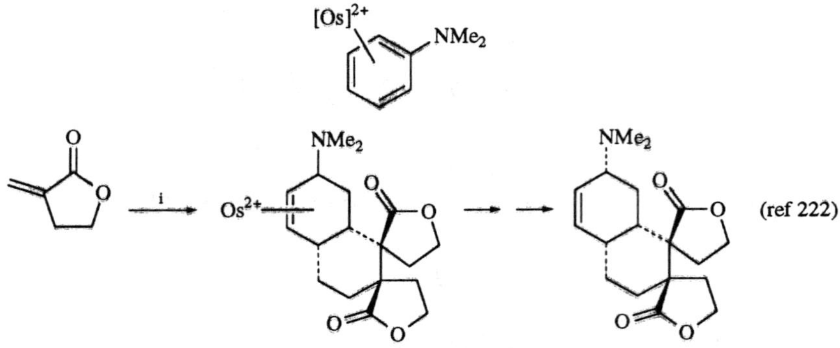

Reagents: i, TBSOTf, Pr^i_2NEt, MeCN

Scheme 99

methodology could have many synthetic applications; further developments are anticipated in this area.

Medium Rings.—The synthesis of medium-sized ring systems has recently been extensively reviewed[223] as have approaches to the highly functionalized enediyne family of antitumour agents.[19]

[223] C. J. Roxburgh, *Tetrahedron*, 1993, **49**, 10749.

6 Aromatic Compounds

By A. P. CHORLTON
ZENECA Specialties, Hexagon House, Blackley, Manchester, M9 3DA, UK

1 General and Theoretical Studies

[1,2-$^{13}C_2$]Benzene automerizes at high temperature to give a 1,2:1,3:1,4 product ratio of 72:24:4. Four plausible mechanisms to explain this have been put forward: the 1,2-carbon shift mechanism; the dyotropic shift mechanism; the benzvalene mechanism; and the 1,2-hydrogen shift mechanism. High-level *ab initio* calculations have been reported that favour the 1,2-hydrogen shift process.[1] The photochemistry of S_1 and S_2 benzene involves the production of benzvalene and Dewar benzene respectively. An MC-SCF study of these processes indicates that the products of benzene photochemistry appear to be controlled by the topology of the S_0/S_1 crossing surface.[2] The X-ray crystal structure and photoelectron spectrum of 3,3′-bicyclopropenyl (1), one of the four valence isomers of benzene, has been determined. The photoelectron spectrum of (1) can be interpreted in terms of about a 2:1 mixture of *anti* and *gauche* conformers, respectively, in the gas phase whereas the X-ray crystal structure of (1) is only present in the *anti* form.[3]

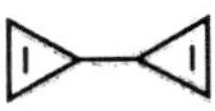
(1)

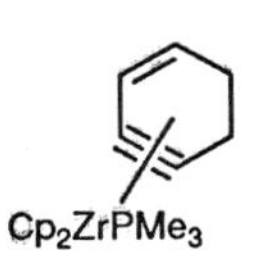

(3)

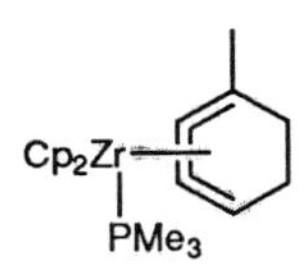

(2)

Zirconocene complexes of the highly strained isomers of benzene, 1,2,3-cyclohexatriene (2) and cyclohexen-3-yne (3), have been synthesized.[4] A new multiphoton fragmentation of benzene at 193 nm has been proposed to proceed *via* the intermediate triplet states and neutral fragments such as C_3H_3.[5] Squires and coworkers have reported values for the heats of formation of *o*-, *m*-, and *p*-benzynes, based on energy-resolved, collision-induced dissociation measurements. The CI calculations carried out independently by the groups of Squires and Borden gave good agreement

[1] K. M. Merz and L. T. Scott, *J. Chem. Soc., Chem. Commun.*, 1993, 412.
[2] I. J. Palmer, I. N. Ragazos, F. Bernardi, M. Olivucci, and M. A. Robb, *J. Am. Chem. Soc.*, 1993, **115**, 673.
[3] R. Boese, D. Blaser, R. Gleiter, K. H. Pfeifer, W. E. Billups, and M. M. Haley, *J. Am. Chem. Soc.*, 1993, **115**, 743.
[4] J. Yin, K. A. Abboud, and W. M. Jones, *J. Am. Chem. Soc.*, 1993, **115**, 8859.
[5] Y. Mori and T. Kitagawa, *Bull. Chem. Soc. Jpn.*, 1993, **66**, 1043.

with the experimental results for the heats of formation of *o*-benzyne, but find that the experimental results for *m*- and *p*-benzyne are too low.[6,7] Photolysis of benzocyclobutenedione (4) gives *o*-benzyne *via* the intermediacy of the bisketene (5). Examination of this process in an argon matrix shows that (5) and (4) can interconvert thermally or photochemically several times before significant photochemical decarbonylation to *o*-benzyne occurs.[8]

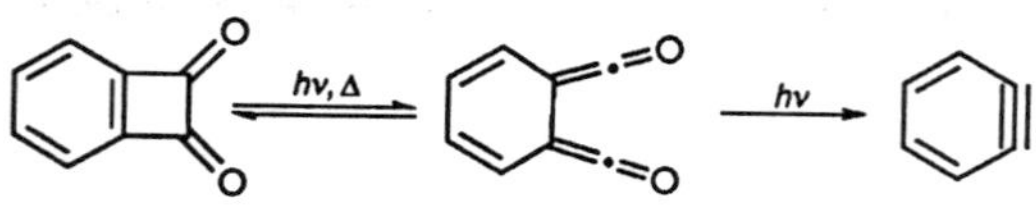

(4) (5)

Evidence has been presented for the presence of a halide–benzyne complex in the gas phase.[9] The generation of arynes in the gas phase has been reviewed.[10]

The role of delocalization in benzene has been the subject of debate; *ab initio* SCF calculations have shown that it is misleading to judge the nature of delocalization based on a σ–π partition. It was therefore concluded that delocalization effects act to stabilize strongly symmetric benzene in accord with concepts of classical resonance theory.[11] Classical valence bond theory has also been applied using quantitative resonance theory (QRT) to access the π-electron transfer in substituted benzenes. These QRT calculations are in agreement with those obtained *via ab initio* MO theory.[12] Messmer has concluded, based on energetic considerations, that the phenomenon of resonance in the classical valence bond sense is best described by a bent-bond model.[13] The results of these calculations indicate that the benzenoid isomer (6) is preferred over the quinoid form (7).[14]

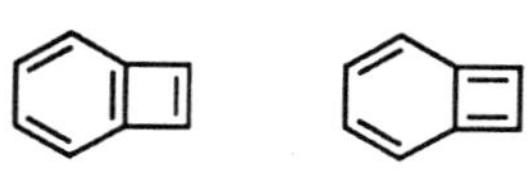

(6) (7)

The structure of 1,3,5-triamino-2,4,6-trinitrobenzene (8) and a number of related compounds have been studied using *ab initio* Hartree–Fock and local density functional computational methods. These calculations predict a planar structure in accord with an *X*-ray crystallographic study; this is a contradiction to semiempirical data, which predict (8) to be in the boat conformation. The planarity of (8) is thought to be due to hydrogen bonding of the flanking groups.[15] Hall has argued that when a benzene ring is

6 S. U. Wierschke, J. J. Nash, and R. R. Squires, *J. Am. Chem. Soc.*, 1993, **115**, 11 960.
7 A. Nicolaides and W. T. Borden, *J. Am. Chem. Soc.*, 1993, **115**, 11 951.
8 T. Mosandl and C. Wentrup, *J. Org. Chem.*, 1993, **58**, 747.
9 H. V. Linnert and J. M. Riveros, *J. Chem. Soc., Chem. Commun.*, 1993, 48.
10 R. F. C. Brown and F. W. Eastwood, *SYNLETT*, 1993, 9.
11 E. D. Glending, R. Faust, A. Streitwieser, K. P. C. Vollhardt, and F. Weinhold, *J. Am. Chem. Soc.*, 1993, **115**, 10 952.
12 Y. Pipeng, *J. Chem. Res. (S)*, 1993, 368.
13 P. A. Schultz and R. P. Messmer, *J. Am. Chem. Soc.*, 1993, **115**, 10 943.
14 J. M. Schulman and R. L. Disch, *J. Am. Chem. Soc.*, 1993, **115**, 11 153.
15 K. K. Baldridge and J. S. Siegel, *J. Am. Chem. Soc.*, 1993, **115**, 10 782.

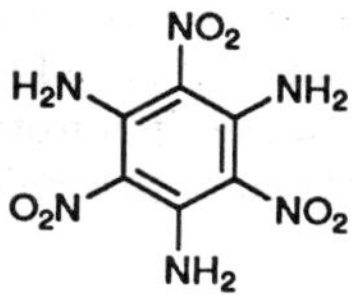

(8)

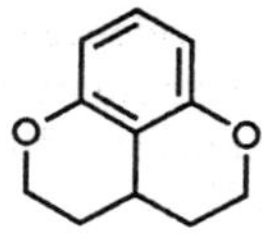

(9)

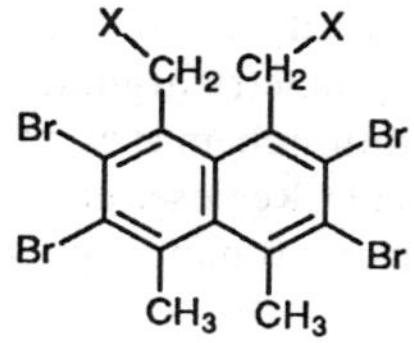

(10), X = H; Y = H.
(11), X = H; Y = Br.
(12), X = Br; Y = Br.

present as an embedded component in a molecule, *e.g.* (9), then the adoption by the benzene ring of the boat form requires the least energy.[16]

X-Ray crystallographic and dynamic ^{1}H NMR spectroscopy of 2,3,6,7-tetrabromo-1,4,5,8-tetramethylnaphthalene (10) and its bromo derivatives (11) and (12) show that these compounds have severely distorted naphthalene nuclei in the solid state. In solution this corresponds to conformations that can have an appreciable barrier to conversion ($\Delta G^{+} \simeq 16\,\text{Kcal mol}^{-1}$), which occurs by both rotation and flipping of the *peri*-substituents.[17] The rotation of the *peri* positions in 1-t-butyl-8-fluoronaphthalene has also been studied by dynamic NMR. The barrier to rotation of the t-butyl group has been calculated[18] at $\Delta G^{+} \simeq 8.6\,\text{Kcal mol}^{-1}$. In a similar study the barrier for the *E–Z* interconversion of atropisomers of naphthyl-1-sulfones having a methyl group at the 2-position has been measured and found to cover the range 10.6–18.4 Kcal mol^{-1}. Low temperature chiral resolution allowed the separation of the two enantiomeric forms and their associated atropisomers.[19] The restricted rotation in 1,8-diarylnaphthalenes (13) has been used to probe the interactions between π-stacked dimers of benzenes. The attractive or repulsive forces between the benzenes should give rise to a higher or lower energy barrier to rotation respectively. The barrier to rotation was found to increase linearly with the increasing electron-withdrawing capacity of the X and Y groups. No extra increase in the rotation barrier was noted when X = OMe and Y = NO_2. These observations point to the domination of polar/π over charge-transfer effects in the interaction of stacked benzene rings.[20]

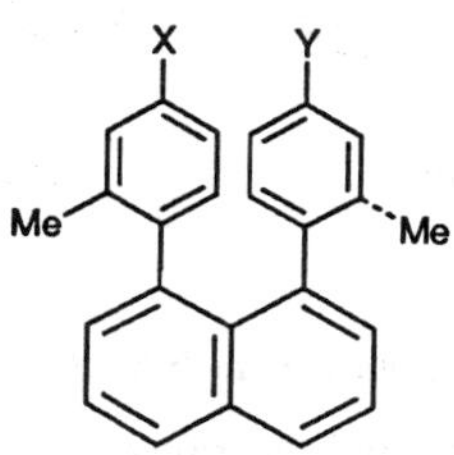

(13)

[16] G. G. Hall, *J. Chem. Soc., Perkin Trans. 2*, 1993, 1491.

[17] P. R. Ashton, G. R. Brown, A. J. Foubister, D. R. Smith, N. Spencer, J. F. Stoddart, and D. J. Williams, *Tetrahedron Lett.*, 1993, **34**, 8333.

[18] G. W. Gribble and E. R. Olson, *J. Org. Chem.*, 1993, **58**, 1631.

[19] D. Casarini, E. Forsti, F. Gasparrini, L. Lunazzi, D. Macciantelli, D. Misiti, and C. Villani, *J. Org. Chem.*, 1993, **58**, 5674.

[20] F. Cozzi, M. Cinquini, R. Annunziata, and J. S. Siegel, *J. Am. Chem. Soc.*, 1993, **115**, 5330.

The aromaticity, or lack of it, in C_{60} is a contentious issue. Fowler has put forward qualitative symmetry arguments, with evidence from bond lengths and energies and from the addition chemistry of C_{60}, which point to a localized description of the π system in this molecule. Regioselective saturation of six double bonds in C_{60} is predicted to give the 48 electron aromatic system (14).[21]

(14)

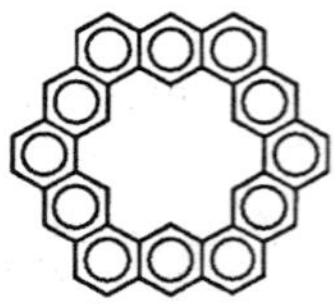

(15)

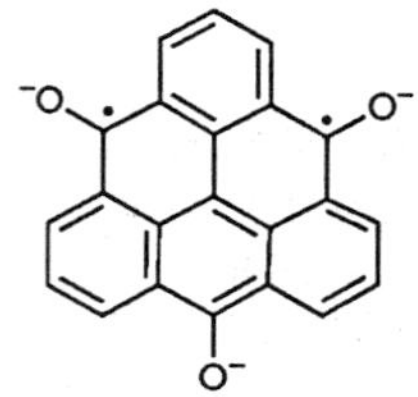

(16)

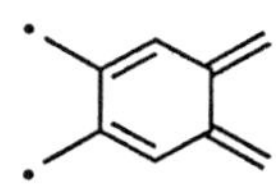

(17)

It has been postulated that large polycyclic conjugated rings such as kekulene (15) with an internal cavity may exhibit superaromaticity. A general graph theory of superaromaticity has been developed to analyse this phenomenon. The results of this work indicate that these systems are essentially non-superaromatic.[22]

The ESR spectrum of a non-Kekulé polynuclear aromatic has been reported for the first time; trioxytriannulene (16) is a remarkably stable triplet biradical.[23] Full characterization of the disjointed non-Kekulé singlet hydrocarbon 1,2,4,5-tetramethylenebenzene biradical (17) has been completed.[24]

Contrary to previous results it has been established by theoretical methods that 1,4-bishomotropylium cation (18) rather than the *cis*-8,9-dihydro-1-indenyl cation (19) corresponds to the minimum energy structure. These calculations, however, are not sensitive enough to confirm the homoaromatic character of (18).[25] Homoaromaticity has been tentatively ascribed to the stabilization shown by arene oxides on their acid-catalysed aromatization.[26]

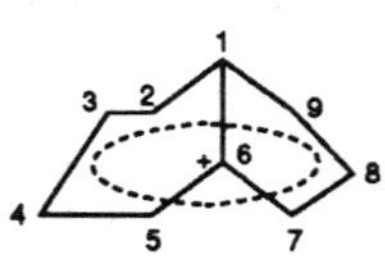

(18)

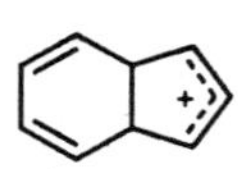

(19)

2 Preparation of Benzenes from Non-aromatic Precursors

The Bergman cyclization, the thermal rearrangement of conjugated *cis*-enediynes to

[21] P. W. Fowler, D. J. Collins, and S. J. Austin, *J. Chem. Soc., Perkin Trans. 2*, 1993, 275.
[22] J. Aihara, *Bull. Chem. Soc. Jpn.*, 1993, **66**, 57.
[23] G. A. Allinson, R. J. Bushby, and J. L. Paillaud, *J. Am. Chem. Soc.*, 1993, **115**, 2062.
[24] J. H. Reynolds, J. A. Berson, K. K. Kumashiro, J. C. Duchamp, K. W. Zilm, J. C. Scaiano, A. B. Berinstain, A. Rubello, and P. Vogel, *J. Am. Chem. Soc.*, 1993, **115**, 8073.
[25] D. Cremer, P. Svensson, E. Kraka, Z. Konkoli, and P. Ahlberg, *J. Am. Chem. Soc.*, 1993, **115**, 7457.
[26] S. Nagaraja Rao, R. A. M. O'Ferrall, S. C. Kelly, D. R. Boyd, and R. Agarwal, *J. Am. Chem. Soc.*, 1993, **115**, 5458.

1,4-benzenoid diradicals, continues to be of immense interest.[27] The incorporation of the *cis*-enediyne in the synthesis of naturally occurring antitumour antibiotics represents the major challenge. This has been achieved by Nicolaou, who has reported the first synthesis of calicheamicin.[28,29] Successful model studies towards the synthesis of dynemicin have also been completed.[30,31] Magriotis has developed a tandem Claisen–Bergman strategy for the stereocontrolled synthesis of tetrahydronaphthalenes (Scheme 1).[32]

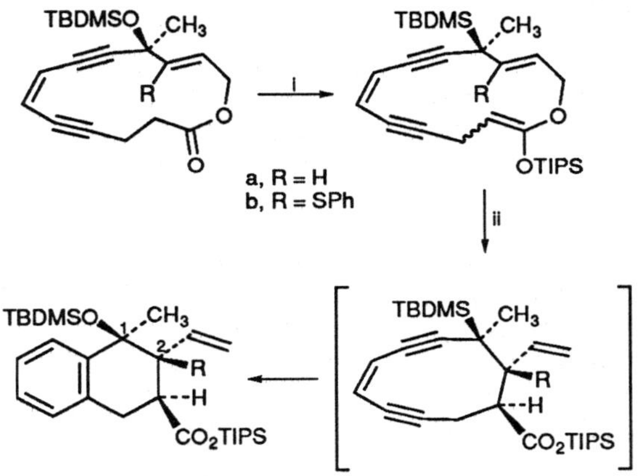

Reagents: i, LHMDS, THF, −78 °C, TIPSOTf

Scheme 1

Grissom has demonstrated that enediynes (20), possessing tethered olefin radical acceptors, can participate in tandem enediyne-radical cyclizations to yield dihydrobenzindenes (21) (Scheme 2).[33–35]

Meyers has shown that 1,6-didehydro[10]annulene (22) undergoes a rapid cyclization to form the biradical 1,5-dehydronaphthalene (23).[36] In a similar study Toshima has cyclized the enyne-allene sulfides (24) and sulfones (25).[37,38]

[27] R. Gleiter and d. Katz, *Angew. Chem., Int. Ed. Engl.*, 1993, **32**, 842.
[28] K. C. Nicolaou, C. W. Hummel, M. Nakada, K. Shibayama, E. N. Pitsinos, H. Saimoto, Y. Mizuno, K. U. Baldenius, and A. L. Smith, *J. Am. Chem. Soc.*, 1993, **115**, 7625.
[29] K. C. Nicolaou, *Angew. Chem., Int. Ed. Engl.*, 1993, **32**, 1377.
[30] K. C. Nicolaou, W. M. Dai, Y. P. Hong, S. C. Tsay, K. K. Baldridge, and J. S. Siegal, *J. Am. Chem. Soc.*, 1993, **115**, 7944.
[31] P. A. Wender, C. K. Zercher, S. Beckham, and E. M. Haubold, *J. Org. Chem.*, 1993, **58**, 5867.
[32] P. A. Magriotis and K. D. Kim, *J. Am. Chem. Soc.*, 1993, **115**, 2972.
[33] J. W. Grissom and T. L. Calkins, *J. Org. Chem.*, 1993, **58**, 5422.
[34] J. W. Grissom, T. L. Calkins, and H. A. McMillen, *J. Org. Chem.*, 1993, **58**, 6556.
[35] J. W. Grissom, T. L. Calkins, and M. Egan, *J. Am. Chem. Soc.*, 1993, **115**, 11 744.
[36] A. G. Myers and P. S. Dragovich, *J. Am. Chem. Soc.*, 1993, **115**, 115, 7021.
[37] K. Toshima, K. Ohta, A. Ohashi, T. Nakamura, M. Nakata, and S. Matsumura, *J. Chem. Soc., Chem. Commun.*, 1993, 1525.
[38] K. Toshima, K. Ohta, A. Ohtsuka, S. Matsumura, and M. Nakata, *J. Chem. Soc., Chem. Commun.*, 1993, 1406.

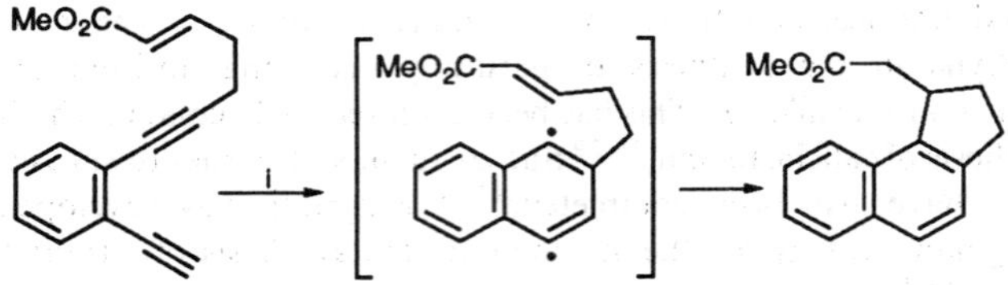

(20) (21)

Reagents: i, PhCl, cyclohexa-1,4-diene, 190° C, 2.5 hr (96%)

Scheme 2

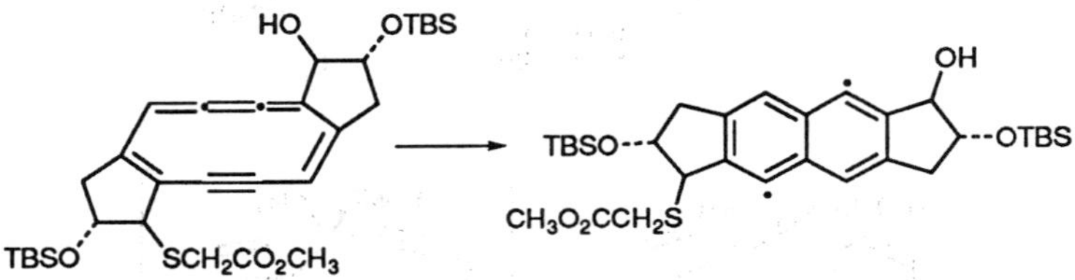

(22) (23)

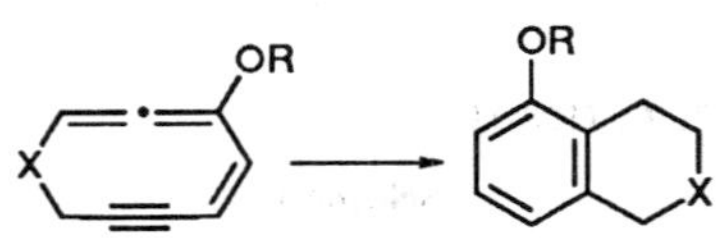

(24) X = S
(25) X = SO_2

This approach has also been utilized by a number of workers to investigate the enyne[3]cumulene cyclization of neocarzinostatin (26) (Scheme 3).[39–42]

Padwa in an investigation of the decomposition of *o*-alkynyl substituted aryl diazo ketones (27) has proposed that the intermediate aryl ketene (28) cyclizes in a Meyers-type process to give the diradical (29), which on quenching gives the *β*-naphthol (30).[43]

The synthesis of benzene rings *via* Fischer carbene complexes continues to be widely utilized. Wulff and coworkers have carried out an in-depth study into this reaction.[44] The results of these experiments suggest that phenol (31) arises from a vinyl carbene complexed intermediate (32) and that phenol (33) and cyclopentenedione (34) arise

[39] S. W. Scheuplein, R. Machinek, J. Suffert, and R. Bruckner, *Tetrahedron Lett.*, 1993, **34**, 6549.
[40] T. Takahashi, H. Tanaka, Y. Hira, T. Doi, H. Yamada, T. Shiraki, and Y. Sugiura, *Angew. Chem., Int. Ed. Engl.*, 1993, **32**, 1657.
[41] M. Tokuda, K. Fujiwara, T. Gomibuchi, and M. Hirama, *Tetrahedron Lett.*, 1993, **34**, 669.
[42] K.-I. Yoshida, Y. Minami, R. Azuma, M. Sacki, and T. Otani, *Tetrahedron Lett.*, 1993, **34**, 2637.
[43] A. Padwa, U. Chiacchio, D. J. Fairfax, J. M. Kassir, A. Litrico, M. A. Semones, and S. L. Xu, *J. Org. Chem.*, 1993, **58**, 6429.
[44] B. A. Anderson, J. Bao, T. A. Brandvold, C. A. Challener, W. D. Wulff, Y.-C. Xu, and A. L. Rheingold, *J. Am. Chem. Soc.*, 1993, **115**, 10 671.

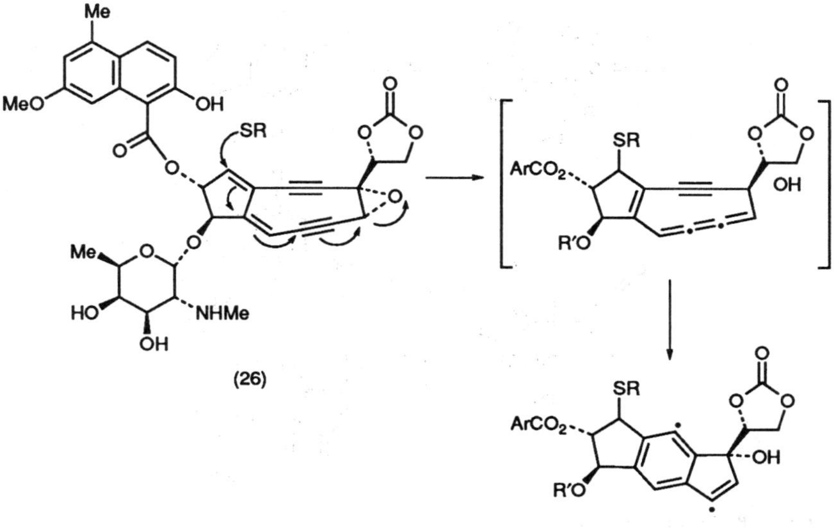

Scheme 3

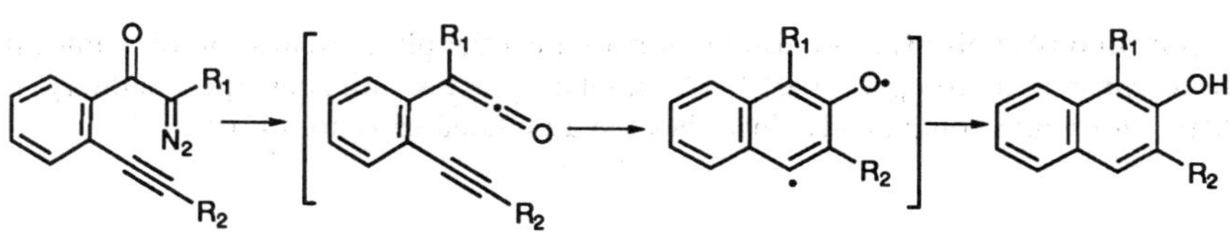

from a vinyl ketone complexed intermediate (35). The product distribution of these products is discussed with relation to solvent, concentration, and nature of metal (Scheme 4). In another study Thomas and coworkers have synthesized metal complexed vinyl ketenes of the type (35), and have shown that their reaction with alkynes leads to phenols (33) if the R group is electron donating, whereas if this substituent is electron withdrawing, cyclopentenediones (34) are formed.[45,46]

Differentially protected hydroquinone–(tricarbonyl)chromium complexes (36) can be prepared directly from alkenyl and alkynyl Fischer carbene complexes by interception with a variety of electrophiles.[47] The first example of benzannulation of alkynes with iron(0) carbene complexes has been reported. Furan annulation is a competing process in this reaction.[48] The reaction of carbene–chromium complex (37) with isocyanides provides a mild and regiospecific access to aromatic amines (38).[49]

[45] K. G. Morris, S. P. Saberi, M. Salter, S. E. Thomas, M. F. Ward, A. M. Z. Slawin, and D. J. Williams *Tetrahedron*, 1993, **49**, 5617.
[46] K. G. Morris, S. P. Saberi, and S. E. Thomas, *J. Chem. Soc., Chem. Commun.*, 1993, 209.
[47] S. Chamberlain, W. D. Wulff, and B. Bax, *Tetrahedron*, 1993, **49**, 5531.
[48] Atta-ur-Rahmann, W. F. K. Schnatter, and N. Manolache, *J. Am. Chem. Soc.*, 1993, **115**, 9848.
[49] R. Aumann, *Chem. Ber.*, 1993, **126**, 1867.

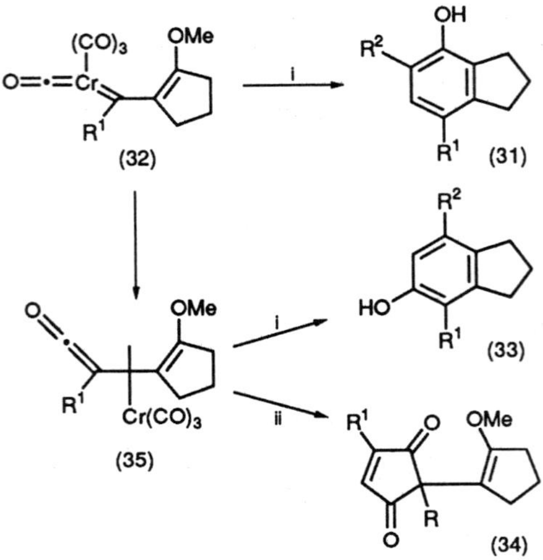

Reagents: i, $R^2C{\equiv}CH$; ii, $R^2C{\equiv}CH$, CO

Scheme 4

o-Amino phenols (39) can also be prepared *via* the photocyclization of aminocarbene–chromium complex (40).[50] A similar process utilizing the methoxycarbene–chromium complex (41) furnishes the annulated catechol (42).[51]

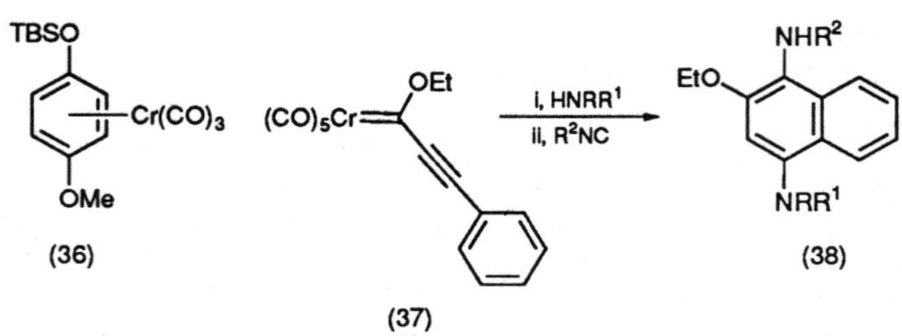

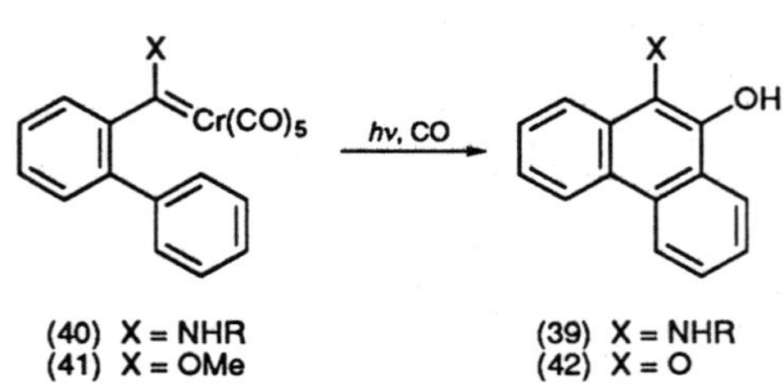

[50] C. A. Merlic, D. Xu, and B. G. Gladstone, *J. Org. Chem.*, 1993, **58**, 538.
[51] C. A. Merlic and W. M. Roberts, *Tetrahedron Lett.*, 1993, **34**, 7379.

Catechols have also been synthesized by the ruthenium-catalysed reaction of 1,6-diynes with carbon monoxide. This transformation is unique in that it represents a rare example of catalytic incorporation of carbon monoxide into a diyne. Also it exhibits a new mode of successive incorporation of two molecules of carbon monoxide (Scheme 5).[52]

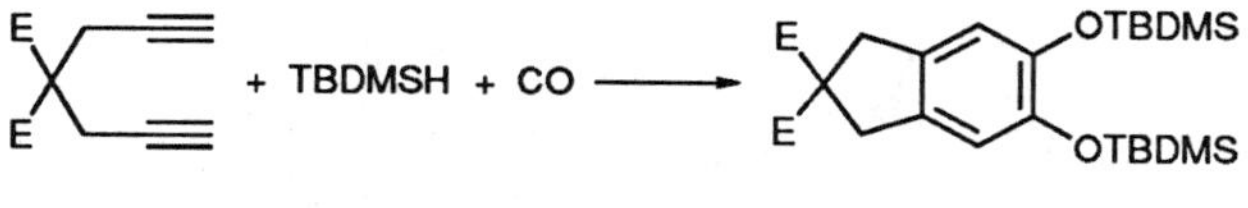

$E = CO_2Et$

Reagents: $Ru(CO)_{12}/PCy_3$, 50 atm

Scheme 5

An alternative regiocontrolled synthesis of substituted catechols has been developed: the products from conjugate addition of vinyl copper reagents to cyclobutenediones, when subjected to thermal rearrangement, yield catechols (Scheme

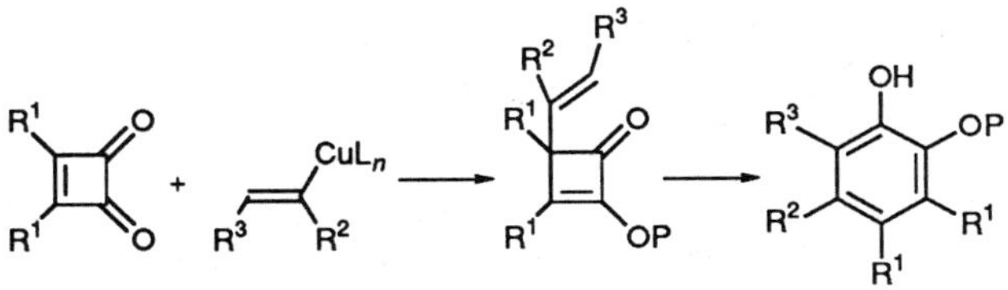

Scheme 6

6).[53] In a similar process, addition of stannylquinones to 4-chlorocyclobutenones followed by thermal rearrangement gives substituted naphthoquinones and anthraquinones.[54]

Liebeskind has developed a general regioselective synthesis of substituted benzocyclobutenedione monoacetals.[55] These reagents can function as important intermediates in the regioselective synthesis of phenols. 3-Alkylidene-4-allenylcyclobutenes undergo thermal ring expansion to give *o*-quinodimethanes, which upon electrocyclic ring closure give benzocyclobutenes (Scheme 7).[56] *o*-Quinodimethanes are of synthetic utility as effective dienes in Diels–Alder reactions for the construction of polycyclic aromatics. A number of novel methods have been developed for the generation of *o*-quinodimethanes (Scheme 7).[57–59]

Benzyne and benzoquinone have been used effectively as 2π components in Diels–Alder reactions to prepare 1,4-naphthoquinones.[60,61] Negishi has developed a

52 N. Chatani, Y. Fukumoto T. Ida, and S. Murai, *J. Am. Chem. Soc.*, 1993, **115**, 11 614.
53 A. Gurski and L. S. Liebskind, *J. Am. Chem. Soc.*, 1993, **115**, 6101.
54 J. P. Edwards, D. J. Krysan, and L. S. Liebeskind, *J. Am. Chem. Soc.*, 1993, **115**, 9868.
55 J. P. Edwards, D. J. Krysan, and L. S. Liebeskind, *J. Org. Chem.*, 1993, **58**, 3942.
56 J. E. Ezcurra and H. W. Moore, *Tetrahedron Lett.*, 1993, **34**, 6177.
57 S. H. Woo, *Tetrahedron Lett.*, 1993, **34**, 7587.
58 H. Fujihara, M. Yabe, and N. Furukawa, *J. Org. Chem.*, 1993, **58**, 5291.
59 G. Kanai, N. Miyaura, and A. Suzuki, *Chem. Lett.*, 1993, 845.

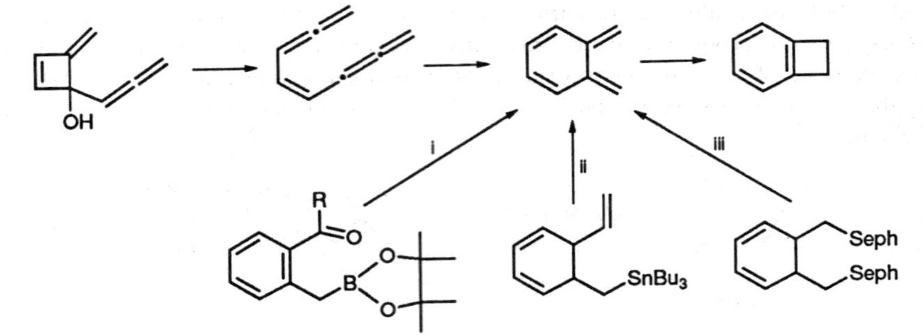

Reagents: i, Δ or *hν*; ii, E^+; iii, *hν*

Scheme 7

selective synthesis of benzene derivatives *via* palladium-catalysed carbometallation of alkynes (Scheme 8).[62]

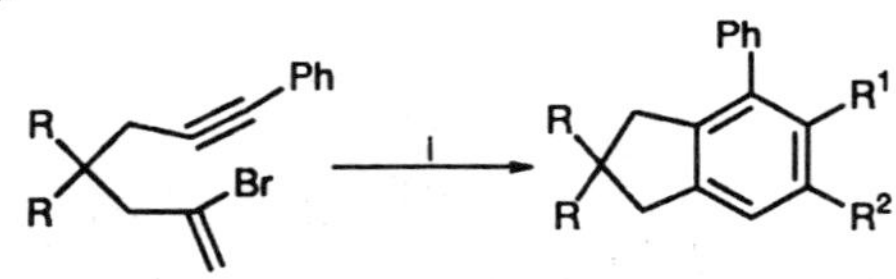

Reagents: i, $R^1C{\equiv}CR^2$, 5% PdL_n, NEt_3, DMF

Scheme 8

The intramolecular Michael–Aldol process has been used in the synthesis of the hexasubstituted aromatic ring present in a key intermediate for the synthesis of pseudopterosin A.[63] Naphthopyrans have been synthesized from a 14-membered polyketide lactone by the intramolecular aldol process.[64] Salicylates have been prepared by thermolytic ring closure of prop-1-en-1-yl substituted malonates (Scheme 9).[65]

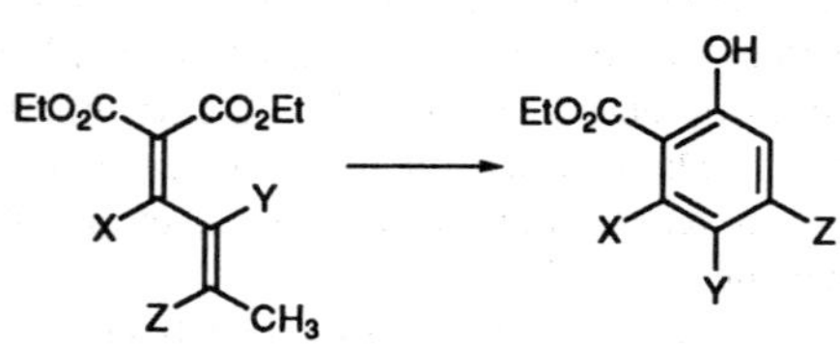

Scheme 9

[60] M. A. Brimble and S. J. Phythia, *Tetrahedron Lett.*, 1993, **34**, 5813.
[61] V. I. Hugo, P. W. Snijman, and I. R. Green, *Synth. Commun.*, 1993, **23**, 577.
[62] E. Negishi, M. Ay, and T. Sugihara, *Tetrahedron*, 1993, **49**, 5471.
[63] M. E. Jung and C. S. Siedem, *J. Am. Chem. Soc.*, 1993, **115**, 3822.
[64] K. Tatsuta, *Tetrahedron Lett.*, 1993, **34**, 4961.
[65] O. E. O. Hormi and L. Hirvela, *Tetrahedron Lett.*, 1993, **34**, 6463.

Frost has made progress in the identification and removal of impediments to the biocatalytic synthesis of aromatics from glucose.[66]

3 Non-aromatic Compounds from Benzene Precursors

Intramolecular anodic oxidation of phenols (43) leads preferentially to spirodienones (44); a competing reaction gives the methanol adduct (45). Swenton has examined the process and has shown that the product ratio is dramatically affected by the olefin structure.[67] A versatile route to quinone *N*-acylimine ketals (46) from *p*-methoxyanilides has been developed utilizing anodic oxidative technology.[68,69] Anodic oxidation of 3,5-dihalogenotyrosines (47) gives a cavernicolin model compound (48). All four possible stereoisomers were formed but this was due to the special characteristics of (48) rather than the methodology.[70] Similar intramolecular oxidative processes using hypervalent iodine have been effected to obtain the key intermediate (49) in the total and partial synthesis of aranorosin (50); the model compound (51), an intermediate towards the synthesis of marine metabolites such as aerothionin, has been synthesized by these methods also.[71–73] A study of phenolic oxidation using phenyliodonium diacetate has also been carried out.[74]

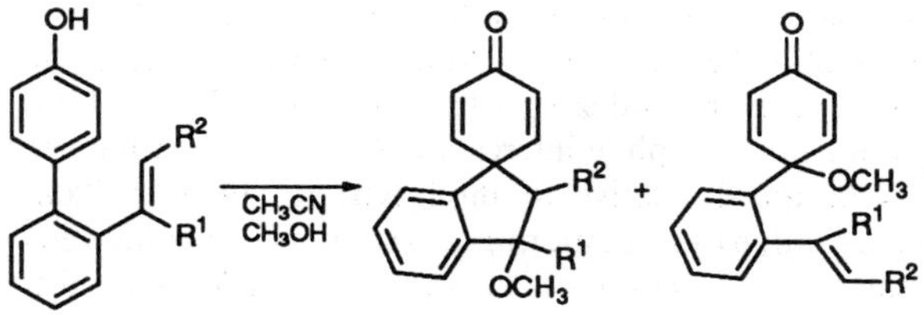

(43) (44) (45)

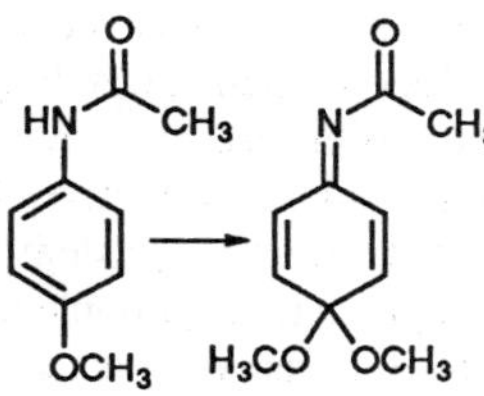

(46)

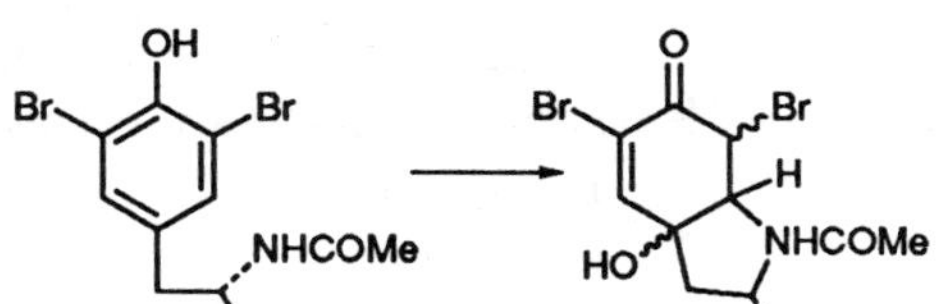

(47) (48)

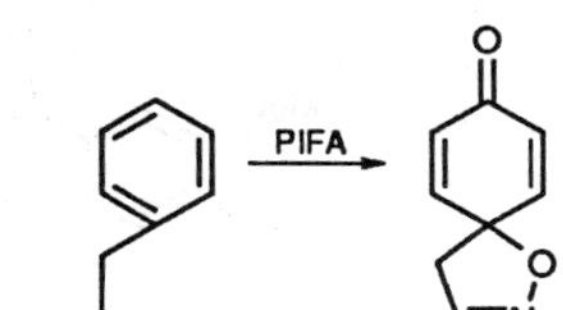

(51)

[66] K. A. Dell and J. W. Frost, *J. Am. Chem. Soc.*, 1993, **115**, 11 581.
[67] J. S. Swenton, K. Carpenter, Y. Chen, M. L. Kerns, and G. W. Morrow, *J. Org. Chem.*, 1993, **58**, 3308.
[68] J. S. Swenton, T. N. Biggs, and W. M. Clark, *J. Org. Chem.*, 1993, **58**, 5607.
[69] M. Novak, J. S. Helmick, N. Oberlies, K. S. Rangappa, W. M. Clark, and J. S. Swenton, *J. Org. Chem.*, 1993, **58**, 867.
[70] M. Cavazza, G. Guella, L. Nucci, F. Pergola, N. Bicchierini, and F. Pietra, *J. Chem. Soc., Perkin Trans. 2*, 1993, 3117.
[71] A. McKillop, L. McLaren, R. J. Watson, R. J. K. Taylor, and N. Lewis, *Tetrahedron Lett.*, 1993, **34**, 5519.
[72] P. Wipf and Y. Kim, *J. Org. Chem.*, 1993, **58**, 1649.
[73] M. Kacan, D. Koyuncu, and A. McKillop, *J. Chem. Soc., Perkin Trans. 1*, 1993, 1771.
[74] A. Pelter and S. M. A. Elgendy, *J. Chem. Soc., Perkin Trans. 1*, 1993, 1891.

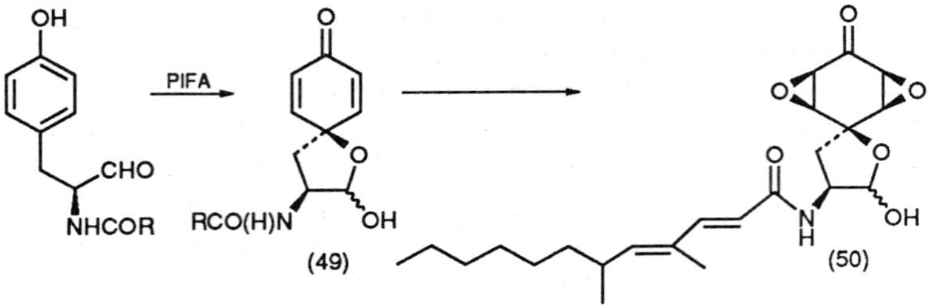

Microbial oxidation of benzenes to *cis*-cyclohexa-3,5-diene diols has been used as the key transformation in the synthesis of inositol stereoisomers and deuteriated conduritol derivatives.[75,76] The bacterial degradation of aromatic compounds has been discussed.[77]

The mechanism of ceric ion oxidation of naphthalene to naphthoquin-1,4-one has been investigated.[78] This oxidation can also be achieved with hydrogen peroxide catalysed by hexafluoroacetone hydrate.[79] The oxidation of hydroquinones to quinones can be effected in the solid state with cerric ammonium nitrate.[80] Hydroxyaromatics can be oxidatively cleaved with superoxide to aromatic 1,2-dicarboxylic acids.[81]

Electrochemical reduction of phthalic acids gives 1,2-dihydrophthalic acids, which have been used as diene components in synthesis.[82] The reduction of disubstituted naphthalene derivatives by potassium–graphite intercalate, C_8K, has been reported. The C_8K–ether system may be an alternative to Birch-type reactions.[83] Oxazoline-mediated 1,6-nucleophilic addition to aromatic rings has continued to be utilized by Meyers, as in the key step in the synthesis of the phytoalexin lancinilene (Scheme 10).[84]

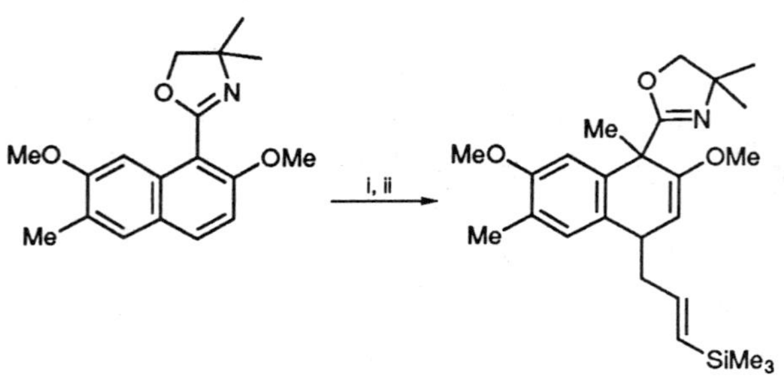

Reagents: i, $(CH_2CHCH_2SiMe_3)Li$; ii, MeI

Scheme 10

[75] H. A. J. Carless, K. Busia, and O. Z. Oak, *SYNLETT*, 1993, 672.
[76] H. A. J. Carless, K. Busia, Y. Dove, and S. S. Malik, *J. Chem. Soc., Perkin Trans. 1*, 1993, 2505.
[77] G. Gottschalk and H.-J. Knackmuss, *Angew. Chem., Int. Ed. Engl.*, 1993, **32**, 1398.
[78] M. V. Bhatt and M. Periasamy, *J. Chem. Soc., Perkin Trans. 2*, 1993, 1811.
[79] W. Adam and P. A. Ganeshpure, *Synthesis*, 1993, 281.
[80] J. Morey and J. M. Saa, *Tetrahedron*, 1993, **49**, 105.
[81] W. Li, C. Sotiriou-Leventis, M. Saha, P. P. Fu, and R. W. Giese, *Synth. Commun.*, 1993, **23**, 97.
[82] T. Ohno, M. Ozaki, A. Inagaki, T. Hirashima, and I. Nishiguchi, *Tetrahedron Lett.*, 1993, **34**, 2629.
[83] I. S. Weitz and M. Rabinovitz, *J. Chem. Soc., Perkin Trans. 1*, 1993, 117.
[84] T. G. Grant and A. I. Meyers, *Tetrahedron Lett.*, 1993, **34**, 3707.

Chiral oxazolines have been prepared from (*S*)-serine; these chiral auxiliaries show favourable asymmetric induction in a process similar to Scheme 10.[85] Tomioka has employed 2,6-di-t-butyl-4-methoxyphenyl esters (52) as an effective inert electron deficient moiety to facilitate 1,4-nucleophilic addition to aromatic rings. Incorporation of the chiral diether (53) in this process affords adducts (54) in high e.e.[86]

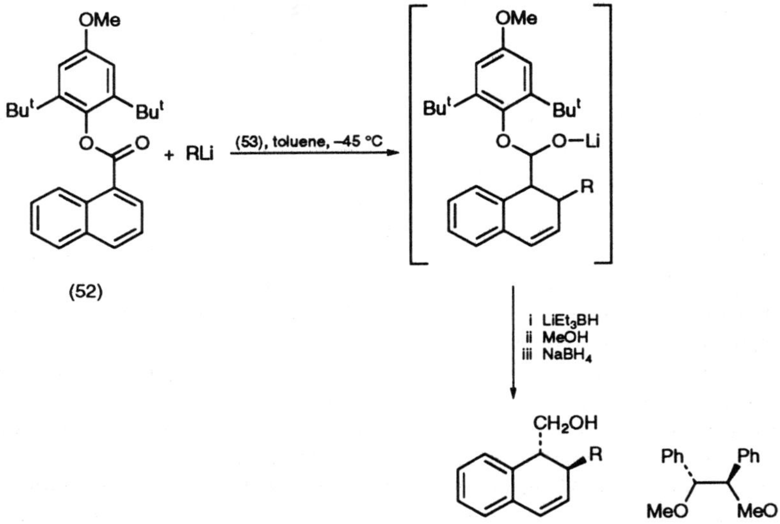

(52)

(54) (53)

The reaction mode and selectivity and regioselectivity of the photocycloaddition of 2,3-dimethylbuta-1,3-diene (DMBD) and benzene are markedly dependent on the arene substituent. In general, mixtures of products are formed. This is in marked contrast to the photoreaction of DMBD and benzonitrile, which gives the (4 + 4) photocycloadduct (55) as the major product (Scheme 11).[87]

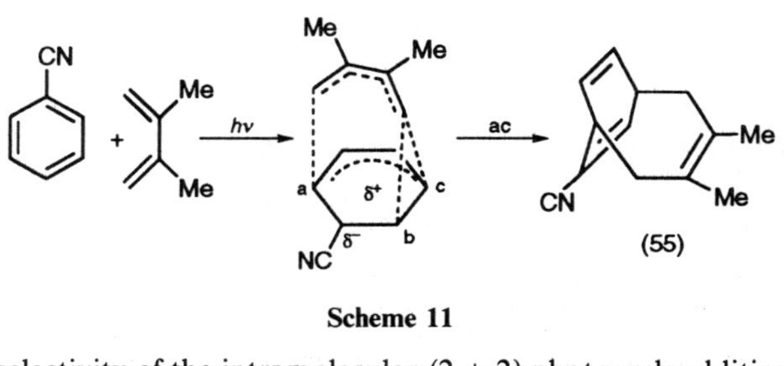

Scheme 11

The stereoselectivity of the intramolecular (2 + 2) photocycloaddition to benzenes

[85] A. I. Meyers, W. Schmidt, and M. J. McKennon, *Synthesis*, 1993, 251.
[86] K. Tomioka, M. Shindo, and K. Koga, *Tetrahedron Lett.*, 1993, **34**, 681.
[87] A. Gilbert and O. Griffiths, *J. Chem. Soc., Perkin Trans. 1*, 1993, 1379.

has been examined.[88] A diastereoselective (4 + 2) cycloaddition of singlet oxygen has been observed in the photooxygenation of a chiral naphthyl alcohol (56) (Scheme 12). These results indicate that the chiral alcohol directs the incoming singlet oxygen.[89]

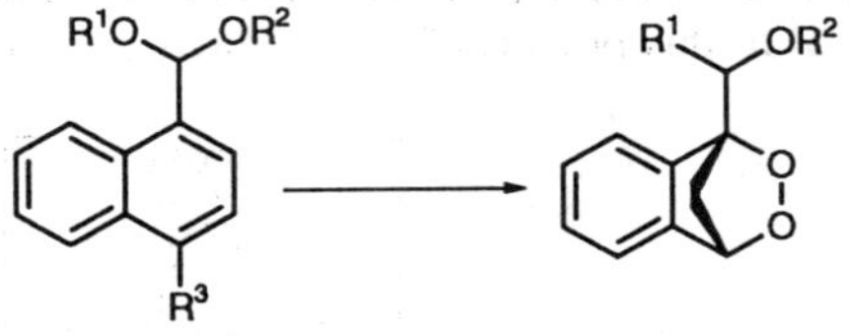

(56)

Reagents: O_2, TPP, *hν*, $CDCl_3$, rt

Scheme 12

4 Substitution in the Benzene Ring

Electrophilic Substitution.—Simple methods to prepare polyiodinated arenes are scarce. Barluenga has developed a method using bis(pyridine)iodonium(I) tetraborate that allows easy access to these iodobenzene derivatives.[90] The introduction of iodine into polyalkylbenzenes has also been achieved using *N*-iodosuccinimide and acidic catalysis.[91] Diaryliodonium salts, $(Ar_2I)^+X$, are a very important class of polyvalent iodine compounds. They are generally synthesized indirectly *via* aryl iodides. However, by the use of iodofluorosulfate they can be obtained directly from aromatic compounds under mild conditions.[92] *Ipso* substitution of sulfonyl chlorides and of the trimethylsilyl group with iodine is an excellent regiospecific method for the preparation of aryl iodides.[93–95]

A number of more selective procedures for introducing bromine into aromatic amines and phenols have been developed. These include the use of hexamethylene tetraamine tribromide, pyridinium hydrobromide, and *N*-bromosuccinimide in the presence of primary or secondary amines.[96–98] Bromination of deactivated aromatics has been achieved using bromine trifluoride without the use of a catalyst.[99]

A facile conversion of arenediazonium salts into arylfluorides using boron trifluoride etherate as a solvent has been developed.[100] Arylfluorides can also be obtained regioselectively by the rapid fluorodesilylation of aryltrimethylsilanes using xenon difluoride.[101]

88 P. J. Wagner and K. Cheng, *Tetrahedron Lett.*, 1993, **34**, 907.
89 W. Adam and M. Prein, *J. Am. Chem. Soc.*, 1993, **115**, 3766.
90 J. Barluenga, J. M. Gonzalez, M. A. Garcia-Martin, and P. J. Campus, *Tetrahedron Lett.*, 1993, **34**, 3893.
91 P. Bovonsombat and E. McNelis, *Synthesis*, 1993, 237.
92 N. S. Zefirov, T. M. Kasumov, A. S. Kozmin, V. D. Sorokin, P. J. Stang, and V. V. Zhdankin, *Synthesis*, 1993, 1209.
93 T. Satoh, K. Itoh, M. Miura, and M. Nomura, *Bull. Chem. Soc. Jpn.*, 1993, **66**, 2121.
94 B. Bennetau and J. Dunogues, *SYNLETT*, 1993, 171.
95 L. A. Jacob, B.-L. Chem, and D. Stec, *Synthesis*, 1993, 611.
96 S. Fujisaka, H. Eguch, A. Omura, A. Okamoto, and A. Nishidas, *Bull. Chem. Soc. Jpn.*, 1993, **66**, 1576.
97 W. P. Reeves and R. M. King, II, *Synth. Commun.*, 1993, **23**, 855.
98 S. C. Bisarya and R. Rao, *Synth. Commun.*, 1993, **23**, 779.
99 S. Rozen and O. Lerman, *J. Org. Chem.*, 1993, **58**, 239.
100 K. Shinhama, S. Aki, T. Farata, and J.-I. Minamikawa, *Synth. Commun.*, 1993, **23**, 1577.
101 A. P. Lothian and C. A. Ramsden, *SYNLETT*, 1993, 753.

In the presence of ozone, nitrogen dioxide exhibits strong nitrating abilities for alkylbenzenes at low temperature, converting them into the corresponding nitro derivatives in high yields.[102] Use of this nitrating agent on alkyl aryl ketones gives high levels of *ortho* substitution along with the *meta* isomer. This is in contrast to traditional nitrating procedures, which lead predominantly to *meta* substitution.[103] The use of calcium nitrate for the nitration of phenols has been studied.[104] An improved procedure for the nitration of benzocylobutene has been developed using acetyl nitrate generated *in situ* by a continuous process in the presence of montmorillonite K10 clay.[105]

Work has continued to render the Friedel–Crafts reaction truly catalytic. The Friedel–Crafts acylation is generally achieved using a stoichiometric amount of aluminium chloride as a Lewis acid catalyst. This can be problematical on a large scale, because the aluminium chloride cannot be reused because of its solubility in the aqueous workup. This problem has been addressed by a number of workers. Friedel–Crafts acylation proceeds smoothly using a catalytic amount of lanthanide triflate, which can be recovered and reused.[106] The Friedel–Crafts acylation can also be carried out in the absence of catalyst using a mixed anhydride. These are prepared by the reaction of a carboxylic acid with trifluoroacetic anhydride in the presence of phosphoric acid or active silicon(IV) catalysts.[107,108]

K10 montmorillonite on its own and impregnated with zinc chloride (clayzic) and other solid acid catalysts have been used for Friedel–Crafts alkylations.[109] These catalysts have a similar activity to aluminium chloride but can be used in smaller quantities and reused. The mild conditions for these catalysts has the added advantage over aluminium chloride that alkyl substituted aromatics can be prepared by the reaction of arenes with allyl alcohol without isomerization of the allyl double bond.[110] Reduced and reversed reactivity has been observed in the Friedel–Crafts alkylation with clays due to differential adsorption of the reactants onto the solid surface.[111–114] The Fries rearrangement of phenyl acetate to *ortho*-hydroxyacetophenone is more selective in the presence of zeolite GaZSM-5.[115] Investigations into the mechanism of the Fries rearrangement in polyphosphoric acid have suggested an intermolecular process.[116] The amino-Claisen rearrangement can be effected at less severe conditions than the thermal process by Lewis acid catalysts.[117,118] The synthetic utility of the Friedel–Crafts reaction has been increased by a number of modifications. A gallium dichloride-mediated reductive Friedel–Crafts reaction has been developed

[102] H. Suzuki, T. Murashima, I. Kogai, and T. Mori, *J. Chem. Soc., Perkin Trans. 1*, 1993, 1591.
[103] H. Suzuki, T. Murashima, A. Tatsumi, and I. Kozai, *Chem. Lett.*, 1993, 1421.
[104] S.C. Bisarya, S.K. Joshi, and A.G. Holkar, *Synth. Commun.*, 1993, **23**, 1125.
[105] P.J. Thomas and R.G. Pews, *Synth. Commun.*, 1993, **23**, 505.
[106] A. Kawada, S. Mitamura, and S. Kobayashi, *J. Chem. Soc., Chem. Commun.*, 1993, 1157.
[107] K. Suzuki, H. Kitagawa, and T. Mukaiyama, *Bull. Chem. Soc. Jpn.*, 1993, **66**, 3729.
[108] C. Galli and S. Fornarini, *J. Chem. Soc., Perkin Trans. 2*, 1993, 1147.
[109] O. Sieskind and P. Albrecht, *Tetrahedron Lett.*, 1993, **34**, 1197.
[110] P.H. Especl, B. Janssens, and P.A. Jacobs, *J. Org. Chem.*, 1993, **58**, 7688.
[111] G.D. Yadav, T.S. Thorat, and P.S. Kumbhar, *Tetrahedron Lett.*, 1993, **34**, 529.
[112] M. Davister and P. Laszlo, *Tetrahedron Lett.*, 1993, **34**, 533.
[113] S.J. Barlow, T.W. Bastock, J.H. Clark, and S.R. Cullen, *Tetrahedron Lett.*, 1993, **34**, 3339.
[114] A. Cornelis, P. Laszlo, and S. Wang, *Tetrahedron Lett.*, 1993, **34**, 3849.
[115] Y.V. Subba Rao, S.J. Kulkarni, M. Subrahmanyam, and A.V. Rama Rao, *Tetrahedron Lett.*, 1993, **34**, 7799.
[116] H. Sharghi and H. Eshghi, *Bull. Chem. Soc. Jpn.*, 1993, **66**, 135.
[117] G. Lai and W.K. Anderson, *Tetrahedron Lett.*, 1993, **34**, 6849.
[118] L.G. Beholz and J.R. Stille, *J. Org. Chem.*, 1993, **58**, 5095.

(Scheme 13).[119] Unconventional regiospecific synthesis of aromatic carboxamides (58) by means of tin-mediated Friedel–Crafts reactions has been achieved (Scheme 13).[120] This methodology can also be used for the preparation of sulfonamides. Friedel–Crafts acylation and alkylation have been shown to take place in a single stage in the presence of aprotic superacids (Scheme 13).[121]

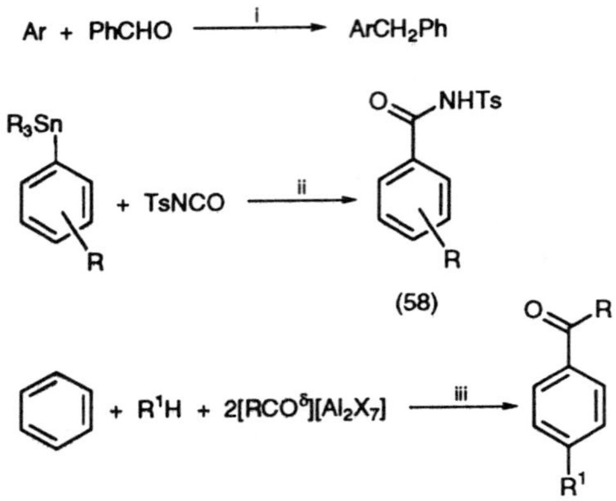

Reagents: i, Ga_2Cl_4;[120] ii, $AlCl_3$;[121] 0–20 °C[122]

Scheme 13

Conjugated dienones add to electron-rich arenes in a Friedel–Crafts annulation process that creates tricyclic compounds containing a central seven-membered ring (Scheme 14).[122] This method has been used as the key step in the synthesis of the diterpene barbatusol (59).[123]

A number of interesting Lewis acid mediated electrophilic substitution reactions related to the Friedel–Crafts reaction have been reported (Scheme 15).[124–128]

Direct aromatic amination has an advantage over the traditional method of reduction of the corresponding nitro compound. A number of novel direct amination methods have been developed. Azides in the presence of trifluoromethane sulfonic acid give the amino diazonium cation $(RNHN_2)^+$, which reacts with arenes to give aromatic amides.[129,130] Electron-rich arenes are regioselectively aminated under mild neutral conditions (Scheme 16).[131]

[119] Y. Hashimoto, K. Hirata, H. Kagoshima, N. Kihara, M. Hasegawa, and K. Saigo, *Tetrahedron*, 1993, **49**, 5969.
[120] M. Arnswald and W. P. Neumann, *J. Org. Chem.*, 1993, **58**, 7022.
[121] I. Akhrem, A. Orlinkov, and M. Volpin, *J. Chem. Soc., Chem. Commun.*, 1993, 257.
[122] G. Majetich, Y. Zhang, T. L. Feltman, and V. Belfoure, *Tetrahedron Lett.*, 1993, **34**, 441.
[123] G. Majetich, Y. Zhang, T. L. Feltman, and S. Duncan, Jr, *Tetrahedron Lett.*, 1993, **34**, 445.
[124] J. N. Kim and E. K. Ryu, *Tetrahedron Lett.*, 1993, **34**, 3567.
[125] A. P. Yakubov, D. V. Tsyganov, L. I. Belenlkii, and M. M. Krayushkin, *Tetrahedron*, 1993, **49**, 3397.
[126] A. R. Katritsky, L. Xie, A. S. Afridi, W.-Q. Fan, and W. Kuzmicrkiewicz, *Synthesis*, 1993, 47.
[127] A. W. van der Made and R. H. van der Made, *J. Org. Chem.*, 1993, **58**, 1263.
[128] G. A. Olah, Q. Wang, X.-Y. Li, and S. Prakash, *SYNLETT*, 1993, 32.
[129] H. Takeuchi, T. Adachi, H. Nishiguchi, K. Itou, and K. Koyama, *J. Chem. Soc., Perkin Trans. 1*, 1993, 867.
[130] G. A. Olah, P. Ramaiah, Q. Wang, and G. K. S. Prakash, *J. Org. Chem.*, 1993, **58**, 6901.
[131] I. Zaltsgendler, Y. Leblanc, and M. A. Bernstein, *Tetrahedron Lett.*, 1993, **34**, 2441.

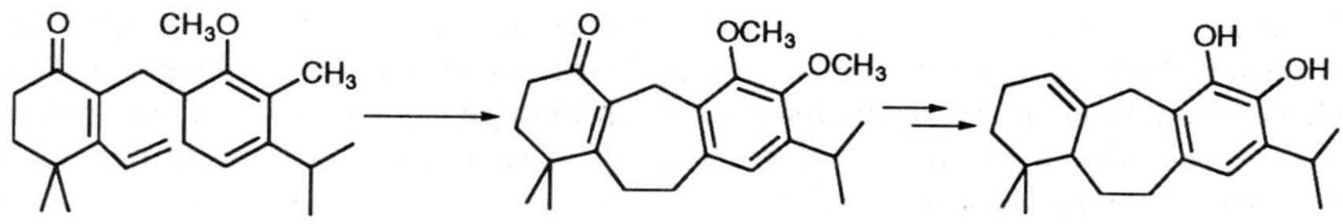

(59)

Reagents: $BF_3 \cdot Et_2O$

Scheme 14

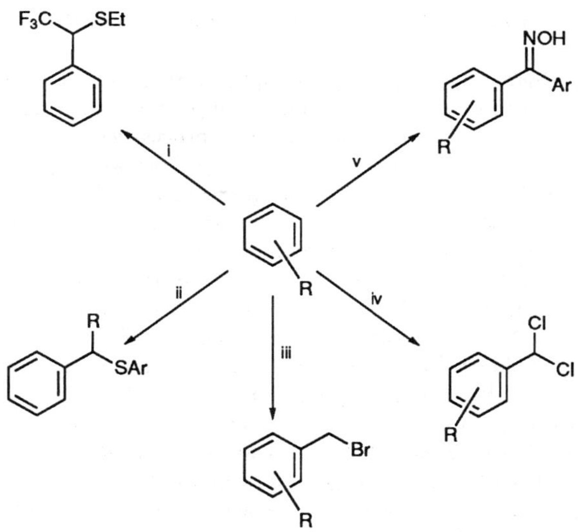

Reagents: i, EtSH, $CF_3CH(OH)_2$, BF_3; ii, CH(Cl)(SAr)R, BF_3; iii, $(CH_2O)_n$, HBr, HOAc; iv, $MeOCHCl_2$, $AlCl_3$; v, $ArCN^+–O^-$, $AlCl_3$

Scheme 15

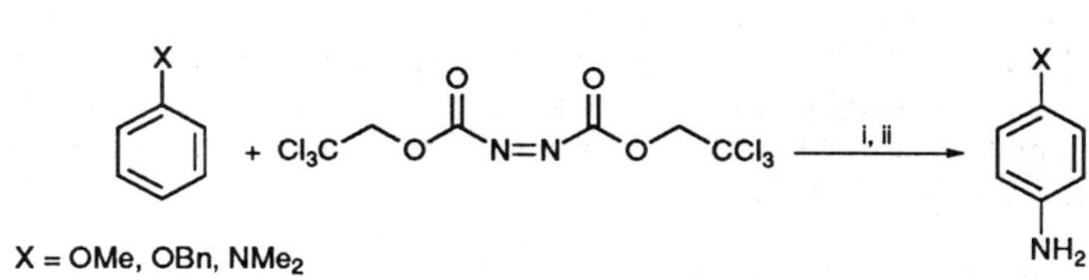

Reagents: i, $LiClO_4$; ii, Zn/HOAc

Scheme 16

Oxygen can also be introduced directly into aromatic systems with the use of HOF–CH_3CN complex or CH_3OF.[132] Tritium can be introduced into activated aromatic compounds using $BF_3 \cdot Et_2O$ and tritiated water.[133]

An unorthodox rate enhancement in the Mannich reaction has been observed.

[132] M. Kol and S. Rozens, *J. Org. Chem.*, 1993, **58**, 1593.
[133] P. McGeady and R. Croteau, *J. Chem. Soc., Chem. Commun.*, 1993, 775.

Under classical conditions the Mannich reaction is dramatically slowed after the formation of the mono-substituted product. Leigh has found that introduction of an electron-withdrawing group into the *para*-position of phenol facilitates the formation of the disubstituted product in high yields.[134] The heating rates associated with microwave heating have been used to control the isomeric ratio in sulfonation of naphthalene.[135] Cerfontain has carried out an in-depth study into the sulfonation of naphthalene containing oxy substituents.[136–138] The thermal decomposition of *meta*-ditoluidinium sulfate to give *ortho*-methylsulfonic acid has been monitored. The rate-determining step for this reaction appears to be proton transfer to give an absorbed phase of *meta*-toluidine and sulfuric acid prior to sulfonation.[139]

Nucleophilic Substitution.—Aromatic nucleophilic substitution has been the subject of debate by a number of workers. The major issue of interest concerns nucleophilic substitution reactions that occur through a chain process with radicals and radical anions as intermediates. The propagation reactions proposed have been, among others, the dissociation of the radical anion intermediate of the substrate to give a radical, which reacts with the nucleophile ($S_{RN}1$ mechanism), or the reaction of such a radical anion with a nucleophile to give the substitution product ($S_{RN}2$ mechanism). Denney has questioned the validity of the $S_{RN}1$ mechanism because these reactions are known to take place with radical anions that are stable to dissociation thus favouring an $S_{RN}2$ mechanism.[140] These claims have been rebuffed by Rossi who favours an $S_{RN}1$ mechanism based on kinetic grounds.[141] Others have also put forward evidence for $S_{RN}1$ mechanisms.[142–144] Alternative mechanisms are also possible for aromatic nucleophilic substitution. Zhang has proposed an S_N2-type non-chain radical process in the reaction of *p*-nitrochlorobenzene with the sodium salt of ethyl α-cyanoacetate.[145,146] In a study of the hydrolysis of ethers of 2,4-dinitrophenols de Rossi has found no evidence for a single-electron transfer mechanism, thus pointing to the classical two-electron addition–elimination reaction.[147] The mechanism of aromatic nucleophilic substitution where amines are the nucleophiles has been intensively investigated.[148–154]

Aromatic nucleophilic substitution with carbon nucleophiles is an important

[134] D. A. Leigh, P. Linnane, and G. Jackson, *Tetrahedron Lett.*, 1993, **34**, 5639.
[135] D. Stuerga, K. Gonon, and M. Lallemant, *Tetrahedron*, 1993, **49**, 6229.
[136] H. R. W. Ansink, E. Zelvelder, and H. Cerfontain, *Recl. Trav. Chim. Pays-Bas*, 1993, **112**, 210.
[137] H. R. W. Ansink, E. Zelvelder, and H. Cerfontain, *Recl. Trav. Chim. Pays-Bas*, 1993, **112**, 216.
[138] H. R. W. Ansink, E. Zelvelder, and H. Cerfontain, *J. Chem. Soc., Perkin Trans. 2*, 1993, 721.
[139] G. Singh, I. P. S. Kapoor, and M. Jain, *J. Chem. Soc., Perkin Trans. 2*, 1993, 1521.
[140] B. Denney, D. Z. Denney, and A. J. Perez, *Tetrahedron*, 1993, **49**, 4463.
[141] R. A. Rossi and S. M. Palacios, *Tetrahedron*, 1993, **49**, 4485.
[142] E. C. Ashby, R. Gurunmurthy, and R. W. Ridlehuber, *J. Org. Chem.*, 1993, **58**, 5832.
[143] S. Montanari, C. Paradisi, and G. Scorrano, *J. Org. Chem.*, 1993, **58**, 5628.
[144] C. Galli and P. Gentili, *J. Chem. Soc., Perkin Trans. 2*, 1993, 1135.
[145] X.M. Zhang, D.-L. Yang, and X. Q.-Jia, and Y.-C. Liu, *J. Org. Chem.*, 1993, **58**, 7350.
[146] X.-M. Zhang, D.-L. Yang, and Y.-C. Liu, *J. Org. Chem.*, 1993, **58**, 224.
[147] E. B. de Vargas, E. L. Setti, M. L. Aimar, and R. H. de Rossi, *J. Org. Chem.*, 1993, **58**, 7364.
[148] N. S. Nadelman, M. Marder, and A. Gurevich, *J. Chem. Soc., Perkin Trans.* 2, 1993, 229.
[149] R. Chamberlin and M. R. Crampton, *J. Chem. Res. (S)*, 1993, 106.
[150] R. Chamberlin and M. R. Crampton, *J. Chem. Soc., Perkin Trans.* 2, 1993, 75.
[151] E. T. Akinyele, D.-F. Crist, and J. Hirst, *J. Chem. Soc., Perkin Trans. 2*, 1993, 905.
[152] S. Sekiguchi, M. Hosokawa, T. Suzuki, and M. Sato, *J. Chem. Soc., Perkin Trans. 2*, 1993, 1111.
[153] T. A. Emokpae, P. U. Uwakwe, and J. Hirst, *J. Chem. Soc., Perkin Trans. 2*, 1993, 125.
[154] T. Ibata, X. Zou, and T. Demura, *Tetrahedron Lett.*, 1993, **34**, 5613.

method for the formation of carbon–carbon bonds; a number of examples are given in Scheme 17.[155-160]

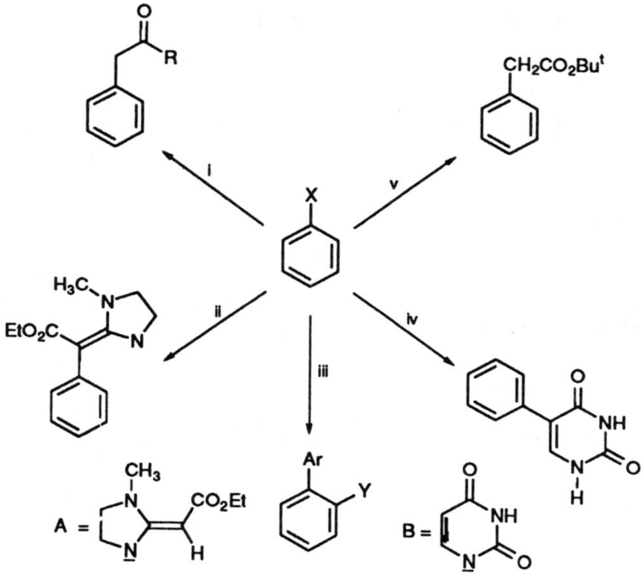

Reagents: i, $H_2C{=}CR(ONa)$; ii, (A); iii, ArMgBr; iv, (B); v, $Na(H_2C{=}CO_2Bu^t)$, $FeSO_4$

Scheme 17

Miyano has demonstrated that Grignard reagents can displace methoxy groups.[156,157] The scope of this reaction has been extended by the utilization of sodium alkoxides and lithium amides as nucleophiles to give alkoxybenzenes and aromatic amines respectively.[161,162] Non-activated aromatic bromides are subject to methoxylation under mild and selective conditions when copper associated with carbon dioxide is used as a catalyst with sodium methoxide.[163] 2,4-Dinitrochlorobenzene undergoes halogen-exchange fluorination with hydrogen fluoride–base solutions to afford the corresponding fluorides.[164] Aryl phosphines can be readily synthesized from aryl fluorides by nucleophilic substitution with potassium diphenylphosphine.[165] Regioselective control can be achieved in aromatic nucleophilic substitution of 2,4-difluorocarbonyl aryls. The carbonyl group of an aldehyde or ketone directs

155 C. Dell'Erba, M. Novi, G. Petrillo, and C. Tavani, *Tetrahedron Lett.*, 1993, **34**, 235.
156 T. Hattori, H. Hotta, T. Suzuki, and S. Miyano, *Bull. Chem. Soc. Jpn.*, 1993, **66**, 613.
157 T. Hattori, T. Suzuki, N. Hayashizaka, N. Koike, and S. Miyamo, *Bull. Chem. Soc. Jpn.*, 1993, **66**, 3034.
158 W.-Y. Zhao and Z.-T. Huang, *Synth Commun.*, 1993, **23**, 2533.
159 H. Medebielle, M. A. Oturan, J. Pinson, and J.-M. Saveant, *Tetrahedron Lett.*, 1993, **34**, 3409.
160 M. van Leeuwen and A. McKillop, *J. Chem. Soc., Perkin Trans. 1*, 1993, 2433.
161 T. Hattori, S. Satoh, and S. Miyano, *Bull. Chem. Soc. Jpn.*, 1993, **66**, 3840.
162 W. ten Hoeve, C. G. Kruse, J. N. Luteyn, J. R. G. Thiecke, and H. Wynberg, *J. Org. Chem.*, 1993, **58**, 5101.
163 D. Nobel, *J. Chem. Soc., Chem. Commun.*, 1993, 419.
164 T. Fukuhara and N. Yoneda, *Chem. Lett.*, 1993, 509.
165 S. J. Coote, G. J. Dawson, C. G. Frost, and J. M. J. Williams, *SYNLETT*, 1993, 509.

nucleophilic displacement to the *ortho*-position in preference to the *para*-position.[166] The application of vicarious nucleophilic substitution (VNS) of hydrogen allows the direct isocyanomethylation of nitroarenes (Scheme 18).[167]

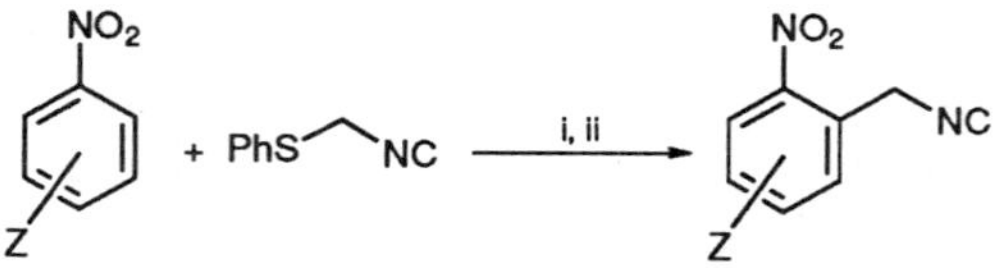

Reagents: i, Bu^{t}OK, DMF; ii, CO_2, AcOEt

Scheme 18

A VNS process is thought to be responsible for the facile monobromination of 1,4-benzodithians (Scheme 19).[168]

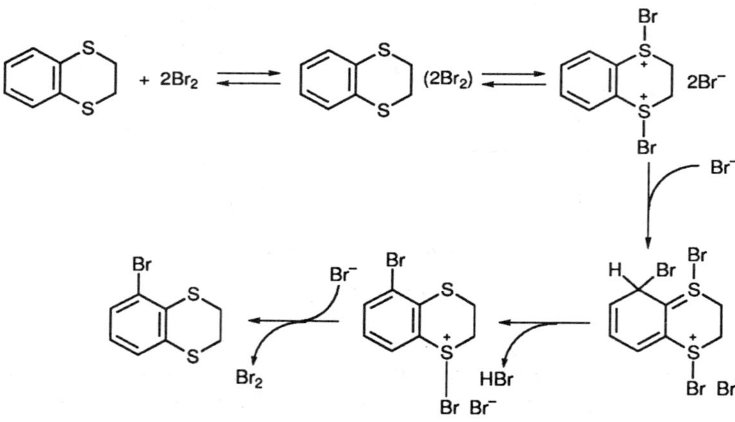

Scheme 19

An unprecedented intramolecular nucleophilic addition process, rather than that of substitution of a diazonium salt, has been reported (Scheme 20).[169]

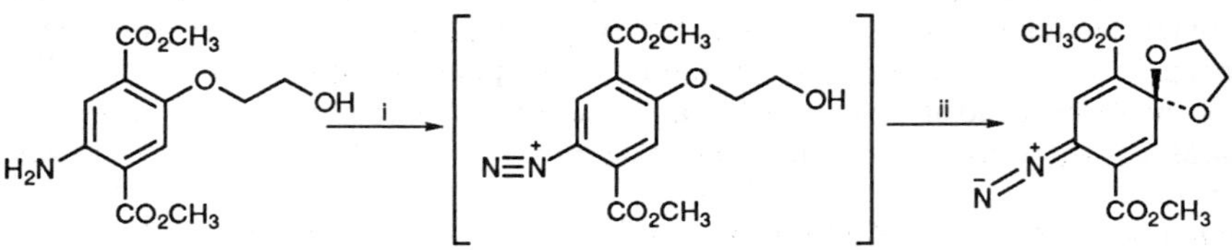

Reagents: i, HNO_2; ii, $Ni(CN)_2$

Scheme 20

[166] D. M. Fink and J. T. Stupczewski, *Tetrahedron Lett.*, 1993, **34**, 6525.
[167] M. Makosza, A. Kimowski, and S. Ostrowski, *Synthesis*, 1993, 1215.
[168] R. Caputo, M. DeNisco, and G. Palumbo, *Tetrahedron* , 1993, **49**, 11 383.
[169] C. A. Panetta, Z. Fang, and N. E. Heimer, *J. Org. Chem.*, 1993, **58**, 6146.

A new carboannulation strategy has been developed that involves an intramolecular nucleophilic addition of silyl enol ethers to photosensitized, electron-transfer generated arene radical cations (Scheme 21).[170]

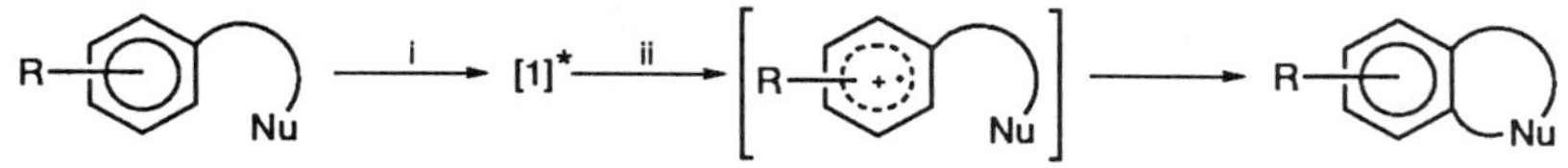

Reagents: i, *hν*, CH_3CN, H_2O; ii, 1,4-dicyanonaphthalene

Scheme 21

Substitution via *Organometallic Intermediates.* The regiospecific nature of directed *ortho*-metallation (DoM) over classical electrophilic aromatic substitution methods continues to be exploited in the synthesis of aromatic compounds. Amides and ethers are still the most common DoM groups to be used for this purpose. The asymmetric synthesis of optically active phthalides *via* DoM and the efficient conversion of alkylphenols into salicyclic acids are examples of their use (Scheme 22).[171,172]

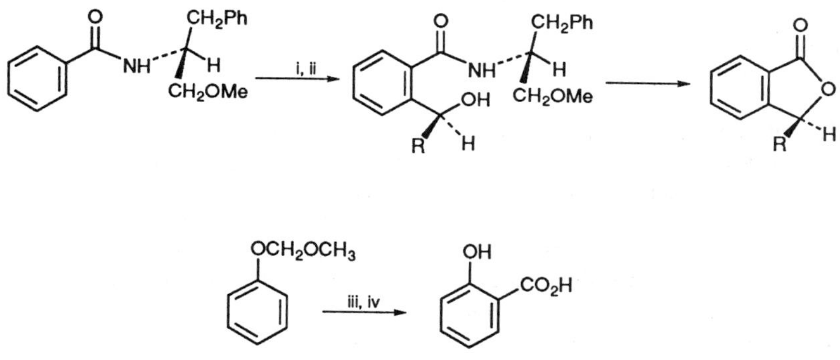

Reagents: i, BuLi; ii, RCHO; iii, CO_2; iv, HOAc

Scheme 22

A number of other DoM groups have been effectively utilized; these include aromatic aldimines (which act as *ortho*-lithiated protected aldehydes) and the thiolate group.[173,174] The silanolate group has been evaluated as a DoM group but it suffers from nucleophilic attack thus reducing its synthetic potential.[175] The regioselectivity of DoM can be effected by manipulation of the side chain adjacent to the amide functionality (Scheme 23).[176] Changing the base used for metallation from an organolithium to a superbase also effects the regioselectivity (Scheme 23).[177]

[170] G. Pandey, A. Krishna, K. Girija, and M. Karthikeyan, *Tetrahedron Lett.*, 1993, **34**, 6631.
[171] S. Matsui, A. Uejima, Y. Suziki, and K. Tanaka, *J. Chem. Soc., Perkin Trans. 1*, 1993, 701.
[172] A. R. Katritzky, H. Lang, and X. Lan, *Synth. Commun.*, 1993, **23**, 1175.
[173] L. A. Flippin, J. M. Muchowski, and D. S. Carter, *J. Org. Chem.*, 1993, **58**, 2462.
[174] S. Masson, J.-F. Saint-Clair, and M. Saquat, *Synthesis*, 1993, 485.
[175] S. McN. Sieburth and L. Fensterbank, *J. Org. Chem.*, 1993, **58**, 6314.
[176] G. Katsoulos and M. Schlosser, *Tetrahedron Lett.*, 1993, **34**, 6263.
[177] S. Cabiddu, C. Fattuoni, C. Floris, G. Gelli, and S. Melis, *Tetrahedron*, 1993, **49**, 4965.

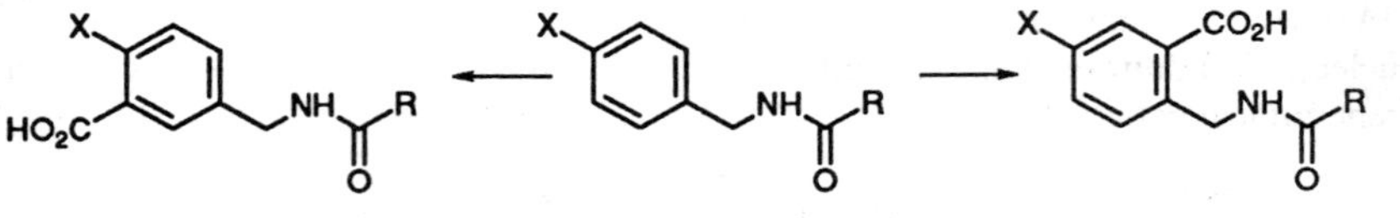

R = OC(CH$_3$)$_3$ X = H$_3$CO or X = F R = – N(CH$_3$)$_2$

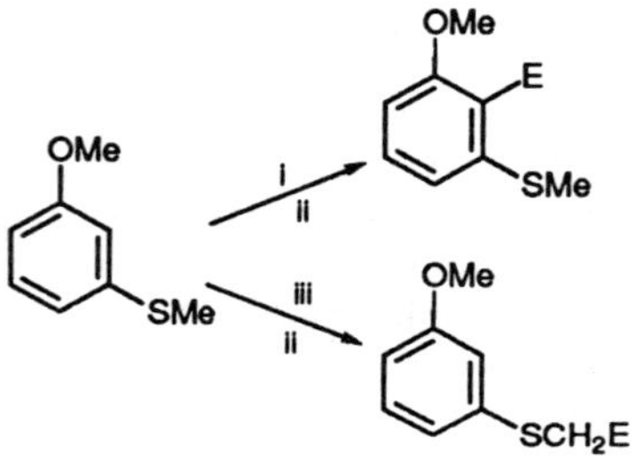

Reagents: i, BuLi; ii, E$^+$; iii, BuLi, BuOK

Scheme 23

The efficiencies of DoM for a series consisting of secondary benzamides, tertiary benzamines, and benzylic alcohols have been reported. It was found for both amidic series that the efficiencies increase as the oxygen and the *ortho* hydrogen in the substrate are more coplanar with the aromatic ring. However, for the alcohol series the opposite order is observed; this is thought to be due to shielding caused by association of the alkoxide with two lithiums as opposed to one in the amidic case.[178]

Palladium-catalysed cross-coupling reactions continue to be widely used for functionalization of aromatic compounds. The utility of the Heck reaction, arylation of olefins, has been extended to the arylation of vinylamines[179] and to reactions in aqueous media.[180] The mode of reaction of the palladium-catalysed reaction of vinyl borate (60) and aryl halides can be varied to favour Heck or Suzuki-type processes by careful control of the reaction conditions (Scheme 24).[181]

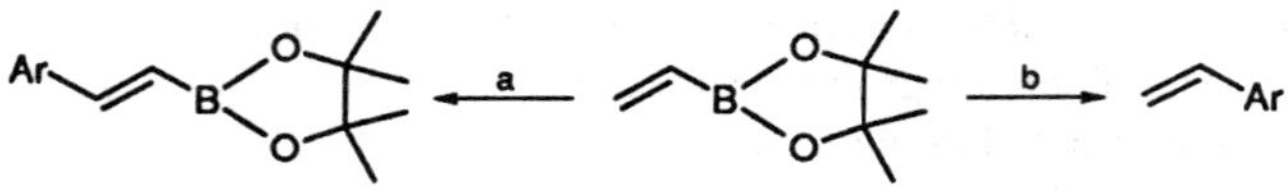

(60)

(a) Heck. (b) Suzuki. Reagents: ArX, Pd(0)

Scheme 24

The palladium-catalysed coupling between organostannanes and unsaturated

[178] P. Beak, S. T. Kerrick, and D. J. Gallagher, *J. Am. Chem. Soc.*, 1993, **115**, 10 268.
[179] C. A. Busacca, R. E. Johnson, and J. Swestock, *J. Org. Chem.*, 1993, **58**, 3299.
[180] S. Sengupta and S. Bhattacharga, *J. Chem. Soc., Perkin Trans. 1*, 1993, 1943.
[181] A. R. Hunt, S. K. Stewart, and A. Whiting, *Tetrahedron Lett.*, 1993, **34**, 3599.

halides or sulfonates, the Stille reaction, has been the subject of a comprehensive study.[182] Aryl alkynes can be synthesized by the palladium-catalysed reaction of aryl triflates with terminal alkynes.[183] The synthetic transformations of aryl triflates has been reviewed.[184]

The biaryl nucleus is present in a number of natural products and is a key feature in chiral catalysts (*e.g.* BINAP) and in bidentate ligands. This has resulted in new developments in the methodology for their formation. Lipshutz has synthesized biaryls *via* higher order cyano cuprates. (Scheme 25).[185]

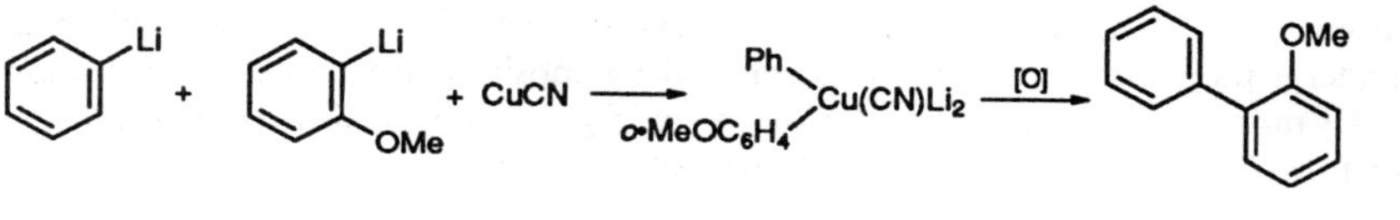

Scheme 25

Meyers has used oxazoline-mediated biaryl coupling and the Ullman reaction to prepare multifunctional chiral biaryls.[186–188] The cross-coupling of arylzinc compounds with aryl halides has been exploited to prepare unsymmetrical biaryls.[189,190] The mechanism of the Ullman coupling reaction on absorbed monolayers has been studied.[191] Barrett has developed a catalytic procedure for the oxidative coupling of phenols (Scheme 26).[192]

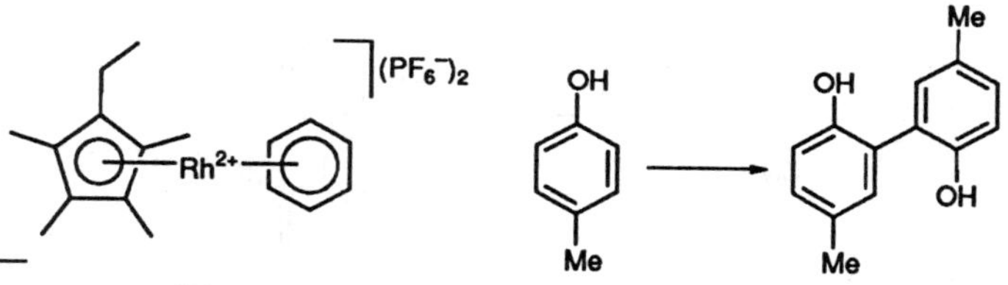

(A)

Reagents: 10 mol % (C), Cs_2CO_3, PhBr

Scheme 26

Organolithium compounds are often formed by bromine–lithium exchange between the appropriate brominated precursors. The use of the more readily available chloroaromatics should be avoided due to their lower reactivity. This process can be made viable by the addition of a naphthalene catalyst.[193] The reaction of *ortho*-

182 V. Farina, B. Krishnan, D. R. Marshall, and G. P. Roth, *J. Org. Chem.*, 1993, **58**, 5434.
183 M. Alami, F. Ferri, and G. Linstrumelle, *Tetrahedron Lett.*, 1993, **34**, 6403.
184 K. Ritter, *SYNLETT*, 1993, 735.
185 B. H. Lipshutz, K. Siegmann, E. Garcia, and F. Kayser, *J. Am. Chem. Soc.*, 1993, **115**, 9276.
186 M. Moorlag and A. I. Meyers, *Tetrahedron Lett.*, 1993, **34**, 6989.
187 H. Moorlag and A. I. Meyers, *Tetrahedron Lett.*, 1993, **34**, 6993.
188 T. P. Nelson and A. I. Meyers, *Tetrahedron Lett.*, 1993, **34**, 3061.
189 K. Takagi, *Chem. Lett.*, 1993, 469.
190 S. Sibille, V. Ratovelomanana, J. V. Nedelec, and J. Perichon, *SYNLETT*, 1933, 425.
191 M. Xi and B. F. Bent, *J. Am. Chem. Soc.*, 1993, **115**, 7426.
192 A. G. M. Barrett, T. Itoh, and E. M. Wallace, *Tetrahedron Lett.*, 1993, **34**, 2233.
193 A. G. Guijarro, D. J. Ramon, and M. Yus, *Tetrahedron Lett.*, 1993, **34**, 469.

haloiodobenzenes with activated copper gives at ambient temperature a stable on *ortho*-halophenylcopper reagent.[194] The use of arene chromium tricarbonyl complexes in organic synthesis has been reviewed.[195,196]

Substitution via *Aryl Radicals*. The production of aryl radicals from *para*-methoxybenzenediazonium tetrafluoroborate is accelerated in the presence of catechol and its derivatives. EPR and spin-trapping investigations indicate that this reduction process is catalysed by semiquinone radicals.[197] Aryl radicals are also formed by the reduction of aryltrimethylammonium cations. Examination of this process demonstrates that the trimethylammonium group can be expelled in much the same manner as chlorine atoms, thus promising application in organic synthesis.[198] Triphenylsulfonium salts undergo photochemical or electrochemical decomposition to generate aryl radicals.[199]

Cyanobenzenes have been shown to undergo photochemical substitution with olefins, allylic organometallics, and alkyl silanes (Scheme 27).[200–202]

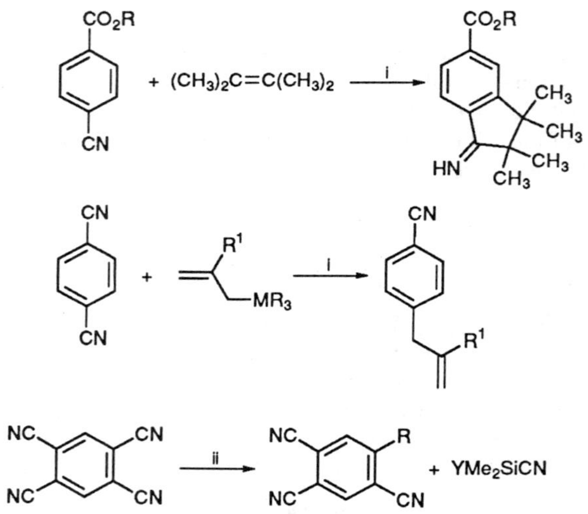

Reagents: i, *hν*, CH_3CN; ii, *hν*, $RSiMe_2Y$

Scheme 27

A number of new photochemical methods have been developed that allow the introduction of fluorinated alkyls into the aryl nucleus. Mallouk has demonstrated that silver triflate can be used as a source of trifluoromethyl radicals, which subsequently

[194] G. W. Ebert, D. R. Pfenning, S. D. Suchen, and T. A. Donovan, Jr., *Tetrahedron Lett.*, 1993, **34**, 2279.
[195] S. G. Davies and T. J. Donohoe, *SYNLETT*, 1993, 323.
[196] M. Sodeoka and M. Shibasaki, *SYNLETT*, 1993, 643.
[197] K. J. Reszka and C. F. Chignell, *J. Am. Chem. Soc.*, 1993, **115**, 7752.
[198] V. V. Konovalov, I. I. Bilkis, B. A. Selivanov, V. D. Shteingarts, and V. D. Tsvetkov, *J. Chem. Soc., Perkin Trans. 2*, 1993, 1707.
[199] A. Stasko, P. Rapta, V. Brezova, O. Nuyken, and R. Vogel, *Tetrahedron*, 1993, **49**, 10917.
[200] K. McMahon and D. R. Arnold, *Can. J. Chem.*, 1992, **71**, 450.
[201] K. Nakanshi, K. Mizuno, and V. Otsuji, *Bull. Chem. Soc. Jpn.*, 1993, **66**, 2371.
[202] M. Mella, N. d'Alessandro, M. Freccero, and A. Albini, *J. Chem. Soc., Perkin Trans. 2*, 1993, 515.

can substitute *C*-hydrogen.[203] Perfluoroalkylation of anilines has been reported; this process involves initial charge-transfer complex formation between the aniline and perfluoroalkyl iodides. The complex then undergoes photo-induced electron transfer to give a radical cation intermediate, which attacks the aniline.[204] Pentafluorophenylation of arenes can be effected *via* photochemical-induced reaction of pentafluorophenylalkane sulfonates with a wide variety of aromatic compounds.[205]

Di- and trihalobenzenes undergo *ipso* substitution on reaction with benzyl radicals to give halo- and dihalodiphenylmethanes respectively.[206] Intramolecular radical *ipso* substitution is thought to be responsible for the conversion of arylpropyl bromides (61) into arylpropanols (62) (Scheme 28).[207]

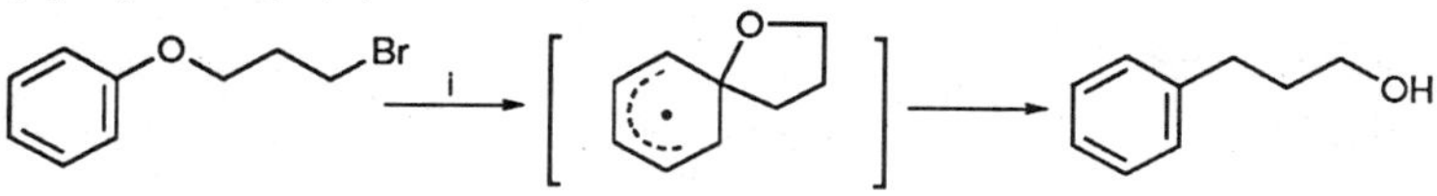

(61) (62)

Reagents: $[CH_3(CH_2)_3]_3SnH$

Scheme 28

The oxidation of benzene to phenol with Underfriend's reagent (Fe^{2+}–EDTA + O_2 + L-absorbic acid) has been examined and evidence is presented for hydroxyl radicals as the activated species.[208]

5 Condensed Polycyclic Aromatic Compounds

Benzenoid Aromatics.—The discovery that buckminsterfullerene, C_{60}, is a stable molecule due to geodesic and electronic properties has generated a renewed interest in hydrocarbons that resemble a portion of the C_{60} surface. The carbon framework of corannulene (63) can be considered to represent the polar cap of C_{60}. In spite of its curvature, corannulene undergoes rapid bowl-to-bowl inversion. Incorporation of the corannulene structure into C_{60} would require the bowl to be in a locked conformation. Rabideau has demonstrated that the bowl becomes rigid with only the addition of two carbons, thus giving the fused five membered ring derivative (64).[209] The corannulene tetraanion has also been subject of debate. The possibility of the tetraanion being viewed as a cyclopentadienyl anion inner core, with the remaining three electrons at the periphery of the molecule producing an 18-π-electron trianion has been raised. Calculations do not lend much support to this hypothesis, and attempted formation of the tetraanion and protonation have also failed.[210]

The bowl shaped hydrocarbon sumanene (65) can also be regarded as a precursor to the synthesis of C_{60}. The synthesis of sumanene has not been reported; Mehta has reported his achievements towards this target, but unfortunately this route fails at the

[203] C. Lai and T. E. Malbuk, *J. Chem. Soc., Chem. Commun.*, 1993, 1359.
[204] Q.-Y. Chem, Z.-T. Li, and C.-M. Zhou, *J. Chem. Soc., Perkin Trans. 1*, 1993, 2457.
[205] Q.-Y. Chen and Z. T. Li, *J. Org. Chem.*, 1993, **58**, 2599.
[206] R. Henriquez and D. C. Nonhebel, *Tetrahedron*, 1993, **49**, 6497.
[207] E. Lee, C. Lee, J. S. Tae, H. S. Whang, and K. S. Li, *Tetrahedron Lett.*, 1993, **34**, 2343.
[208] S. Ito, K. Ueno, A. Mitarai, and K. Sasaki, *J. Chem. Soc., Perkin Trans. 2*, 1993, 255.
[209] A. H. Abdourazak, A. Sygula, and P. W. Rabideau, *J. Am. Chem. Soc.*, 1993, **115**, 3010.
[210] P. W. Rabideau, Z. Marcinow, R. Sygyula, and A. Sygyla, *Tetrahedron Lett.*, 1993, **34**, 6351.

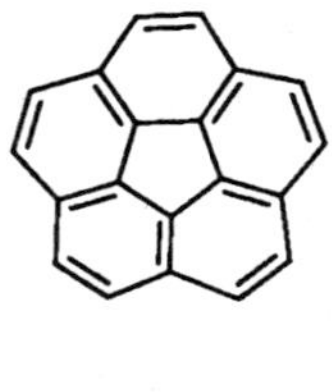

(63)

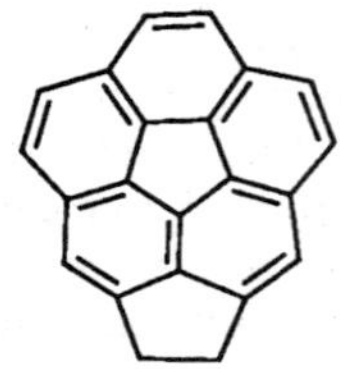

(64)

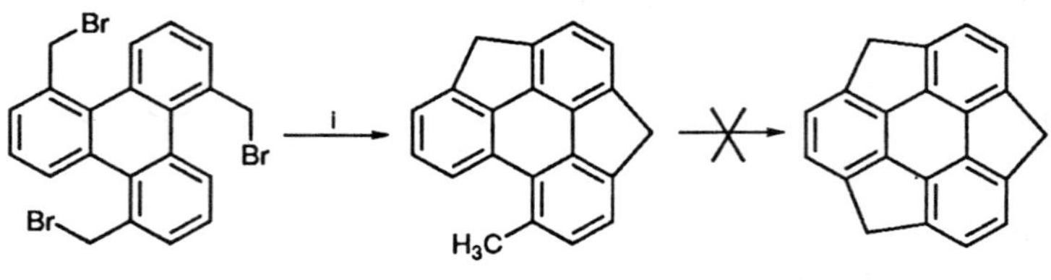

(65)

Reagents: FVP, 850 °C, 10^{-4} Torr

Scheme 29

last step[211] (Scheme 29).

Plater has suggested the synthesis of C_{60} from suitable polycyclic aromatic hydrocarbon (PAH) precursors. A number of appropriate PAHs have been synthesized towards this aim, the most complex being 1,3,5-tri(benzol[c]phenanthren-9-yl)benzene (66), which could be envisaged under suitable gas-phase dehydrogenation conditions to close up like the petals of a flower to yield C_{60}.[212]

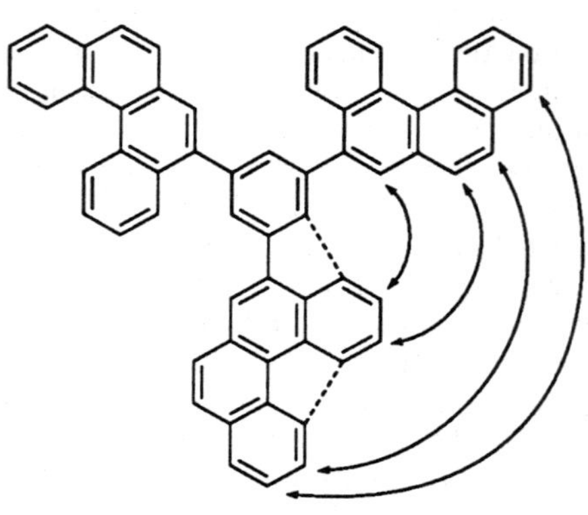

(66)

Cyclopentene-fused PAHs are related to C_{60} in that they are constructed of fused hexagonal and pentagonal arrays of carbon atoms. The reactivity of these nonalternate

[211] G. Mehta, S. R. Shtah, and K. Ravikumar, *J. Chem. Soc., Chem. Commun.*, 1993, 1007.
[212] M. J. Plater, *SYNLETT*, 1993, 405.

PAHs is frequently different from alternate PAHs. A detailed study of the electronic behaviour of a series of nonalternate PAHs has been accomplished using heavy atom perturbation effects.[213]

Methylene bridged PAHs are ubiquitous environmental pollutants often formed by incomplete combustion of fossil fuels. These PAHs have been shown to exhibit potent tumourigenic activity. However, surprisingly little is known concerning their chemistry or biological properties. This situation has been remedied by the synthesis of these compounds by standard techniques.[214–217] The electrophilic substitution and the catalytic reduction of a number of these PAHs has been studied.[218–220] The carcinogenic properties of PAHs are thought to be due to oxidation of the bay region to give diol epoxides (67), which are implied as the ultimate carcinogenic metabolites. Evidence for this is the identification of its metabolic precursor, *trans*-3,4-dihydrodiol (68). To permit biological study of these systems a number of workers have disclosed synthetic routes to these metabolites.[221–224]

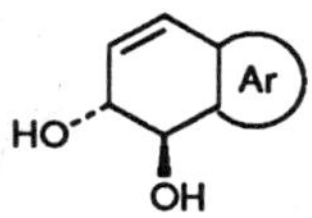
(68)

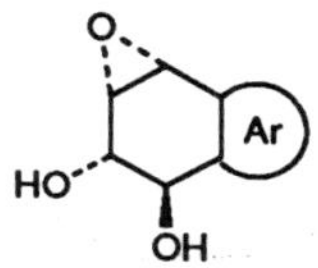
(67)

Flash vacuum pyrolysis has proved to be useful in the synthesis of acenaphthylene and its derivatives (Scheme 30).[225–227]

The simple and rapid production of complex PAHs has been demonstrated; a number of the more interesting examples are highlighted in Scheme 31.[228–230]

Non-benzenoid Aromatics.—1,6-Methano[10]annulene (69), a 10π-electron system, can also exist in the less stable bisnorcaradiene form (70). Neidlein has demonstrated that bridging between the C-2 and C-10 positions shifts the equilibrium in favour of the bisnorcaradiene structure (71).[231]

213 B. F. Plummer, L. K. Steffen, T. L. Braley, W. G. Reese, K. Zych, G. Van Dyke, and B. Tully, *J. Am. Chem. Soc.*, 1993, **115**, 11 542.
214 C. Yang and R. G. Harvey, *J. Org. Chem.*, 1993, **58**, 4155.
215 P. P. J. Mulder, B. B. Boere, J. Cornelisse, and Lugtenburg, *Recl. Trav. Chim. Pays-Bas*, 1993, **112**, 255.
216 T. Gimisis, J. W. Kampt, and M. Koreeda, *J. Org. Chem.*, 1993, **58**, 5858.
217 M. J. Tanya and J. E. Bupp, *J. Org. Chem.*, 1993, **58**, 4173.
218 B. P. Cho, M. Kim, and R. G. Harvey, *J. Org. Chem.*, 1993, **58**, 5788.
219 M. Minabe, R. Nishimura, T. Kimura, and M. Tsubota, *Bull. Chem. Soc. Jpn.*, 1993, **66**, 1248.
220 M. Minabe, S. Urushibara, F. Mishina, T. Kimura, and M. Tsubota, *Bull. Chem. Soc. Jpn.*, 1993, **66**, 670.
221 P. K. Sharma, *Synth. Commun.*, 1993, **23**, 389.
222 R. J. Young, C. Cortez, E. Luna, H. Lee, and R. G. Harvey, *J. Org. Chem.*, 1993, **58**, 356.
223 R. Agarwal and D. R. Boyde, *J. Chem. Soc., Perkin Trans. 1*, 1993, 2869.
224 Y.-S. Wu, J.-S. Lai, and P. P. Fu, *J. Org. Chem.*, 1993, **58**, 7283.
225 R. F. C. Brown, F. W. Eastwood, and N. R. Wong, *Tetrahedron Lett.*, 1993, **34**, 1223.
226 R. F. C. Brown, F. W. Eastwood, and N. R. Wong, *Tetrahedron Lett.*, 1993, **34**, 3607.
227 U. E. Wiersum and L. W. Jeneskens, *Tetrahedron Lett.*, 1993, **34**, 6615.
228 G. Dyker, J. Korning, P. G. Jones, and P. Bubenitschek, *Angew. Chem., Int. Ed. Engl.*, 1993, **32**, 1733.
229 F. M. Raymo, M. F. Paris, and F. H. Kohnke, *Tetrahedron Lett.*, 1993, **34**, 5331.
230 A. Ghanimi and J. Simonet, *Tetrahedron Lett.*, 1993, **34**, 6893.
231 R. Neidlein and U. Kux, *Angew. Chem., Int. Ed. Engl.*, 1993, **32**, 1324.

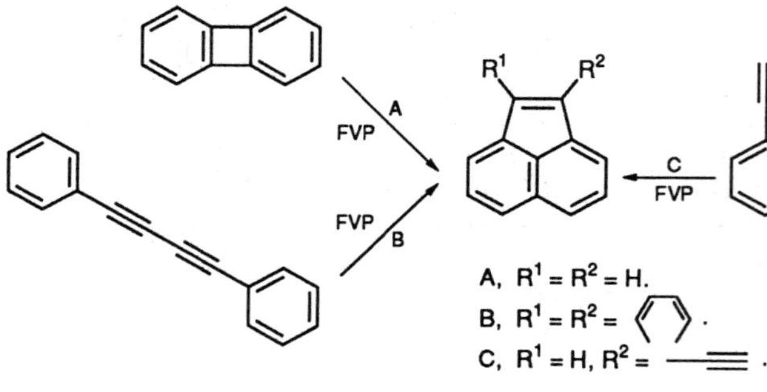

Scheme 30

(69) (70)

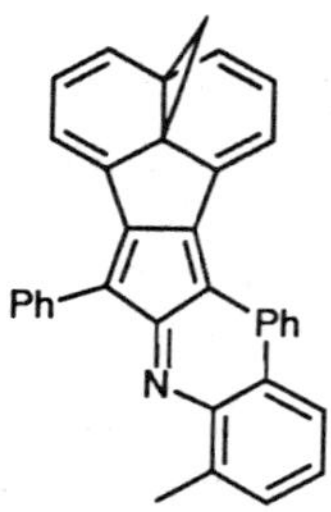

(71)

Azulenes with systems that have a peripheral 14π-electron (72) and an 18π-electron (73) conjugation have been synthesized.[232,233] These systems show little contribution from the extra conjugation, but rather are composed of two separate azulene-type π systems, as in (73), and azulene and fulvene systems, as in (72).

A series of methano-bridged-tetrahydro[20]-, -[24]-, -[26]-, -[28]-, -[30]-, and -[32]annulenediones, (74a)–(74f), have been synthesized. These compounds were expected to show the ring-current effect for the $[4n + 2]\pi$ or $[4n]\pi$-electron system arising from polarization of the two carbonyl groups. However, contrary to these expectations, the annulenediones showed weak tropicity, presumably due to a low planarity of their molecular skeleton. The dication from the tetrahydromethano[24]-annulendione (74b) is strongly diatropic due to possession of a 22π-electron aromatic system.[234] A similar dication of the methano[24]annulenedione (75) exhibits the strongest diatropicity recorded for a 22π-electron system.[235] In an analogous study, methanohomo[15]annulene (76) and methanohomo[19]annulene (77) were prepared. Protonation of these species should yield the corresponding 14π and 18π-electron

232 M. Yasunami, J. Hoioki, Y. Kitamori, I. Kikuchs, and K. Takase, *Bull. Chem. Soc. Jpn.*, 1993, **66**, 2273.
233 M. Nitta, K. Nishimura, and Y. Ino, *Tetrahedron Lett.*, 1993, **34**, 2157.
234 H. Higuchi, K. Asano, K. Nakafuku, Y. Takai, J. Ojima, and G. Yamamoto, *J. Chem. Soc., Perkin Trans. 1*, 1993, 89.
235 G. Yamamoto, H. Higuchi, K. Asano, and J. Ojima, *Chem. Lett.*, 1993, 1829.

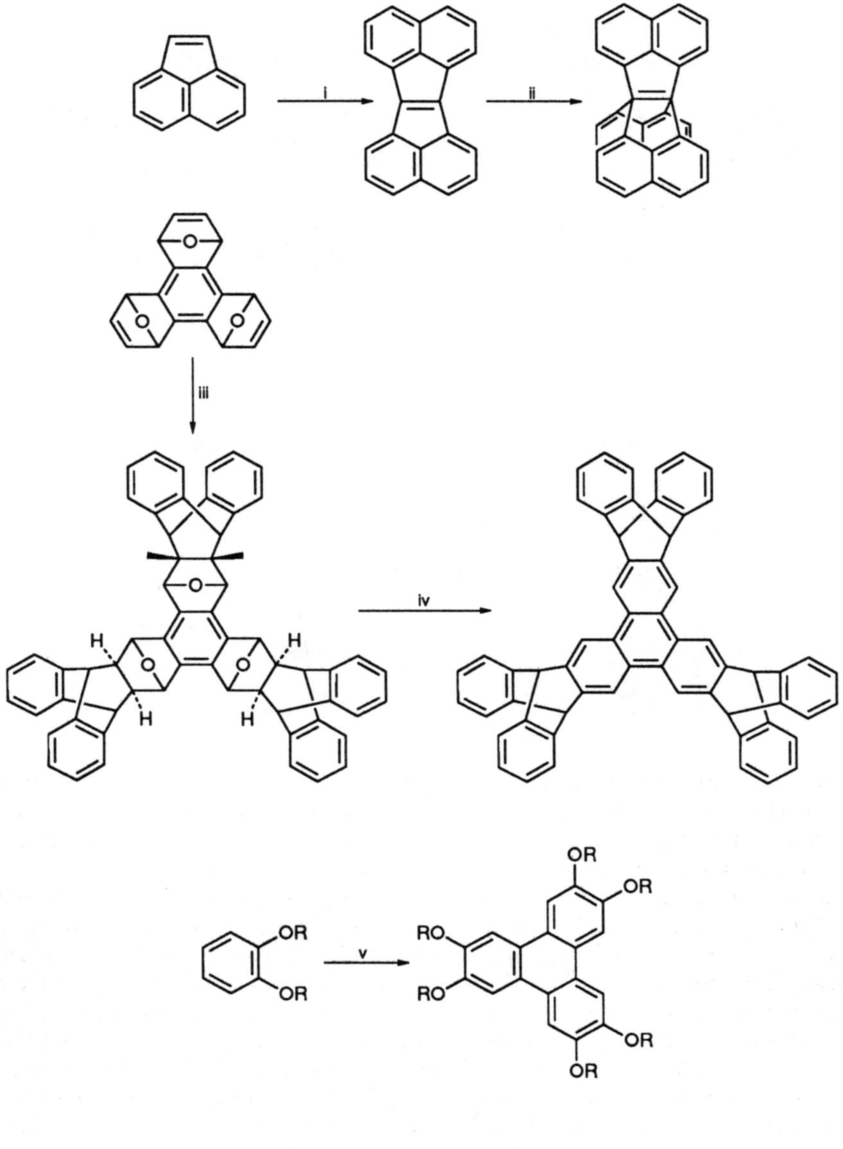

Reagents: i, 1,8-diiodonaphthalene; ii, 1-iodonaphthalene; iii, anthracene, xylene; iv, $HClO_4$, EtOH; v, $-6e^-$, $-6H^+$

Scheme 31

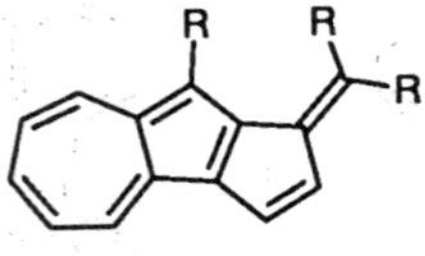

(72)

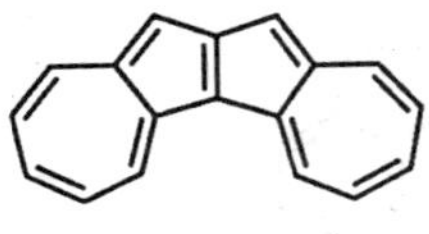

(73)

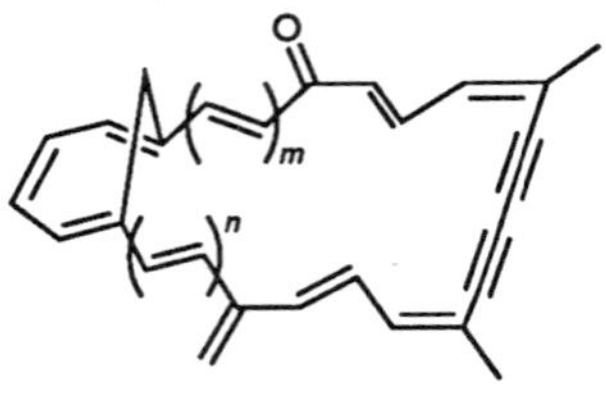

(74a) [20]; $m = n = 0$
(74b) [24]; $m = n = 1$
(74c) [26]; $m = 1; n = 2$
(74d) [28]; $m = n = 2$
(74e) [30]; $m = 2; n = 3$
(74f) [32]; $m = n = 3$

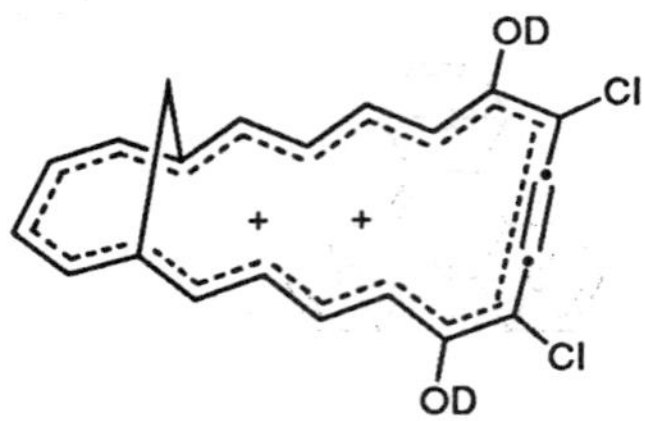

(75)

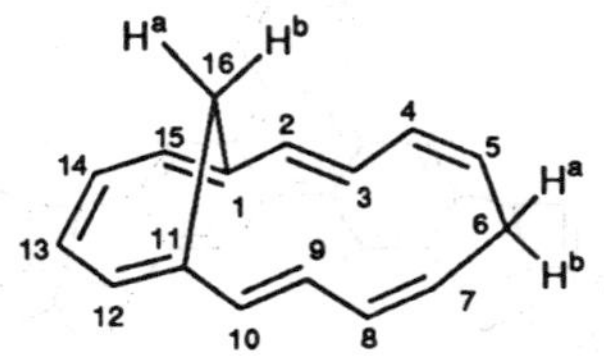

(76)

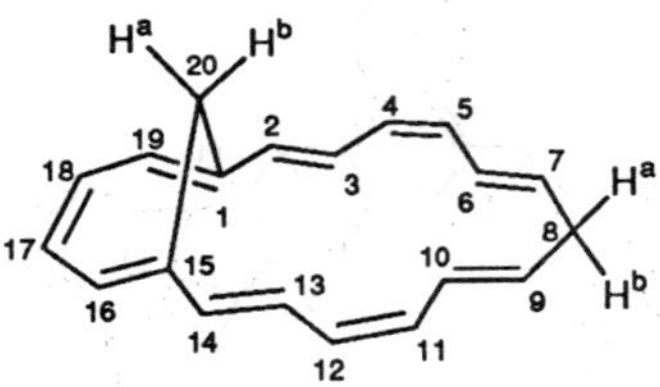

(77)

systems respectively. Attempts at the protonation of (76) and (77) were unsuccessful.[236]

Hay coupling of the tetraethynylethene (78) gives the [12]annulene (79) and the [18]annulene (80); spectral examination of (79) and (80) confirms the expected antiaromatic character of the 12π-electron system and the aromatic character of the 18 π-electron system.[237]

It has been demonstrated that resonance stabilization in [26]annulene (81) is not sufficient to stabilize the molecule in a planar conformation with aromatic properties. However, if this 26π-electron perimeter incorporates four pyrrole units, as in the [26]porphyrin (82), then sufficient stabilization is present and the expected aromatic character is exhibited.[238]

6 Cyclophanes

The strain and distortion imposed upon the benzene ring in cyclophanes imparts an

[236] H. Higuchi, T. Hashimoto, K. Nakafuku, J. Ojima, and G. Yamamoto, *Bull. Chem. Soc. Jpn.*, 1993, **66**, 294.
[237] J. Anthony, C. B. Knobler, and F. Diederich, *Angew. Chem., Int. Ed. Engl.*, 1993, **32**, 406.
[238] T. Wessel, B. Franck, M. Moller, U. Rodewald, and M. Lage, *Angew. Chem., Int. Ed. Engl.*, 1993, **32**, 1148.

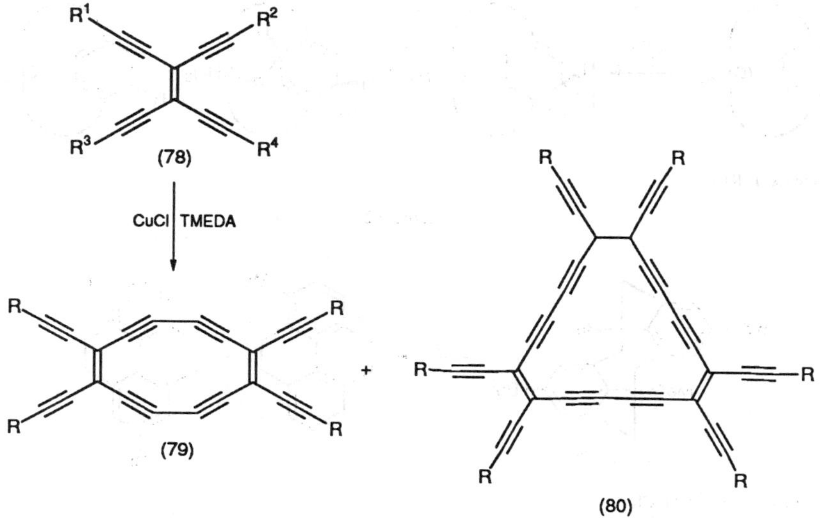

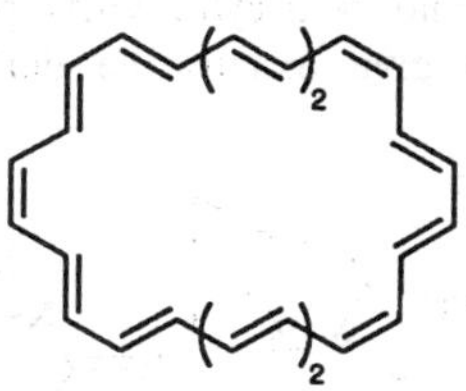

(81)

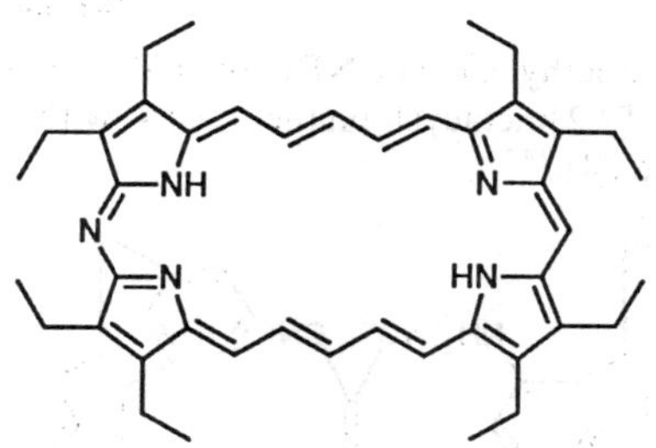

(82)

enhanced reactivity, which can subsequently lead to unusual reactions. [6]Paracyclophanes undergo nucleophilic addition with Bu^nLi and Bu^sLi and formal nucleophilic substitution with Bu^tLi. This is an example of nucleophilic substitution being effected under mild conditions, considering that the benzene ring is not activated by strongly electron-withdrawing groups. This result is ascribed to the release in strain on the formation of the initial adduct (Scheme 32).[239]

[2.2]Metacyclophanes undergo a novel Lewis-acid catalysed transannular cyclization. The cyclization is attributable to the different π-electron densities of the two benzene rings, which are separated from each other by two ethylene units (Scheme 33).[240]

Similar transannular cyclizations have been observed in the bromination of 8-substituted [2]metacyclo[2](1,3)pyrenophanes.[241] Transannular interactions in [2.2]metacyclophanes also affect the selectivity of the tricyanovinylation reaction with

[239] Y. Tobe, M. Jimbo, S. Saiki, K. Kakiuchi, and K. Naemura, *J. Org. Chem.*, 1993, **58**, 5883.
[240] T. Yamato, J.-I. Matsumoto, M. Shigekuni, T. Ishili, and M. Tashiro, *J. Chem. Res.(S)*, 1993, 272.
[241] T. Yamato, A. Miyazawa, and M. Tashiro, *J. Chem. Soc., Perkin Trans. 1*, 1993, 3127.

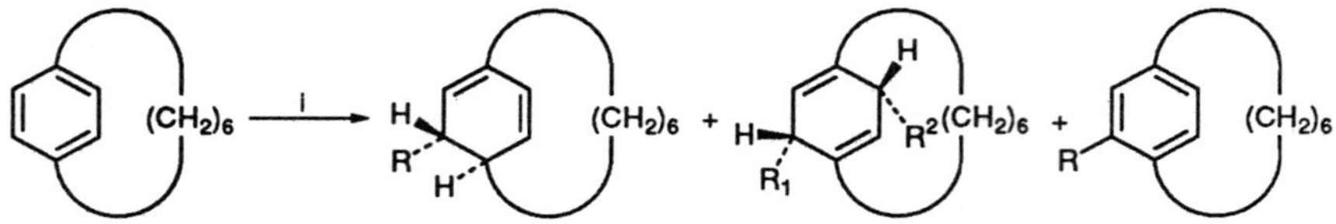

Reagents: i, RLi

Scheme 32

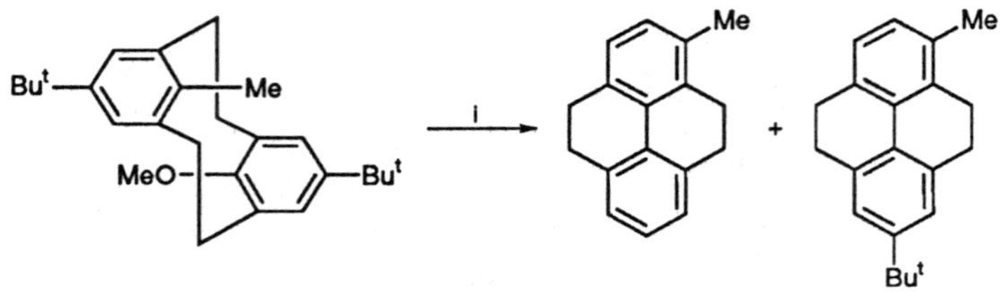

Reagents: i, BBr_3, CH_2Cl_2

Scheme 33

tetracyanoethylene (TCNE). In addition to the normal *N*-tricyanovinyl product, 4-amino[2.2]metacyclophane reacts with TCNE to give oxaziridine derivatives (Scheme 34).[242]

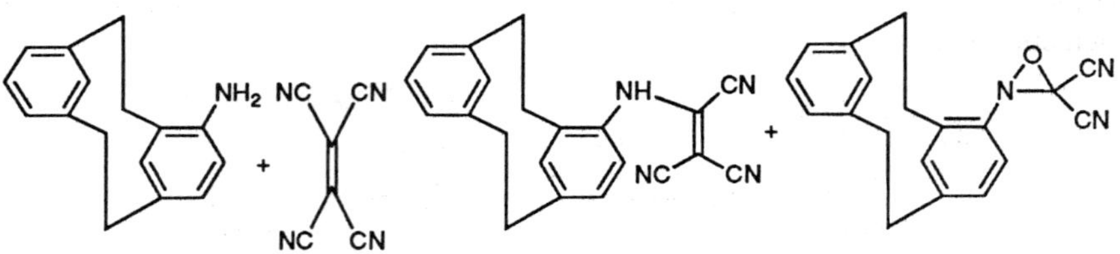

Scheme 34

The highly strained Adamantanophane (83) renders the benzene ring susceptible to dienophile addition in a Diels–Alder reaction.[243] Similar Diels–Alder reactions with the mono–(Dewar benzene) isomer of [1.1]metacyclophane have also been observed (Scheme 35).[244]

The intramolecular interaction between the aromatic rings of small cyclophanes has been the subject of a number of investigations. 8,16-Dimethoxy[2.2]metacyclophane (84) undergoes the diazo coupling reaction easily in one ring, but never in both. The absorption maxima of a series of azo compounds was found to be bathochromically shifted from those of model reference compounds. When the λ_{max} values of the azo compound that carries a functional group in the couter (C-13) position are plotted

[242] A. A. Aly, A. A. Hassan, and A.-F. E. Mourad, *Can. J. Chem.*, 1993, **71**, 1845.
[243] R. Lemmerz, M. Nieger, and F. Vogtle, *J. Chem. Soc., Chem. Commun.*, 1993, 1168.
[244] G. W. Wijsman, D. S. Van Es, W. H. de Wolf, and F. Bickelhaupt, *Angew. Chem., Int. Ed. Engl.*, 1993, **32**, 726.

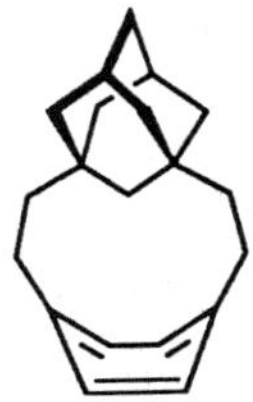

(83)

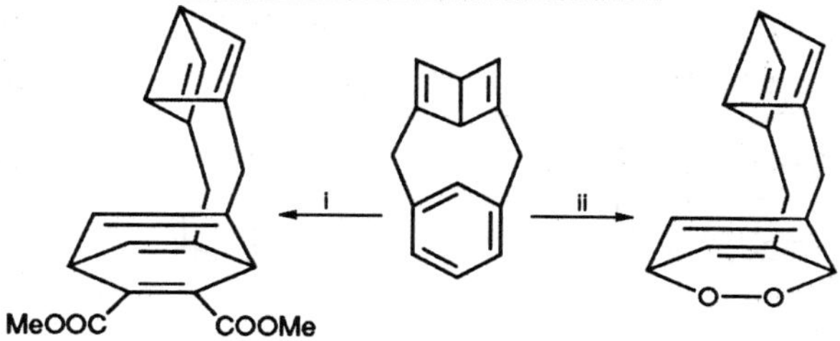

Reagents: $MeO_2CC{\equiv}CCO_2Me$; ii, O_2

Scheme 35

against the substituent Hammett constants (σ,ρ), a straight line is obtained. These results establish that the electronic effects of the functional group are transmitted between the two aromatic rings in [2.2]metacyclophanes by a through-space interaction (Scheme 36).[245]

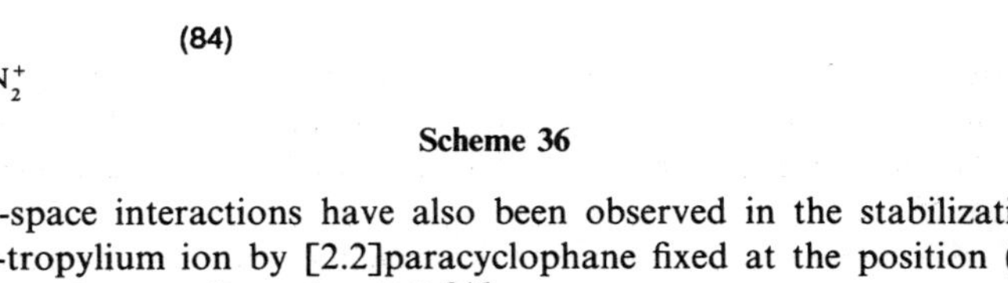

Reagents: PhN_2^+

Scheme 36

Through-space interactions have also been observed in the stabilization of the naphthyl-1-tropylium ion by [2.2]paracyclophane fixed at the position (C-8–C-5′) face-to-face to the tropylium ring (85).[246]

The electronic properties of benzoannelated [2.2]paracyclophane (86) have been studied; such molecules are found to be highly efficient electron acceptors. The unusual geometry of the carbon framework allows reversible acceptance of four to six electrons, thus making it an attractive model for a charge-storage system.[247] The last and most

245 A. Tsuge, T. Moriguchi, S. Mataka, and M. Tachiro, *J. Chem. Soc., Perkin Trans.* 1, 1993, 2211.

246 K. Komatsu, R. Tsuji, Y. Inoue, and K. Takeuchi, *Tetrahedron Lett.*, 1993, **34**, 99.

247 O. Reiser, B. Konig, K. Meerholz, J. Heinze, T. Wellauer, F. Gerson, R. Frim, M. Ravinovitz, and A. de Meijere, *J. Am. Chem. Soc.*, 1993, **115**, 3511.

strained [2.2]cyclophane, (87), has been synthesized; the thermal isomerization of the Dewar benzene valence isomer (88) is the key step.[248]

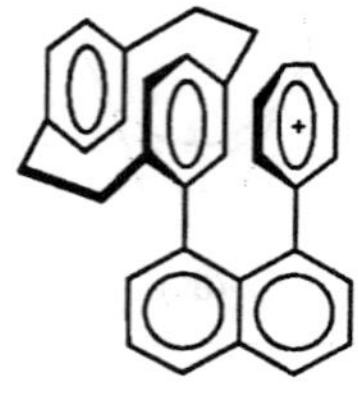

(85)

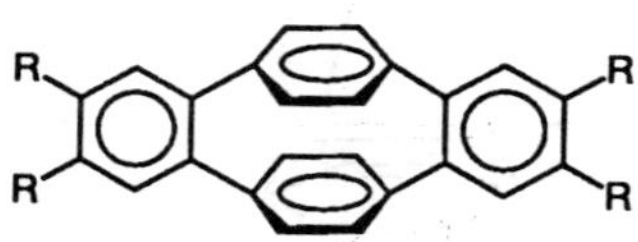

(86)

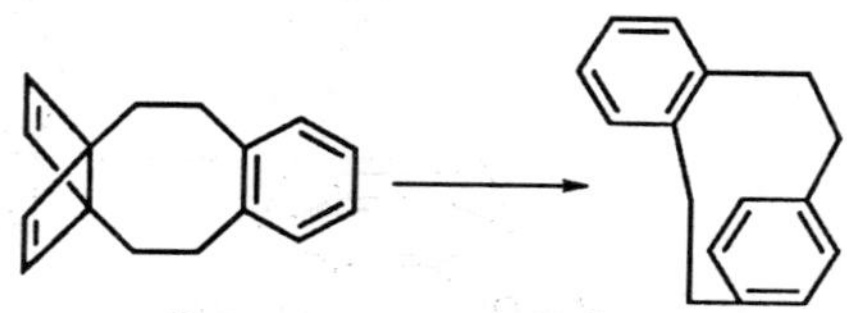

(88) (87)

[248] Y. Tobe, M. Kawaguchi, K. Kakiuchi, and K. Naemura, *J. Am. Chem. Soc.*, 1993, **115**, 1173.

7 Heterocyclic Compounds

By P.W. SHELDRAKE
SmithKline Beecham Pharmaceuticals, Old Powder Mills, nr Leigh, Tonbridge, Kent, TN11 9AN, UK

1 Three-membered Rings

Oxiranes are valuable intermediates standing at the interface of acyclic and heterocyclic chemistry and provide interest to both areas. Oxiranes, exemplified by (1), can be prepared stereoselectively, and rearrange on treatment with stannic chloride to give keto-alcohols (2), again stereoselectively and in excellent yield (Scheme 1).[1]

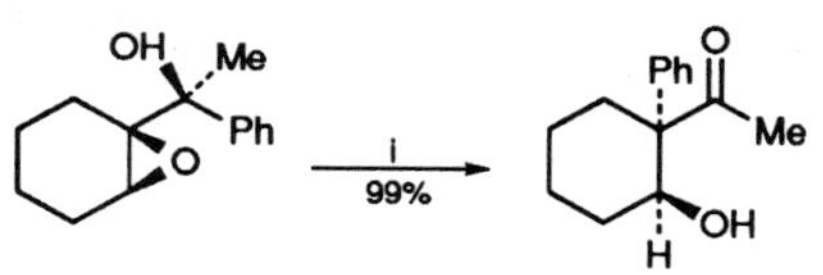

(1) (2)

Reagents: i, $SnCl_4$

Scheme 1

Lithium tetrachlorocuprate(II) has been shown to be efficient at opening epoxides to give chloro-alcohols.[2] For example, the bis-oxirane (3) gives (4) in 65% yield (Scheme 2).

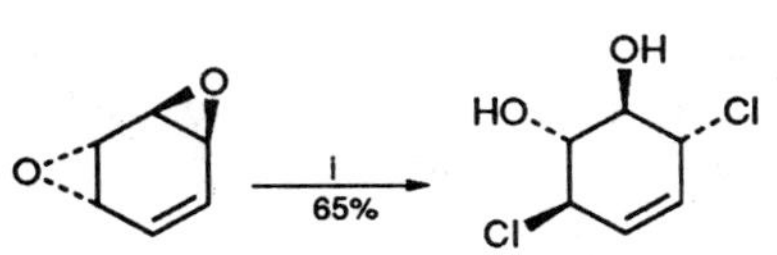

(3) (4)

Reagents: i, Li_2CuCl_4, THF

Scheme 2

When the racemic oxirane (5) is treated with acid the product is a (racemic)

[1] C.M. Marson, A.J. Walker, J. Pickering, A.D. Hobson, R. Wrigglesworth, and S.J. Edge, *J. Org. Chem.*, 1993, **58**, 5994.
[2] Z.-X. Guo, A.H. Haines, and R.J.K. Taylor, *SYNLETT.*, 1993, 607.

tetrahydrofuran (6). However, on treatment with antibodies raised against (7) the product is predictably optically active, but is the tetrahydropyran (8) (Scheme 3).[3]

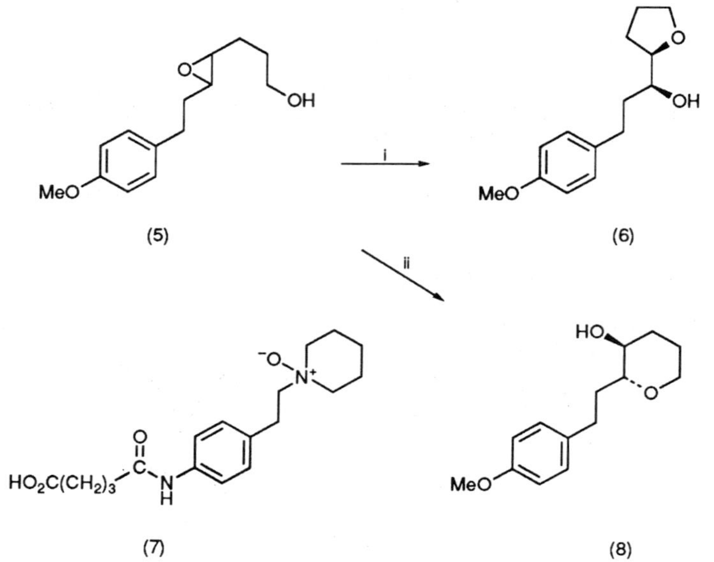

Reagents: i, H^+; ii, antibodies

Scheme 3

The biosynthetically-patterned cascade oxirane cyclization of (9) has been used as a key step to produce (10), in 30% yield, in studies directed towards the synthesis of etheromycin (Scheme 4).[4]

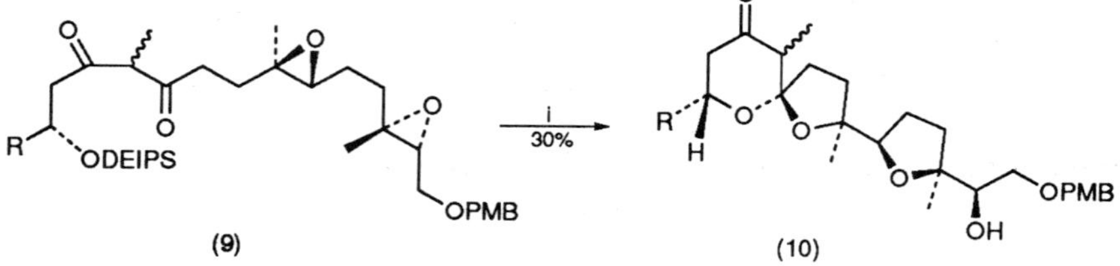

Reagents: i, 0.5 M HCl, THF

Scheme 4

Treatment of hydroxylamine derivatives (11) with sodium hydride in the presence of suitable olefins (12) gives aziridines (13) in yields of up to 95% (Scheme 5).[5]

[3] K. D. Jandu, C. G. Shevlin, and R. A. Lerner, *Science*, 1993, **259**, 490.
[4] I. Paterson, R. D. Tillyer, and J. B. Smaill, *Tetrahedron Lett.*, 1993, **34**, 7137.
[5] M. M. Pereira, P. P. O. Santos, L. V. Reis, A. M. Lobo, and S. Prabhaker, *J. Chem. Soc., Chem. Commun.*, 1993, 38.

Reduction of 6*H*-1,2-oxazines (14) with lithium aluminium hydride gives aziridines (15).[6] The *cis*-isomer predominates; for example, where $R^1 = R^2 = Me$ the yield is 64% with a 94:6 ratio of *cis*:*trans* (Scheme 6).

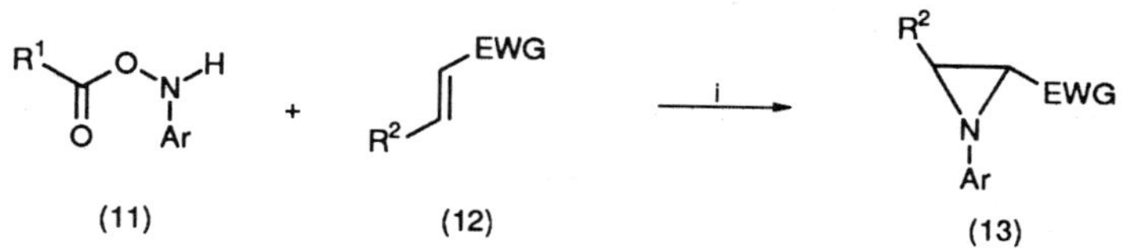

Reagents: i, NaH

Scheme 5

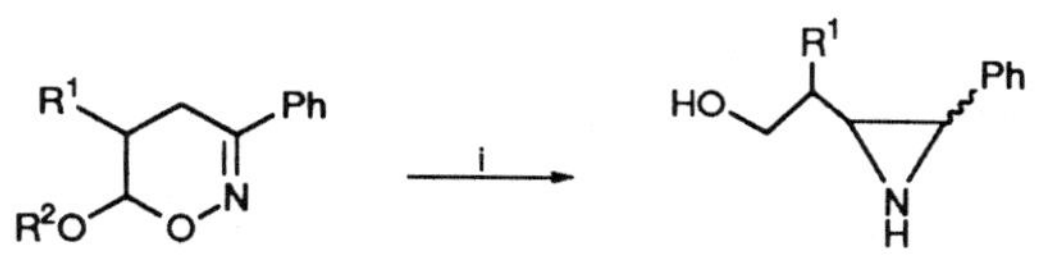

Reagents: i, $LiAlH_4$, Et_2O

Scheme 6

The chiral auxiliary (16) has been used in the asymmetric synthesis of aziridines (18) from olefins (17).[7] Enantiomeric excesses up to 96% were observed, with yields up to 76% (Scheme 7).

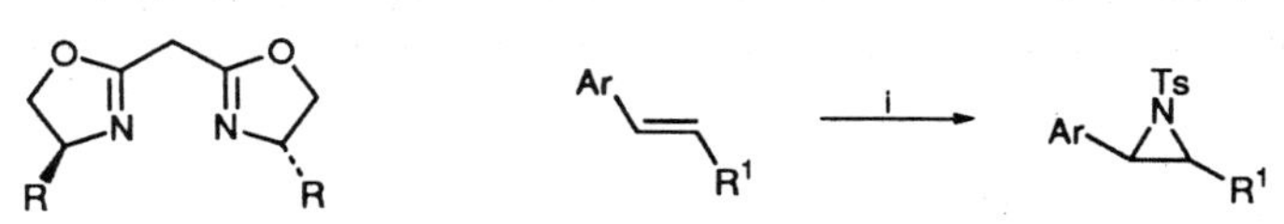

Reagents: i, PhI=NTs, CuOTf, (16)

Scheme 7

Dimethyldioxirane has been used to oxidize furans (including furan itself) to malealdehydes.[8] These were reacted *in situ*, *e.g.* with Wittig reagents, and the products isolated in good yields (Scheme 8).

Methyl(trifluoromethyl)dioxirane oxidizes unactivated carbon–hydrogen bonds in

[6] R. Zimmer, K. Hommann, and H.-U. Reissig, *Liebigs Ann. Chem.*, 1993, 1155.
[7] D. A. Evans, M. M. Faul, M. T. Bilodeau, B. A. Anderson, and D. M. Barnes, *J. Am. Chem. Soc.*, 1993, **115**, 5328.
[8] B. J. Adger, C. Barrett, J. Brennan, P. McGuigan, M. A. McKervey, and B. Tarbit, *J. Chem. Soc., Chem. Commun.*, 1993, 1220.

amine salts with remarkable selectivity.[9] The ultimate products from derivatives of pentylamine salts (19) are 3,4,5,6-tetrahydropyridines (20) (Scheme 9).

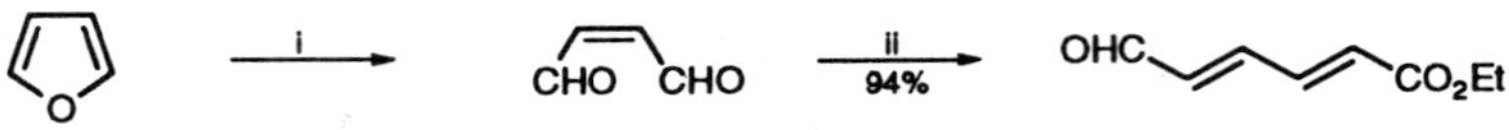

Reagents: i, dimethyldioxirane; ii, $Ph_3P{=}CHCO_2Et$ (1 equiv.)

Scheme 8

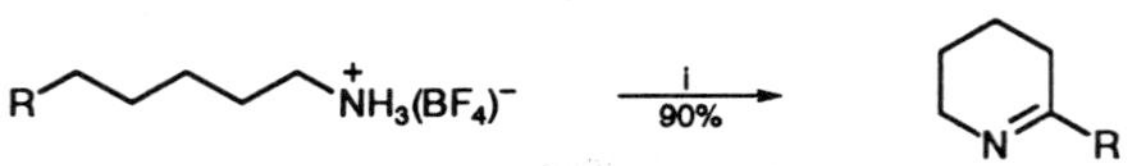

(19) (20)

Reagents: i, methyl(trifluoromethyl)dioxirane

Scheme 9

The oxaziridines (21) transfer the *N*-Boc group to nitrogen or carbon nucleophiles.[10]

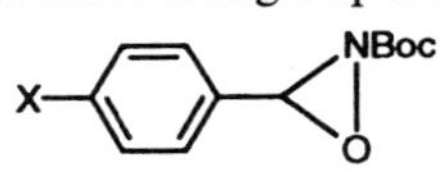

(21) X = H, CN

2 Four-membered Rings

There are reviews of industrial transformations of penicillins and cephalosporins[11] and of oxo and imino-functionalized 1,2-oxazetidines.[12]

On treatment with silver triflate the silyl enol ethers (22) cyclize onto the oxonium ion intermediate and an oxetane is annulated to the existing ring giving the unusual bicycles (23) (Scheme 10).[13]

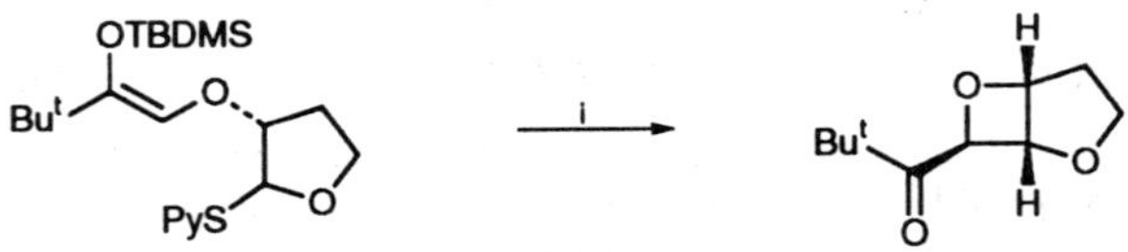

(22) (23)

Reagents: i, AgOTf, CH_2Cl_2, 4Å Sieves

Scheme 10

Tin bis(hexamethyldisilazide) was studied as a reagent for converting esters into

[9] G. Asensio, M. E. Gonzalez-Nunez, C. B. Bernardini, R. Mello, and W. Adam, *J. Am. Chem. Soc.*, 1993, **115**, 7250.
[10] J. Vidal, L. Guy, S. Sterin, and A. Collett, *J. Org. Chem.*, 1993, **58**, 4791.
[11] J. Verweij and E. de Vroom, *Recl. Trav. Chim. Pays-Bas*, 1993, **112**, 66.
[12] D. Monderhack, *J. Heterocycl. Chem.*, 1993, **30**, 579.
[13] D. Craig and V. R. N. Munasinghe, *J. Chem. Soc., Chem. Commun.*, 1993, 901.

secondary amides[14] and was found to be efficient in cyclizing β-alkylamino esters to β-lactams (Scheme 11).[15]

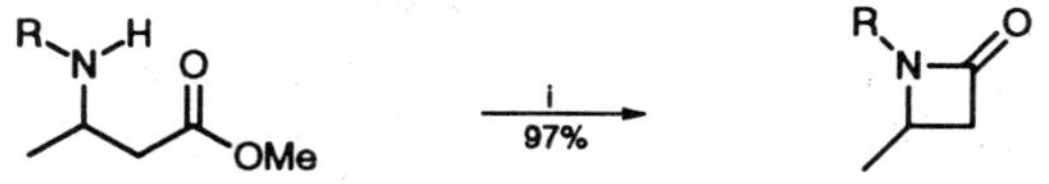

Reagents: i, $Sn[N(SiMe_3)_2]_2$

Scheme 11

Allyl diethyl phosphate was used to form an allyl palladium species from which β-lactams (24) were constructed[16] by carbonylation followed by reaction with an imine (Scheme 12).

The 2-aza-1,3-dienes (25) were prepared in good yield.[17] Staudinger reaction with acid chlorides gave β-lactams (26) in which the *cis* isomer predominated and from which the nitrogen substituent was removed using potassium permanganate (Scheme 13).

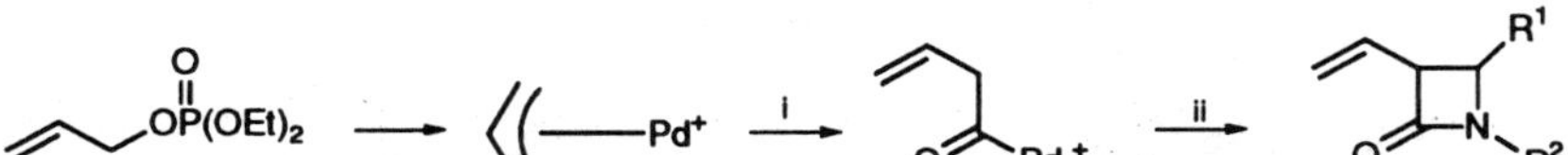

(24)

Reagents: i, CO; ii, *trans*-R'CH=NR^2

Scheme 12

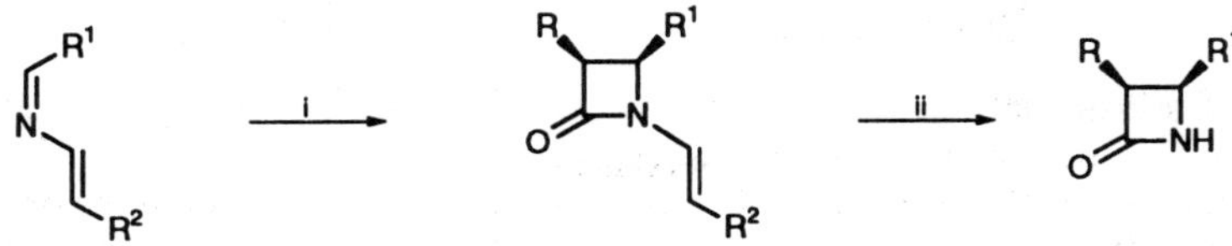

(25) (26) (27)

Reagents: i, RCH_2COCl, Et_3N; ii, $KMnO_4$

Scheme 13

A new synthesis of the important intermediate (29) was reported[18] starting with the furanyl β-lactam (28). Degradation of the furan was brought about using singlet oxygen and the resulting peroxy-compounds rearranged thermally (Scheme 14).

A radical-mediated ring closure formed the carbapenam (31) from the dithiane (30) (Scheme 15).[19] Another radical-mediated cyclization[20] demonstrated formation of the

[14] W. B. Wang, J. A. Restituyo, and E. J. Roskamp, *Tetrahedron Lett.*, 1993, **34**, 7217.
[15] W.-B. Wang and E. J. Roskamp, *J. Am. Chem. Soc.*, 1993, **115**, 9417.
[16] S. Torii, H. Okumoto, M. Sadakane, A. K. M. A. Hai, and H. Tanaka, *Tetrahedron Lett.*, 1993, **34**, 6553.
[17] G. I. Georg, P. He, J. Kant, and Z. J. Wu, *J. Org. Chem.*, 1993, **58**, 5771.
[18] J. E. Lynch, W. L. Laswell, R. P. Volante, R. A. Reamer, D. M. Tschaen, and I. Shinkai, *Heterocycles*, 1993, **35**, 1029.
[19] J. Anaya, D. H. R. Barton, S. D. Gero, M. Grande, N. Martin, and C. Tachdijian, *Angew. Chem., Int. Ed. Engl.*, 1993, **32**, 867.
[20] H. Ishibashi, C. Kameoka, A. Yoshikawa, R. Ueda, K. Kodama, T. Sato, and M. Ikeda, *SYNLETT.*, 1993, 649.

β-lactam ring itself in the conversion of (32) into (33) (Scheme 16).

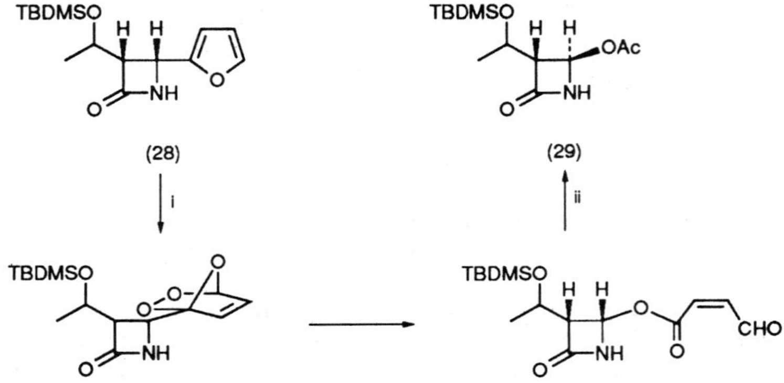

Reagents: i, $^1O_2$; ii, KOAc

Scheme 14

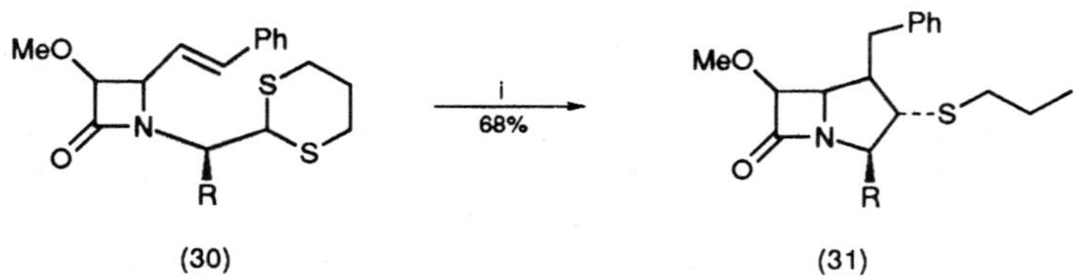

Reagents: i, Bu_3SnH, AIBN

Scheme 15

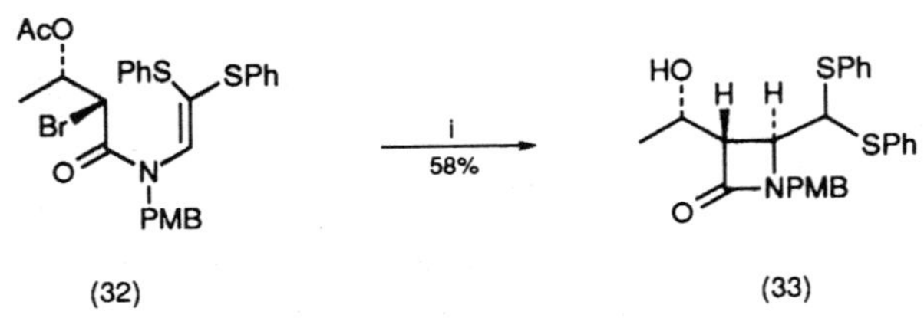

Reagents: i, Bu_3SnH, AIBN

Scheme 16

Titanium tetrachloride was used to cyclize (34) to the carbapenem (35) (Scheme 17),[21] which was taken on to thienamycin.

Irradiation of a powdered inclusion complex of *N*-(aroylmethyl)lactams (36) in the chiral auxiliary (37) produces the azetidino-lactams (38) in yields up to 59% and up to 98% enantiomeric excess (Scheme 18).[22]

[21] G. B. Feigelson, *Tetrahedron Lett.*, 1993, **34**, 4747.
[22] F. Toda, K. Tanaka, O. Kakindii, and T. Kawakami, *J. Org. Chem.*, 1993, **58**, 3783.

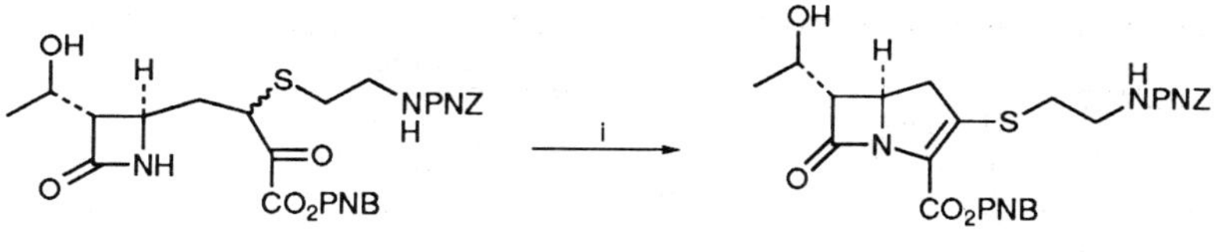

(34) (35)

Reagents: i, $TiCl_4$, CH_2Cl_2

Scheme 17

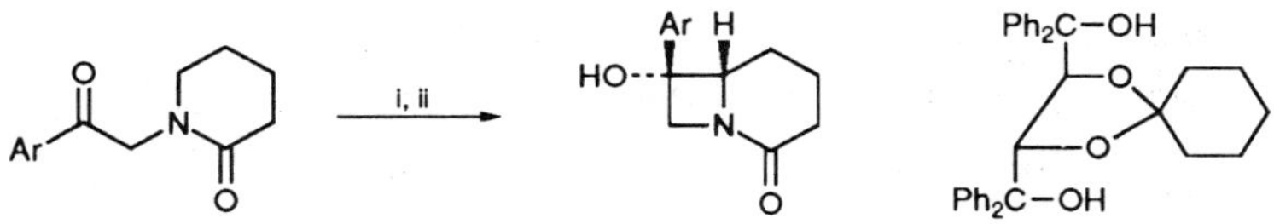

(36) (38) (37)

Reagents: i, (37); ii, *hν*

Scheme 18

Rhodium(II) acetate mediated closure of the diazo-esters (39) has been studied (Scheme 19).[23] When $n = 0$ there is no ring closure because the sulfur interferes with the carbene intermediate; for the sulfoxide or sulfone the reaction is fairly efficient.

The chemistry of the 3-methylidene-β-lactams (41) has been investigated:[24] the double bond can be epoxidized or cyclopropanated and is a good Michael acceptor and dienophile. On reaction with a hydrazine (Scheme 20) the primary product is the spiro compound (42) from which it is possible to obtain the pyrazole (43).

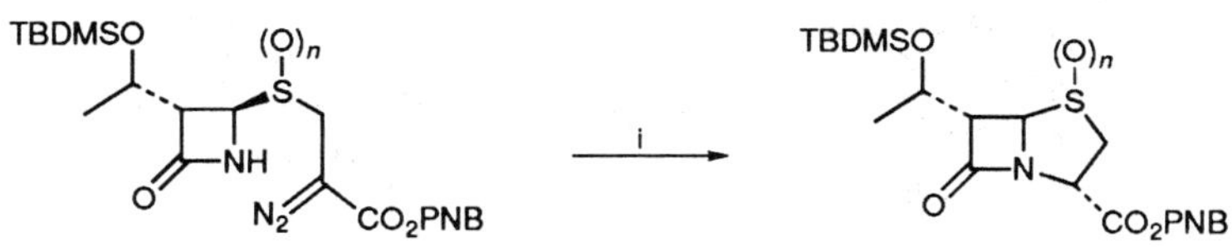

(39) n = 0,1,2 (40)

Reagents: i, $Rh_2(OAc)_4$

Scheme 19

Deprotection of β-lactam (44) with ceric ammonium nitrate proceeds as expected.[25] However, on similarly treating its diastereoisomer (45) there is an unprecedented oxidative ring transformation giving (46) in high yield (Scheme 21).

[23] P. Bissolino, M. Alpegiani, D. Borghi, E. Perrone, and G. Franceschi, *Heterocycles*, 1993, **36**, 1529.
[24] S. Gurthler, M. Johner, S. Ruf, and H.-H. Otto, *Helv. Chim. Acta*, 1993, **76**, 2959.
[25] F. Bertha, J. Fetter, M. Kajtar-Peredy, G. M. Kesura, K. Lempot, L. Parkanji, and J. Tamas, *Tetrahedron.*, 1993, **49**, 7803.

The β-lactams (47a) and (47b) on treatment with stannous chloride give rise to a pyrazine dione (48a) or 1,4-oxazine dione (48b) (Scheme 22).[26]

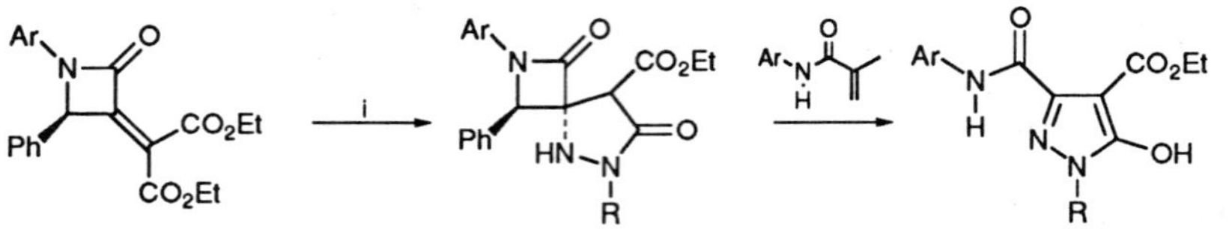

(41) (42) (43) R = Me or $PhCH_2$

Reagents: i, $RNHNH_2$

Scheme 20

The β-lactam (49) is readily opened with dimethyloxosulfonium methylide and the product cyclizes on treatment with rhodium(II) acetate to give the 3-ketopyrrolidine (50) (Scheme 23).[27]

The silacyclobutane (51) reacts with lithium carbenoids, thus undergoing ring expansion to give 2-iodo-1,1-dimethyl-1-silacyclopentane (52) (Scheme 24).[28]

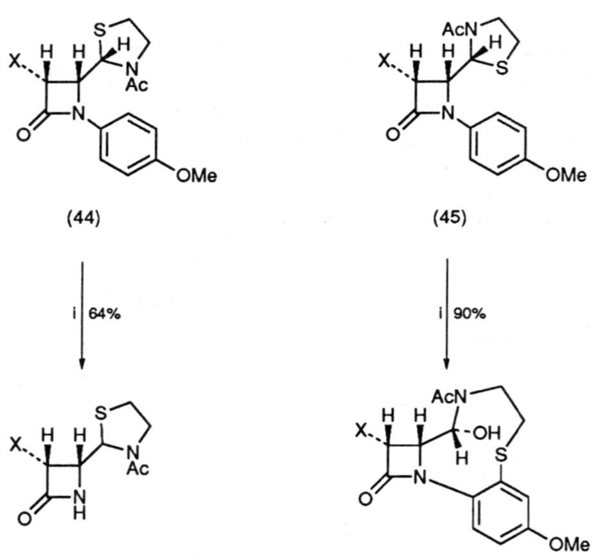

(44) (45) (46)

Reagents: i, $(NH_4)Ce(NO_3)_6$

Scheme 21

[26] B. Alcaide, Y. Martin-Cantelejo, J. Rodriquez-Lopez, and M. A. Sierra, *J. Org. Chem.*, 1993, **58**, 4767.
[27] J. E. Baldwin, R. M. Adlington, C. R. A. Godfrey, D. W. Gollins, and J. G. Vaughan, *J. Chem. Soc., Chem. Commun.*, 1993, 1434.
[28] K. Matsumoto, Y. Aoki, K. Oshima, K. Utimoto, and N. A. Rahman, *Tetrahedron*, 1993, **49**, 8487.

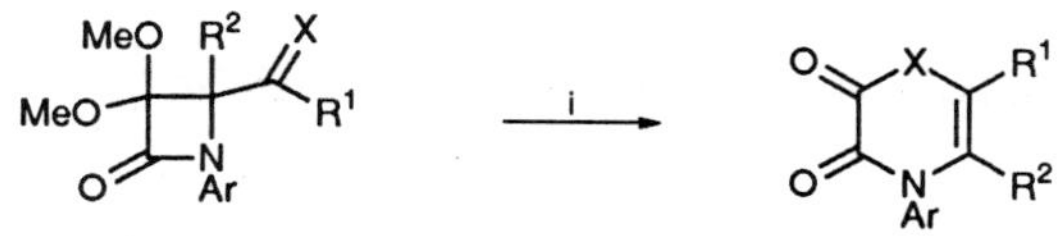

(47a) X = NAr
(47b) X = O

(48a) X = NAr
(48b) X = O

Reagents: i, $SnCl_2 \cdot 2H_2O$

Scheme 22

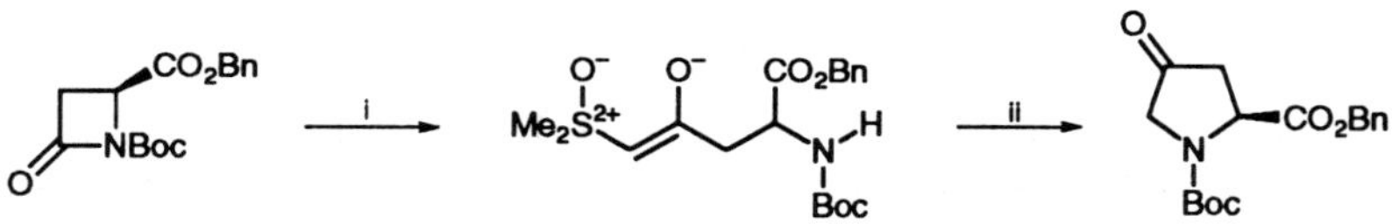

(49) (50)

Reagents: i, Me_2SOCH_2, DMSO; ii, Rh_2 $(OAc)_4$

Scheme 23

Bis(diisopropylamino)phosphanylazide (53) reacts with dimethyl acetylenedicarboxylate to give the six-membered ring (54),[29] from which on heating nitrogen is lost producing the 1,2-λ^5-azaphosphate (55) (Scheme 25). The product is air-stable and melts (mp 109–110 °C) without decomposition!

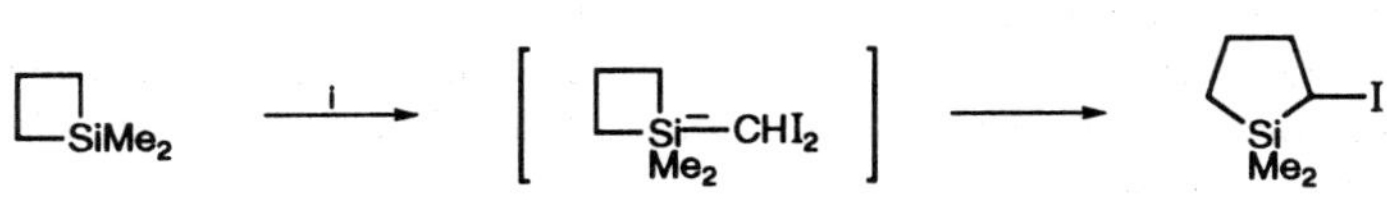

(51) (52)

Reagents: i, $LiCHI_2$

Scheme 24

3 Five-membered Rings

A review of the use of isocyanides in heterocyclic synthesis[30] predominantly covers five-membered rings. More specifically there are reviews of new chemistry of oxazoles,[31] recent progress in preparation and synthetic uses of 3*H*-pyrazoles,[32] and the preparation and reactions of indolin-2(3*H*)-ones (oxindoles).[33]

[29] J. Tejeda, R. Reau, F. Dahan, and G. Bertrand, *J. Am. Chem. Soc.*, 1993, **115**, 7880.
[30] S. Marcaccini and T. Torroba, *Org. Prep. Proced. Int.*, 1993, **25**, 143.
[31] A. Hassner and B. Fischer, *Heterocycles*, 1993, **35**, 1441.
[32] T. Nagai and M. Hamaguchi, *Org. Prep. Proced. Int.*, 1993, **25**, 403.
[33] G. M. Karp, *Org. Prep. Proced. Int.*, 1993, **25**, 483.

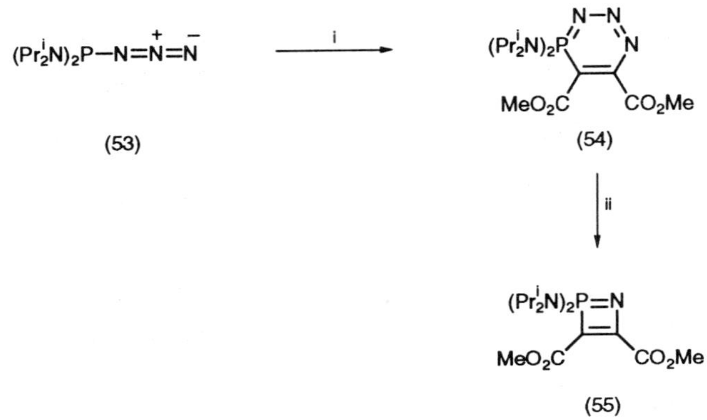

Reagents: i, $MeO_2CC{\equiv}CCO_2Me$; ii, Δ

Scheme 25

There is a review of the lateral metallation of isoxazoles.[34] Metallation of heterocycles continues to grow as a powerful synthetic tool complementary to 'classical' synthetic routes. A definitive paper[35] on the metallation of benzo-fused nitrogen heterocycles covers reaction both adjacent to nitrogen, with formamidine derivatives (56), and in the aromatic ring, with t-butoxycarbonyl derivatives (58) (Scheme 26). The t-butoxycarbonyl derivatives of pyrrolidine, piperidine, and hexamethyleneimine can also be lithiated efficiently (Scheme 27).[36] Dilithiation of β-bromo-β-trimethylsilyl styrene[37] (60) gives an intermediate (61) into which can be inserted as the heteroatom sulfur, selenium, tellurium, phosphorus, arsenic, antimony, bismuth, silicon, germanium, or tin (Scheme 28).

Both enantiomers of imidazoline (63) are available *via* resolution. Lithiation–alkylation gives (64) from which amino acids are obtained in high yield and high optical purity (Scheme 29).[38]

1-Substituted-1,2,4-triazoles, where the substituent is an alkyl or diethoxymethyl group, have been lithiated at the 5-position and reacted with a range of electrophiles.[39] 1-Substituted-4-iodoimidazoles treated with ethyl magnesium bromide followed by an electrophile react at the 4-position.[40] No products from isomerization of the 4-metallated imidazole to the 2-metallated species were detected. The intermediates (66) and (67) have both been formed[41] by direct deprotonation of the corresponding acids.

[34] N. R. Natale and Y. R. Mirzaei, *Org. Prep. Proced. Int.*, 1993, **25**, 517.
[35] A. I. Meyers and G. Milot, *J. Org. Chem.*, 1993, **58**, 6538.
[36] P. Beak and W. K. Lee, *J. Org. Chem.*, 1993, **58**, 1109.
[37] J. Kurita, M. Ishii, S. Yasuika, and T. Tsuchiya, *J. Chem. Soc., Chem. Commun.*, 1993, 1309.
[38] S. Blank and D. Seebach, *Angew. Chem., Int. Ed. Engl.*, 1993, **32**, 1765.
[39] S. Ohta, I. Kaawasaki, A. Fukuno, M. Yamashita, T. Tada, and T. Kawabata, *Chem. Pharm. Bull.*, 1993, **41**, 1226.
[40] R. M. Turner, S. V. Ley, and S. D. Lindell, *SYNLETT.*, 1993, 748.
[41] C. D. Buttery, R. G. Jones, and D. W. Knight, *J. Chem. Soc., Perkin Trans. 1*, 1993, 1425.

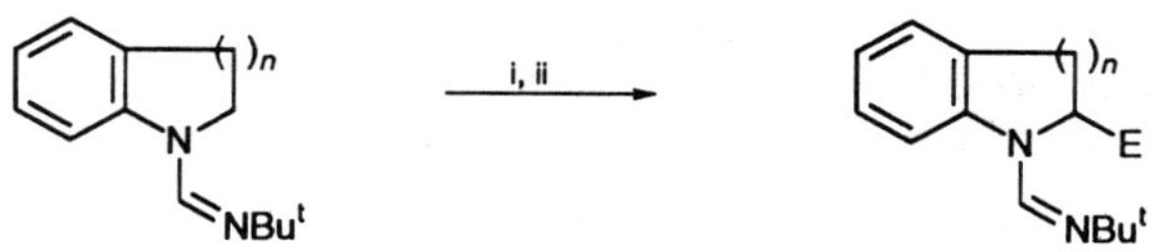

(56) n = 1,2,3 (57)

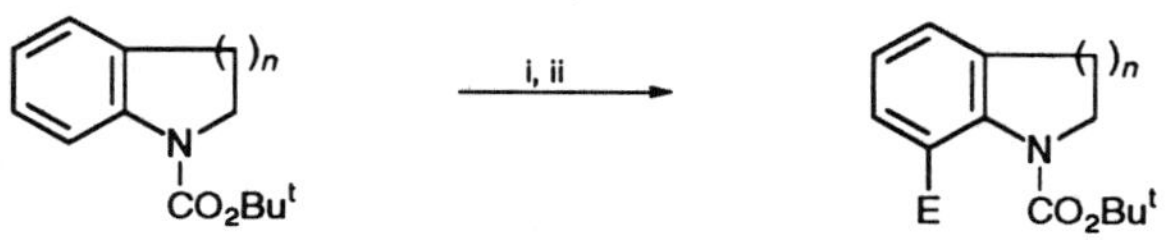

(58) n = 1,2,3 (59)

Reagents: i, Bu^tLi; ii, electrophile, E

Scheme 26

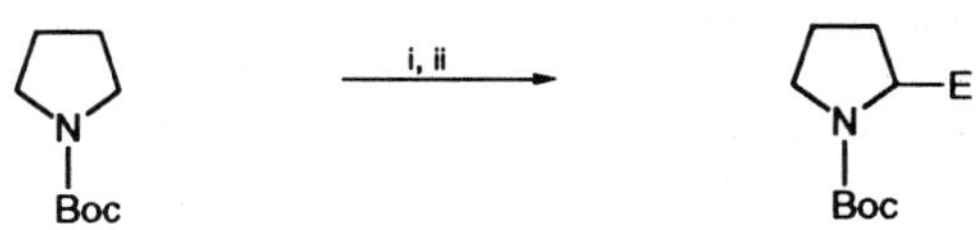

Reagents: i, RLi, TMEDA; ii, electrophile, E

Scheme 27

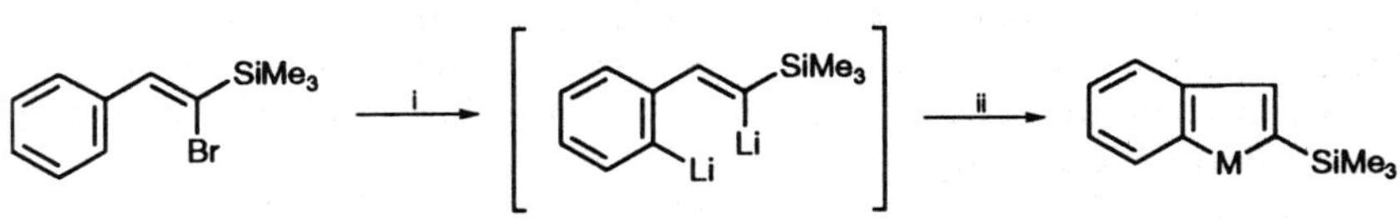

(60) (61) (62)

Reagents: i, Bu^nLi, Et_2O; ii, for M = S, $(PhSo_2)_2S$; for M = Se, Se; for M = AsPh, $PhAsCl_2$; for M = $SiMe_2$, Me_2SiCl_2

Scheme 28

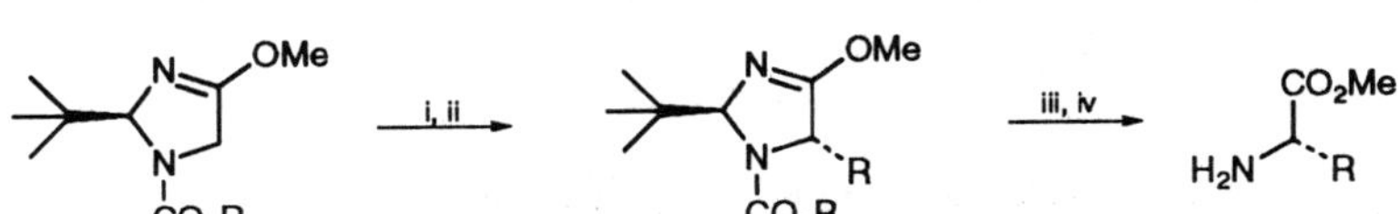

(63) (64) (65)

Reagents: i, LDA, THF, −78 °C; ii, RX; iii, CF_3CO_2H, CH_2Cl_2; iv, H_3O^+

Scheme 29

The mesylate and tosylate (68) undergo intramolecular cyclization involving the oxygen of the ether linkage to give the tetrahydrofuran (69).[42] The reaction occurs in warm DMF or in acetone or acetonitrile, provided the last two solvents contain lithium bromide (Scheme 30).

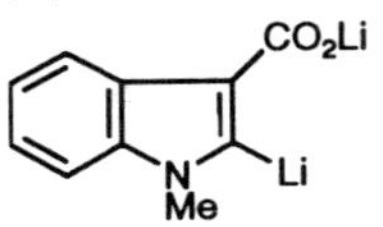

(66)

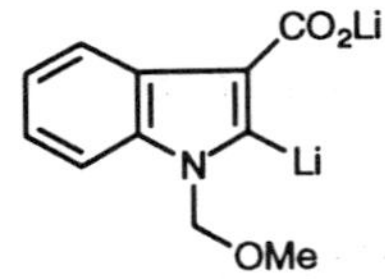

(67)

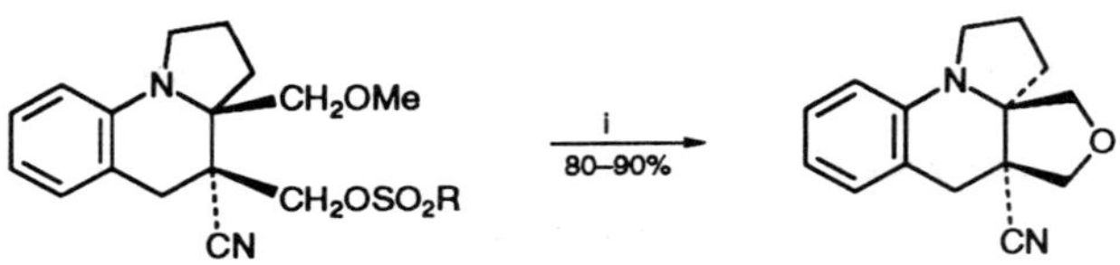

(68) R = Me, Tol (69)

Reagents: i, $Me_2C{=}O$, LiBr or MeCN, LiBr

Scheme 30

On treatment with AIBN/tributyltin hydride the bromide (70a) forms a radical which ring closes to give the tetrahydrofuran (71a).[43] A tetrahydropyran (71b) is available by the same methodology (Scheme 31).

4-Substituted furan-2-ylacetic acid esters are available by the route shown in Scheme 32.[44]

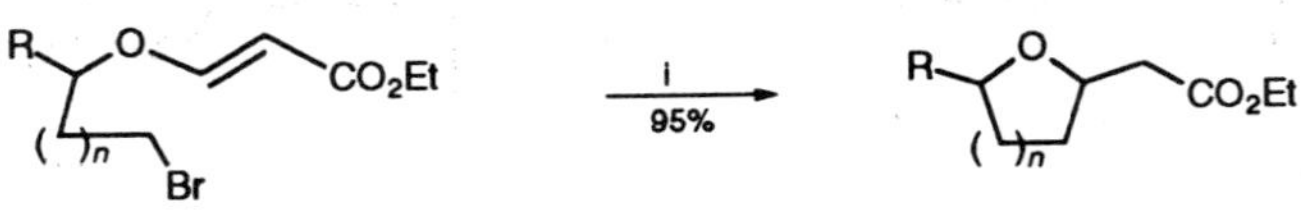

(70a) $n = 1$
(70b) $n = 2$

(71a) $n = 1$
(71b) $n = 2$

Reagents: i, Bu_3SnH, AIBN

Scheme 31

Di-, tri-, or tetrasubstituted furans are available[45] by base-catalysed cyclization of hydroxyenynes (72); some variations and simplifications[46] of the idea are shown in Scheme 33.

[42] W. Verbroom, Y. Morzherin, E. Kelderman, J. F. J. Engbersen, G. I. van Hummel, S. Harkema, and D. N. Reinhoudt, *Recl. Trav. Chim. Pays-Bas*, 1993, **112**, 549.
[43] E. Lee, J. S. Tae, C. Lee, and C. M. Park, *Tetrahedron Lett.*, 1993, **34**, 4831.
[44] R. W. Carling and P. D. Leeson, *SYNLETT.*, 1993, 40.
[45] J. A. Marshall and W. J. DuBay, *J. Org. Chem.*, 1993, **58**, 3435.
[46] J. A. Marshall and W. J. DuBay, *J. Org. Chem.*, 1993, **58**, 3602.

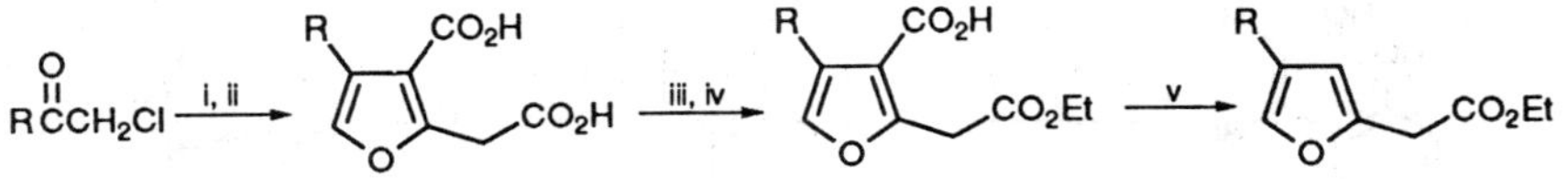

Reagents: i, $(EtO_2CCH_2)_2C{=}O$; ii, KOH; iii, CH_3COCl; iv, EtOH; v, Cu_2O, quinoline

Scheme 32

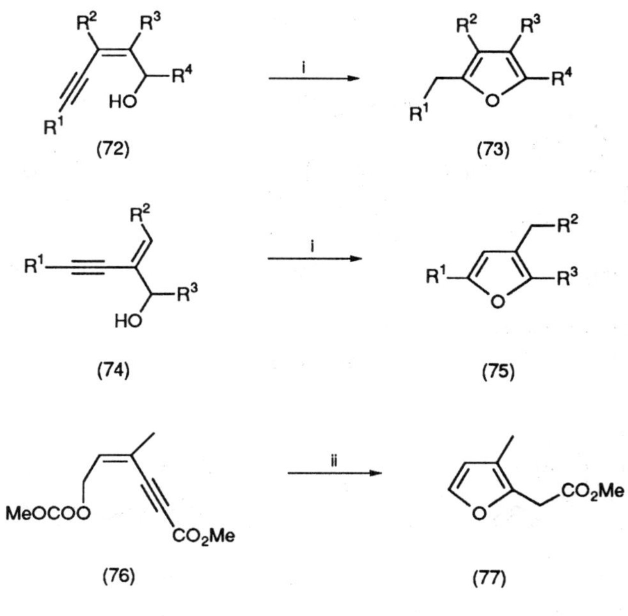

Reagents: i, KOBu^t, Bu^{t}OH, 18-crown-6; ii, NaOMe

Scheme 33

Benzotriazole is a versatile reagent. Just one of its new uses involves setting up the intermediate (78), which by cyclization and elimination of benzotriazole gives benzofurans (79) (Scheme 34).[47]

The sugar-derived triflate (80) is converted into (81) in high yield by acid,[48] but in base the more unusual bicycle (82) is produced with almost equal efficiency (Scheme 35).

The unsaturated ketal (83) gives diol (84) in high yield and enantiomeric excess by Sharpless asymmetric dihydroxylation.[49] Treatment with acid completes an efficient synthesis of (+)-*exo*-brevicomin (Scheme 36).

[47] A. R. Katritzky, X. Lan, and Z. Zhang, *J. Heterocycl. Chem.*, 1993, 381.
[48] J. R. Wheatley, C. J. F. Bichard, S. J. Mantell, J. C. Son, D. J. Hughes, G. W. J. Fleet, and D. Brown, *J. Chem. Soc., Chem. Commun.*, 1993, 1065.
[49] J. A. Sodequist and A. M. Rane, *Tetrahedron Lett.*, 1993, **34**, 5031.

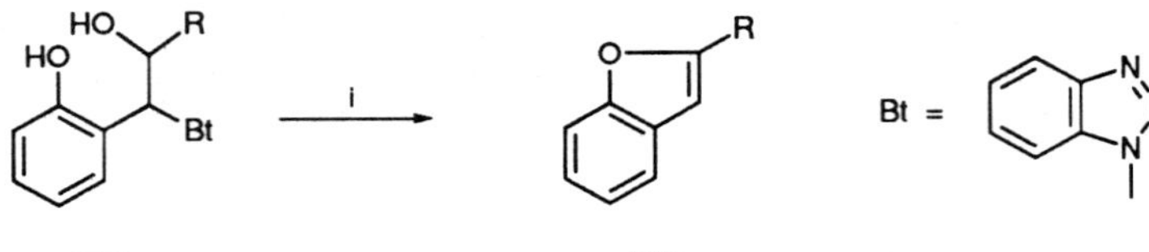

(78) (79)

Reagents: i, TsOH, AcOH

Scheme 34

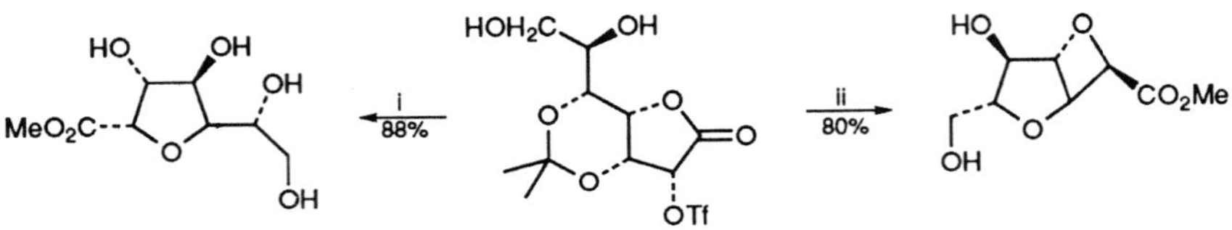

(81) (80) (82)

Reagents: i, HCl, MeOH; ii, K_2CO_3, MeOH

Scheme 35

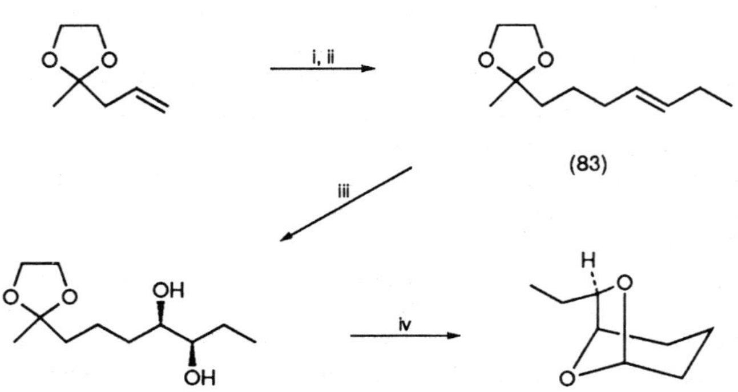

(84)

Reagents: i, 9BBN; ii, EtCH=CHBr, $Pd(PPh_3)_4$, NaOH; iii, OsO_4, DHQD-PHAL; iv, TsOH

Scheme 36

The tertiary alcohol (85) when treated with potassium hydride provides the only example of an anionic oxy-Cope reaction on a simple furan by rearranging to (86) in good yield (Scheme 37).[50]

Iodolactonization of unsaturated carboxylic acids using iodine and potassium

[50] D. Martin, J. A. Wurster, M. J. Boylan, R. M. Borzilleri, G. T. Engel, and E. J. Walsh, *Tetrahedron Lett.*, 1993, **34**, 8395.

iodide is well known. Generating the iodine using potassium iodide and sodium persulfate is claimed[51] to be more convenient, faster, and higher yielding.

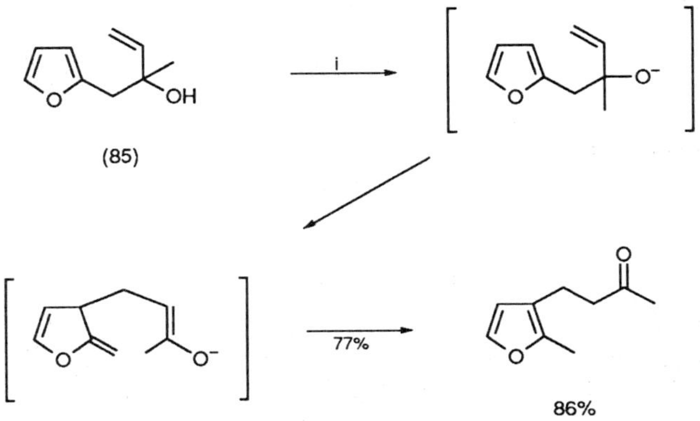

Reagents: i, KH, 18-crown-6, THF

Scheme 37

The oxidative cyclization of aldehyde (87) is the key step in a synthesis[52] of aranorosin (88), which is also the subject of an independent study.[53] In a related cyclization[54] oxime (89) gives isoxazoline (90) (Scheme 38).

The β-keto-esters (91a) and (91b) react with the propargyl carbonate (92) under palladium(0) catalysis to give furans (93) (Scheme 39).[55]

The 1-aminobenzotriazole (94) is a precursor of the dihydrobenzofuran (95) when treated with lead tetraacetate.[56] When *N*-bromosuccinimide is used as the oxidant the bromo-derivative (96) is formed (Scheme 40).

Oxonium ylids are formed from cyclic ethers bearing pendant diazoketones (97) when treated with rhodium(II). Either [1,2] or [2,3] shifts give oxygen-bridged ring systems (98) (Scheme 41).[57]

A variety of systems (99) are substituted when treated with 2-haloesters, triethylborane, and air.[58] The products (100) are postulated to form *via* a radical pathway (Scheme 42).

The bis(thiophene) (101) exhibits a pronounced solvatochromism;[59] yellow in hexane, the hue traverses the spectrum to blue in formamide–water.

On treatment with fluoride the thiophenes (102a) and (102b) give 2,3-dimethylene-2,3-dihydrothiophene (103).[60] Non-thermal preparations of this reactive intermediate are rare (Scheme 43).

[51] A.C. Royer, R.C. Mebane, and A.M. Swafford, *SYNLETT.*, 1993, 899.
[52] A. McKillop, L. McLaren, R.J. Watson, R.J.K. Taylor, and N. Lewis, *Tetrahedron Lett.*, 1993, **34**, 5519.
[53] P. Wipf, Y. Kim, and P.C. Fritch, *J. Org. Chem.*, 1993, **58**, 7195.
[54] M. Kacun, D. Koyuncu, and A. McKillop, *J. Chem. Soc., Perkin Trans. 1*, 1993, 1771.
[55] N. Greeves and J.S. Torode, *Synthesis*, 1993, 1109.
[56] M.A. Birkett, D.W. Knight, and M.B. Mitchell, *Tetrahedron Lett.*, 1993, **34**, 6939.
[57] F.G. West, T.H. Eberlein, and R.W. Tester, *J. Chem. Soc., Perkin Trans. 1*, 1993, 2857.
[58] E. Baciocchi and E. Muraglia, *Tetrahedron Lett.*, 1993, **34**, 5015.
[59] F. Effenberger and F. Wurthner, *Angew. Chem., Int. Ed. Engl.*, 1993, **32**, 719.
[60] K.J. van den Berg and A.M. van Leusen, *Recl. Trav. Chim. Pays-Bas*, 1993, **112**, 7.

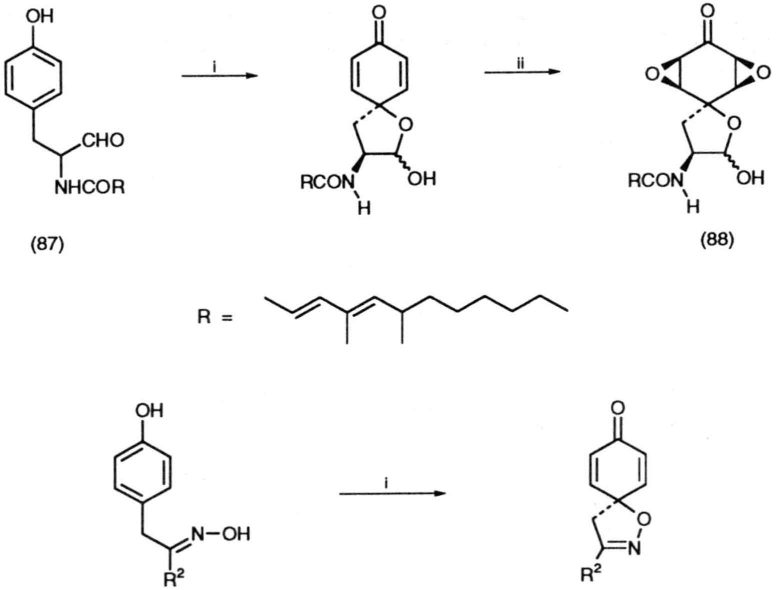

(87) (88)

(89) (90)

Reagents: i, $PhI(OCOCF_3)_2$; ii, H_2O_2, LiOH

Scheme 38

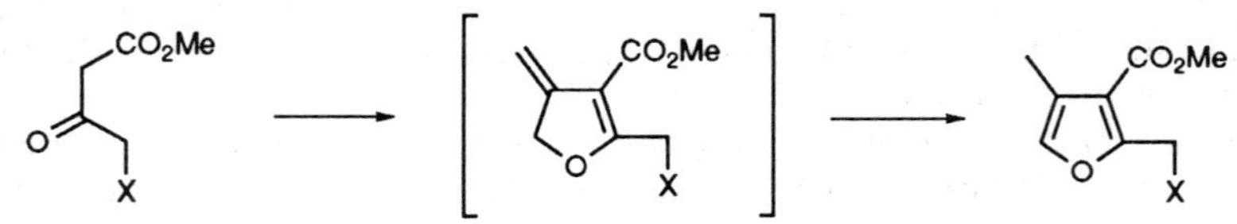

(91a) X = OMe
(91b) X = SPh

(93)

Reagents: i, $HC{\equiv}CCH_2OCO_2Me$ (92), $Pd(dba)_3 \cdot CHCl_3$, dppe

Scheme 39

Benzothiophenes (105) having a mercaptopropyl substituent are formed[61] when the dithianes (104) are treated with AIBN/tributyltin hydride (Scheme 44).

Cyclic sulfates prepared from 1,2-diols are known to be useful for further transformation of such diols. It has been found that the corresponding cyclic sulfites (often precursors of the sulfates) can be as efficient.[62]

[61] D. C. Harrowven, *Tetrahedron Lett.*, 1993, **34**, 5653.
[62] P. H. J. Carlsen and K. Aase, *Acta Chim. Scand.*, 1993, **47**, 737.

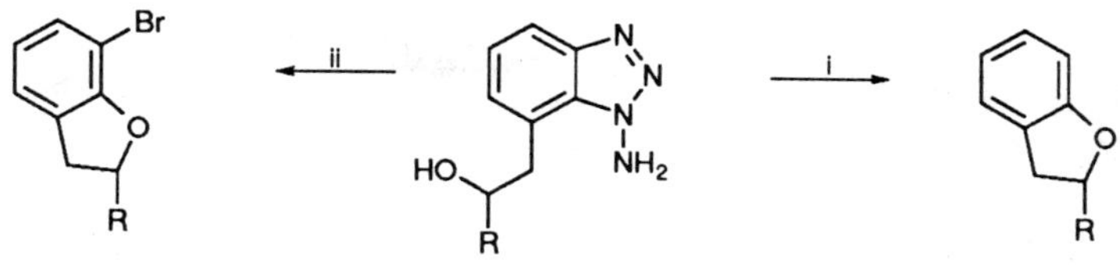

(96) (94) (95)

Reagents: i, $Pb(OAc)_4$; ii, NBS

Scheme 40

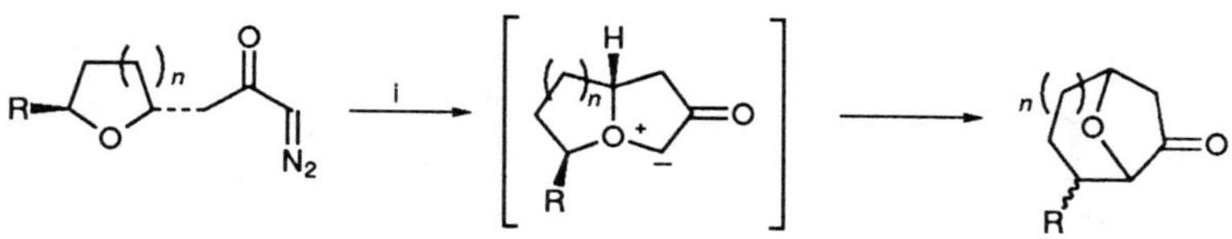

(97) $n = 1, 2$

(98)

Reagents: i, $Rh_2(OAc)_4$

Scheme 41

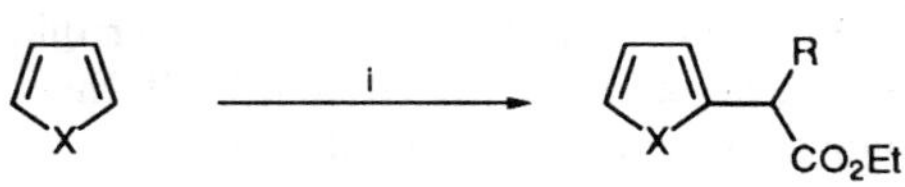

(99) X = O, S, NMe,NH

(100) R = H, Me

Reagents: i, $RCHBrCO_2Et$, BEt_3, air, DMSO

Scheme 42

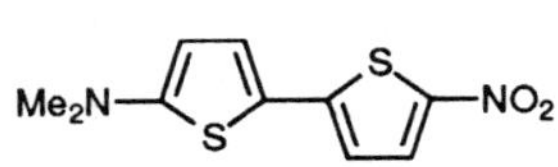

(101)

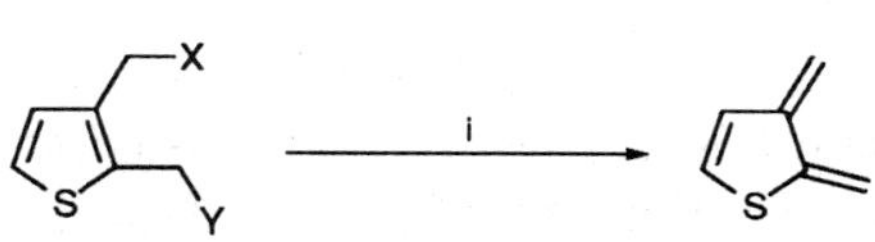

(102a) X = $^+NMe_3$; Y = $SiMe_3$

(102b) X = $SiMe_3$; Y = $^+NEt_2Me$

(103)

Reagents: i, Bu_4NF

Scheme 43

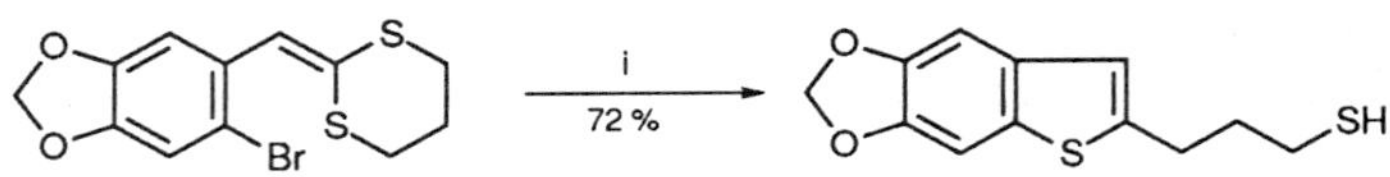

(104) (105)

Reagents: i, Bu_3SnH, AIBN

Scheme 44

A review of progress in the Fischer indole reaction has appeared.[63] A regioselective Fischer indole synthesis is also reported.[64] The *Z* form of hydrazone (106) on treatment with diethylaluminium tetramethylpiperidide gives 1,3-dimethyl-2-(2-methylbutyl)-indole (107) in 93% yield, whereas the *E* isomer gave the 2-ethyl-1-methyl-3(1-methyl)-propyl product (108) in lower yield (42%), but of 93% purity (Scheme 45).

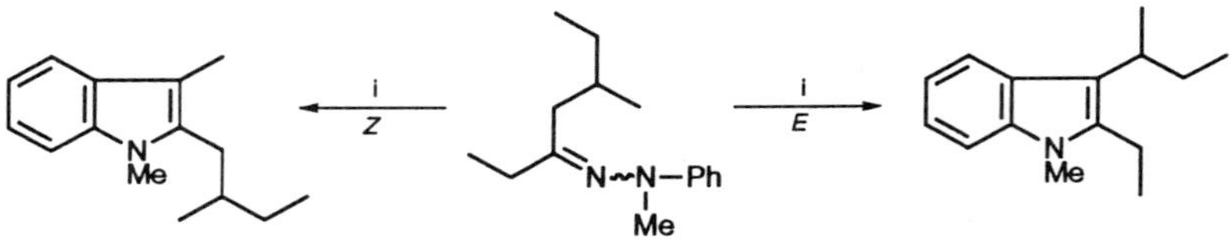

(107) (106) (108)

Reagents: i, diethylaluminium tetramethylpiperidide

Scheme 45

Keto-amides (109) readily prepared from *N*-vinylpyrrolidinone yield cyclic imines (110) on successive acid and base treatment.[65] If they are first alkylated, the disubstituted products (111) are obtained. (Scheme 46).

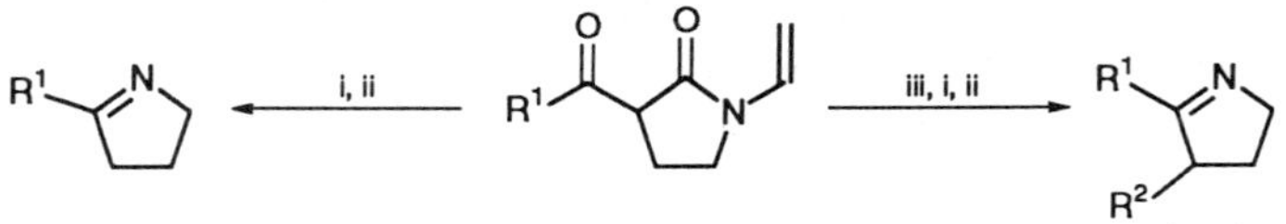

(110) (109) (111)

Reagents: i, H_3O^+; ii, NaOH; iii, NaH, R^2X

Scheme 46

The iminophosphorane (112) can be reacted with isocyanates either to isolate a pyridine (113) or, on heating, to obtain an aza-indole (114) (Scheme 47).[66]

The secondary amide (115) cyclizes under the action of phosphorus pentachloride to give 3-phenylimidazo[1,5-*a*]pyridine (116) (Scheme 48). The reaction also works when the amide moiety is attached at the 2-position of a thiazole.[67] Some mechanistic studies

[63] D. L. Hughes, *Org. Prep. Proced. Int.*, 1993, **25**, 607.
[64] E. Maruoka, M. Oishi, and H. Yamamoto, *J. Org. Chem.*, 1993, **58**, 7638.
[65] M. L. Haslego, C. A. Maryanoff, L. Scott, and K. L. Sorgi, *Heterocycles*, 1993, **35**, 643.
[66] P. Molina, E. Aller, and M. A. Lorenzo, *Synthesis*, 1993, 1239.
[67] T. Benincori, E. Brenna, and F. Sannicolo, *J. Chem. Soc., Perkin Trans. 1*, 1993, 675.

were undertaken supporting a nitrile ylid intermediate in a transformation related to the Wallach imidazole synthesis.

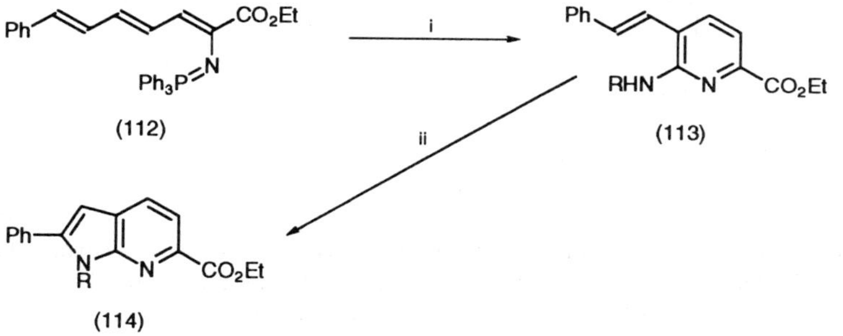

Reagents: i, RN=C=O; ii, Δ, $PhNO_2$

Scheme 47

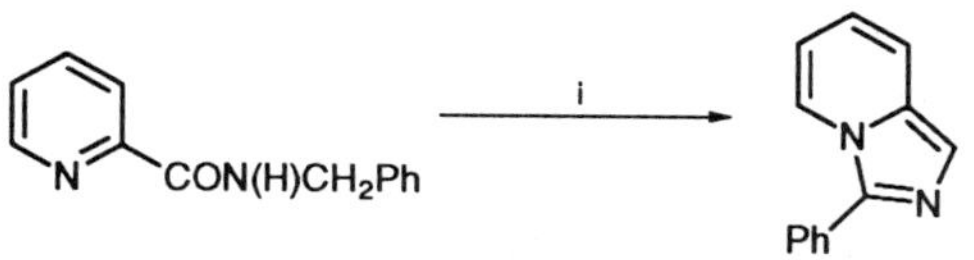

(115) (116)

Reagents: i, PCl_5

Scheme 48

The novel analgesic epibatidine (117) discovered in the skin of an Ecuadorian frog has been synthesized. Approaches include a Diels–Alder strategy from 1-methoxycarbonylpyrrole,[68] and a Heck reaction pyridinating a 7-azabicyclo[2.2.1]heptene.[69]

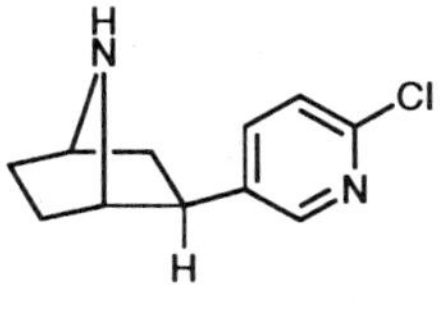

(117)

Regeneration of a nitrone from bicycles such as (118) is known. However, control of which nitrone is formed can now be obtained[70] from variation of the solvent in which the oxidation is carried out (Scheme 49).

Isoxazoles (121) can be prepared from acetylenes and nitric acid using a gold catalyst (Scheme 50). Trapping experiments using alkenes confirmed that there is an acyl nitrile oxide intermediate.[71]

[68] D. F. Huang and T. Y. Shen, *Tetrahedron Lett.*, 1993, **34**, 4477.
[69] S. C. Clayton and A. C. Regan, *Tetrahedron Lett.*, 1993, **34**, 7493.
[70] S. A. Ali and M. I. M. Wazeer, *Tetrahedron*, 1993, **49**, 4339.
[71] F. Gasparrini, M. Giovannoli, D. Misiti, G. Natile, G. Palmieri, and L. Maresca, *J. Am. Chem. Soc.*, 1993, **115**, 4401.

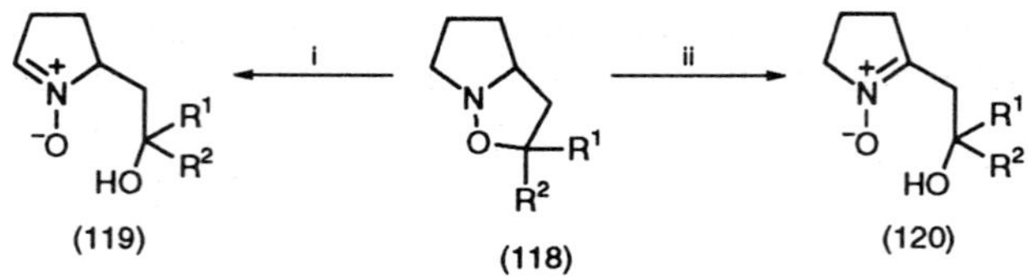

Reagents: i, *m*CPBA, CH_2Cl_2; ii, *m*CPBA, AcOH

Scheme 49

Oximes such as (122) or acyclic examples on treatment with dimethyl carbonate and potassium carbonate give oxazolin-2-ones (123) as indicated (Scheme 51).[72]

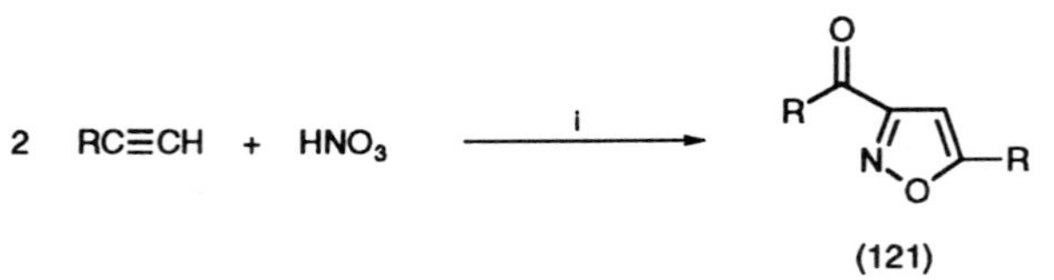

Reagents: i, Bu_4NAuCl_4

Scheme 50

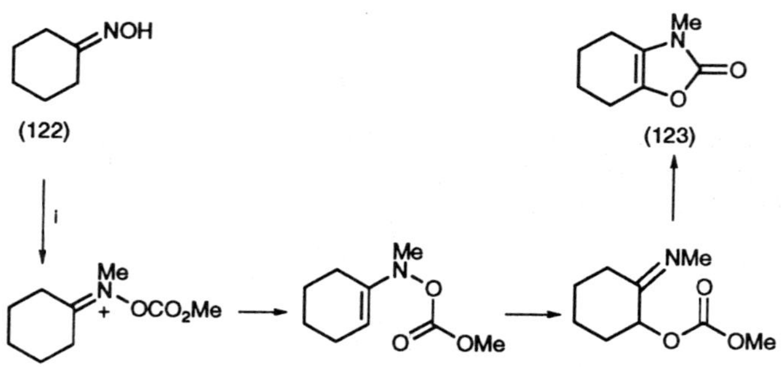

Reagents: i, $(MeO)_2C{=}O$, K_2CO_3, 190 °C

Scheme 51

Diazo compounds (124) in the presence of a nitrile and rhodium(II) acetate produce oxazoles (125) (Scheme 52).[73]

Homochiral oxazolines are useful chiral auxiliaries. The phosphine (126) has been studied in the palladium-catalysed reaction of allylic acetates and malonic esters; high yields and enantiomeric excesses up to 99% are reported.[74,75] The thiazolidine (127) is

[72] C. A. Marques, M. Sehra, P. Tunde, and F. Montanari, *J. Org. Chem.*, 1993, **58**, 5765.
[73] R. D. Connell, M. Tebbe, A. R. Ganghoff, P. Helquist, and B. Akermark, *Tetrahedron*, 1993, **49**, 5445.
[74] P. von Matt and A. Pfaltz, *Angew. Chem., Int. Ed. Engl.*, 1993, **32**, 566.
[75] G. J. Dawson, C. G. Frost, J. M. J. Williams, and S. J. Coote, *Tetrahedron Lett.*, 1993, **34**, 3149.

remarkable as the product of a seven-component condensation.[76] The necessary ingredients are sodium hydrogen sulfide, 2-bromo-2-methylpropanal, ammonia, 2-methylpropanal, carbon dioxide, methanol, and t-butylisonitrile. The yield is 43%.

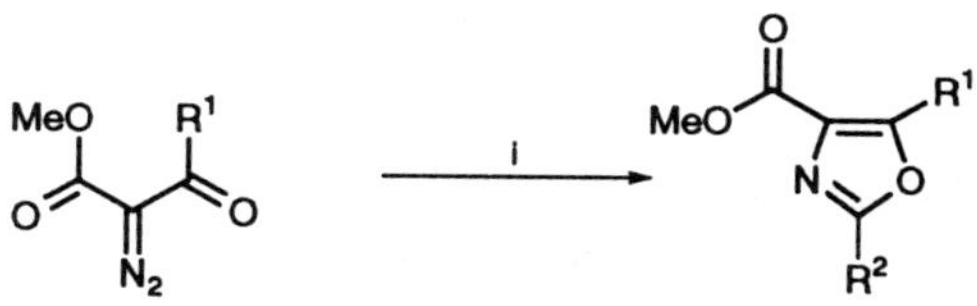

(124) (125)

Reagents: i, $Rh_2(OAc)_4$, R^2CN

Scheme 52

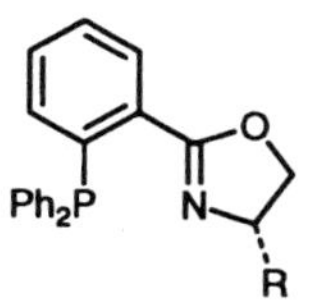

(126)

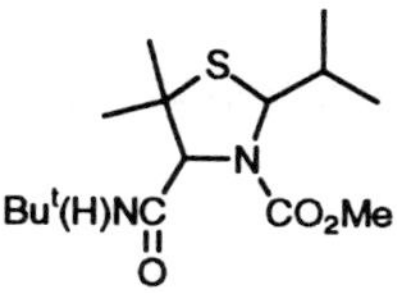

(127)

Oximes (128) react with an imidoyl chloride to produce isolable amides (129) (Scheme 53). On heating these amides are transformed to imidazoles (130).[77]

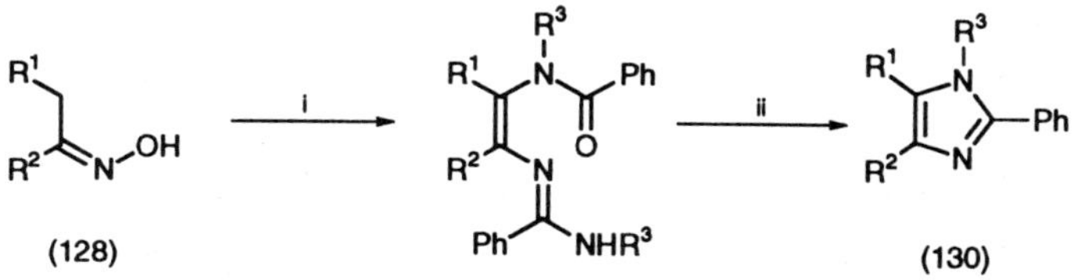

(128) (129) (130)

Reagents: i, $PhC(=NR^3)Cl$; ii, H^+, toluene, Δ

Scheme 53

Tributyltin-radical-induced cyclization of the amide (131) gives 5-butyl-3-methyl-3*H*-imidazo[4,5-*c*]quinolin-4[5*H*]-one (132) *via* a spirocyclic radical intermediate and [1,2]acyl shift (Scheme 54). Cyclization by a palladium-catalysed coupling gives the 1-methyl isomer (133).[78]

Treatment of the aryl triazolyl propanol (134) with triphenylphosphine/carbon tetrachloride gives the salt (135) containing the novel 5*H*-6-arylpyrazolo[1,2-*a*][1,2,4]triazol-4-ium ion (Scheme 55).[79]

[76] A. Domling and I. Ugi, *Angew. Chem., Int. Ed. Engl.*, 1993, **32**, 563.
[77] I. Lantos, W.-Y. Zhang, X. Shui, and D. S. Eggleston, *J. Org. Chem.*, 1993, **58**, 7092.
[78] F. Suzuki and T. Kuroda, *J. Heterocycl. Chem.*, 1993, **30**, 811.
[79] Y. Arredondo, R. Pleixats, and M. Moreno-Manas, *Synth. Comm.*, 1993, **23**, 1245.

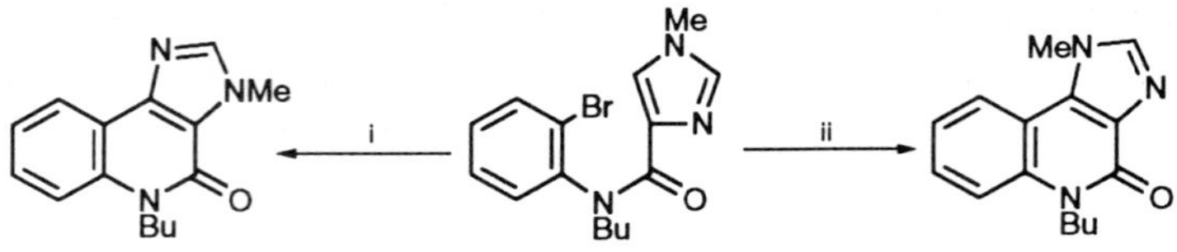

(132) (131) (133)

Reagents: i, Bu_3SnH, AIBN; ii, $Pd(OAc)_2$

Scheme 54

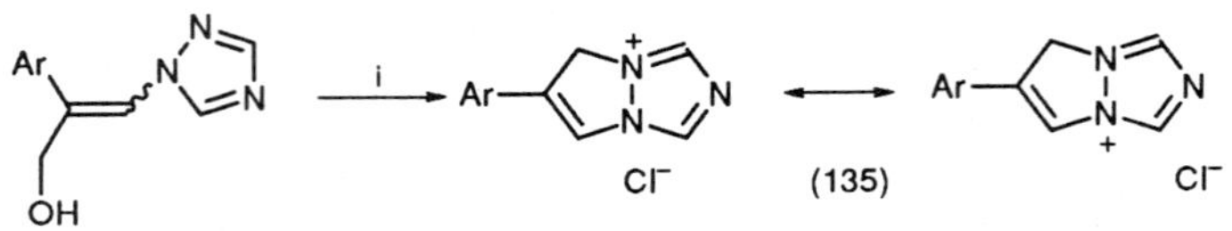

(134) (135)

Reagents: i, PPh_3/CCl_4, CH_2Cl_2

Scheme 55

4 Six-membered Rings

Among the reviews there is an account of the reactivity of 6-unsubstituted 2*H*-pyran-2-ones;[80] the synthesis of six- and seven-membered phosphorus heterocycles by ring enlargement is covered,[81] as is the synthesis of 9-substituted guanines.[82] The intramolecular double Michael reaction has been surveyed[83] and includes heterocyclic examples; the principle is exemplified in Scheme 56.

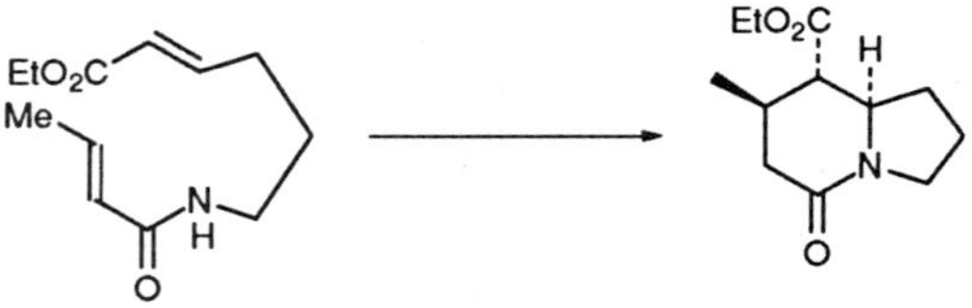

Scheme 56

There is coverage of the metallation and metal-assisted bond formation in π-electron-deficient heterocycles[84] and there are details of the lithiation of several halogen-substituted pyridines using lithium diisopropylamide.[85]

Lithiation of the pyrimidines (136a) and (136b) has been found[86] to occur at the 5-position using lithium diisopropylamide, but significant formation of products from 6-lithiation is found with lithium tetramethylpiperidide. With two equivalents of

[80] V. Kvita and W. Fischer, *Chimia*, 1993, **47**, 3.
[81] G. Keglevich, *Synthesis*, 1993, 931.
[82] F. P. Clausen and J. Juhl-Christensen, *Org. Prep. Proced. Int.*, 1993, **25**, 373.
[83] M. Ihara and K. Fukumoto, *Angew. Chem., Int. Ed. Engl.*, 1993, **32**, 1010.
[84] K. Undheim and T. Benneche, *Acta Chem. Scand.*, 1993, **47**, 102.
[85] G. W. Gribble and M. G. Saulnier, *Heterocycles*, 1993, **35**, 151.
[86] N. Ple, A. Turck, P. Martin, S. Barbey, and G. Quéguiner, *Tetrahedron Lett.*, 1993, **34**, 1605.

s-butyllithium a proton can be abstracted from the methyl group of 5-hydroxy-2-methylpyridine (137).[87] If the hydroxy substituent is first protected the approach fails.

(136a) X = Cl
(136b) X = SMe

(137)

The tactic of introducing temporarily a complexing/directing moiety has been applied to the lithiation of pyridine-2-aldehyde.[88] Electrophiles are introduced in the 3-position *via* the (presumed) complex (139) (Scheme 57). The metallation of iodopyridines is a little more complex.[89] Lithiation (lithium diisopropylamide was used) is *ortho*-directed by the iodine, but a fast migration occurs to give a stabilized iodolithiopyridine from which the product is derived. Thus, for example, 2-chloro-3-iodopyridine (141) produces 2-chloro-4-iodo-3-substituted products (142) (Scheme 58).

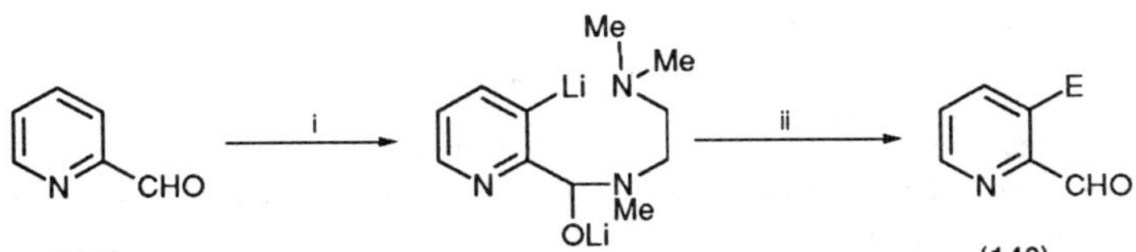

(138) (139) (140)

Reagents: i, BuLi, $Me_2NCH_2CH_2N$ (Me)Li; ii, electrophile, E

Scheme 57

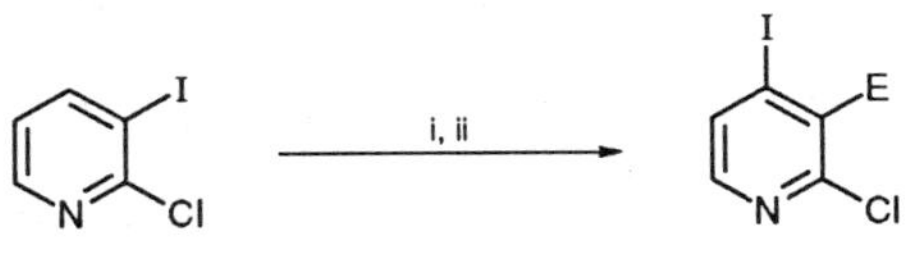

(141) (142)

Reagents: i, LDA, THF, −75 °C; ii, electrophile, E

Scheme 58

1,2,4-Triazines (143) have been metallated at the 6-position using lithium tetramethylpiperidide,[90] the first examples of metallated triazines. The homochiral stannane (144) on treatment with n-butyllithium at −78 °C gives with retention of configuration the corresponding organolithium,[91] which is configurationally stable up to −40 °C.

87 A. K. Saksena, R. G. Lovey, V. M. Girijavallabhan, H. Guzik, and A. K. Ganguly, *Tetrahedron Lett.*, 1993, **34**, 3267.
88 T. R. Kelly, W. Xu, and J. Sundaresan, *Tetrahedron Lett.*, 1993, **34**, 6173.
89 P. Rocca, C. Cochennac, F. Marsais, L. Thomas-dit-Dumond, M. Mallet, A. Godard, and G. Quéguiner, *J. Org. Chem.*, 1993, **58**, 7832.
90 N. Ple, A. Turck, G. Quéguiner, B. Glassl, and H. Neunhoeffer, *Liebigs Ann. Chem.*, 1993, 583.
91 R. E. Gawley and Q. Zhang, *J. Am. Chem. Soc.*, 1993, **115**, 7515.

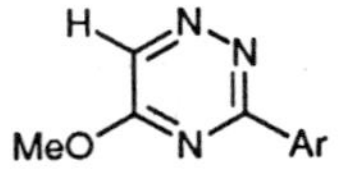

(143)

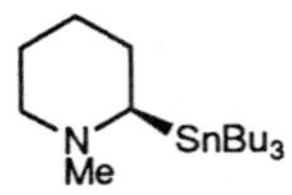

(144)

N,N-Dimethylcinnamide has been reacted photochemically with (145), the first example of a photochemical addition to a flavone (Scheme 59).[92]

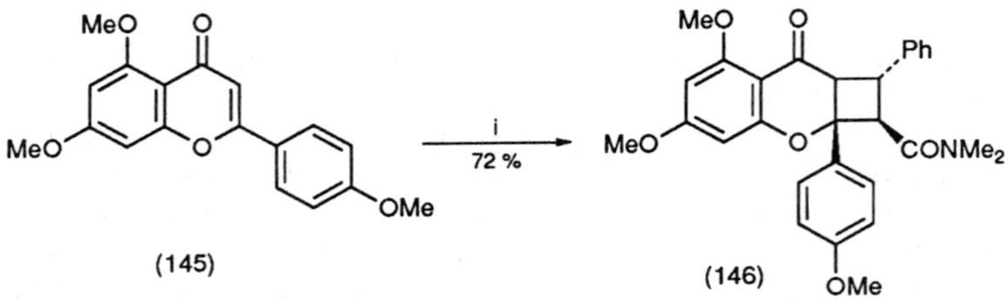

(145) (146)

Reagents: i, PhCH=CHCONMe$_2$, *hv*

Scheme 59

Pyran-2-ones (149) are produced in moderate to good yields by the reaction of trimethylsilylketene (147) and 1,3-dienes (148), the first examples of [4 + 2] addition of such components. (Scheme 60).[93]

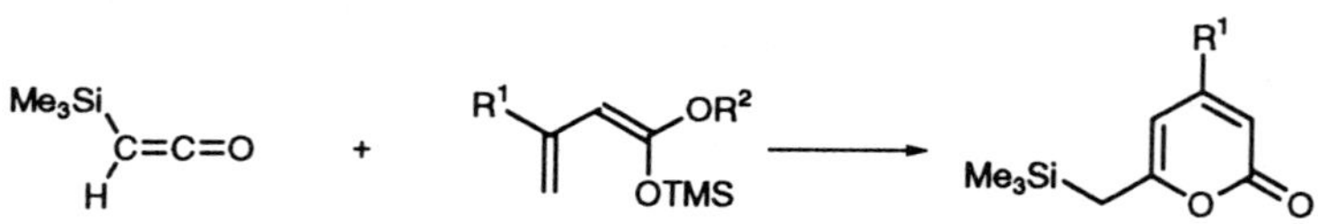

(147) (148) R^1 = Me, MeO, TMSO; R^2 = Me, TMS (149)

Scheme 60

Isochroman (150) is converted into an acetal (151) by the action of dichlorodicyanoquinone in water or methanol (Scheme 61). There are other more complex examples.[94]

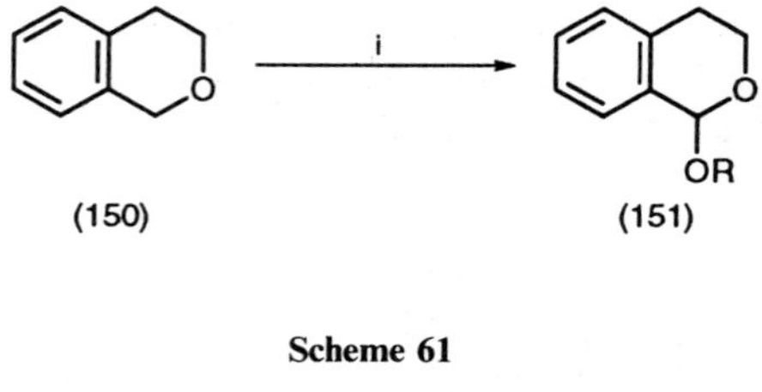

(150) (151)

Reagents: i, DDQ, ROH

Scheme 61

[92] H. C. Hailes, R. A. Raphael, and J. Staunton, *Tetrahedron Lett.*, 1993, **34**, 5313.
[93] T. Ito, T. Aoyama, and T. Shirori, *Tetrahedron Lett.*, 1993, **34**, 6583.
[94] Y. C. Xu, E. Lebeau, J. W. Gillard, and G. Attardo, *Tetrahedron Lett.*, 1993, **34**, 3841.

Treatment of diazo-ketones (152) with copper complexes (the hexafluoroacetylacetonate being the most efficient) gives rise to usefully substituted and functionalized cyclic ethers (153).[95] The mechanism involves oxygen insertion of a metal carbenoid followed by rearrangement. Seven- and eight-membered rings are available by the same methodology (Scheme 62).

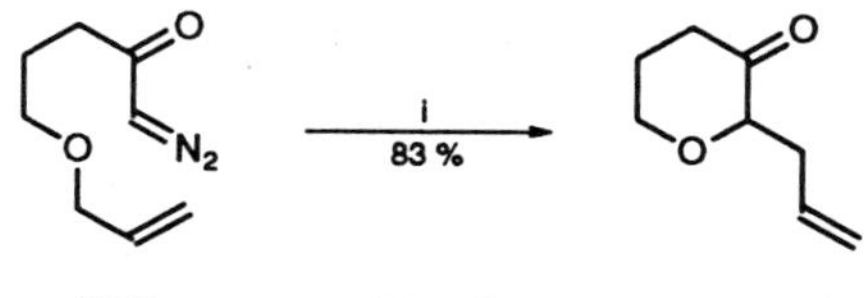

(152) (153)

Reagents: i, $Cu(hfacac)_2$, CH_2Cl_2, reflux

Scheme 62

Reaction of pyrylium tetrafluoroborate (154) with methylenetriphenylphosphorane followed by deprotonation (n-butyllithium/TMEDA) gives a phosphorane (155), which will give six-carbon homologation of aromatic, but not enolizable aliphatic aldehydes (Scheme 63).[96]

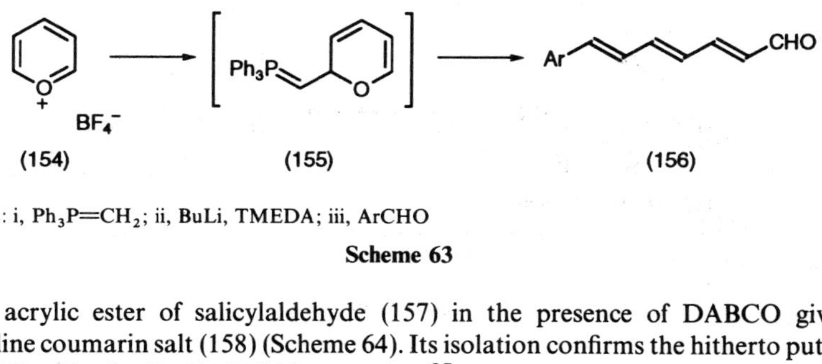

(154) (155) (156)

Reagents: i, $Ph_3P{=}CH_2$; ii, BuLi, TMEDA; iii, ArCHO

Scheme 63

The acrylic ester of salicylaldehyde (157) in the presence of DABCO gives a crystalline coumarin salt (158) (Scheme 64). Its isolation confirms the hitherto putative intermediate in the Baylis–Hillman reaction.[97]

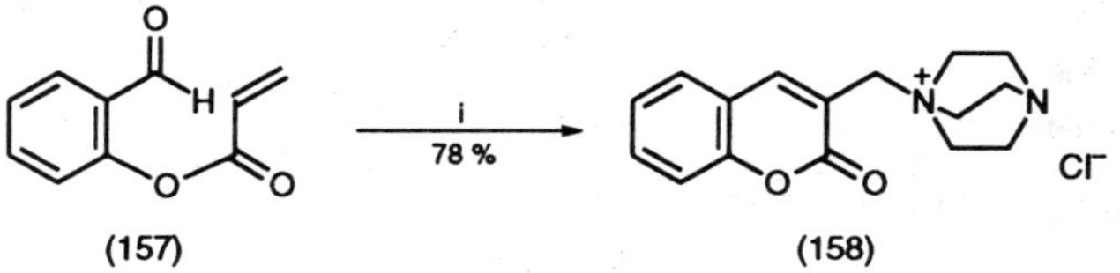

(157) (158)

Reagents: i, DABCO, CH_2Cl_2

Scheme 64

Diazo-ketones (159) on treatment with rhodium(II) acetate give cyclic amino-

95 J. S. Clark, S. A. Krowiak, and L. J. Street, *Tetrahedron Lett.*, 1993, **34**, 4385.
96 K. Henning and R. J. K. Taylor, *J. Chem. Soc., Chem. Commun.*, 1993, 1409.
97 S. E. Drewes, O. L. Njamela, N. D. Emslie, N. Ramesar, and J. S. Field, *Synth. Commun.*, 1993, **23**, 2807.

ketones (161) by nitrogen insertion of the carbenoid followed by rearrangement of the intermediate (160) (Scheme 65);[98] compare with the analogous Scheme 62.

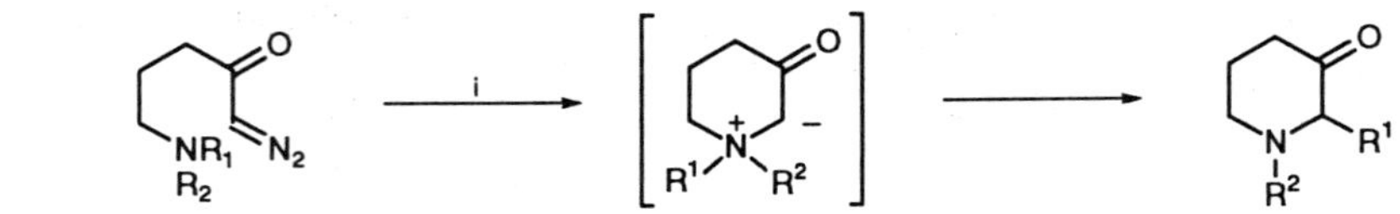

(159) $R^1 = CO_2Et$, Ar; R^2 = Me, Et (160) (161)

Reagents: i, $Rh_2(OAc)_4$

Scheme 65

Oximes of 5-oxoalkanenitriles [(162), R^1 = Me, Et)] have been cyclized with acetyl chloride/acetic anhydride mixtures to give 2-aminopyridine derivatives (163) (Scheme 66). Oximes of 5-oxopentanenitriles [(162), R^1 = H] gave only glutaronitriles.[99]

In phosphorus oxychloride at 80 °C *N*-methylformanilide (164) reacts with tertiary amides (165) to give 3-substituted-4-chloroquinolinium salts (166), from which 4-quinolones (167) are obtained by hydrolysis.[100] Yields are generally good (Scheme 67). The formanilide reacts as its Vilsmeier salt and the alkanoamide as an α-chloroenamine.

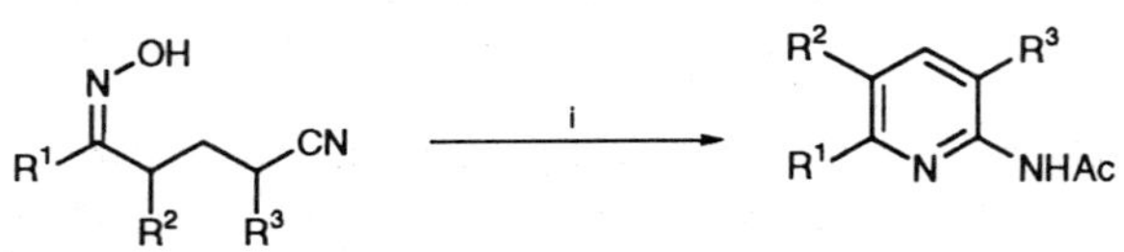

(162) R^2, R^3 = H, alkyl (163)

Reagents: i, AcCl, Ac_2O

Scheme 66

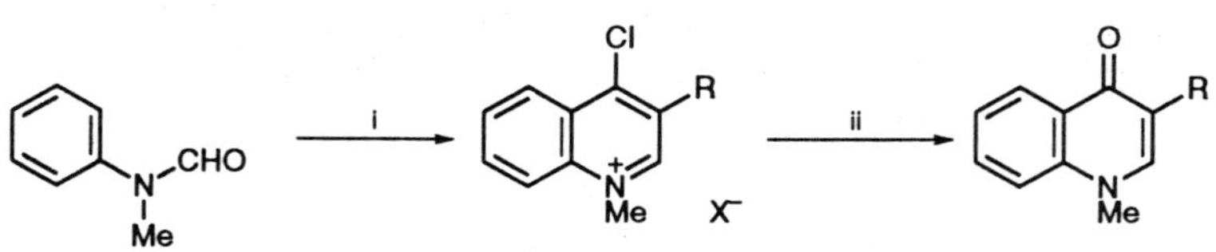

(164) (166) (167)

Reagents: i, RCH_2CONMe_2 (165), $POCl_3$; i, NaOH

Scheme 67

5-Phenylpyrroline-2,3-diones (168) (available from an imine and oxalyl chloride) react with benzyne (from benzenediazonium-2-carboxylate) to give 3-phenyl-

[98] F. G. West and B. N. Naidu, *J. Am. Chem. Soc.*, 1993, **115**, 1177.
[99] R. J. Vijn, H. J. Arts, P. J. Maas, and A. M. Castelijns, *J. Org. Chem.*, 1993, **58**, 887.
[100] O. Meth-Cohn and D. L. Taylor, *Tetrahedron Lett.*, 1993, **34**, 3629.

isoquinolones (169) (Scheme 68). Mechanistic hypotheses are discussed.[101]

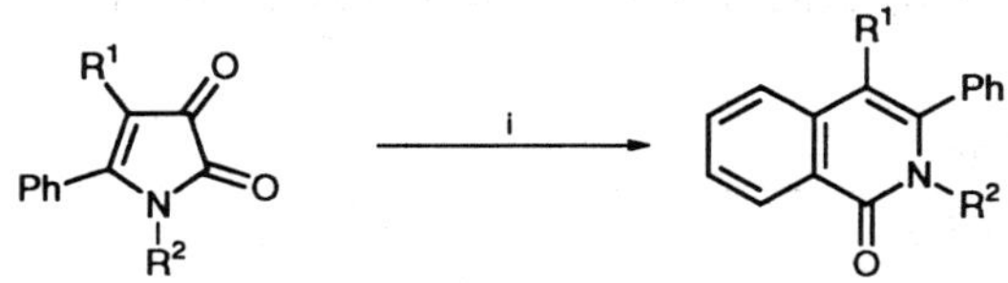

(168) R¹ = H, Me, Et; R² = alkyl, phenyl (169)

Reagents: i, benzyne

Scheme 68

Both *R* and *S* forms of enaminoester (170) are available, thus allowing the preparation of (*R*)- or (*S*)-2-alkylpiperidines (171) by the protocol[102] of Scheme 69.

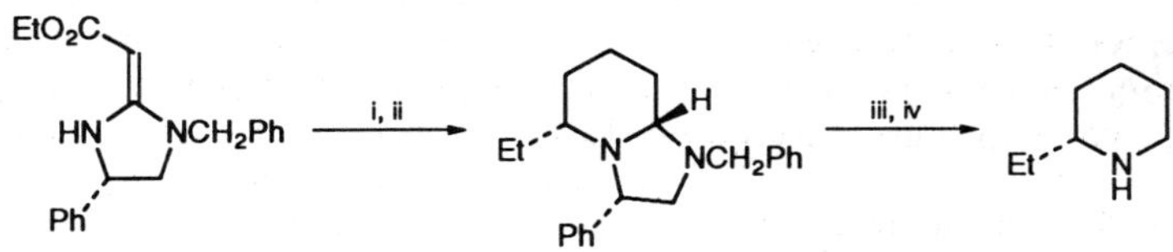

(170) (171)

Reagents: i, EtC(O)CH=CH_2; ii, BH_3, 50% aq. H_2SO_4; iii, $NaBH_3CN$; iv, H_2, catalyst

Scheme 69

Lithiated 2-chloro-*N*-cycloalkylidene-3-pyridinimines derived from (172) react with suitable *O*-ethyl thiocarboxylates (173) to give pyrido[2,3-*b*][1,5]thiazepines (174) (Scheme 70). A thermally induced ring contraction then gives the 6-arylcycloalka[*b*][1,5]naphthyridines (175) in moderate to good yield.[103]

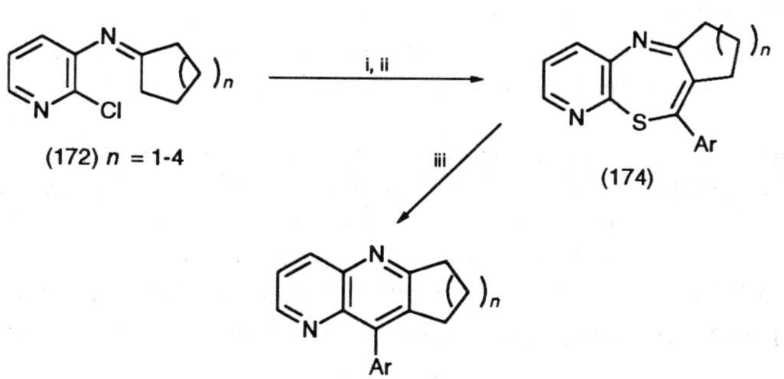

(175)

Reagents: i, LDA, −78 °C; ii, ArCSOEt (173); iii, Δ

Scheme 70

[101] A. Cobes, E. Guitian, and L. Castedo, *J. Org. Chem.*, 1993, **58**, 3113.
[102] R. C. F. Jones, I. Turner, and K. J. Howard, *Tetrahedron Lett.*, 1993, **34**, 6329.
[103] A. Couture, E. Deniau, P. Grandclaudron, and C. Simion, *Synthesis*, 1993, 1227.

The nitrobenzene (176) forms a delocalized coloured anion on treatment with sodium hydroxide. As the appropriately localized anion cyclizes onto the nitro group, quinoline *N*-oxides (177) are formed (Scheme 71).[104]

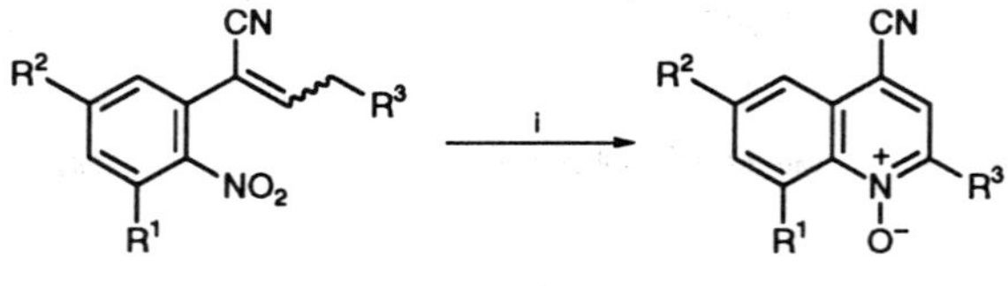

(176) (177)

Reagents: i, NaOH, MeOH

Scheme 71

6,8-Dioxabicyclo[3.2.1]octanes (178) are readily prepared from methyl vinyl ketone and are efficiently transformed into 2,6-disubstituted pyridines (179) by the action of hydroxylamine and aluminium chloride (Scheme 72).[105]

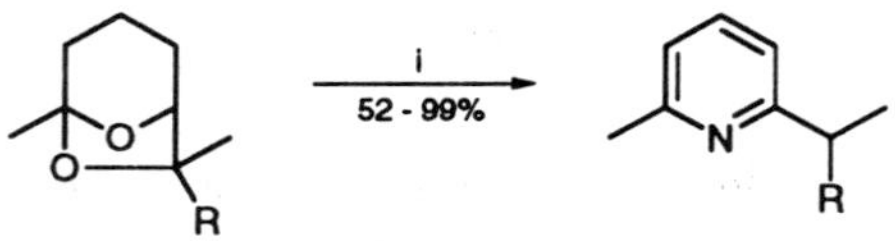

(178) (179)

Reagents: i, $NH_2OH \cdot HCl$, $AlCl_3$, AcOH

Scheme 72

The iminophosphoranes (180) react with α,β-unsaturated aldehydes and the resulting azatriene system gives the pyridines (181) by cyclization and dehydrogenation. By contrast, if the iminophosphorane is based on furan or thiophene (182) the product is a mixture of the (181) analogue together with a furyl or thienylpyridine (183) (Scheme 73).[106]

In the course of a quinocarcin synthesis the phosphonium salt (184) was treated with potassium t-butoxide to produce dihydroisoquinoline (185), a synthesis claimed to be the first of a dihydroisoquinoline by Wittig methodology (Scheme 74).[107]

2-Chloro-3-oxiranylmethoxypyridine (186) is readily available. Various alcohol, amine, or hydride nucleophiles open the oxirane ring and subsequent base treatment gives the 3-substituted 2,3-dihydro-1,4-dioxino[2,3-*b*]pyridine system (187) (Scheme 75).[108]

The conformationally restricted Claisen rearrangement of lactone (188) to pipecolic ester (189) forms the key step of a (+)-monomorine synthesis (Scheme 76).[109]

The pyrimidine-2,4-diones (190a) and (190b) react with certain α,β-unsaturated

[104] Z. Wrobel, A. Kwast, and M. Makosga, *Synthesis*, 1993, 31.
[105] J.-G. Jun, H. S. Shin, and S. H. Kim, *J. Chem. Soc., Perkin Trans. 1*, 1993, 1815.
[106] P. Molina, A. Pastor, and M. J. Vilaplanu, *Tetrahedron*, 1993, **49**, 7769.
[107] P. Garner, W. B. Ho, and H. Shin, *J. Am. Chem. Soc.*, 1993, **115**, 10 742.
[108] A. Benareb, P. Poirot, and G. Guillaumet, *Heterocycles*, 1993, **36**, 1589.
[109] S. R. Angle and J. G. Breitenbacher, *Tetrahedron Lett.*, 1993, **34**, 3985.

acyl-cyanides in a [4 + 2] manner.[110] Acid-catalysed elimination provides a new entry to 5-substituted uracils (192) (Scheme 77).

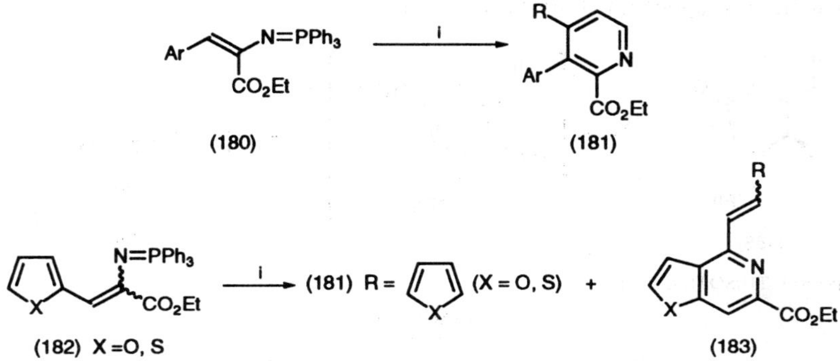

Reagents: i, RCH=CHCHO

Scheme 73

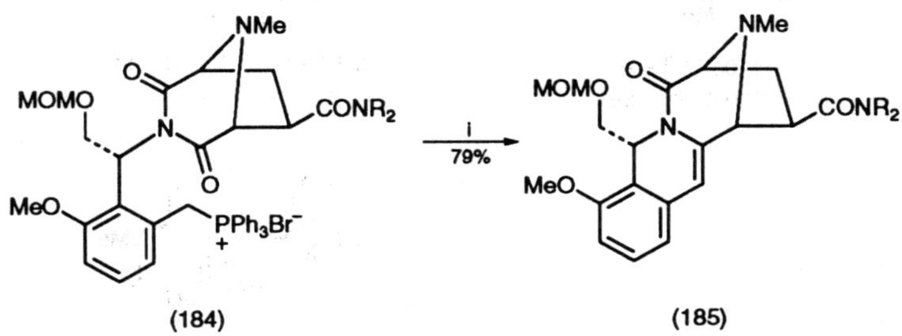

Reagents: i, KOBu^t, DMF

Scheme 74

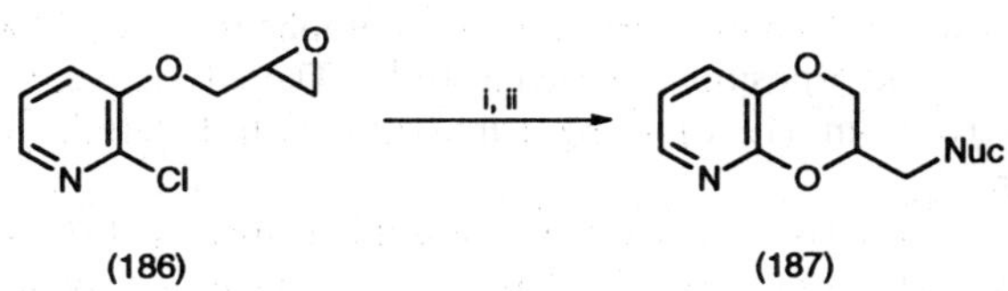

Reagents: i, nucleophile; ii, NaH, DME

Scheme 75

The pyrido[2,3-*d*]pyridazino[2,3-*a*]indole (194) is a new heterocyclic system.[111] It is prepared by an intramolecular Heck coupling in excellent yield despite the constraints of the system precluding an intermediate that can undergo a *cis* elimination of the palladium species. A base-catalysed *trans* elimination is postulated (Scheme 78).

[110] J.-C. Zhuo and H. Wyler, *Helv. Chim. Acta*, 1993, **76**, 1916.
[111] P. Melnyk, J. Gasche, and C. Thal, *Tetrahedron Lett.*, 1993, **34**, 5449.

The fused 1,2,5-oxadiazinone (195) reacts with enamines in [4 + 2] fashion, followed by elimination of carbon dioxide and amine, thus providing a route to unsymmetrical pyrazines (196) (Scheme 79).[112]

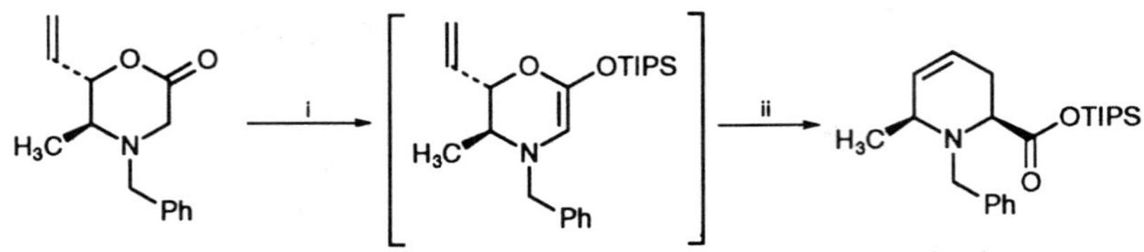

(188) (189)

Reagents: i, TIPSOTf; ii, 25 °C

Scheme 76

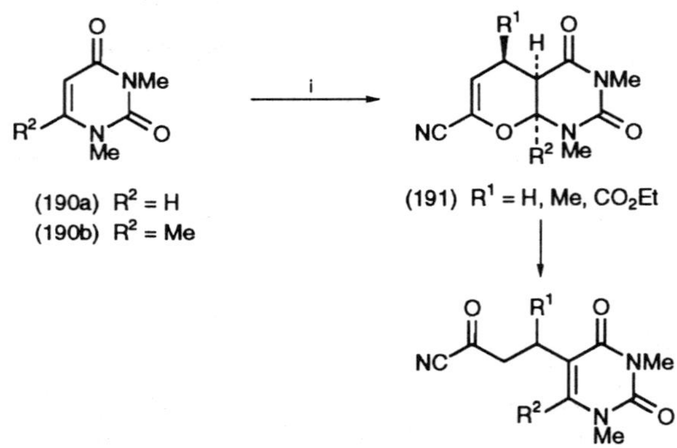

(190a) R^2 = H
(190b) R^2 = Me

(191) R^1 = H, Me, CO_2Et

(192)

Reagents: i, $R^1CH{=}CHCOCN$

Scheme 77

The 2,4,6-tribromophosphinines (197a) and (197b) undergo palladium-catalysed coupling with 2-trimethylstannylpyrrole, -furan, -thiophene, or -pyridine, all of which substitute at the 2- and 6-positions (Scheme 80).[113] The 2-bromo substituent can be selectively reduced from (197b) using tributyltin hydride/tetrakis(triphenylphosphine)palladium.

Flash vacuum pyrolysis of dialkylvinylphosphine gives a 40% yield of phosphinine.[114] The synthesis of strychnine has been a popular endeavour;[115,116,117] in particular the application of aza-Cope–Mannich methodology is elegant.[118,119]

[112] A. Ganesan and C. H. Heathcock, *J. Org. Chem.*, 1993, **58**, 6155.
[113] P. LeFloch, D. Carmichael, L. Ricard, and F. Mathey, *J. Am. Chem. Soc.*, 1993, **115**, 10 665.
[114] P. LeFloch and F. Mathey, *J. Chem. Soc., Chem. Commun.*, 1993, 1295.
[115] P. Magnus, M. Giles, R. Bonnert, G. Johnson, L. McQuire, M. Deluca, A. Merritt, C. S. Kim, and N. Vicker, *J. Am. Chem. Soc.*, 1993, **115**, 8116.
[116] M. E. Kuehne and F. Xu, *J. Org. Chem.*, 1993, **58**, 7490.
[117] P. Magnus and M. Giles, *Tetrahedron Lett.*, 1993, **34**, 6355.
[118] S. R. Angle, J. M. Fevig, S. D. Knight, R. W. Marquis, Jr, and L. E. Overman, *J. Am. Chem. Soc.*, 1993, **115**, 3966.
[119] S. D. Knight, L. E. Overman, and G. Pairaudeau, *J. Am. Chem. Soc.*, 1993, **115**, 9293.

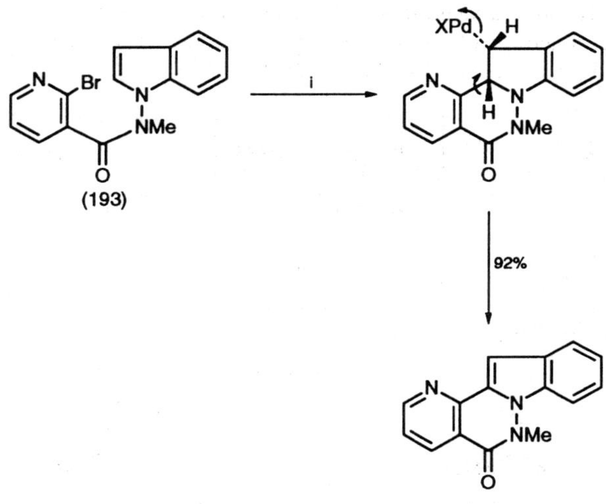

(193)

(194)

Reagents: i, $Pd(OAc)_2$, PPh_3, K_2CO_3

Scheme 78

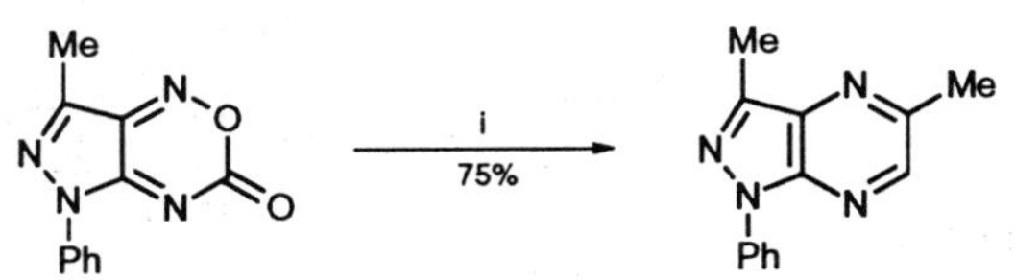

(195)

(196)

Reagents: i, $R^1R^2NCH{=}CHMe$

Scheme 79

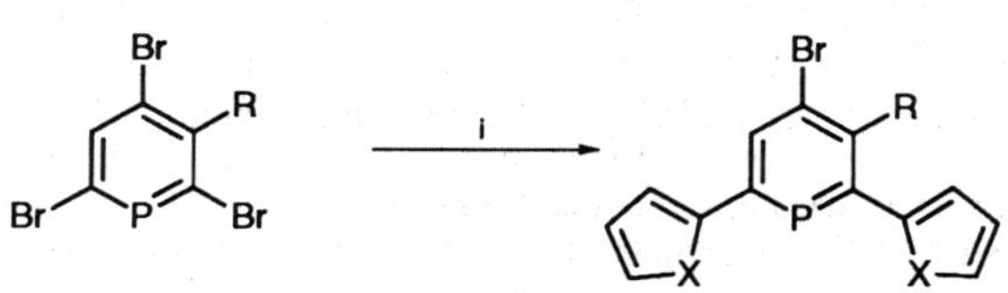

(197a) R = H
(197b) R = Me

(198) X = O, S, NMe

Reagents: i, $Pd(dba)_3$, Ph_3P

Scheme 80

5 Seven-membered Rings

A review of the syntheses of medium-sized rings by ring-expansion reactions includes heterocyclic examples.[120]

The diene (199) is converted into a dilithium compound with t-butyllithium and serves as precursor to a variety of unusual heterocycles,[121] including the first arsepines and bismepines (200) (Scheme 81), the latter of which (M = Bi) has a half-life of only seven minutes.

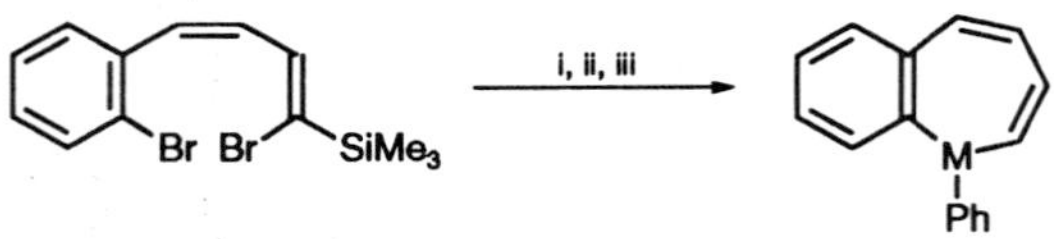

(199) (200) M = P, As, Sb, Bi

Reagents: i, Bu^tLi, $-80\,^{\circ}C$; ii, $PhMCl_2$; iii, TBAF

Scheme 81

The bis-cyclopropanated furan (201) isomerizes to 2,3-dihydrooxepine (202) on heating at 130 °C. The corresponding *anti* isomer requires a temperature of 368 °C. Dienophiles react with (201) to give 9-oxabicyclo[4.2.1]nonanes (203) (Scheme 82).[122]

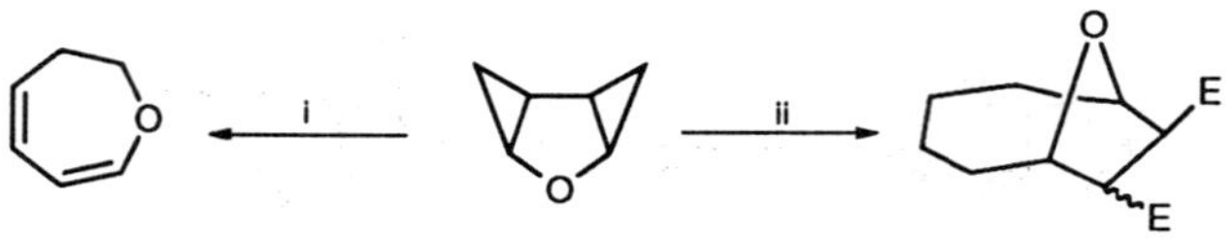

(202) (201) (203)

Reagents: i, 130 °C; ii, EHC=CHE

Scheme 82

The naturally occurring, antifungal benzopentathiapin varacin (205) has been synthesized.[123] Intermediate (204) is derived from a benzyne and carbon disulfide and subsequent reaction with disulfur dichloride completes construction of the pentathiapin ring (Scheme 83).

The utility of the displacement reaction of aromatic fluorides is demonstrated by the closure of the oxepine ring to form 6,7-dihydro[1]benzoxepino[4,5-*c*]quinolin-8(9*H*)-one (207) from the alcohol (206) (Scheme 84).[124]

An intramolecular Heck reaction closes (208) to the dialkylmethylidene-substituted 2,3,4,5-tetrahydro-1*H*-3-benzazepine (209) (Scheme 85).[125]

[120] C. J. Roxburgh, *Tetrahedron Lett.*, 1993, **49**, 10 749.
[121] S. Yasuike, H. Ohta, S. Shiratori, J. Kurita, and T. Tsuchiya, *J. Chem. Soc., Chem. Commun.*, 1993, 1817.
[122] T. Golz, S. Hammes, and F.-G. Klarner, *Chem. Ber.*, 1993, **126**, 485.
[123] V. Behar and S. Danishefsky, *J. Am. Chem. Soc.*, 1993, **115**, 7017.
[124] M. Anzini, A. Cappelli, and S. Vomero, *Heterocycles*, 1993, **36**, 1065.
[125] L. F. Tietze and R. Schimpf, *Synthesis*, 1993, 876.

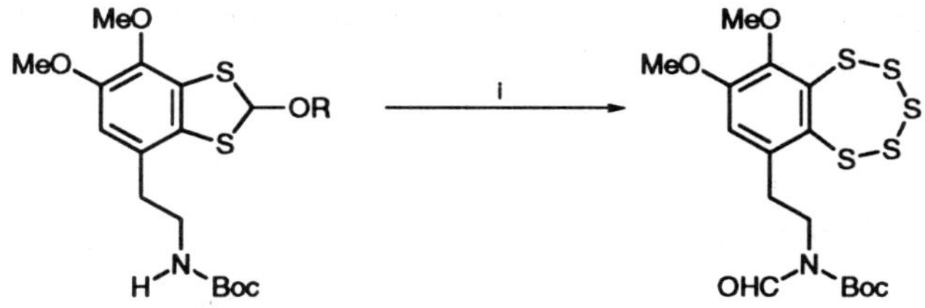

(204) (205)

Reagents: i, S_2Cl_2, THF, MeOH, HCl

Scheme 83

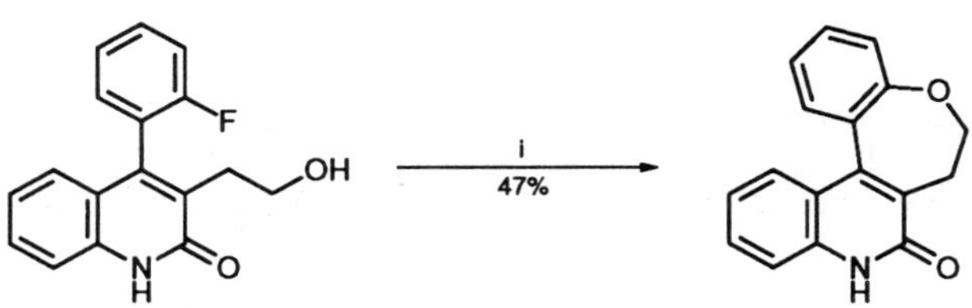

(206) (207)

Reagents: i, NaH, DMF, 155 °C, 5 min

Scheme 84

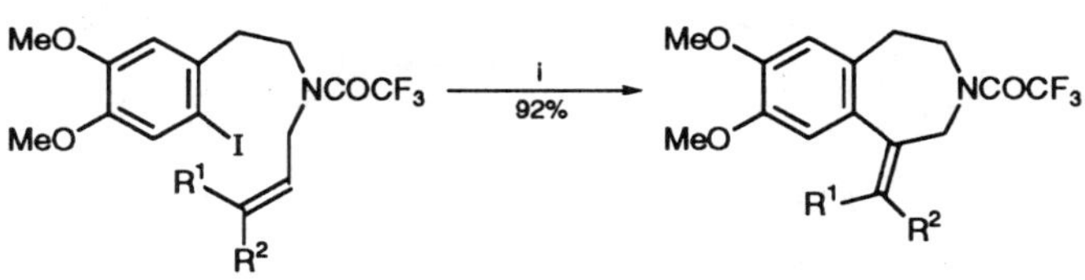

(208) R^1, R^2 = H, Me (209)

Reagents: i, $Pd(OAc)_2$, Ph_3P, KOAc

Scheme 85

Radical-mediated ring expansion of (210) gives the benzazepine (211) (Scheme 86).[126] The technique was also applied to the preparation of dibenzazepines.

Base treatment of (212) was expected to induce ring closure to (213), but this turned out to be the minor (20%) product;[127] a higher yield (39%) was obtained of (214), resulting from a novel Smiles rearrangement, a mechanism confirmed by the isolation of intermediate (215) in 34% yield (Scheme 87).

The functionalized azepinone (216) on treatment with methyl propiolate gives the aryl pyrrole (217) in high yield.[128] The mechanism involves cycloaddition, an

[126] Z. B. Zheng and P. Dowd, *Tetrahedron Lett.*, 1993, **34**, 7709.
[127] J. R. Poudfoot, U. R. Patel, and S. J. Campbell, *J. Org. Chem.*, 1993, **58**, 6997.
[128] E. Cartmell, J. E. Mayo, H. McNab, and I. H. Sadler, *J. Chem. Soc., Chem. Commun.*, 1993, 1417.

aromatization that liberates an α-aminoketone, and condensation with a second molecule of methyl propiolate (Scheme 88).

Reagents: i, Bu_3SnH, AIBN

Scheme 86

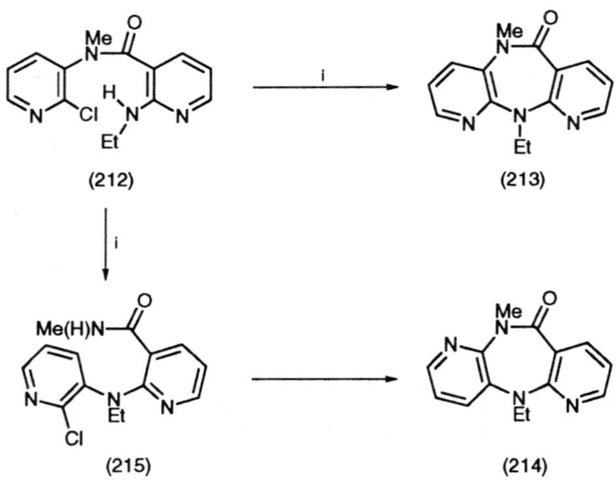

Reagents: i, LiHMDS, THF

Scheme 87

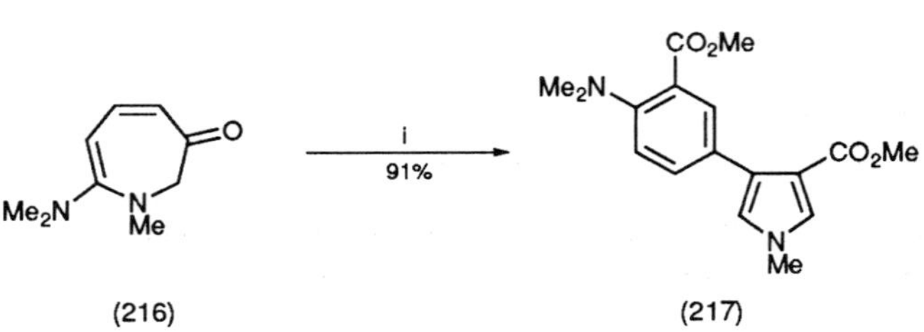

Reagents: i, $HC{\equiv}CCO_2Me$

Scheme 88

The nitrilimine precursor (218) is readily prepared. Cyclization gives a 3,3a-dihydro-4*H*,6*H*-pyrazolo[1,5-*a*][4,1]benzoxazepine (219) (Scheme 89).[129] This could be oxidized to the corresponding pyrazole with DDQ.

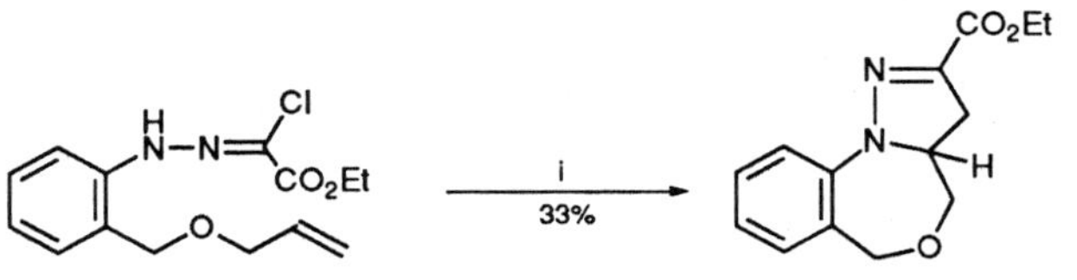

(218) (219)

Reagents: i, NEt_3, toluene, reflux

Scheme 89

The reaction of 2-aminothiophenol (220) with various 2-(1-haloalkyl)oxiranes (221) provides a convenient synthesis of 1,5-benzothiazepins (222) (Scheme 90).[130]

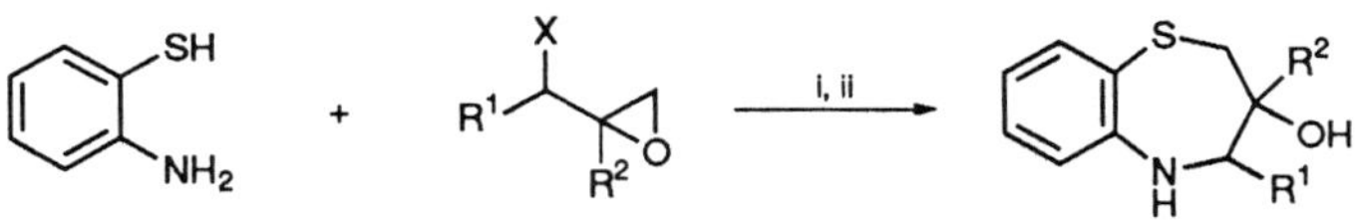

(220) (221) (222)

Reagents: i, NEt_3; ii, KOH

Scheme 90

On exposure to Mitsunobu conditions dithioketals (223) give ring-expanded products (224) *via* 1,3-sulfur migration (Scheme 91).[131]

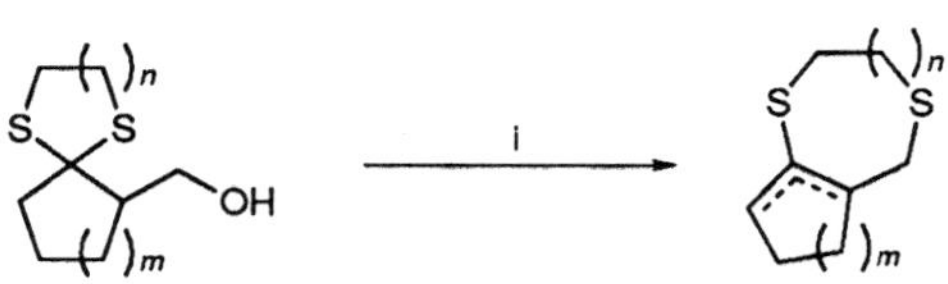

(223) $m = n = 1,2$ (224)

Reagents: i, DEAD, Ph_3P, THF

Scheme 91

Dihydrothiepin-1,1-dioxide (225) can be lithiated and reacted with electrophiles.[132] The products (226) are convenient precursors to trienes (227) (Scheme 92).

[129] L. Garanti, G. Zecchi, and L. Bruche, *J. Heterocycl. Chem.*, 1993, 559.
[130] M. Karikomi, S. Yamori, and T. Toda, *Heterocycles*, 1993, **35**, 619.
[131] S. Takano, H. Iida, and K. Ogasawara, *Heterocycles*, 1993, **36**, 2203.
[132] J.H. Rigby and A.C. Kruger, *SYNLETT.*, 1993, 829.

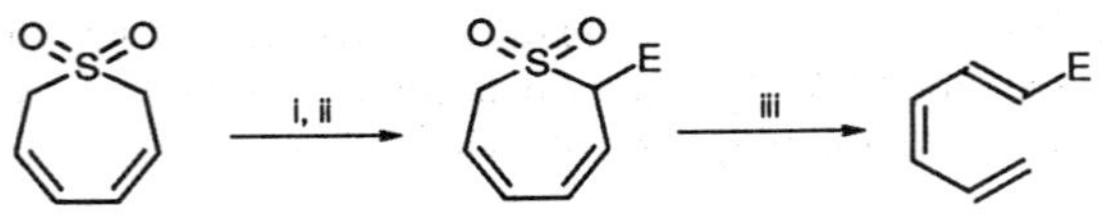

(225) (226) (227)

Reagents: i, Bu^{n}Li, HMPA, −105 °C; ii, electrophile, E; iii, 175 °C

Scheme 92

6 Larger Rings

There is a review of improved methods for the synthesis of aza-crown macrocycles and cryptands.[133] Two accounts[134,135] of the biosynthesis of vitamin B_{12} have appeared.

B-Methoxyborinanes (228) react with α-chloroallyllithium to give, after treatment of the initial adduct with boron trifluoride, the *B*-methoxyboracyclenes (229) (Scheme 93).[136]

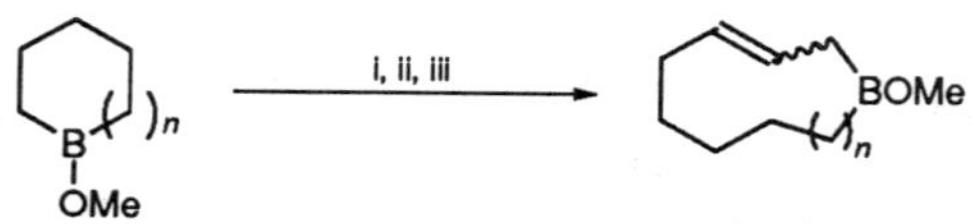

(228) $n = 0–7$ (229)

Reagents: i, CH_2=$CHCH_2Cl$, $LiNR_2$, −78 °C; ii, $BF_3·Et_2O$; iii, −78 °→r.t

Scheme 93

Pyrido[1,2-*a*]azepinone (230) is deprotonated in the pyridine ring by LDA and the lithiated species will react with a variety of electrophiles. When an aldehyde is used the intermediate alkoxide attacks the lactam giving [7](2,6)pyridinophane (231) (Scheme 94).[137]

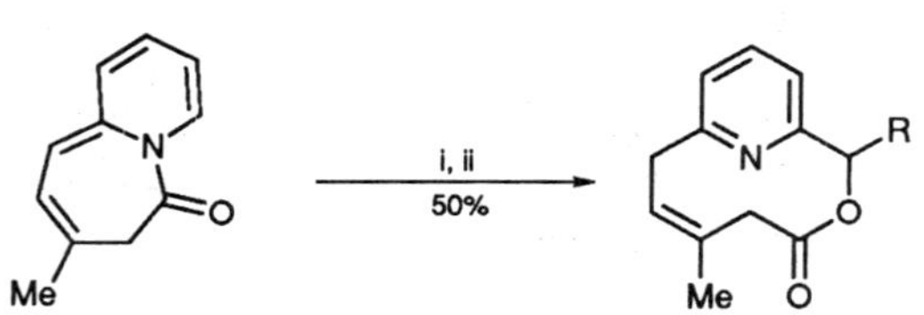

(230) (231)

Reagents: i, LDA; ii, RCHO

Scheme 94

[133] K. E. Krakowiak, J. S. Bradshaw, and R. M. Izatt, *SYNLETT.*, 1993, 611.
[134] A. I. Scott, *Angew. Chem., Int. Ed. Engl.*, 1993, **32**, 1223.
[135] A. R. Battersby, *Acc. Chem. Res.*, 1993, **26**, 15.
[136] H. C. Brown and S. Jayaraman, *Tetrahedron Lett.*, 1993, **34**, 3997.
[137] W. Maier, M. Keller, and W. Eberbach, *Heterocycles*, 1993, **35**, 817.

The *N*-oxide (232) is converted in methanol under reflux into 3,6-epoxyhexahydroazocino[5,4-*b*]indole (233) (Scheme 95). A radical pathway is suggested.[138]

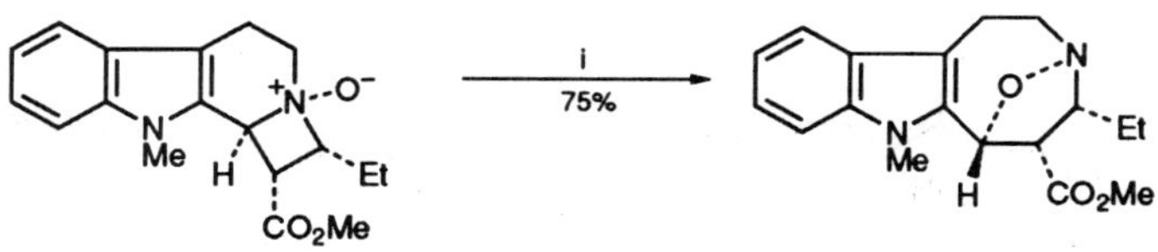

(232) (233)

Reagents: i, 90% MeOH

Scheme 95

Irradiation of the *β*-ketovinylogous amide (234) leads to formation of the acetylpyrrole (237). Formation of the stabilized diradical (235), which can form keto-imine (236), is a likely mechanism (Scheme 96).[139]

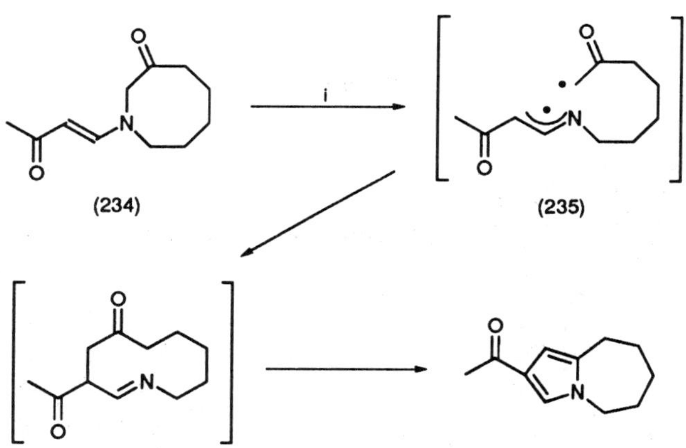

(234) (235)

(236) (237)

Reagents: i, *hv*

Scheme 96

The thionocarbonate (238) was treated with base to form the cyclic thionocarbonate (239), which by [3,3] sigmatropic ring expansion gave thiolcarbonate (240) (Scheme 97).[140] This material was converted into yellow-scale pheromone.

The subtlety of the experiments needed to elucidate the biosynthesis of haem and vitamin B_{12} (see Chapter 11) tends to obscure the underlying synthetic work. The key step in the preparation of the spiro compound (244) was the combination of

[138] T. Kurihara, Y. Sakamoto, M. Takai, K. Ohuchi, S. Harasawa, and R. Yoneda, *Chem. Pharm. Bull.*, 1993, **41**, 1221.
[139] J. D. Winkler and M. G. Siegel, *Tetrahedron Lett.*, 1993, **34**, 7697.
[140] S. Harusawa, S. Takemura, H. Osaki, R. Yoneda, and T. Kurihara, *Tetrahedron*, 1993, **49**, 7657.

iododihydrodipyrrin (241) with the pyrrole (242) to give the lactam (243) (Scheme 98);[141] methodology developed specifically for this target. The ultimate target was the octaacid corresponding to (244) required for use in enzyme inhibition studies.

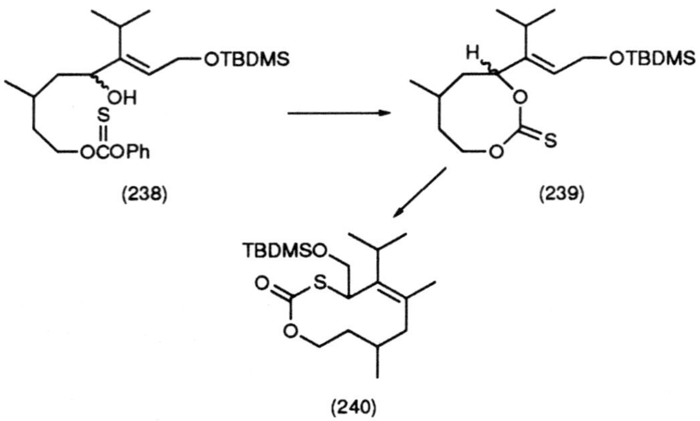

Reagents: i, LiHMDS

Scheme 97

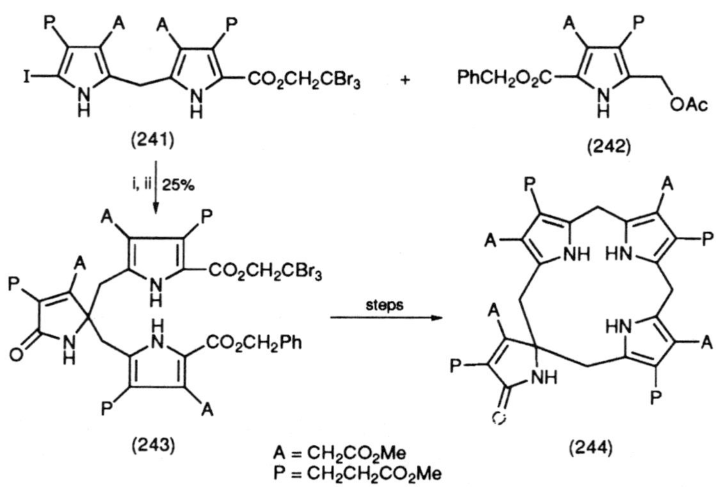

Reagents: i, $SnCl_4$, CH_2Cl_2; ii, AgOAc, TsOH, THF, H_2O

Scheme 98

[141] W. M. Stark, C. J. Hawker, G. J. Hart, A. Philippides, P. M. Petersen, J. D. Lewis, F. J. Leeper, and A. R. Battersby, *J. Chem. Soc., Perkin Trans. 1*, 1993, 2875.

8 Organometallic Chemistry
The Transition Elements

By G. R. STEPHENSON
School of Chemical Sciences, University of East Anglia, Norwich, NR4 7TJ, UK

1 Introduction

Not every year produces results which point to new directions in organometallic chemistry. During 1993, however, significant advances have been made in the use of supercritical fluids as solvents for preparative-scale organometallic synthesis. At Nottingham, Bannister, Howdle, and Poliakoff[1] have developed a flow reactor to perform photochemistry under supercritical conditions. In the Nottingham set-up, an infrared flow cell is placed in series with the reactor vessel to allow on-line study of the reaction process (Figure 1). In this way, the pentacarbonylchromium complex of ethene was isolated for the first time. This experiment used supercritical ethene as the solvent in the flow reactor.

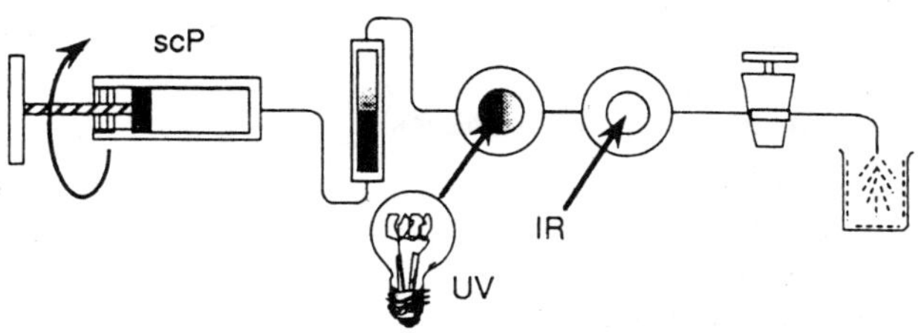

Figure 1
(Reproduced with permission from *J. Chem. Soc., Chem. Commun.*, 1993, 1814)

At Clemson University in the USA, supercritical water has been used as a reaction solvent. Dicarbonyl(cyclopentadienyl)cobalt-catalysed cyclotrimerization of alkynes was studied under these conditions. At 374 °C good yields of aromatic products were obtained, with regioisomer distributions comparable to those observed in the organic solvents more customarily employed for this type of reaction. The temperature conditions were important. At 140 °C a diverse mixture of products was formed.[2]

The cyclooligomerization and cooligomerization of alkynes has always been an

[1] J. A. Banister, S. M. Howdle, and M. Poliakoff, *J. Chem. Soc., Chem. Commun.*, 1993, 1814.
[2] K. S. Jerome and E. J. Parsons, *Organometallics*, 1993, **12**, 2991.

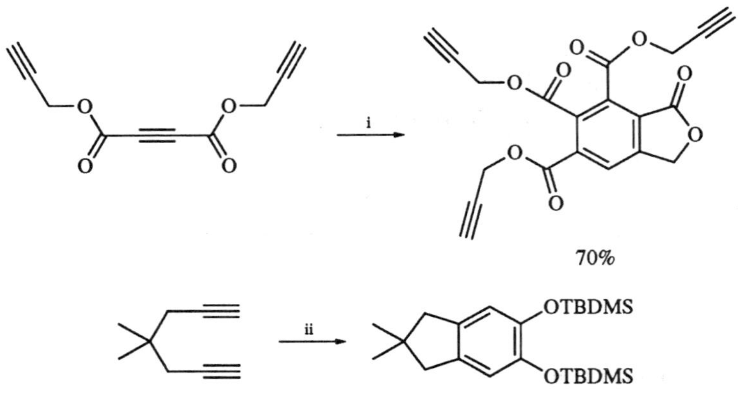

Reagents: i, $[Pd(dba)_2]$; ii, TBDMSCl, CO, $[Ru_3(CO)_{12}]$

Scheme 1

important topic in organometallic chemistry. Alkynes are highly reactive starting materials and so participate readily in a great variety of reactions. During the year, there has been particular progress with the development of the organometallic chemistry of alkynes for controlled carbon–carbon bond formation. Few of the reactions under study are new in type, but, looking back over the year, it is clear that this continues to be an exciting and rapidly developing area of organometallic chemistry. In this year's Report, we will look in detail at the most recent developments in this field.

2 Alkynes in C–C Bond Formation

Dicarbonyl(cyclopentadienyl)cobalt is the most common reagent used for the cyclotrimerization of alkynes. An alternative, based on organopalladium chemistry, is remarkable for its chemical selectivity (Scheme 1). Palladium catalysis produces a dimer from a bispropargyl ester of an alkyne-containing dicarboxyclic acid.[3] Combination of two alkynes with two molecules of carbon monoxide can produce a 1,2-dioxygenated aromatic ring. $Ru_3(CO)_{12}$ and tricyclohexylphosphine are used to promote this cyclization reaction.[4] When two alkynes are joined within a larger ring, these can be dimerized using a (cyclopentadienyl)cobalt system to form a superphane containing η^4-cyclobutadiene CoCp complexes at each end. Yields are low (less than 1%), but this interesting product was characterized by X-ray crystallography.[5]

Enyne cyclization.—When an alkene and an alkyne are combined by a palladium catalyst, the typical product is a 1,3-diene. An example from the Trost group forms a six-membered ring in this way.[6] When an extra alkyne is present in the starting material, further carbon–carbon bond formation forms a triene which cyclizes

[3] C. Stephan and H. Tom Dieck, *J. Mol. Catal.*, 1993, **81**, 313.
[4] N. Chatani, Y. Fukumoto, T. Ida, and S. Murai, *J. Am. Chem. Soc.*, 1993, **115**, 11 614.
[5] R. Gleiter, G. Pflästerer, and B. Nuber, *J. Chem. Soc., Chem. Commun.*, 1993, 454.
[6] B. M. Trost and O. J. Gelling, *Tetrahedron Lett.*, 1993, **34**, 8233.

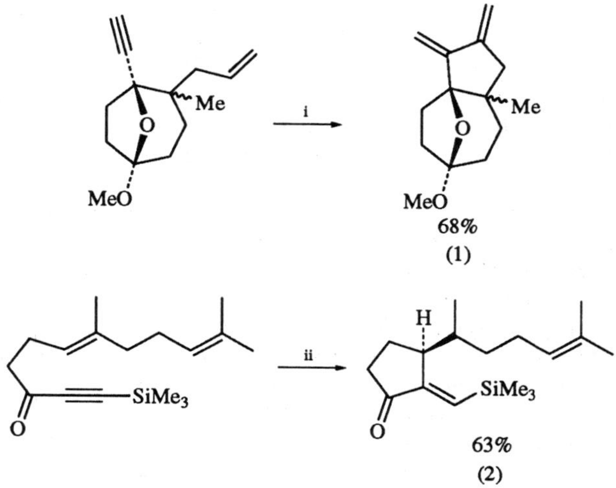

Reagents: i, $[Pd(OAc)_2]$, Py; ii, $Pd[(dba)_2]$, CH_3CO_2H,$P(O\text{-tol})_3$

Scheme 2

thermally thus providing a cyclohexa-1,3-diene synthesis.[7] If the alkene is a vinylbromide,[8] cocyclization with an additional alkyne affords an aromatic product.

Enyne cyclization can be performed on elaborate substrates, as evidenced by the formation of (1) (Scheme 2).[9] Inclusion of acetic acid among the reagents allows the formation of an alkene [*e.g.* (2)] rather than a diene. The product (2) was used as an intermediate in the total synthesis of chokol C.[10]

There are now many variants on the enyne cyclization. A combination of the cyclization reaction with a metathesis step is of particular note. Palladium-catalysed cyclization of (3) affords (4) in 85% yield. The CHMe unit, originating in the alkene, has been transferred to the alkyne-end of the carbon chain.[11] In some cases a cyclobutene, for example (5), can be isolated.[12] Use of a palladium chloride catalyst, in the presence of five equivalents of copper(II) dichloride, allows the introduction of two

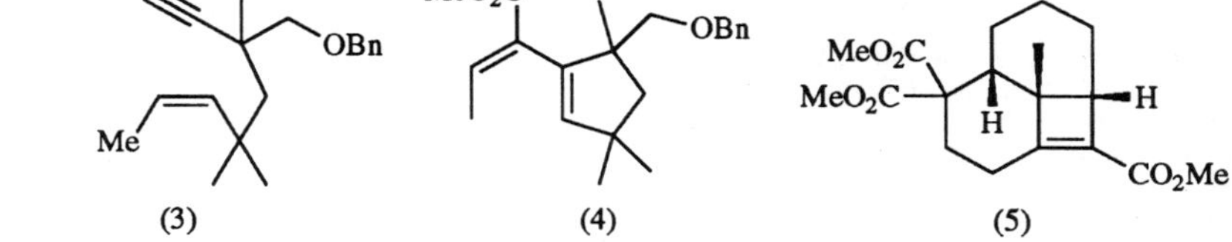

(3) (4) (5)

[7] B. M. Trost and Y. Shi, *J. Am. Chem. Soc.*, 1993, **115**, 12491.
[8] E.-i. Negishi, M. Ay, and T. Sugihara, *Tetrahedron*, 1993, **49**, 5471.
[9] F.-H. Wartenberg, B. Hellendahl, and S. Blechert, SYNLETT, 1993, 539.
[10] B. M. Trost and L. T. Phan, *Tetrahedron Lett.*, 1993, **34**, 4735.
[11] B. M. Trost and V. K. Chang, *Synthesis*, 1993, 824.
[12] B. M. Trost, M. Yanai, and K. Hoogsteen, *J. Am. Chem. Soc.*, 1993, **115**, 5294.

chlorine atoms, one to the alkene and one to the alkyne.[13] The potential for further elaboration of the alkenyl and the alkyl chloride groups in such products will make them attractive as intermediates in organic synthesis. To gain access to a precursor of isohinokinin, hydrogenation conditions were employed to remove both halides and the carbon–carbon double bond.[14] In a further variant, β-ketoester and alkyne coreactants in an intermolecular process combine in a new synthesis of 2,3,4-trisubstituted furans.[15]

Allylic alcohols, acetates, and chlorides have been examined as starting materials in intramolecular cyclizations with terminal alkynes. Lithium chloride is included as a coreactant. In the case of the allylic alcohol, depending on the substitution at the CHOH centre, either hydrogen or OH transfer to the metal is observed. Bulky substituents favour hydrogen transfer forming an enol ether and hence, ultimately, a ketone. With a methyl group, the only product is that of oxygen transfer. In both cases, a chlorine atom becomes attached, forming an alkenyl chloride.[16] Allylic acetates show more general selectivity for leaving group displacement in this type of reaction.[17] Allylic chlorides have also been used as starting materials.[18] In the presence of carbon monoxide, the reaction takes a different course. This is a palladium-catalysed version of the Pauson–Khand reaction. The dienone (6) was formed in 43% yield.[19]

Progress continues to be made in the development of the organocobalt Pauson–Khand reaction. Dimethyl sulfoxide and other polar solvents have been shown to promote cyclopentenone formation from alkyne hexacarbonyldicobalt complexes.[20] The dry-state adsorption conditions developed by Smit and Caple have been employed in a version of the Pauson–Khand cyclization of N-protected allyl propargyl amines, which, in the absence of oxygen, give rise to saturated bicyclic cyclopentanone products, rather than the usual cyclopentenones.[21] The use of coordinating ligands has also been evaluated,[22] and a 1-alkynylcyclopropanol has been shown[23] to be an alternative starting material for this type of cyclization. In this case, a catalytic amount of octacarbonyldicobalt and triphenylphosphite was employed.

In the patent literature, too,[24] reports of catalytic versions of the Pauson–Khand cyclization have appeared. An intermolecular reaction between hept-1-yne and ethene has been performed. Pentacarbonyliron may be used in combination with the dinuclear cobalt catalyst (Scheme 3). The reaction is described as regioselective and useful on an industrial scale. Another catalytic version of the reaction used dicyclopentadienyltitanium(trimethylphosphine) in place of the cobalt complex, and an isocyanide in place of carbon monoxide.[25] Since the recently reported first catalytic

[13] S. Ma and X. Lu, *J. Org. Chem.*, 1993, **58**, 1245.
[14] X. Lu and G. Zhu, *SYNLETT*, 1993, 68.
[15] N. Greeves and J. S. Torode, *Synthesis*, 1993, 1109.
[16] S. Ma and X. Lu, *J. Organomet. Chem.*, 1993, **447**, 305.
[17] G. Zhu, S. Ma, and X. Lu, *J. Chem. Res.* (*S*), 1993, 366.
[18] S. Ma, G. Zhu, and X. Lu, *J. Org. Chem.*, 1993, **58**, 3692.
[19] N. C. Ihle and C. H. Heathcock, *J. Org. Chem.*, 1993, **58**, 560.
[20] Y. K. Chung, B. Y. Lee, N. Jeong, M. Hudecek, and P. L. Pauson, *Organometallics*, 1993, **12**, 220.
[21] D. P. Becker and D. L. Flynn, *Tetrahedron Lett.*, 1993, **34**, 2087.
[22] M. E. Krafft, I. L. Scott, R. H. Romero, S. Feibelmann, and C. E. Van Pelt, *J. Am. Chem. Soc.*, 1993, **115**, 7199.
[23] N. Iwasawa and T. Matsuo, *Chem. Lett.*, 1993, 997.
[24] W. Keim, *Patentschrift* (*Switz.*), CH 681224, 1993 (*Chem. Abstr.*, 1993, **119**, 95004).
[25] S. C. Berk, R. B. Grossman, and S. L. Buckwald, *J. Am. Chem. Soc.*, 1993, **115**, 4912.

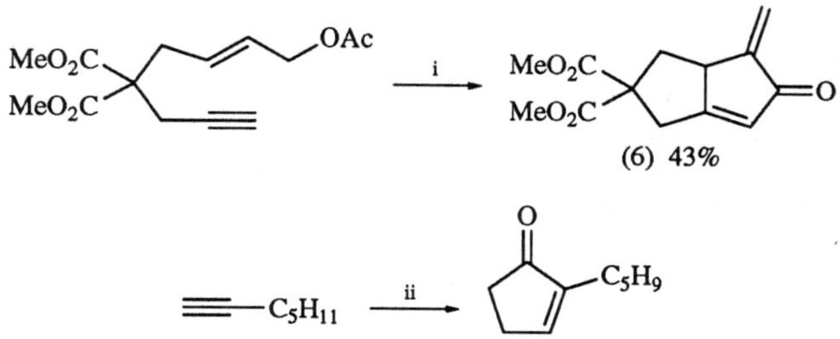

(6) 43%

51%

Reagents: i, $[Pd_2(dba)_3]$, PPh_3,LiCl,CO; ii, ethene, CO, $[Co_2(CO)_8]$ (cat.), $[Fe(CO)_5]$

Scheme 3

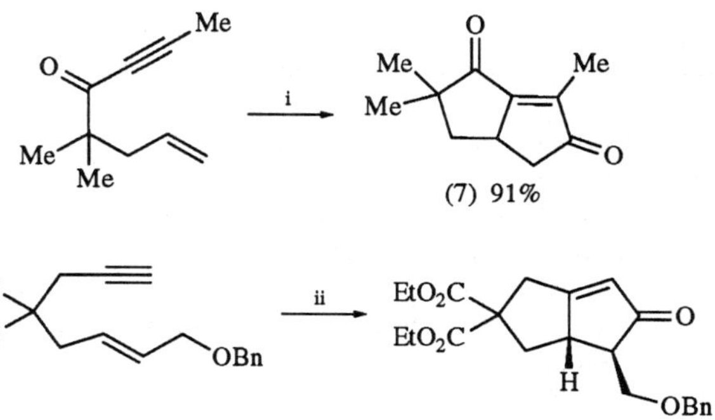

(7) 91%

62%

Reagents: i, $[Co_2(CO)_8]$; ii, $[Mo(CO)_6]$, DMSO, toluene

Scheme 4

version of a Pauson–Khand cyclization, considerable progress has been made.

Under stoichiometric conditions, cyclizations of electron deficient alkynes have been examined. The enedione product (7) was obtained in excellent yields[26] (Scheme 4). With three equivalents of a simple propargylketone and (octacarbonyl)dicobalt, a compound with a dinuclear 'flyover bridge-type' structure has been obtained and characterized by *X*-ray crystallography.[27] The use of other transition metal systems in cyclopentenone formation continues to be developed. Both tungsten[28] and molybdenum[29] complexes have been used to effect the intramolecular cyclization of enynes to form cyclopentenones. In the tungsten case, a pentacarbonyltungsten–THF complex was used. The hexacarbonylmolybdenum method, on the other hand, employed the recently developed DMSO promoter conditions. The second example shown in Scheme 4 includes a potential allylic leaving group, but, unlike the reaction leading to

[26] T. R. Hoye and J. A. Suriano, *J. Org. Chem.*, 1993, **58**, 1659.
[27] G. Gervasio, E. Sappa, and L. Marko, *J. Organomet. Chem.*, 1993, **444**, 203.
[28] T. R. Hoye and J. A. Suriano, *J. Am. Chem. Soc.*, 1993, **115**, 1154.
[29] N. Jeong, S. J. Lee, B. Y. Lee, and Y. K. Chung, *Tetrahedron Lett.*, 1993, **34**, 4027.

(6) (Scheme 3), in this benzyl ether case, the ether is retained in the product. Indenones can be prepared with palladium catalysts from alkynes and substituted benzaldehydes.[30]

There have been further developments with cyclopentadienone formation.[31] Cyclization of diynes with carbonyl insertion initially formed an η^2-complex, from which the metal can be removed by oxidation with trimethylamine *N*-oxide. The tetrasubstituted cyclopentadienone (8) can be prepared in 46% yield in this way from

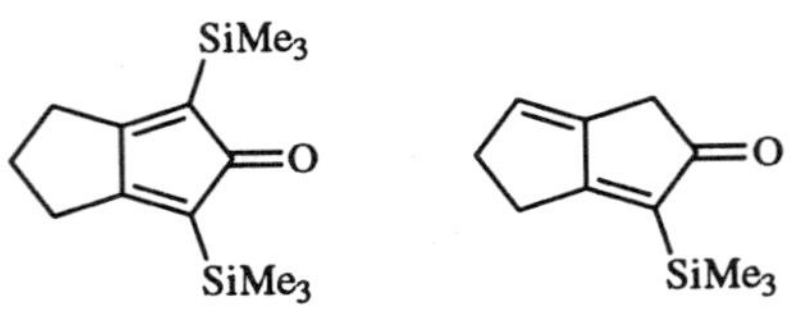

(8) (9)

1,7-bistrimethylsilylhept-1,6-diyne. Four equivalents of the amine oxide reagent are used. When a large excess of amine oxide is employed, (9) is obtained in an improved yield of 67%. It has been shown that (8) is converted into (9) by the action of trimethylamine *N*-oxide. In the case of the thioether (10), cyclopentadienone formation affords the η^4-complex in 76% yield. On metal removal, both double bonds move to produce the substituted thiophene (11) (Scheme 5).

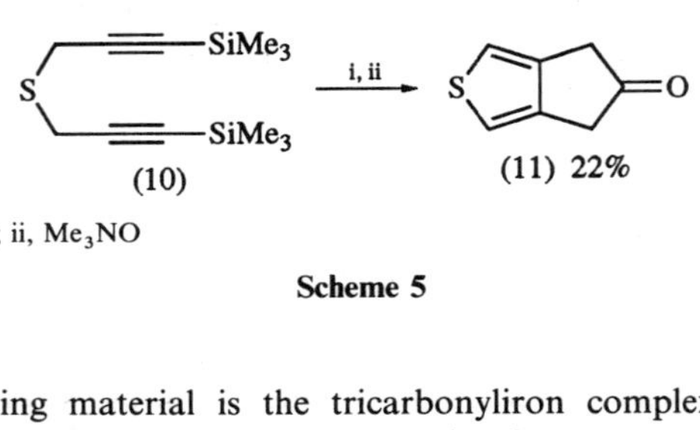

(10) (11) 22%

Reagents: i, $[Fe(CO)_5]$; ii, Me_3NO

Scheme 5

When the starting material is the tricarbonyliron complex of a vinylketene, intermolecular reaction with an alkyne can, under the correct conditions, produce a 2,2-disubstituted cyclopent-4-ene-1,3-dione. Phenolic products are also obtained in this way, particularly when electron-rich alkynes are used.[32] An example of phenol production is given in Scheme 6. This type of cyclization is related to the Dötz–Wulff cyclization of vinylcarbenes with alkynes, which is also believed to involve coordinated ketene-type intermediates. In the case of (12), however, the usual 1,4-oxygenation pattern on the aromatic ring is replaced by a 1,3-regiochemistry.

The Dötz–Wulff cyclization has itself been undergoing further development. The direct preparation of differentially protected hydroquinone complexes has been demonstrated,[33] and the dry-state adsorption conditions discussed above for the

[30] R. C. Larock, M. J. Doty, and S. Cacchi, *J. Org. Chem.*, 1993, **58**, 4579.
[31] H.-J. Knölker and J. Heber, *SYNLETT*, 1993, 924.
[32] K. G. Morris, S. P. Saberi, and S. E. Thomas, *J. Chem. Soc., Chem. Commun.*, 1993, 209; K. G. Morris, S. P. Saberi, M. W. Salter, S. E. Thomas, and M. F. Ward, *Tetrahedron*, 1993, **49**, 5617.
[33] S. Chamberlin, W. D. Wulff, and B. Bax, *Tetrahedron*, 1993, **49**, 5531.

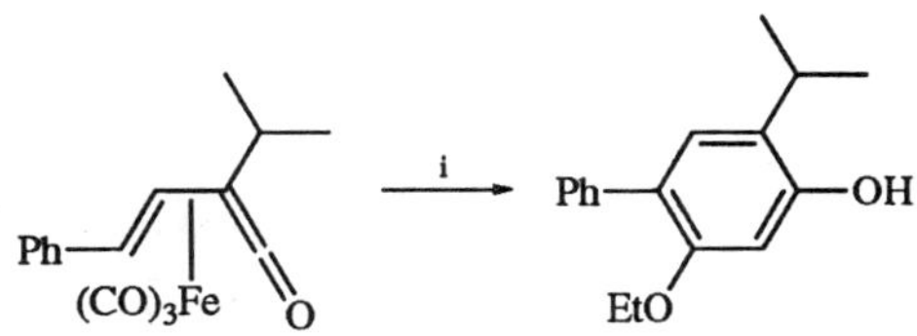

71%
(12)

Reagents: i, EtOC≡CH, Δ

Scheme 6

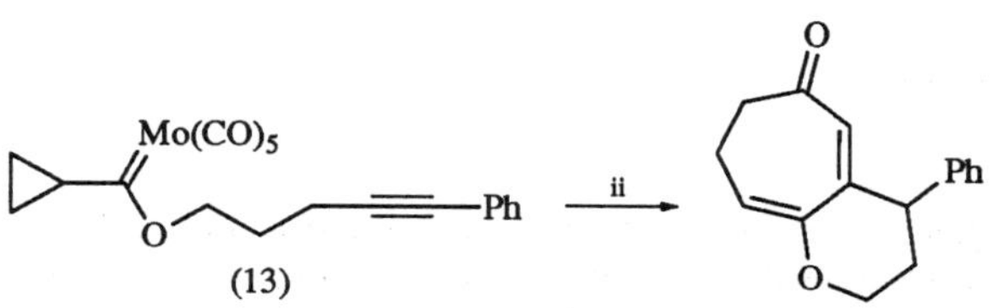

(13)

(14) 55%

Reagents: i, Cerium(IV); ii, PPh_3

Scheme 7

Pauson–Khand reaction have been shown to be effective.[34] In a variant under examination in the Herndon group, a cyclopropylcarbene complex (13) is used (Scheme 7). Opening of the cyclopropyl ring forms a cycloheptadienone product (14).[35] Fluorinated vinylcarbene complexes afford fluorinated hydroquinones.[36] 1,4-Dioxygenated 1,3-dienes can be obtained from molybdenum carbene complexes and alkynes.[37] With aminocarbenes bearing a pendant alkyne, tricyclic five-membered lactams are obtained.[38]

3 Palladium-catalysed Cross Coupling

There are many palladium-catalysed procedures to effect coupling reactions. Most involve aryl or vinyl halides (or equivalents) with alkenes or hetero-substituted vinyl compounds or arenes. Heck and Stille/Suzuki (organotin/organoboron) coupling reactions are among the most widely used. In the Heck procedure, an alkene receives the incoming group. After carbon–carbon bond formation, β-elimination detaches the

[34] J. P. A. Harrity, W. J. Kerr, and D. Middlemiss, *Tetrahedron*, 1993, **49**, 5565.
[35] J. W. Herndon and M. Zora, *SYNLETT*, 1993, 363.
[36] K. H. Dötz and J. Glänzer, *J. Chem. Soc., Chem. Commun.*, 1993, 1036.
[37] D. F. Harvey and D. A. Neil, *Tetrahedron*, 1993, **49**, 2145.
[38] E. Chelain, A. Parlier, M. Audouin, H. Rudler, J.-C. Daran, J. Vaissermann, *J. Am. Chem. Soc.*, 1993, **115**, 10 568.

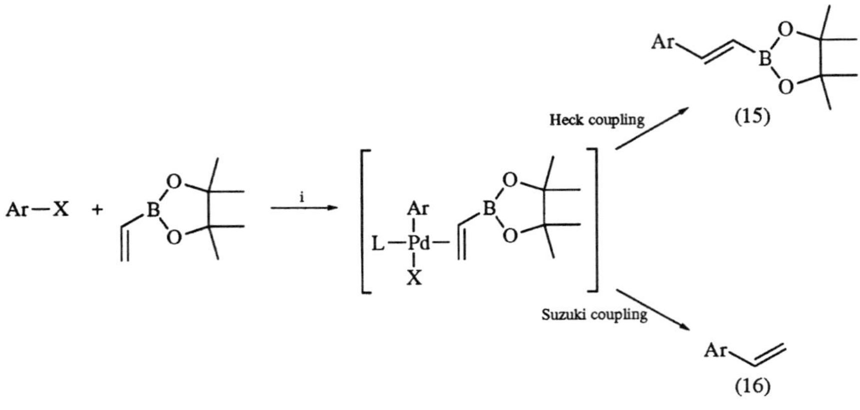

Reagents: 1, Palladium(0), L, 65 °C

Scheme 8

metal and replaces the unsaturation in the product. In the Stille/Suzuki cross-couplings, the alkene is substituted with a heteroatom (tin in the Stille case, boron in the Suzuki case) to allow oxidative addition to transfer an alkenyl substituent to the metal. In this case, carbon–carbon bond formation occurs by reductive elimination. When the substrate is a simple vinylboronate ester, it is an open question whether reaction with an aryl halide will proceed by a Heck or a Suzuki process (Scheme 8).

A study of the selectivity of this reaction[39] has shown that the nature of the aryl halide exerts a considerable influence. With iodobenzene, the ratio of (15):(16) was greater than 95:5. With 4-iodoanisole, on the other hand, reaction was slower and selectivity was reversed (10:90) to favour product (16).

Acrylic esters are commonly used substrates for the Heck coupling. Bates and Gabel have applied this reaction to good effect in a short synthesis of plicatin B (17) (Scheme 9).[40] Styrenes are also typical receiving groups. The effect of chelating ligands in Heck coupling to styrenes has been studied in detail under base-free conditions.[41] A recent trend is to apply Heck coupling in heterocyclic synthesis. 4-Vinylpyridine has been combined with 4-bromoiodobenzene under the normal palladium acetate, triethylamine conditions.[42] Indoles have been coupled to acrylates.[43] Oxidative addition to an iodopyridine by a triphenylphosphine complex, formed from palladium acetate, begins a short total synthesis of (±)-epibatidine. Coupling to the alkene (18) is followed by a reductive procedure using formic acid to complete the substituted ring system in 35% yield (Scheme 10). Deprotection at nitrogen completed the synthesis of epibatidine (19) (R = H).[44]

[39] A. R. Hunt, S. K. Stewart, and A. Whiting, *Tetrahedron Lett.*, 1993, **34**, 3599.
[40] R. W. Bates and C. J. Gabel, *Tetrahedron Lett.*, 1993, **34**, 3547.
[41] M. Portnoy, Y. Ben-David, and D. Milstein, *Organometallics*, 1993, **12**, 4734.
[42] J. Burdeniuk and D. Milstein, *J. Organomet. Chem.*, 1993, **451**, 213.
[43] Y. Yokoyama, M. Takashima, C. Higaki, K. Shidori, S. Moriguchi, C. Ando, and Y. Murakami, *Heterocycles*, 1993, **36**, 1739.
[44] S. C. Clayton and A. C. Reagan, *Tetrahedron Lett.*, 1993, **34**, 7493.

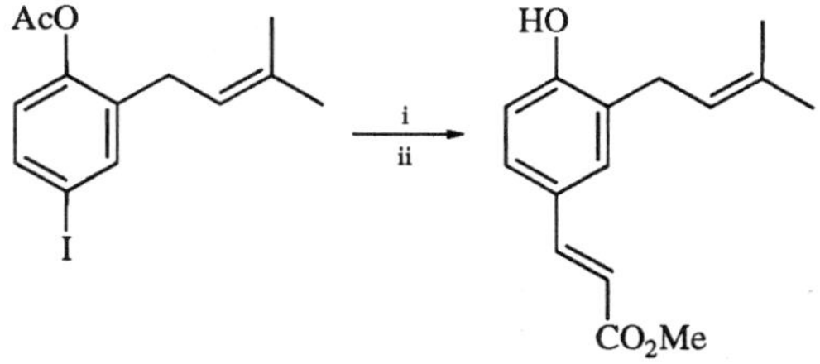

89%
(17)

Reagents: i, methyl acrylate, $[Pd(OAc)_2]$, PR_3, NEt_3; ii, K_2CO_3, MeOH

Scheme 9

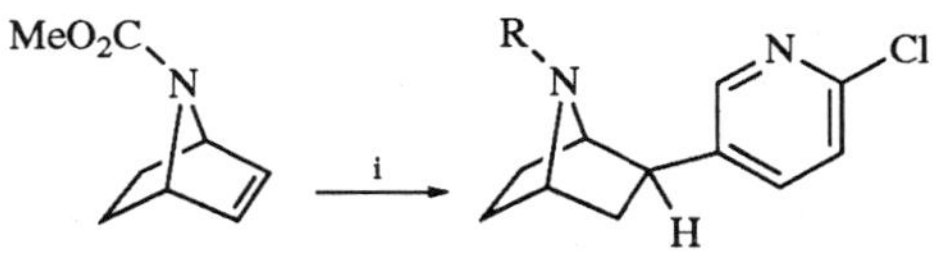

(18) 35%
(19)

Reagents: i, $[Pd(OAc)_2(PPh_3)_2]$, piperidine, formic acid, 2-chloro-5-iodopyridine

Scheme 10

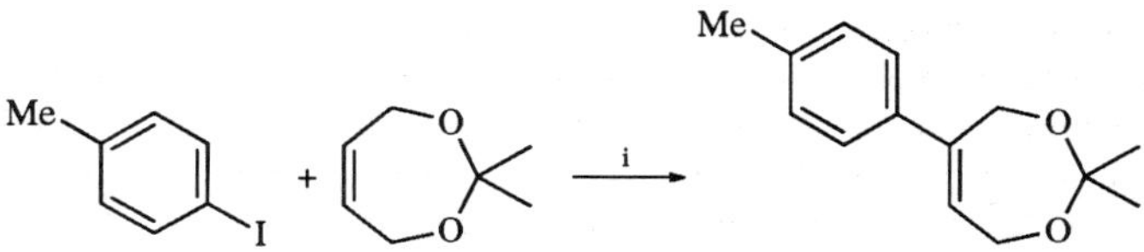

(20) 72%

Reagents: i, $[Pd(OAc)_2]$, Ag_2CO_3, PPh_3, DMF

Scheme 11

Attention is currently turning to more unusual substitution on the alkene. Examples involving vinylphosphine oxides[45] and enamines[46] illustrate this trend. The alkene linkage in a 1,3-dioxep-5-ene is also suitable (Scheme 11).[47] In this case, silver carbonate was used as the base in place of the more normal triethylamine. The product (20) was employed in a synthesis of 4-arylbutyrolactones.

[45] K. M. Petrusiewicz, M. Kuznikowski, and M. Koprowski, *Tetrahedron: Asymmetry*, 1993, **4**, 2143.
[46] C. A. Busacca, R. E. Johnson, and J. Swestock, *J. Org. Chem.*, 1993, **58**, 3299.
[47] T. Sakamoto, Y. Kondo, and H. Yamanaka, *Heterocycles*, 1993, **36**, 2437.

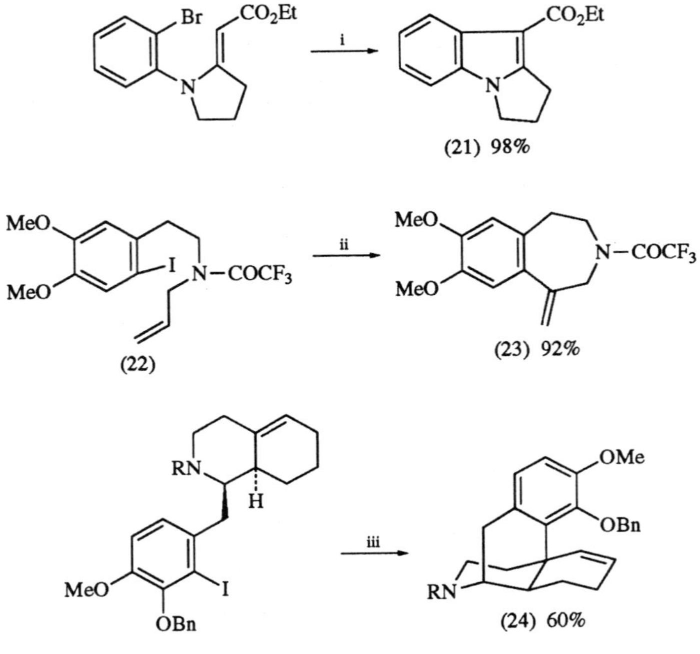

Reagents: i, [$Pd(OAc)_2$], NEt_3, CH_3CN; ii, [$Pd(OAc)_2$], PPh_3, KOAc; iii, [$Pd(OTf)_2(PPh_3)_2$], toluene

Scheme 12

Intramolecular Heck reactions work well (Scheme 12). Indoles can be made by cyclization of *N*-allyl-2-iodoanaline derivatives. The β-elimination step completes the methylene substituent at C-3. The reaction has been used in the synthesis of the CC-1065/duocarmycin pharmacophore.[48] Cyclization of an *N*-arylenamine also works. The highly efficient formation of (21) illustrates that ester substituents on the receiving alkene are not restricted to the far end of the double bond.[49] Polycyclic indole derivatives are accessible by use of the C-2, C-3 double bond as the receiving unit.[50] Acrylic amides obtained from iodoanilines offer similar coupling possibilities.[51] Larger rings [*e.g.* (23)] can be formed by attachment of the allylamine portion *via* a spacer unit as in (22).[52] More elaborate polycyclic structures are obtained in a similar way.[53] The example leading to (24) employs the relatively unusual palladium trifluoroacetate catalyst. The alkene links in the products (23) and (24) illustrate the importance of the β-elimination step at the conclusion of the coupling process. A highly functionalized substrate has been used in an intramolecular Heck arylation to give rapid access to

[48] T. Sakamoto, Y. Kondo, M. Uchiyama, and H. Yamanaka, *J. Chem. Soc., Perkin Trans.*1, 1993, 1941.
[49] J. P. Michael, S.-F. Chang, and C. Wilson, *Tetrahedron Lett.*, 1993, **34**, 8365.
[50] P. Melnyk, J. Gasche, and C. Thal, *Tetrahedron Lett.*, 1993, **34**, 5449.
[51] A. Arcadi, S. Cacchi, F. Marinelli, and P. Pace, *SYNLETT*, 1993, 734.
[52] L. F. Tietze and R. Schimpf, *Synthesis*, 1993, 876.
[53] C. Y. Hong, N. Kado, and L. E. Overman, *J. Am. Chem. Soc.*, 1993, **115**, 11028.

congeners of FR900482.[54] An allylic leaving group, an N–O bond, and an aziridine were all present as structural features in the substrate in this case, yet side reactions did not prevent the effective application of the process.

Regiocontrol has been studied. Depending on the reaction conditions, coupling to vinyl ethers can occur at either end, but pendant amines in the alkyl group of the enol ether promote far-end attack by chelation control.[55] The regiocontrol of the β-elimination step in the Heck coupling of allylamines has been studied.[56] When a vinyl epoxide is the receiving group, carbon–oxygen bond cleavage replaces β-elimination at the metal-removal step.[57] A reductive base-free version of the Heck reaction has been reported. Half an equivalent of zinc is included in the reaction mixture.[58]

The reaction conditions shown in Schemes 9–11 are typical of those currently employed in Heck coupling reactions. Normally, palladium acetate and triethylamine are used in polar aprotic solvents such as DMF or acetonitrile. Common variants on the base include other alkylamines, silver carbonate, or an acetate. The palladium complex need not be the diacetate. Other palladium(II) and (0) complexes are also employed.

The use of aryl triflates in place of aryl halides is common. An example with an enol ether again employs a coordinating group in the alkyl side chain of the receiving unit. In this case, an alkyldiphenylphosphine was used. The enol ether couples with the triflate of 1-naphthol.[59] An intramolecular coupling involving an aryl triflate has been used to build benzofluoranthenes.[60]

Turning to cross-coupling, similar trends are clear. Aryl bromides are again typical building blocks, but are now combined with tin- or boron-containing alkenes. Scheme 13 shows a typical example.[61] Bromobenzene couples with the vinyltin reagent (25) to afford (26) in 95% yield. The reaction has been applied to the preparation of seven-membered azaheterocycles, *e.g.* (27), using bromobenzaldehydes. The palladium(0) catalyst $[Pd(PPh_3)_4]$ is typical for this type of coupling. Product (26) is formed efficiently in the presence of the free amine, but a version of the reaction using a bistrimethylsilyl-protected amine has also been reported.[62] An advantage of the cross-coupling is that regiocontrol in the coupling step can be dictated by the location of the carbon–tin bond. Coupling 4-iodonitrobenzene with an α-stannyl acrylate affords a branched product in 76% yield.[63] The reaction has been applied to 2-bromo-6-methoxynaphthalene, again to afford a branched coupling product. This requires only reduction and ester hydrolysis to complete a short synthesis of naproxen, a commercially important inhibitor of cyclooxygenase. An intramolecular version of vinyltin coupling has been reported.[64] This type of coupling reaction is useful in heterocyclic synthesis. 2,6-Dibromoaniline couples with a tributyltin derivative of an

[54] K. F. McClure and S. J. Danishefsky, *J. Am. Chem. Soc.*, 1993, **115**, 6094.
[55] M. Larhed, C.-M. Andersson, and A. Hallberg, *Acta Chem. Scand.*, 1993, **47**, 212.
[56] L. Filippini, M. Gusmeroli, and R. Riva, *Tetrahedron Lett.*, 1993, **34**, 1643.
[57] R. C. Larock and S. Ding, *J. Org. Chem.*, 1993, **58**, 804.
[58] M. Portnoy, Y. Ben-David, and D. Milstein, *Organometallics*, 1993, **12**, 4734.
[59] D. Badone and U. Guzzi, *Tetrahedron Lett.*, 1993, **34**, 3603.
[60] J. E. Rice and Z.-W. Cai, *J. Org. Chem.*, 1993, **58**, 1415.
[61] R. J. P. Corriu, B. Geng, and J. J. E. Moreau, *J. Org. Chem.*, 1993, **58**, 1443.
[62] R. J. P. Corriu, G. Bolin, and J. J. E. Moreau, *Bull. Soc. Chim. Fr.*, 1993, 273.
[63] J. I. Levin, *Tetrahedron Lett.*, 1993, **34**, 6211.
[64] H. Finch, N. A. Pegg, and B. Evans, *Tetrahedron Lett.*, 1993, **34**, 8353.

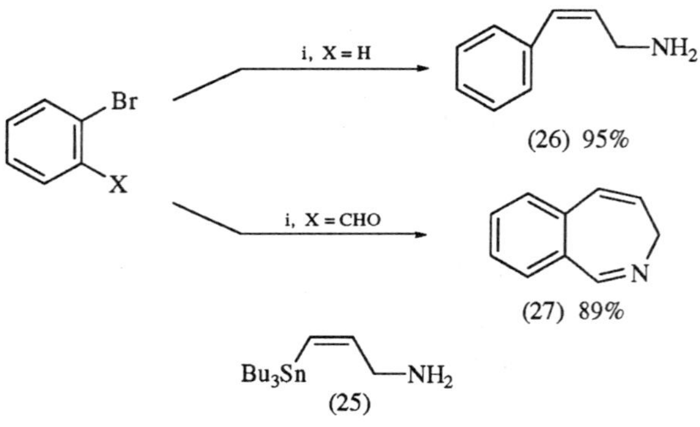

Reagents: i, (25), $[Pd(PPh_3)_4]_2$

Scheme 13

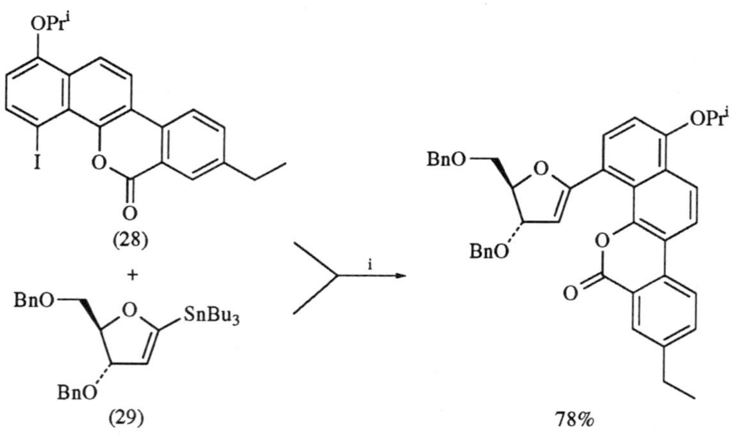

Reagents: i, $[Pd(OAc)_2]$, $AsPh_3$

Scheme 14

enol ether providing bromoindoles for use in a short synthesis of hippadine.[65] With tetramethyltin, alkyl groups are transferred. Methylation of an iodo derivative of a purine nucleoside has been effected in this way.[66] Cross-coupling reactions can be performed on highly elaborate intermediates. The combination of (28) with the tin-substituted chiral dihydrofuran derivative (29) provides a good example (Scheme 14).[67] The catalysts in these cases were palladium(II) complexes. Stannylquinones can

[65] T. Sakamoto, A. Yasuhara, Y. Kondo, and H. Yamanaka, *Heterocycles*, 1993, **36**, 2597.
[66] A. A. Van Aerschot, P. Mamos, N. J. Weyns, S. Ikeda, E. De Clerq, and P. A. Herdewijn, *J. Med. Chem.*, 1993, **36**, 2938.
[67] H.-C. Zhang, M. Brakta, and G. D. Daves, Jr., *Tetrahedron Lett.*, 1993, **34**, 1571.

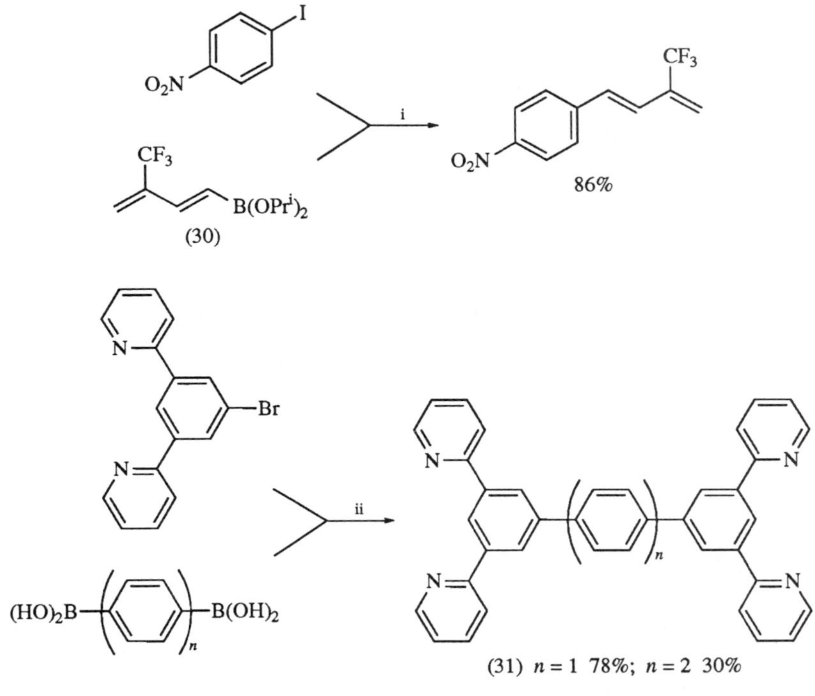

Reagents: i, $[PdCl_2(PPh_3)]$, NaOEt; ii, $[Pd(PPh_3)_4]$, $NaCO_3$

Scheme 15

be coupled with aryl iodides,[68] and heteroaryltin derivatives couple similarly.[69] Tin derivatives of BOC-protected anilines couple with bromochlorobenzenes in the first step of a new carbazole synthesis. The bromine in the haloarene takes part in the coupling step and the chlorine is eliminated with strong base to complete the carbazole by a benzyne mechanism.[70]

The use of carbon–boron bonds in cross-coupling is now common (Scheme 15). A trifluoromethyl-substituted diene (30) has been linked to 4-iodonitrobenzene in this way.[71] In this case, a boronate ester was used, though often it is a simple boronic acid that is employed. An azacarbazole synthesis is based on this approach, using iodofluoropyridines in combination with N-protected boronic acid derivatives of aniline.[72] The reaction has been employed in a synthesis of the antimicrobial arene sponge pigment fascaplysin.[73] A bisboronic acid can be used to produce symmetrical,

[68] L. S. Liebeskind and S. W. Riesinger, *J. Org. Chem.*, 1993, **58**, 408.
[69] S. Gronwitz, P. Björk, J. Malm, and A.-B. Hörnfeldt, *J. Organomet. Chem.*, 1993, **460**, 127.
[70] M. Iwao, H. Takehara, S. Furukawa, and M. Watanabe, *Heterocycles*, 1993, **36**, 1483.
[71] F. Jin, Y. Xu, and B. Jiang, *J. Fluorine Chem.*, 1993, **65**, 111.
[72] P. Rocca, F. Marsais, A. Godard, and G. Quéguiner, *Tetrahedron*, 1993, **49**, 49.
[73] P. Rocca, F. Marsais, A. Godard, and G. Quéguiner, *Tetrahedron Lett.*, 1993, **34**, 7917.

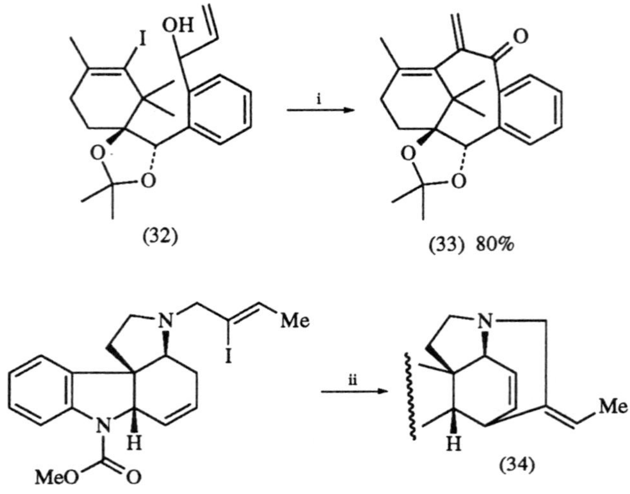

Reagents: i, [$Pd(OAc)_2$], K_2CO_3; ii, [$Pd(OAc)_2$], K_2CO_3, DMF

Scheme 16

doubly-extended structures such as the tetrapyridine-substituted triphenyl derivative (31) ($n = 1$).[74] Aryl triflates have been employed in the same way as aryl halides. The reaction can couple a porphyrin-substituted aryl triflate with 2,4-dimethoxyphenylboronic acid, providing an alternative access to elaborated quinones.[75]

Coupling with Vinyl Halides and Triflates.—Vinyl halides work well in Heck couplings (Scheme 16). The reaction proceeds in the presence of unprotected OH groups, even in allylic alcohols, and has been performed without racemization on optically active allyl alcohol in an enantioselective direct synthesis of dienols.[76] Palladium acetate and silver acetate or carbonate were used in this coupling. In a similar reaction (using palladium acetate and potassium carbonate) intramolecular cyclization of (32) to form the taxol intermediate (33) proceeded with concomitant oxidation of the allylic alcohol in the product.[77] Another intramolecular Heck coupling[78] is notable because of the inversion of geometry in the exocyclic alkene in the product (34).

Vinyl halides are useful in cross-coupling. Combination with vinyltin coupling partners provides an attractive route to conjugated polyenes (Scheme 17). The tetraene building block for calyculins has been prepared in this way.[79] Intramolecular coupling of the same type closes the macrocyclic ring in (35).[80] A still more ambitious

[74] M. Beley, S. Chodorowski, J.-P. Collin, and J.-P. Sauvage, *Tetrahedron Lett.*, 1993, **34**, 2933.
[75] C.-S. Chan, C. C. Mak, and K. S. Chan, *Tetrahedron Lett.*, 1993, **34**, 5125.
[76] T. Jeffery, *Tetrahedron Lett.*, 1993, **34**, 1133.
[77] J. J. Masters, D. K. Jung, W. G. Bornmann, S. J. Danishefsky, and S. de Gala, *Tetrahedron Lett.*, 1993, **34**, 7253.
[78] V. H. Rawal and C. Michoud, *J. Org. Chem.*, 1993, **58**, 5583.
[79] F. Yokokawa, Y. Hamada, and T. Shioiri, *Tetrahedron Lett.*, 1993, **34**, 6559.
[80] G. Pattenden and S. M. Thom, *SYNLETT*, 1993, 215.

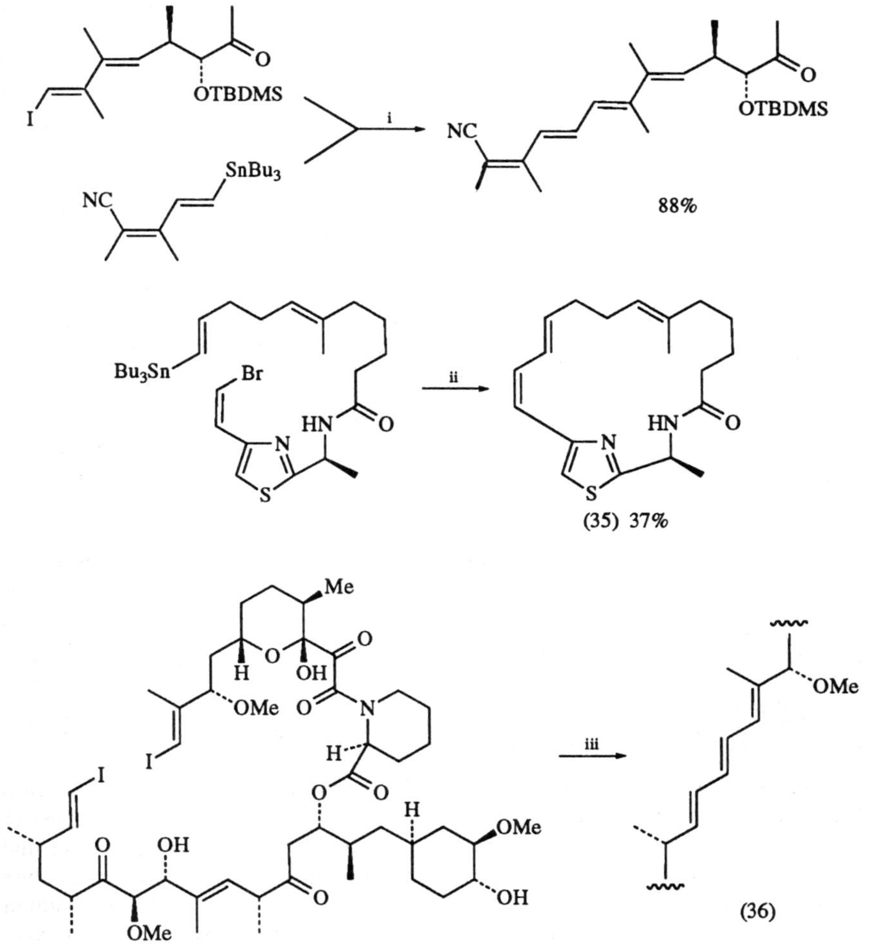

Reagents: i, $[PdCl_2(CH_3CN)_2]$, DMF; ii, $[Pd(AsPh_3)_4]$; iii, *trans*-1,2-bis(tributylstannyl)ethyne, $[PdCl_2(CH_3CN)_2]$, DMF–THF

Scheme 17

cross-coupling constitutes the key step in a total synthesis of rapamycin (36). Here 1,2-di(tributylstannyl)ethene is coupled with a pair of vinyl iodide end-groups to complete a triene and close the macrocycle in the target molecule.[81] Palladium dichloride or triphenylphosphine complexes are generally used in these reactions, though in the case of the formation of (35) triphenylphosphine proved ineffective and the triphenylarsine analogue of the catalyst was required. Enantiomerically pure

[81] K. C. Nicolaou, T. K. Chakraborty, A. D. Piscopio, N. Minowa, and P. Bertinato, *J. Am. Chem. Soc.*, 1993, **115**, 115, 4419.

dienylsulfoxides can be obtained by coupling tributylvinyltin with a sulfoxide-substituted vinyl bromide.[82] An unusual coupling of distannylethene was used to link two bromocyclooctatetraenes.[83] Boronic acids are often used in coupling with vinyl halides. In a synthesis of belamcandaquinones, a substituted arylboronic acid is coupled with a bromoquinone derivative by $[Pd(PPh_3)_4]$ catalysis and sodium carbonate.[84] This reaction is the converse of the quinone elaboration illustrated in the preceding section.

Vinyl triflates are used in coupling reactions with vinyltin reagents.[85] This type of coupling has been applied to polyene synthesis[86] and the elaboration of aromatic rings[87] by organotin reagents. Vinyltriflates couple with alkyl-, alkenyl- and aryl-boronic acid derivatives, and other organoboron compounds.[88] 3-Indoleboronic acids have been used in similar couplings to form 3-substituted indoles.[89] Carbapenems have been obtained by Suzuki cross-coupling between boronic acid derivatives and vinyltriflate functionality in the carbapenem unit.[90]

Nucleophilic Reagents in Palladium-catalysed Cross-coupling.—Nucleophilic species are commonly used in cross-coupling reactions. One of the most general procedures employs acetylide reagents. Typical conditions involve, besides the palladium catalyst, the presence of copper(I) iodide and an amine base in the reaction mixture. Coupling with aryl halides[91] and with heterocyclic derivatives[92] illustrate this reaction. Considerable attention has been paid to the elaboration of uracils and pyridines. Protected uracil derivatives couple with propargyl alcohols and ketones under the typical copper iodide conditions.[93] This type of coupling can be performed with enol triflate derivatives of pyrimidinucleosides.[94] Propargyl alcohols couple with aryl halides in a reaction that has been used in the early stages of the synthesis of water-soluble phthalocyanine derivatives destined for use in the photosensitized inactivation of cancer cells and viruses.[95] Chiral propargyl alcohols (37) have been employed,[96] and retain their stereointegrity during the coupling process (Scheme 18). The product was taken on in another palladium-catalysed alkyne addition step to complete a synthesis of lipoxin B_4.

The use of a coupling step to circumvent the need for protecting groups on the alcohols in this synthesis of (38) (of which only reduction is required to complete the tetraene heart of lipoxin B_4) is an important advantage. The synthesis of polyenynes in a fashion similar to that illustrated in Scheme 17 is a typical application of this type of

[82] R. S. Paley and A. de Dios, *Tetrahedron Lett.*, 1993, **34**, 2429.
[83] D. A. Siesel and S. W. Staley, *Tetrahedron Lett.*, 1993, **34**, 3679.
[84] Y. Fukuyama, Y. Kiriyama, and M. Kodama, *Tetrahedron Lett.*, 1993, **34**, 7637.
[85] I. N. Houpis, L. DiMichele, and A. Molina, *SYNLETT*, 1993, 365.
[86] B. H. Lipshutz and M. Alami, *Tetrahedron Lett.*, 1993, **34**, 1433.
[87] V. Farina, B. Krishnan, D. R. Marshall, and G. P. Roth, *J. Org. Chem.*, 1993, **58**, 5434.
[88] T. Oh-e, N. Miyaura, and A. Suzuki, *J. Org. Chem.*, 1993, **58**, 2201.
[89] Q. Zheng, Y. Yang, and A. R. Martin, *Tetrahedron Lett.*, 1993, **34**, 2235.
[90] N. Yasuda, L. Xavier, D. L. Rieger, Y. Li, A. E. DeCamp, and U.-H. Dolling, *Tetrahedron Lett.*, 1993, **34**, 3211.
[91] B. V. Nguyen, Z.-Y. Yang, and D. J. Burton, *J. Org. Chem.*, 1993, **58**, 7368.
[92] G. Galambos, C. Szántay, J. Tamás, and C. Szántay, *Heterocycles*, 1993, **36**, 2241.
[93] N. G. Kundu and S. K. Dasgupta, *J. Chem. Soc., Perkin Trans.* 1, 1993, **21**, 2657.
[94] G. T. Crisp and B. L. Flynn, *J. Org. Chem.*, 1993, **58**, 6614.
[95] R. W. Boyle and J. E. van Lier, *SYNLETT*, 1993, 351.
[96] M. Alami, B. Crousse, G. Linstrumelle, L. Mambu, and M. Larchevêque, *SYNLETT*, 1993, 217.

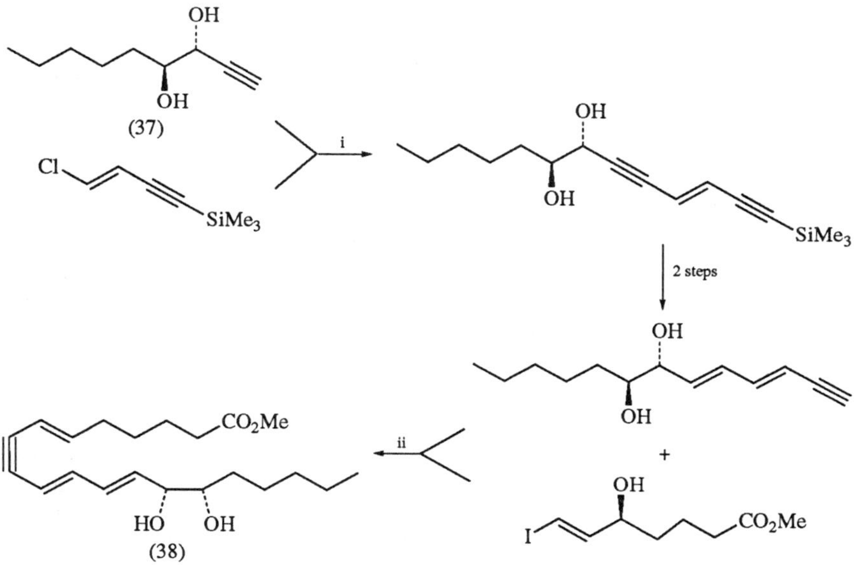

Reagents: i, $[PdCl_2(PhCN)_2]$, CuI, piperidine; ii, $[Pd(PPh_3)_4]$, CuI, piperidine

Scheme 18

reaction. The procedures to form dienediyne systems[97,98] are effective, and provide the starting point[98] for work related to the neocarzinostatin series of natural products. Tribenzocyclotriynes[99] and higher polybenzopolydehydroannulenes have been prepared by sequences of coupling reactions (Scheme 19) between arylalkynes and aryl halides. The formation[100] of (39) ($n = 5$) provides another fine example of large ring closure using palladium-catalysed coupling methods.

In these reactions, it has always been thought that the combination of organic base and copper salt produced traces of copper acetylides as the active nucleophile. In a route to building blocks for functionalized butenolides, coupling at the alkyne was performed in the presence of a 1,3-dicarbonyl compound that could also be easily deprotonated.[101] Coupling of alkynes with aryl halides, however, can be efficient in the presence of a typical amine base such as pyrrolidine, without the presence of a copper salt.[102] In a synthesis of aryl- and alkynyl-linked pyridine systems, a key coupling between a functionalized alkyne and a similarly functionalized aryl halide was found to proceed efficiently (71–79% yield) in the presence of n-propylamine. When copper salts were used, side reactions were observed.[103]

97 J. Suffert, A. Eggers, S. W. Scheuplein, and R. Brückner, *Tetrahedron Lett.*, 1993, **34**, 4177.
98 K. Nakatani, K. Arai, and S. Terashima, *Tetrahedron*, 1993, **49**, 1901.
99 J. D. Kinder, C. A. Tessier, and W. J. Youngs, *SYNLETT*, 1993, 149.
100 K. P. Baldwin, J. D. Bradshaw, C. A. Tessier, and W. J. Youngs, *SYNLETT*, 1993, 853.
101 A. Arcadi, S. Cacchi, G. Fabrizi, and F. Marinelli, *SYNLETT*, 1993, 65.
102 M. Alami, F. Ferri, and G. Linstrumelle, *Tetrahedron Lett.*, 1993, **34**, 6403.

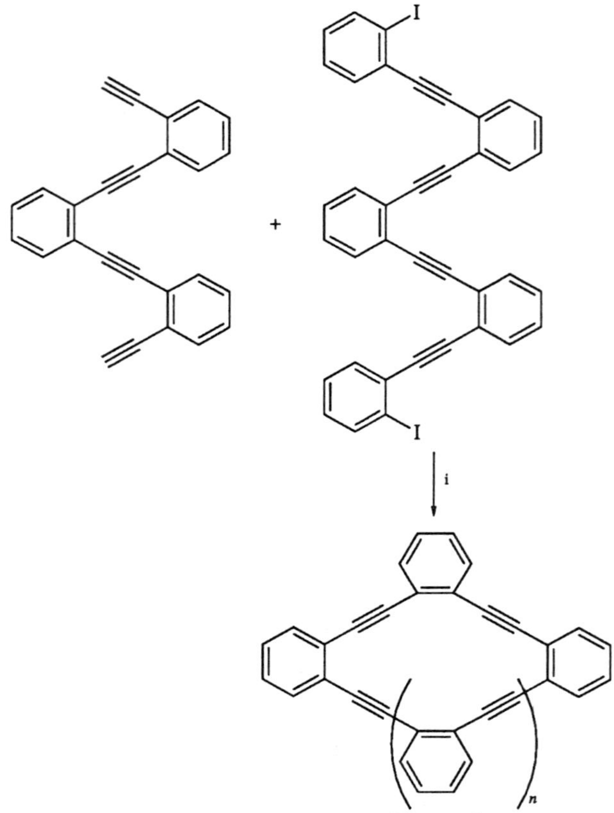

(39) $n = 5$

Reagents: i, $[Pd(PPh_3)_4]$, CuI, NEt_3

Scheme 19

Not all aryl halides participate in coupling with alkynes. Efficient homocoupling of the acetylide (to form a diyne) has been reported in the presence of the iodoarene portion of 2-nitro-3-hydroxy-4-iodophenol.[104] Coupling of acetylides to a second alkyne carrying a propargylic leaving group, has proved to be an efficient route to 1,2-dien-4-ynes.[105] Two examples have appeared where silylalkynes participate in coupling with 2-iodoanilines (and related heterocyclic systems) to provide effective syntheses of heterocondensed pyrroles[106] and substituted tryptophans.[107] In these

[103] V. Grosshenny and R. Ziessel, *J. Organomet. Chem.*, 1993, 453, C19.
[104] N. G. Kundu, M. Pal, and C. Chowdhury, *J. Chem. Res.* (*S*), 1993, 432.
[105] S. Gueugnot and G. Linstrumelle, *Tetrahedron Lett.*, 1993, **34**, 3853.
[106] D. Wensbo, A. Eriksson, T. Jeschke, U. Annby, S. Gronowitz, and L. A. Cohen, *Tetrahedron Lett.*, 1993, **34**, 2823.
[107] T. Jeschke, D. Wensbo, U. Annby, S. Gronowitz, and L. A. Cohen, *Tetrahedron Lett.*, 1993, **34**, 6471.

examples, the silicon is retained during the coupling step, which involves palladium transfer to the alkyne and ring closure with the nitrogen atom.

Frequently, more potent nucleophiles are used in coupling reactions. In many cases, a carbon–zinc bond initiates reaction with a palladium complex. This has been employed in the elaboration of serine-derived iodozinc derivatives of amino acids,[108] and imidazole-derived organozinc[109] and vinylzinc[110,111] reagents. Vinylzincs are often prepared *in situ*, for example, from a vinyl halide,[112] or vinyltin reagent.[113] Their use thus provides a direct alternative to the combination of the halides and tin compounds themselves in cross-coupling steps. Heterocyclic systems have been used. Deprotonation of N-protected indoles followed by addition of zinc chloride prepares the way for coupling with iodobenzene.[114] In this reaction, the protecting group is an electron-withdrawing group. Similar zinc chloride derivatives of indoles have been used in coupling with pyridines.[115] Grignard reagents have also been used, even in the presence of alcohol functional groups.[116] Boron is a less-common precursor for metallation, but there are examples of the use of trivinylboron derivatives in transmetallation with diethylzinc, prior to coupling under palladium-catalysed conditions.[117] Electrochemically prepared arylzinc halide derivatives have also been successfully employed.[118]

Double Heteroatom Attachment.—The use of palladium-catalysed coupling conditions to form bonds to heteroatoms, rather than to carbon, is of growing importance (Scheme 20). Coupling of 2-bromotrifluoromethylbenzene with hexamethyldisilane affords (40) in 50% yield.[119] Double silylation, with concomitant dimerization of 1,3-dienes to form (41), has also been reported.[120] In the case of formation of (42), a dimethylsilyl group is introduced during the cyclodimerization of two alkynes.[121] An intramolecular process adds an O-tethered $R_2Si–SiR_3$ unit across an alkene.[122] With a mixed trimethyltin–trimethylsilyl reagent, tin and silicon add across norbornene.[123] Hexamethylditin can be used to replace an enol triflate by a trimethyltin unit.[124] Thioboration of alkynes proceeds under palladium-catalysed conditions.[125] Iodine can be introduced in place of an SO_2Ph group by the use of zinc iodide with $[PdCl_2(PhCN)_2]$.[126] The examples given in Scheme 20 indicate that all the typical

108 M. J. Dunn, R. F. W. Jackson, J. Pietruszka, N. Wishart, D. Ellis, and M. J. Wythes, *SYNLETT*, 1993, 499.
109 R. M. Turner, S. V. Ley, and S. D. Lindell, *SYNLETT*, 1993, 748.
110 T. L. Gilchrist and M. A. M. Healy, *Tetrahedron*, 1993, **49**, 2543.
111 R. Rossi, A. Carpita, F. Bellina, and P. Cossi, *J. Organomet. Chem.*, 1993, **451**, 33.
112 E.-i. Negishi, M. Ay, Y. V. Gulevich, and Y. Noda, *Tetrahedron Lett.*, 1993, **34**, 1437.
113 B. H. Lipshutz, M. Alami, and R. B. Susfalk, *SYNLETT*, 1993, 693.
114 T. Sakamoto, Y. Kondo, N. Takazawa, and H. Yamanaka, *Heterocycles*, 1993, **36**, 1993.
115 M. Amat, S. Hadida, and J. Bosch, *Tetrahedron Lett.*, 1993, **34**, 5005.
116 R. W. Hoffman, V. Giesen, and M. Fuest, *Liebigs Ann. Chem.*, 1993, 629.
117 K. A. Agrios and M. Srebnik, *J. Organomet. Chem.*, 1993, **444**, 15.
118 S. Sibille, V. Ratovelomanana, J. Y. Nédélec, and J. Périchon, *SYNLETT*, 1993, 425.
119 P. Babin, B. Bennetau, M. Theurig, and J. Dunoguès, *J. Organomet. Chem.*, 1993, **446**, 135.
120 Y. Obora, Y. Tsuji, and T. Kawamura, *Organometallics*, 1993, **12**, 2853.
121 K. Ikenaga, K. Hiramatsu, N. Nasaka, and S. Matsumoto, *J. Org. Chem.*, 1993, **58**, 5045.
122 M. Murakami, M. Suginome, K. Fujimoto, H. Nakamura, P. G. Andersson, and Y. Ito, *J. Am. Chem. Soc.*, 1993, **115**, 6487.
123 Y. Obora, Y. Tsuji, M. Asayama, and T. Kawamura, *Organometallics*, 1993, **12**, 4697.
124 T. A. Tius, G. S. K. Kannangara, M. A. Kerr, and K. J. S. Grace, *Tetrahedron*, 1993, **49**, 3291.
125 T. Ishiyama, K.-i. Nishijima, N. Miyaura, and A. Suzuki, *J. Am. Chem. Soc.*, 1993, **115**, 7219.
126 T. Satoh, K. Itoh, M. Miura, and M. Nomura, *Bull. Chem. Soc. Jpn.*, 1993, **66**, 2121.
T. Satoh, K. Itoh, M. Miura, and M. Nomura, *Bull. Chem. Soc. Jpn.*, 1993, **66**, 2121.

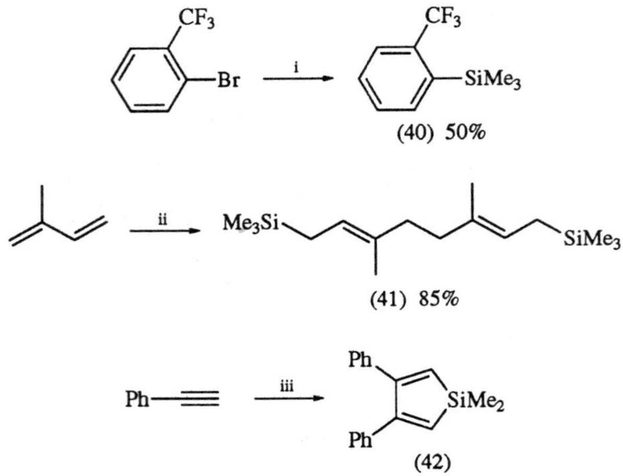

Reagents: i, $Me_3SiSiMe_3$, [$Pd(PPh_3)_4$], HMPT; ii, $Me_3SiSiMe_3$, [$Pd(dba)_2$], DMF; iii, $HMe_2SiSnBu_3$, [$PdCl_2(PPh_3)_3$]

Scheme 20

types of palladium(0) and palladium(II) catalysts can be employed.

Multi-step Coupling Reactions.—A simple two-step process adds two coupling partners across an alkene. Norbornadiene couples with 4-methoxyiodobenzene and a tributyltin acetylide. Effectively, this is a Heck coupling reaction intercepted by a Stille coupling. Pyrolysis of the product effects a retro Diels–Alder reaction to place a *cis*-alkene between the acetylene and the aromatic ring.[127] The second double bond in norbornadiene can also participate in coupling. A new carbon–carbon bond is formed across the base of the norbornadiene framework, and a second coupling reaction completes the process. A combination of zinc, zinc dichloride, and an alkyl iodide have been used in the final stage of this reaction.[128] Reaction of norbornadiene with 3-iodocyclohexenone attached two enone units to the same alkene. In the presence of triethylamine, these products then cyclized in a normal aldol reaction.[129] A vinylzinc reagent provides the starting point for an unusual double coupling (Scheme 21) to a diiodobiaryl coreactant with axial chirality and C_2 symmetry. The product (43) arises from a sequence of metal-mediated steps ending in β-elimination to place the alkene in the five-membered ring. Since a quaternary centre has been formed, the alkene moves to the next place around the ring.[130] An iodocinnamate ester couples with norbornadiene by a double Heck process. In this reaction, too, an alkene is put back into the

[127] M. Kosugi, T. Kimura, H. Oda, and T. Migita, *Bull. Chem. Soc. Jpn.*, 1993, **66**, 3522.
[128] C. S. Li, D. C. Jou, and C. H. Cheng, *Organometallics*, 1993, **12**, 3945.
[129] C.-S. Li, D.-C. Jou, C.-H. Cheng, F.-L. Liso, and S.-L. Wang, *Organometallics*, 1993, **12**, 3553.
[130] T. J. Katz, A. M. Gilbert, M. E. Huttenloch, G. Min-Min, and H. H. Brintzinger, *Tetrahedron Lett.*, 1993, **34**, 3551.

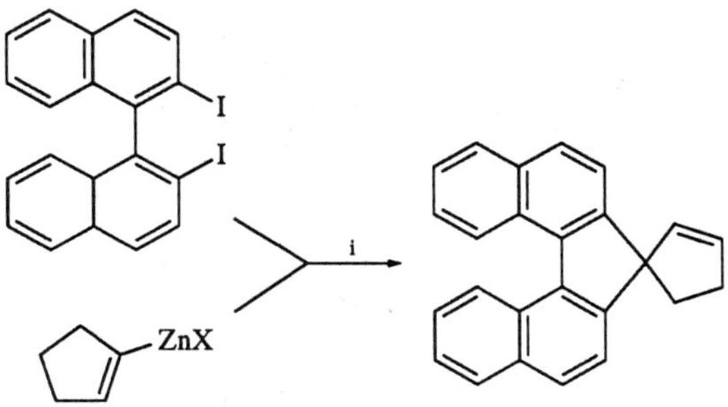

(43) 66%

Reagents: i, $[Pd(PPh_3)_4]$, THF

Scheme 21

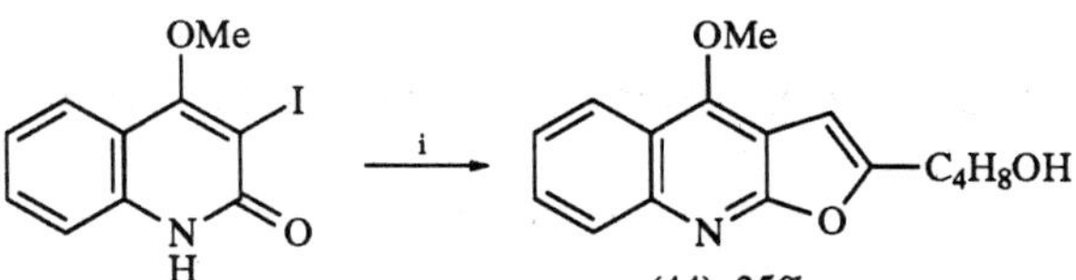

(44) 35%

Reagents: i, $HC{\equiv}C_4H_8OH$, $[PdCl_2(PPh_3)_2]$, CuI, NEt_3

Scheme 22

product by β-elimination.[131] Heteroatoms can participate in the final bond formation. Coupling of a monosubstituted alkyne with 3-bromoacrylic acid under the copper iodide–triethylamine conditions affords butenolides by a sequence of carbon–carbon and carbon–oxygen bond formations.[132] 2-Iodobenzoic acid couples with alkynes in a similar fashion.[133] The quinolinofuren (44) can be prepared by attachment of an acetylide to a vinylhalide and trapping of the oxygen atom after tautomerization of the amide (Scheme 22). Benzofurans can also be obtained.[134] 1,3-Dicarbonyl compounds are precursors to furan ring systems in a tandem coupling process in which tautomerization to the enol form completes the product structure.[135] In all these reactions, an intermolecular initial coupling is followed by an intramolecular process. With a 1-en-6-yne, a sequence of three carbon–carbon bond formations is possible to effect a [2 + 2 + 2] cycloaddition.[136]

An alternative way to combine inter- and intramolecular steps is to place both the initial site of reaction and the first receiving carbon–carbon multiple bond within the

[131] R. Grigg, P. Kennewell, A. J. Teasdale, and V. Sridharan, *Tetrahedron Lett.*, 1993, **34**, 153.
[132] X. Lu, X. Huang, and S. Ma, *Tetrahedron Lett.*, 1993, **34**, 5963.
[133] N. G. Kundu and M. Pal, *J. Chem. Soc., Chem. Commun.*, 1993, 86.
[134] J. Reisch, P. Nordhaus, and T. Pflug, *J. Heterocyclic Chem.*, 1993, **30**, 1161; I. Candiani, S. DeBernardinis, W. Cabri, M. Marchi, A. Bedeschi, and S. Penco, *SYNLETT*, 1993, 269.
[135] A. Arcadi, S. Cacchi, R. C. Larock, and F. Marinelli, *Tetrahedron Lett.*, 1993, **34**, 2813.
[136] S. Brown, S. Clarkson, R. Grigg, and V. Sridharan, *Tetrahedron Lett.*, 1993, **34**, 157.

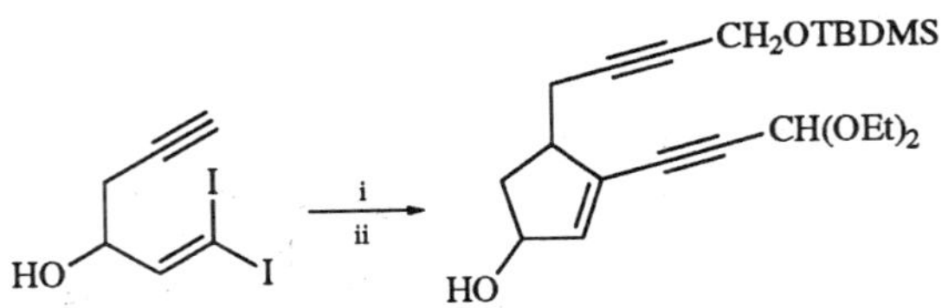

(45) 50%

Reagents: i, $Bu_3SnC{\equiv}CCH_2OTBDMS$, $[Pd(PPh_3)_4]$, THF; ii, $[PdCl_2(PPh_3)_2]$,CuI,Et_2NH, $HC{\equiv}CCH(OEt)_2$

Scheme 23

same molecule. Reaction in the presence of an external coreactant traps the final organometallic intermediate. A dienediyne motif has been obtained in this way from a 1,1-diiodoalkene equipped with a terminal alkyne group (Scheme 23). These two unsaturated portions become the 1,3-diene moiety in the diene/diyne. The acetylides became attached in different ways. One by direct coupling with the vinyl iodide, and the other in the trapping step following the closure to the pendent alkyne to complete (45).[137] The control of stereochemistry at the exocyclic alkene in this reaction is an important feature, and is determined by the *syn* stereochemistry of metal-mediated double addition to the alkyne. In a variant of the reaction a 1,1-dibromoalkene was used, with palladium acetate and triphenylphosphine as the catalyst system. One equivalent of a tributyltin derivative of a protected propargyl alcohol was employed to complete the coupling reaction. The intermediate obtained in this way contained one remaining vinylbromide unit and was coupled with a second propargyl alcohol derivative using the copper iodide method.[138] The final trapping can also use a copper-mediated acetylide addition. An example that starts through transmetallation to produce an alkylzinc reagent from an alkyl iodide proceeds by intramolecular cyclization to a pendent alkene. This produces an intermediate that couples to a propargylic ester.[139] A similar strategy, but in a different ring size and with the vinyl halide at an internal position (Scheme 24), gives access to an exocyclic 1,3-diene which can couple to a vinyltin derivative. In this way a route to the diene portion of vitamin D can be achieved.[140] In a similar cyclization, an exocyclic vinylbromide derived from a steroidal CD fragment has been used to terminate the coupling process.[141]

Since the metal moves through the molecule with each successive carbon–carbon bond formation, an extensive series of new bonds can be made in a single reaction. This forms the basis of the palladium zipper polycyclization process (Scheme 25).[142] Three new carbon–carbon bonds and three rings are formed in the cyclization leading to (46). In the route to (47), reaction is initiated at the vinyl halide with cyclization occurring, in

137 J. M. Nuss, R. A. Rennels, and B. M. Levine, *J. Am. Chem. Soc.*, 1993, **115**, 6991.
138 S. Torii, H. Okumoto, T. Tadokoro, A. Nishimura, and M. A. Rashid, *Tetrahedron Lett.*, 1993, **34**, 2139.
139 H. Stadtmüller, C. E. Tucker, A. Vaupel, and P. Knochel, *Tetrahedron Lett.*, 1993, **34**, 7911.
140 J. M. Nuss, M. M. Murphy, R. A. Rennels, B. H. Heravi, and B. J. Mohr, *Tetrahedron Lett.*, 1993, **34**, 3079.
141 B. M. Trost and J. Dumas, *Tetrahedron Lett.*, 1993, **34**, 19.
142 B. M. Trost and Y. Shi, *J. Am. Chem. Soc.*, 1993, **115**, 9421.

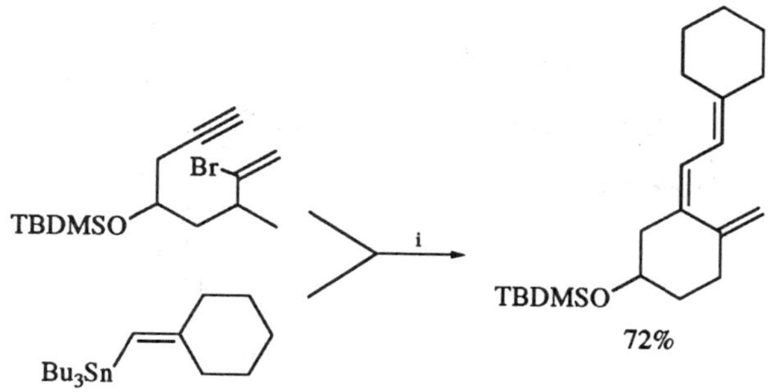

Reagents: i, [Pd(PPh$_3$)$_4$], THF

Scheme 24

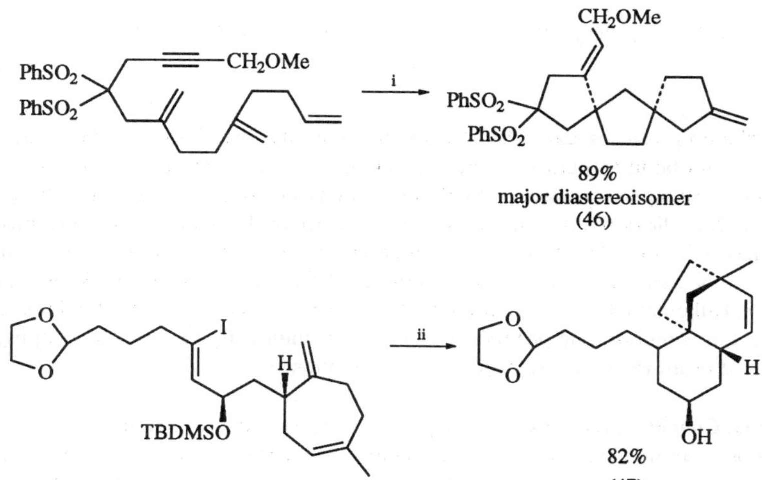

Reagents: i, palladium(0), HOAc; ii, [Pd(OAc)$_2$], PPh$_3$, Ag$_2$CO$_3$, THF

Scheme 25

turn, at each of the two alkene links in the right-hand portion of the substrate.[143] This type of cyclization has been used in the first total synthesis of (±)-scopadulcic acid B.[144] In this case, oxidative addition to an aryl halide initiated the reaction sequence. Heteroatom groups can become involved in the termination of tandem palladium coupling reactions. Intramolecular Heck cyclization of a vinyl bromide to an alkene has been combined with carbon–nitrogen bond formation to a terminal NHTs group.[145]

143 D. J. Kucera, S. J. O'Connor, and L. E. Overman, *J. Org. Chem.*, 1993, **58**, 5304.
144 L. E. Overman, D. J. Ricca, and V. D. Tran, *J. Am. Chem. Soc.*, 1993, **115**, 2042.
145 G. D. Harris, Jr., R. J. Herr, and S. M. Weinreb, *J. Org. Chem.*, 1993, **58**, 5452.

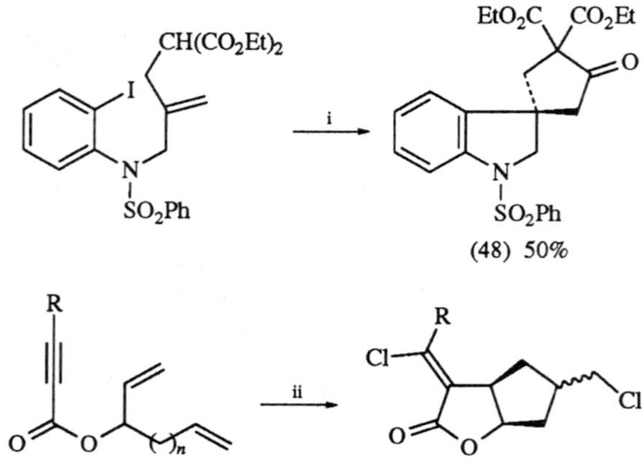

(48) 50%

(49) 73–82%

Reagents: i, $Pd(OAc)_2$, PPh_3, CH_3CN, CO(1atm), TlOAc; ii, $PdCl_2$, $CuCl_2$, LiCl, CH_3CN

Scheme 26

Carbanion centres can also terminate coupling reactions (Scheme 26). Carbon–carbon bond formation by cyanide ion capture provides a good example.[146] In another case, carbonylation is combined with Heck coupling and anion capture to form spirocyclic derivatives of indole (49).[147] Capture by halide ions is also possible. An unusual example, employing copper(II) chloride and lithium chloride in the presence of a catalytic amount of palladium(II) chloride, couples an alkyne and two terminal alkenes with the introduction of two halide ions to form (49). Under different conditions a monocyclic product predominates, indicating that halide trapping had occurred at an earlier stage in the reaction sequence.[148]

Unusual Coupling Processes.—In the previous sections, developments over the year in the most common types of coupling reactions have been surveyed. Clear trends are emerging for the pairing of coupling partners. The same patterns apply with the more unusual coupling processes. For example, a diazonium salt takes the place of the iodide or triflate in a Heck coupling process performed using palladium acetate in aqueous media.[149] Vinyltin reagents have been coupled with acid chlorides,[150] and with dienes with concomitant silylation by a hexaalkyldisilane.[151] Coupling of acid chlorides and aromatic halides with heterocyclic tributyltin reagents has been examined.[152]

Several investigations have examined cyclobutane-1,2-diones, and their haloketone analogues. Both vinyltin–acid chloride coupling with a palladium(II) catalyst in the

146 R. Grigg, V. Santhakumar, and V. Sridharan, *Tetrahedron Lett.*, 1993, **34**, 3163.
147 R. Grigg and V. Sridharan, *Tetrahedron Lett.*, 1993, **34**, 7471.
148 J. Ji and X. Lu, *SYNLETT*, 1993, 745.
149 S. Sengupta and S. Bhattacharya, *J. Chem. Soc., Perkin Trans.* 1, 1993, **17**, 1943.
150 J.-L. Parrain, I. Beaudet, A. Duchêne, S. Watrelot, and J. P. Quintard, *Tetrahedron Lett.*, 1993, **34**, 5445.
151 Y. Obora, Y. Tsuji, and T. Kawamura, *J. Am. Chem. Soc.*, 1993, **115**, 10414.
152 W. Shihua, Y. Shenggang, H. Xinquan, and G. Hefu, *Huaxue Xuebao*, 1993, **51**, 393.

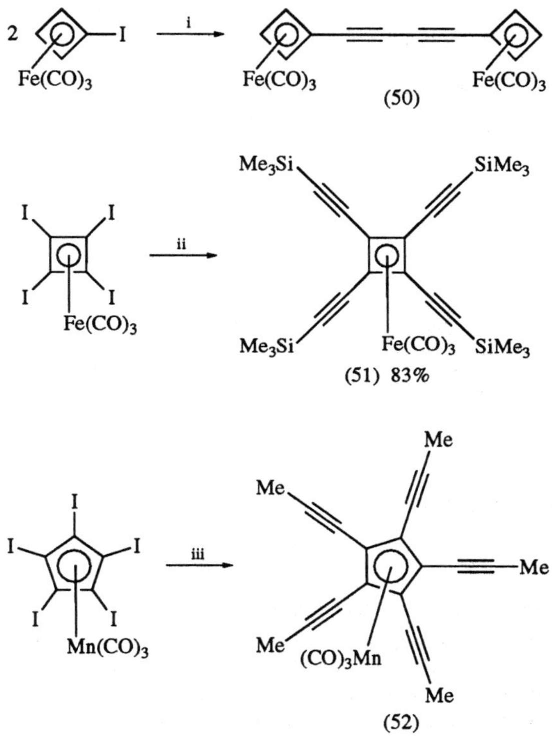

Reagents: i, $Me_3SnC{\equiv}CC{\equiv}CSnMe_3$; ii, $Me_3SiC{\equiv}CSiMe_3$, $[Pd_2(dba)_3]$, $AsPh_3$, DMF; iii, $MeC{\equiv}CSnMe_3$

Scheme 27

presence of copper iodide[153] and the reverse reaction with a chlorocyclobutenone combined with a tin derivative of a quinone[154] succeed. A tin–silicon disubstituted furan has been combined with a chlorocyclobutenone in a new synthesis of benzofurans. Benzothiophenes are also accessible.[155] Hydrosilylation and palladium-catalysed cross-coupling have been combined in a one-pot reaction sequence.[156] During 1993 there have been a series of investigations of palladium-catalysed coupling to heteroatom-substituted π-bound ligands (Scheme 27). The bimetallic product (50)[157] and the tetrayne (51)[158] were prepared by coupling reactions of iodocyclobutadiene complexes. Similar coupling with a (pentaiodocyclopentadienyl)manganese complex leads to (52).[159] Products (50) and (51) have been characterized by *X*-ray

[153] L. S. Liebeskind, M. S. Yu, and R. W. Fengl, *J. Org. Chem.*, 1993, **58**, 3543.
[154] J. P. Edwards, D. J. Krysan, and L. S. Liebeskind, *J. Am. Chem. Soc.*, 1993, **115**, 9868.
[155] L. S. Liebeskind and J. Wang, *J. Org. Chem.*, 1993, **58**, 3550.
[156] K. Takahashi, T. Minami, Y. Ohara, and T. Hiyama, *Tetrahedron Lett.*, 1993, **34**, 8263.
[157] J. E. C. Wiegelmann and U. H. F. Bunz, *Organometallics*, 1993, **12**, 3792.
[158] U. H. F. Bunz and V. Enkelmann, *Angew. Chem., Int. Ed. Engl.*, 1993, **32**, 1653.
[159] U. H. F. Bunz, V. Enkelmann, and J. Rader, *Organometallics*, 1993, **12**, 4745.

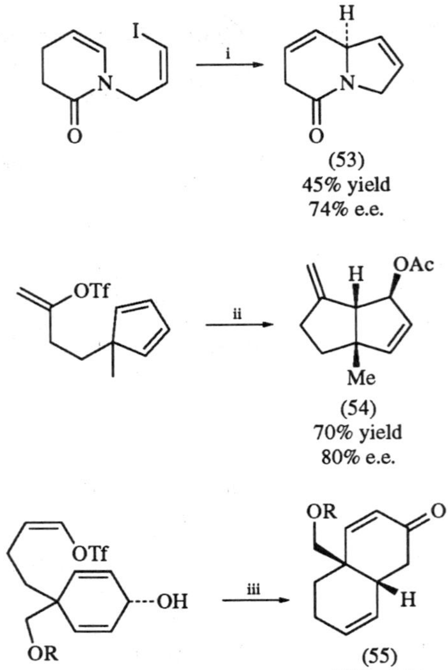

(53) 45% yield 74% e.e.

(54) 70% yield 80% e.e.

(55) 76% yield 86% e.e.

Reagents: i, $[Pd_2(dba)_3]$, (*R*,*S*) – BPPFOH, $CaCO_3$, $AgPO_4$; ii, $Pd(OAc)_2$, (*S*)-BINAP, Bu_4NOAc, DMSO; iii, $[Pd_2(dba)_3]$, (R)-BINAP, Bu^tOH, $ClCH_2CH_2Cl$

Scheme 28

crystallography. Cross-coupling to tricarbonylchromium arene complexes is well precedented. An asymmetric version has been reported in which the symmetrical tricarbonylchromium complex of 1,2-dichlorobenzene is coupled in the presence of a chiral auxiliary to introduce a vinyl group in place of one of the chlorines. The resulting metal complex is chiral, and is formed in 44% enantiomeric excess.[160]

Asymmetric Palladium-catalysed Cross-coupling reactions.—At the end of the preceding section, an example of palladium-catalysed cross-coupling induced asymmetry to form a product with planar chirality. The asymmetric modification of palladium-catalysed cross-coupling has been a topic of growing importance. Most work in this area concentrates on the induction of asymmetry at chiral centres at carbon. Further examples of enantiofacial differentiation in coupling with dihydrofurans have appeared.[161,162] In both cases, BINAP was used as the chiral auxiliary. An

[160] M. Uemura, H. Nishimura, and T. Hayashi, *Tetrahedron Lett.*, 1993, **34**, 107.
[161] S. Hillers and O. Reiser, *Tetrahedron Lett.*, 1993, **34**, 5265.
[162] F. Ozawa, Y. Kobatake, and T. Hayashi, *Tetrahedron Lett.*, 1993, **34**, 2505.

intermediate in an efficient synthesis of (−)-eptazocine also relied on BINAP to differentiate the enantiofaces of a silyl-protected allylic alcohol. After the β-elimination step, a silylenol ether was produced. Enantioselectivity in the coupling step was 91%.[163] A similar strategy, employing a pendant allyl alcohol and BINAP, was used in an enantioselective synthesis of physostigmine. The enol ether product was hydrolysed to position an aldehyde correctly for a condensation reaction with methylamine in the following step.[164]

Enantiofacial differentiation leading to the asymmetric formation of (53) (Scheme 27) has been studied in detail. Variation of the chiral auxiliary, and the use of silver additives, produced a pronounced variation in the efficiency of asymmetric induction. Enantiomeric excesses ranged from 5–86% and the chemical yields from 6–94%. An additional complication was the possibility of formation of mixtures of α,β- and β,γ-unsaturated lactams. In the first example illustrated in Scheme 28, complete selectivity for (53) was achieved.[165] Enantiotopic group differentiation holds the key to the asymmetric synthesis of (54);[166] the same strategy afforded (55),[167] which was required as a key intermediate for a synthetic route to vernolepin.[168] An example of enantioselective Grignard cross-coupling employs a chiral ferrocenylaminophosphine ligand to effect an asymmetric transformation during the coupling of a chiral Grignard reagent with a vinyl bromide.[169]

4 Use of Other Metals in Coupling Reactions

Zirconium complexes are effective at forming carbon–carbon bonds between alkenes and alkynes. Because of the stoichiometric nature of these reactions, functional groups can be introduced by exploiting the carbon–metal bonds in the final metallocyclic intermediate. Protonation can simply replace the metal by a hydrogen. The use of iodine produces vinyl halides.[170]

The chemoselectivity of functionalization of zirconacyclopentenes has been examined.[171] When a pendent alkyl halide is included in the alkene, cyclization to form cyclopropanes can occur.[172] Two alkenes can also be coupled using dicyclopentadienylzirconium dichloride. A study with an unsymmetrical dialkene has demonstrated that the cyclization step is reversible at 75 °C. An initial kinetic product (dimerization) is converted into an intramolecular cycloaddition product over a period of time.[173] Non-conjugated dienes have been converted into conjugated dienezirconocenes.[174] Organometallic nucleophiles can be combined with alkynes by the reaction of dicyclopentadienylzirconium dichloride. Carboalumination can be effected even in the presence of water.[175]

163 T. Takemoto, M. Sodeoka, H. Sasai, and M. Shibasaki, *J. Am. Chem. Soc.*, 1993, **115**, 8477.
164 A. Ashimori, T. Matsuura, L. E. Overmann, and D. J. Poon, *J. Org. Chem.*, 1993, **58**, 6949.
165 S. Nukui, M. Sodeoka, and M. Shibasaki, *Tetrahedron Lett.*, 1993, **34**, 4965.
166 K. Kagechika, T. Ohshima, and M. Shibasaki, *Tetrahedron*, 1993, **49**, 1773.
167 K. Kondo, M. Sodeoka, M. Mori, and M. Shibasaki, *Synthesis*, 1993, 920.
168 K. Kondo, M. Sodeoka, M. Mori, and M. Shibasaki, *Tetrahedron Lett.*, 1993, **34**, 4219.
169 B. Jedicka, C. Kratky, W. Weissensteiner, and M. Widhalm, *J. Chem. Soc., Chem. Commun.*, 1993, 1329.
170 T. Takahashi, N. Suzuki, M. Kageyama, D. Y. Kondakov, and R. Hara, *Tetrahedron Lett.*, 1993, **34**, 4811.
171 T. Takahashi, K. Aoyagi, R. Hara, and N. Suzuki, *J. Chem. Soc., Chem. Commun.*, 1993, 1042.
172 T. Takahashi, D. Y. Kondakov, and N. Suzuki, *Tetrahedron Lett.*, 1993, **34**, 6571.
173 D. F. Taber, J. P. Louey, and J. A. Lim, *Tetrahedron Lett.*, 1993, **34**, 2243.
174 J. P. Maye and E.-i. Negishi, *Tetrahedron Lett.*, 1993, **34**, 3359.
175 P. Wipf and S. Lim, *Angew. Chem., Int. Ed. Engl.*, 1993, **32**, 1068.

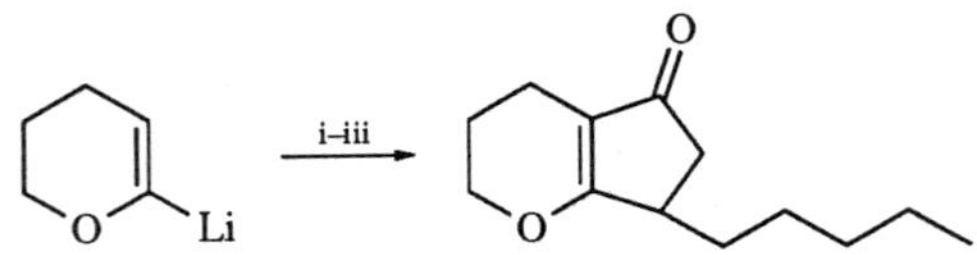

(56)

Reagents: i, $Cp_2ZrMeCl$; ii, hept-1-ene; iii, CO

Scheme 29

Catalytic amounts of dicyclopentadienylzirconium dichloride have been used to promote the addition of alkyl Grignard reagents to alkynes.[176] Oxygenated vinyllithium reagents can also be linked to alkynes. The resulting metallocycles are cleaved to dienes in protic solvents, or can be carbonylated in a cyclopentenone synthesis. The bicyclic product (56) was obtained in this way (Scheme 29).[177]

Dimethyltitanocene reacts with alkynes to form metallocyclobutenes. Subsequent addition to a ketone provides a route to homoallylic alcohols.[178] Organic nitriles can replace the alkyne in this reaction sequence. There has been a structural study of a titanium ester homoenolate derivative.[179] Dicyclopentadienyltitanium derivatives of arenes also react well with alkynes. After protonation, substituted styrenes such as (57) are obtained (Scheme 30). This reaction sequence proceeds through a titanium aryne intermediate (58) and the metallocycle (59).[180]

Nickel catalysts couple aryl triflates to form biaryls in the presence of zinc or potassium iodides.[181] A combination of nickel dichloride and chromium dichloride has been used to convert vinyl triflates into nucleophilic species for reaction with carbonyl groups, in work that completes the tricyclic framework needed for the taxane ring system.[182]

The dimerization of alkynes using a rhodium catalyst can proceed in two different fashions. Under acidic conditions, followed by exposure to carbon monoxide, enynes are produced. On the other hand, if reaction with carbon monoxide occurs prior to acidification, a disubstituted butatriene is obtained.[183] A dicationic benzene(ethyltetramethylcyclopentadienyl)rhodium complex promotes oxidative coupling of phenols.[184] (Pentamethylcyclopentadienyl)ruthenium complexes are used to codimerize conjugated dienes with 1,5-dienes. Intermediate complexes can be obtained, or catalytic conditions afford direct access to the free polyenes.[185] A mechanistic study has been made of reactions between cyclopentadienyl- and (pentamethylcyclopentadienyl)ruthenium cations and conjugated dienes and ethyne.[186]

[176] T. Takahashi, K. Aoyagi, V. Denisov, N. Suzuki, D. Choueiry, and E.-i. Negishi, *Tetrahedron Lett.*, 1993, **34**, 8301.
[177] M. C. J. Harris, R. J. Whitby, and J. Blagg, *SYNLETT*, 1993, 705.
[178] N. A. Petasis and D.-K. Fu, *Organometallics*, 1993, **12**, 3776.
[179] P. G. Cozzi, T. Carofiglio, and C. Floriani, *Organometallics*, 1993, **12**, 2845.
[180] J. Cámpora and S. L. Buchwald, *Organometallics*, 1993, **12**, 4182.
[181] A. Jutand and A. Mosleh, *SYNLETT*, 1993, 568.
[182] M. H. Kress, R. Ruel, W. H. Miller, and Y. Kishi, *Tetrahedron Lett.*, 1993, **34**, 6003.
[183] M. Schäfer, N. Mahr, J. Wolf, and H. Werner, *Angew. Chem., Int. Ed. Engl.*, 1993, **32**, 1315.
[184] A. G. M. Barrett, T. Itoh, and E. M. Wallace, *Tetrahedron Lett.*, 1993, **34**, 2233.
[185] K. Masuda, H. Ohkita, S. Kurumatani, and K. Itoh, *J. Organomet. Chem.*, 1993, **454**, C13.
[186] K. Masuda, H. Ohkita, S. Kurumatani, and K. Itoh, *Organometallics*, 1993, **12**, 2221.

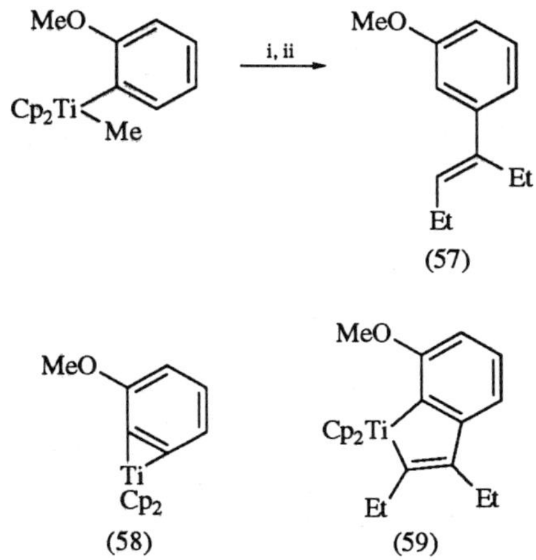

Reagents: i, hex-3-yne; ii, MeOH

Scheme 30

Unusual 1,1-dizinc carbenoid organometallics have been under investigation in Paris.[187] These species can be combined with Grignard reagents and, after copper modification, propargylic esters to produce coupling products. Full details have appeared of homologations of copper reagents using (iodomethyl)zinc iodide.[188] Boron–zinc transmetallation has also been investigated.[189] Zinc/copper chemistry gives rise to new α-amino acid γ-amine equivalents and so to direct methods for the synthesis of enantiomerically pure protected amino acids.[190]

5 The Nicholas Reaction

The chemistry of dicobalt-stabilized propargyl cations continues to develop. The main step forward comes from the laboratory of Nicholas himself, where chiral metal clusters [(HC≡CR)CoL(CO)$_2$Co(CO$_3$)] have been used in stereoselective propargyl addition reactions.[191] Chiral allylboron reagents have also been employed with propargyl aldehyde complexes.[192] Diastereoselectivity is good in Lewis-acid promoted aldol reactions between trimethylsilylenol ethers and dicobalt complexes of propynals.[193] Stereoselectivity is also good in the addition of boron enolates to (propargyl ether)dicobalt complexes.[194] This reaction (Scheme 31) has been employed in a formal total synthesis of thienamycin.[195] After the enolate addition step, oxidative removal of

[187] F. Chemla, I. Marek, J.-F. Normant, *SYNLETT*, 1993, 665.
[188] A. Sidduri, M. J. Rozema, and P. Knochel, *J. Org. Chem.*, 1993, **58**, 2694.
[189] F. Langer, J. Waas, and P. Knochel, *Tetrahedron Lett.*, 1993, **34**, 5261.
[190] R. F. W. Jackson, N. Wishart, and M. J. Wythes, *SYNLETT*, 1993, 219.
[191] A. J. M. Caffyn and K. M. Nicholas, *J. Am. Chem. Soc.*, 1993, **115**, 6438.
[192] P. Ganesh and K. M. Nicholas, *J. Org. Chem.*, 1993, **58**, 5587.
[193] C. Mukai, O. Kataoka, and M. Hanoaka, *J. Chem. Soc., Perkin Trans.* 1, 1993, 563.
[194] P. A. Jacobi and W. Zheng, *Tetrahedron Lett.*, 1993, **34**, 2581.
[195] P. A. Jacobi and W. Zheng, *Tetrahedron Lett.*, 1993, **34**, 2585.

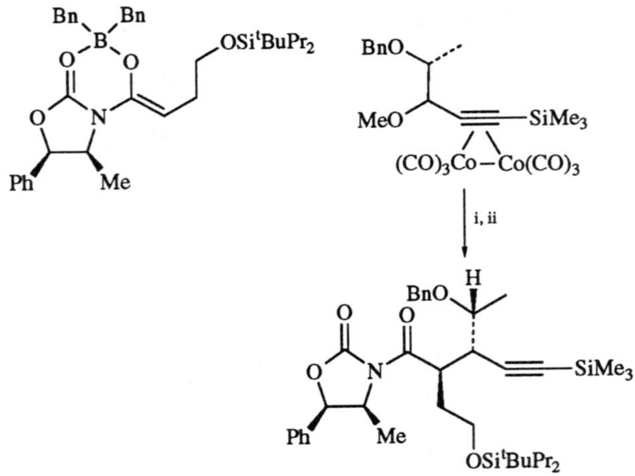

(60) 79%

Reagents: i, Bu_2BOTf; ii, cerium(IV)

Scheme 31

the metal with ceric ammonium nitrate afforded the key building block (60).

Intramolecular reactions between silyl enol ethers and propargyl cations have been examined.[196] Amines[197] and indoles[198] react well with propargyl cations. A crystal structure study of metal-cluster-stabilized propargyl cations has been performed.[199] Alkynyl derivatives of pyranoses easily invert their stereochemistry when converted into hexacarbonyldicobalt complexes.[200] Carbon–oxygen bond cleavage under acidic conditions must be combined with the interconversion of stereoisomeric propargyl cation complexes to account for this observation. The effect of modification of the alkyne portion of estradiol on the binding of organometallic units in the estradiol receptor provides evidence for the irreversible covalent attachment of the metal-labelled steroid to its specific receptor.[201]

The Pauson–Khand reaction was discussed in Section 2. A formal total synthesis of loganine has made use first of stabilization of positive charge α to the cobalt complexed alkyne, and second a Pauson–Khand step to complete a bicyclic cyclopentenone.[202]

[196] E. Tyrrell, P. Heshmati, and L. Sarrazin, *SYNLETT*, 1993, 769.
[197] K.-D. Roth and U. Müller, *Tetrahedron Lett.*, 1993, **34**, 2919.
[198] K.-D. Roth, *SYNLETT*, 1993, 529.
[199] M. Gruselle, H. El Hafa, M. Nikolski, G. Jaouen, J. Vaissermann, L. Li, and M. J. McGlinchey, *Organometallics*, 1993, **12**, 4917.
[200] S. Tanaka, T. Tsukiyama, and M. Isobe, *Tetrahedron Lett.*, 1993, **34**, 5757.
[201] D. Osella, G. Dutto, G. Jaouen, A. Vessiéres, P. R. Raithby, L. De Benedetto, and M. J. McGlinchey, *Organometallics*, 1993, **12**, 4545.
[202] N. Jeong, B. Y. Lee, S. M. Lee, Y. K. Chung, and S.-G. Lee, *Tetrahedron Lett.*, 1993, **34**, 4023.

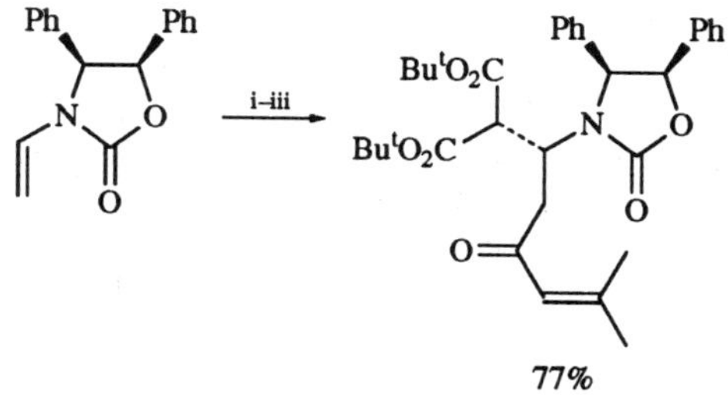

Reagents: i, $NaCH(CO_2Et)(CO_2Bu^t)$, NEt_3, $[PdCl_2(CH_3CN_2)]$; ii, CO; iii, $Me_3SnCH{=}CMe_2$

Scheme 32

6 Transition Metal Alkene and Arene π-Complexes

η^2-Complexes.—Palladium-catalysed nucleophilic addition to alkenes and palladium-catalysed coupling reactions are closely related by the similarity of their reaction products, though mechanistically the reactions are distinct. Intramolecular versions are popular. An example leading to 2-alkenyltetrahydrofurans uses one double bond link of an allene as the metal-bound electrophilic centre.[203] Combining nucleophile addition with a subsequent cross-coupling step provides a good use for the metal–carbon σ-bond that is formed when a π-bound alkene acts as an electrophile. Intramolecular closure of a pendent amine to an allene has been combined with coupling to an alkyl halide.[204] A similar procedure, again using an allene but in the presence of carbon monoxide and methanol, combines nucleophilic attack by a NHTs group with carbonyl insertion to place an ester at the end of the double bond to promote the site of nucleophile addition.[205] Unlike the preceding examples which used palladium(0) catalysts and a base (typical conditions for coupling reactions), this carbonyl insertion example employs palladium dichloride with copper(II) chloride as an oxidant. These are typical conditions for a Wacker oxidation modified for carbonylation. Intermolecular nucleophilic addition to an enone carrying a cyclic chiral auxiliary at nitrogen is followed by carbonyl insertion in the early steps of an enantioselective total synthesis of (+)-negamycin (Scheme 32).[206]

An intramolecular closure to a 1,3-diene combined with chlorination by the action of lithium chloride under palladium acetate catalysis illustrates further extensions of the methods for 1,4-difunctionalization of dienes under investigation in Bäckvall's group.[207] Hexacarbonylmolybdenum–TMNO has been reported to affect cyclization of pendent OH groups to alkynes to afford substituted dihydrofuran complexes.[208]

203 R. D. Walkup, L. Guan, M. D. Mosher, S. W. Kim, and Y. S. Kim, *SYNLETT*, 1993, 88.
204 I. W. Davies, D. I. C. Scopes, and T. Gallagher, *SYNLETT*, 1993, 85.
205 M. Kimura, N. Saeki, S. Uchida, H. Harayama, S. Tanaka, K. Fugami, and Y. Tamaru, *Tetrahedron Lett.*, 1993, **34**, 7611.
206 J. J. Masters and L. S. Hegedus, *J. Org. Chem.*, 1993, **58**, 4547.
207 J. E. Bäckvall, K. L. Granberg, P. G. Andersson, R. Gatti, and A. Gogoll, *J. Org. Chem.*, 1993, **58**, 5445.
208 F. E. McDonald, C. B. Connolly, M. M. Gleason, T. B. Towne, and K. D. Treiber, *J. Org. Chem.*, 1993, **58**, 6952.

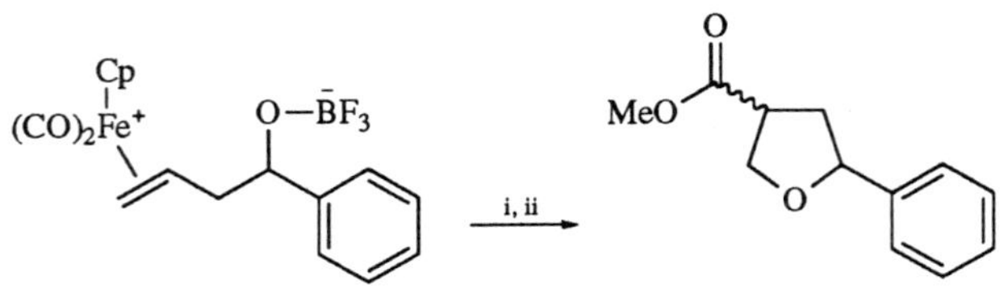

(61) 45%

Reagents: i, KOBu^t, CH_2Cl_2; ii, cerium(IV), CO, MeOH

Scheme 33

In stoichiometric examples, development of the chemistry of cationic dicarbonyl(cyclopentadienyl)iron–alkene complexes continues. Again, intramolecular nucleophile attack has been the focus of attention (Scheme 33). Suitable electrophiles can be obtained by the reaction of ketones with allyliron complexes in the presence of BF_3. Treatment of the resulting η^2-complex with base effects cyclization which, under oxidative carbonylation conditions, can ultimately replace the iron atom by an ester group. In this way (61) can be obtained.[209]

Cyclopentadienylruthenium–nitrosyl–phosphine complexes of 1,3-dienes and enynes provide interesting η^2-complexes that undergo linkage isomerism.[210] Work continues on η^2-styrene complexes[211] and side-on aldehyde complexes.[212] η^2-Enone complexes have also been investigated.[213] Following work on tricarbonyliron complexes of enones, the Thomas group has now extended this investigation to β,γ-unsaturated esters and amides.[214] New examples of cyclic alkyne complexes have employed the dicarbonyl(cyclopentadienyl)manganese group to form an η^2-adduct which rearranges to a 1,2-cyclooctadiene complex.[215]

η^3-Complexes.—The regioselectivity of nucleophilic addition to η^3-transition metal complexes is a topical issue. Further results with cationic iridium complexes from the Stryker group illustrate the generality of reaction pathways for nucleophilic addition at the centre of the η^3-unit.[216] Examples with neutral η^3-palladium chloride dimers have now also been reported.[217] Both examples use ketone enolates as the nucleophile, though in the palladium case the presence of TMEDA is important, and the use of a carbon monoxide atmosphere improves yields, which can be as high as 70%. With an allyl bromide as the starting material, central attack has been observed in a

209 S. Jiang and E. Turos, *Organometallics*, 1993, **12**, 4280.
210 T.-S. Peng, Y. Wang, A. M. Arif, and J. A. Gladysz, *Organometallics*, 1993, **12**, 4535.
211 M. A. Dewey, Y. Zhou, Y. Liu, and J. A. Gladysz, *Organometallics*, 1993, **12**, 3924.
212 D. P. Klein, N. Q. Méndez, J. W. Seyler, A. M. Arif, and J. A. Gladysz, *J. Organomet. Chem.*, 1993, **450**, 157.
213 Y. Wang, F. Agbossou, D. M. Dalton, Y. Liu, A. M. Arif, and J. A. Gladysz, *Organometallics*, 1993, **12**, 2699.
214 C. Alt, S. L. Griffiths, C. F. Marcos, M. M. Salter, A. M. Z. Slawin, S. E. Thomas, and D. J. Williams, *J. Chem. Soc., Chem. Commun.*, 1993, 201.
215 S. M. Coughlan and G. K. Yang, *J. Organomet. Chem.*, 1993, **450**, 151.
216 K. E. Schwiebert and J. M. Stryker, *Organometallics*, 1993, **12**, 600.
217 A. Wilde, A. R. Otte, and H. M. R. Hoffmann, *J. Chem. Soc., Chem. Commun.*, 1993, 615.

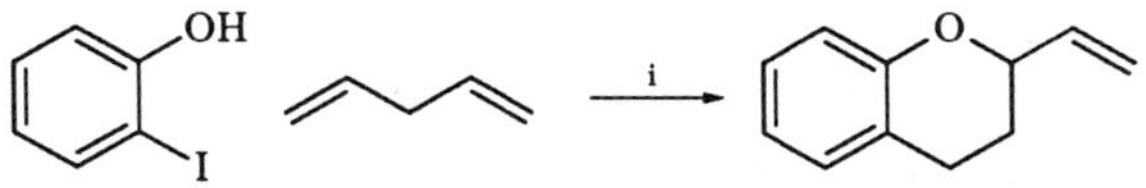

(62) 70%

Reagents: i, $Pd(OAc)_2$, PPh_3, $NaHCO_3$

Scheme 34

palladium-*catalysed* process; selectivity to favour central attack can be as high as 97 : 3 under the correct conditions, though in other cases, exclusive terminal attack was observed. The nucleophile was a trimethylsilylketene acetal and the reactions were performed in the presence of thallium acetate.[218] The centre of an η^3-dienyl complex corresponds to the vinyl-substituted *end* of the η^3 unit. Stabilized-enolate addition under palladium acetate catalysis can be manipulated to favour central attack by the presence of tributylphosphine.[219] The position of the acetate leaving group in the starting material has also been found to influence regiochemistry, showing that the approach of the nucleophile can be influenced by the nature of the ion pair after departure of the leaving group.[220]

Manipulation of *syn*/*anti* ratios continues to receive attention. There has been a detailed investigation of dynamic processes in η^3-palladium complexes,[221] and molecular mechanics parameters for the π-allyl palladium moiety have been reported.[222]

As with η^2-chemistry, so too with η^3-examples; combination of π-allyl chemistry with coupling reactions is an important progression (Scheme 34). Silylation of an allylic acetate with hexamethyldisilane can be achieved with a palladium catalyst.[223] In another example, coupling to an alkene followed by a hydrogen shift forms a η^3-allyl complex which is picked up by an internal nucleophile. In this way, the cyclic ether products (62) can be obtained.[224] Allylic acetates and alcohols can be cyclized to pendent alkenes. In the first case,[225] the reaction is a metallo–ene process, and was combined with coupling to a vinyltin species. In the second example, the tether between the allyl moiety and the alkene was short and a substituted cyclopentadiene was obtained.[226]

Ring opening of vinyloxetanes[227] and β-lactams[228] has been combined with palladium-catalysed coupling reactions. Allylic displacement can be accompanied by

218 M. Formica, A. Musco, and R. Pontellini, *J. Mol. Catal.*, 1993, **84**, 239.
219 Y. I. M. Nilsson, P. G. Andersson, and J. E. Bäckvall, *J. Am. Chem. Soc.*, 1993, **115**, 6609.
220 B. M. Trost and R. C. Bunt, *Tetrahedron Lett.*, 1993, **34**, 7513.
221 S. Hansson, P.-O. Norrby, M. P. T. Sjögren, and B. Åkermark, M. E. Cucciolito, F. Giordano, and A. Vitagliano, *Organometallics*, 1993, **12**, 4940.
222 P.-O. Norrby, B. Åkermark, F. Haeffner, S. Hansson, and M. Blomberg, *J. Am. Chem. Soc.*, 1993, **115**, 4859.
223 Y. Tsuji, S. Kajita, S. Isobe, and M. Futano, *J. Org. Chem.*, 1993, **58**, 3607.
224 R. C. Larock, N. G. Berrios-Peña, C. A. Fried, E. K. Yum, C. Tu, and W. Leong, *J. Org. Chem.*, 1993, **58**, 4509.
225 W. Oppolzer and J. Ruiz-Montes, *Helv. Chim. Acta*, 1993, **76**, 1266.
226 T. Zair, C. Santelli-Rouvier, and M. Santelli, *Tetrahedron*, 1993, **49**, 3313.
227 R. C. Larock, S. Ding, and C. Tu, *SYNLETT*, 1993, 145.
228 R. C. Larock and S. Ding, *Tetrahedron Lett.*, 1993, **34**, 979.

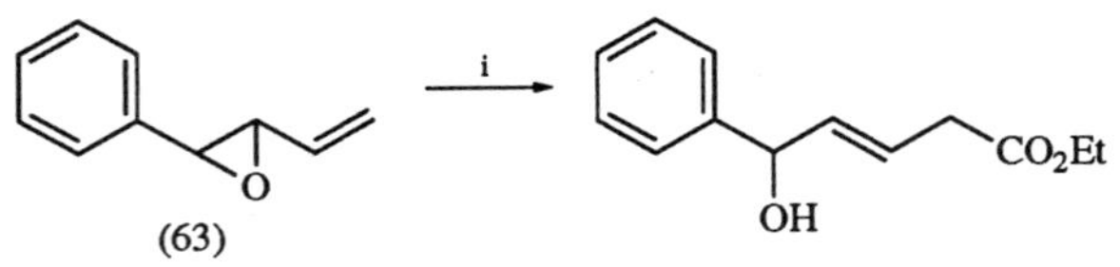

Reagents: i, $[PdCl_2(C_4H_7)_2]$, EtOH, CO, Pr^i_2NEt, NaBr, maleic anhydride

Scheme 35

carbonylation, as illustrated for the opening of the vinyl epoxide (63) (Scheme 35).[229] In this case palladium catalysis was used, but allylic carbonylation (with a phosphate leaving group) employing catalysis by a rhodium carbonyl cluster has also been reported.[230]

Work continues to extend the range of nucleophiles that can be employed in palladium-catalysed allylic displacement. Trimethylsilylcyanide,[231] triphenylsilanol (as a water surrogate),[232] and an aluminate formed *in situ* from an aluminium acetylide and dimethyl ethylidenemalonate,[233] have all been used. Interest in bisphosphonates as enzyme inhibitors has revived work on their use as nucleophiles in palladium-catalysed allylic displacement.[234] N-protected amines carrying two Boc groups have been employed as nucleophiles in a reaction that forms intermediates which afford esters of N-protected amino acids, after ozonolysis of the alkene formed by allylic substitution.[235] This approach has also been used in a synthesis of (+)-N^6-hydroxylysine.[236] By the use of BPh_3, an oxyborane leaving group has been brought into play to provide a new way to use allylic alcohols directly in palladium-catalysed allylic substitution.[237] Less-common substitution patterns have been introduced on stoichiometric η^3-ligands in palladium chloride dimers. Trialkylsilyl[238] and formyl and acetal substituents[239] have been examined.

Synthetic applications this year (Scheme 36) have featured some of the less common types of nucleophiles. Palladium-catalysed allylic rearrangement of propenyl-3-pyrimidinyl carbonates[240] have been used in the preparation of pyrimidine nucleoside analogues (64).[241] The Boc-protected hydroxylamines (referred to earlier) have been used in a synthesis of (+)-laminine.[242] In syntheses of buspirone (65) and gepirone, a 1,4-disubstituted but-2-ene is used to introduce a simple C_4H_8 strap by a sequence of two palladium-catalysed allylic displacements, followed by hydrogenation to remove

229 I. Shimizu, T. Maruyama, T. Makuta, and A. Yamamoto, *Tetrahedron Lett.*, 1993, **34**, 2135.
230 Y. Imada, O. Shibata, and S.-I. Murahashi, *J. Organomet. Chem.*, 1993, **451**, 183.
231 Y. Tsuji, N. Yamada, and S. Tanaka, *J. Org. Chem.*, 1993, **58**, 16.
232 B. M. Trost, N. Ito, and P. D. Greenspan, *Tetrahedron Lett.*, 1993, **34**, 1421.
233 B. M. Trost and C.-J. Li, *Tetrahedron Lett.*, 1993, **34**, 2271.
234 R. Sulsky and D. R. Magnin, *SYNLETT*, 1993, **12**, 933.
235 R. Jumnah, J. M. J. Williams, and A. C. Williams, *Tetrahedron Lett.*, 1993, **34**, 6619.
236 J. P. Genêt, S. Thorimbert, and A.-M. Touzin, *Tetrahedron Lett.*, 1993, **34**, 1159.
237 I. Starý, I. G. Stará, and P. Kočovský, *Tetrahedron Lett.*, 1993, **34**, 179.
238 S. Ogoshi, W. Yoshida, K. Ohe, and S. Murai, *Organometallics*, 1993, **12**, 578.
239 S. Ogoshi, K. Hirako, J. Nakanishi, K. Ohe, and S. Murai, *J. Organomet. Chem.*, 1993, **445**, C13.
240 M. L. Falck-Pedersen, T. Benneche, and K. Undheim, *Acta Chem. Scand.*, 1993, **47**, 63.
241 M. L. Falck-Pedersen, T. Benneche, and K. Undheim, *Acta Chem. Scand.*, 1993, **47**, 72.
242 J. P. Genêt, S. Thorimbert, S. Mallart, and N. Kardos, *Synthesis*, 1993, 321.

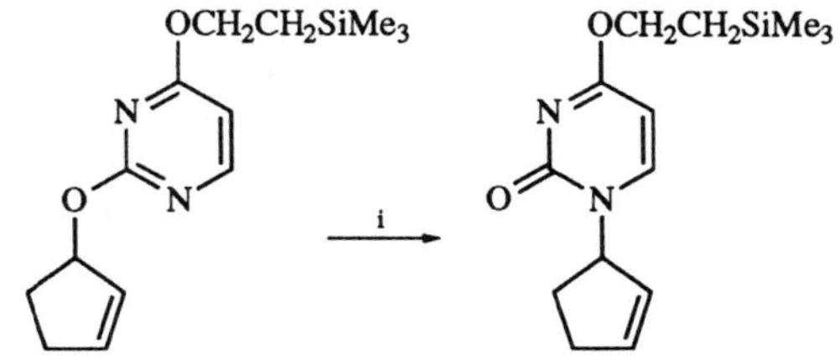

(64) 95%

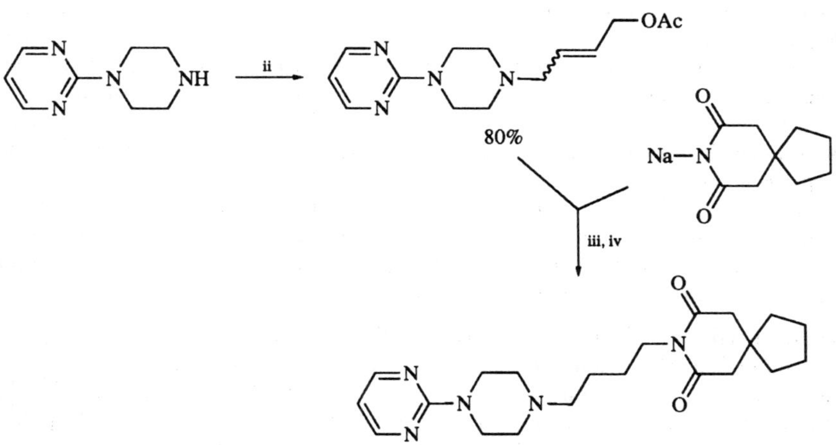

80%

(65) 70%

Reagents: i, $[PdCl_2(PhCN)_2]$, THF or DMF; ii, 4-chlorobut-2-enyl acetate, $[Pd(PPh_3)_4]$; iii, $[Pd(PPh_3)_4]$; iv, H_2, Pd, EtOH

Scheme 36

the central alkene.[243] An intramolecular *N*-alkylation[244] has made use of the selectivity between halide and oxygen leaving groups at the two ends of an alkene.

An example (Scheme 37) that illustrates the use of η^3-complexes to prepare optically active silylallylamines makes use of the trimethylsilyl directing group in a catalytic process.[245] Similar silicon-directed displacements of $(EtO)_2PO_2^-$ from chiral allylphosphates with methylmagnesium bromide proceed in a highly regiocontrolled manner.[246] In another example, with optically active allylic acetates, manipulation of leaving-group selectivity gives access to both *R* and *S* products from a biologically derived monoester of a *meso*-2-ene-1,4-diol.[247] Since biotechnology is used to gain

243 D. L. Kuo, *Heterocycles*, 1993, **36**, 1463.

244 K. Tadano, K. Takao, Y. Nigawara, E. Nishino, I. Takagi, K. Maeda, and S. Ogawa, *SYNLETT*, 1993, 565.

245 H. Inami, T. Ito, H. Urabe, F. Sato, *Tetrahedron Lett.*, 1993, **34**, 5919.

246 H. Urabe, H. Inami, and F. Sato, *J. Chem. Soc., Chem. Commun.*, 1993, 1595.

247 J.-E. Bäckvall, R. Gatti, and H. E. Schink, *Synthesis*, 1993, 343.

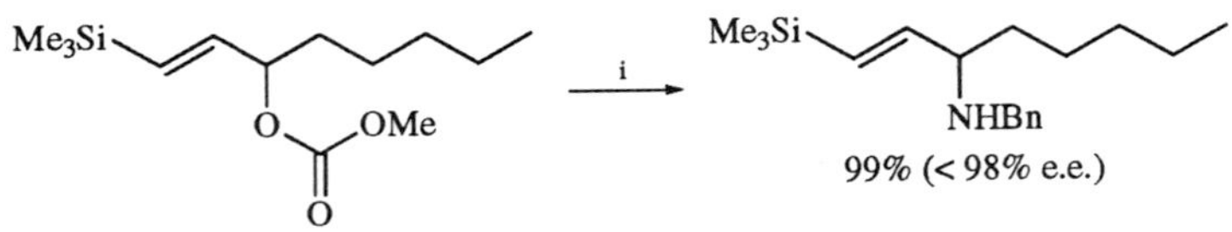

Reagents: i, $BnNH_2$, $[Pd_2(dba)_3]$, PBu_3

Scheme 37

entry to the optically active allylic acetate, the stereodivergence of the two routes is an important feature.

Further advances have been made in reactions employing trimethylenemethane complexes. Cycloaddition to *N*-tosylimines,[248] to ynones,[249] and to sugar-derived acrylates[250] are developments this year. The chemistry of (oxodimethylenemethane)palladium and -platinum complexes has been reported.[251] The complexes form metallacyclobutanones.

A mild and selective method for palladium-catalysed deallylation has been described.[252] Allyloxycarbonyl protecting groups have been removed by palladium-catalysed reactions[253] and the procedure (performed in aqueous media) has been applied to liberation of N-protected amino acid derivatives.[254]

Enantioselective Palladium-catalysed Allylic Displacement.—The use of chiral auxiliary ligands to direct palladium-catalysed allylic displacement is a major growth area. A range of chelating phosphine auxiliaries have been applied to the differentiation of enantiotopic allylic leaving groups. By use of a 1,2-aminoalcohol as the nucleophile, the asymmetric construction of chiral morpholines and piperazines has been achieved.[255] A more common substrate is a 1,3-diphenyl substituted allylic alcohol which becomes prochiral once the leaving group has been displaced. A range (66)–(74) of heteroatom-functionalized chiral bisphosphine ligands has now been employed in this reaction.[256] The best optical yield was obtained with (66) (BHMP-COOH) which proved much more effective (49% e.e.) than the corresponding t-butyl ester. This indicates a role for the pendent carboxylic acid group. The 1,2-diamine derivative (67), with a C_2-symmetric pyrrolidine at each end, has also proved effective with an e.e as high as 91%.[257] A more unusual nitrogen-containing chelating ligand (68) has been prepared from menthone. Asymmetric allylic substitution proceeded in 84% e.e.[258]

[248] B. M. Trost and C. M. Marrs, *J. Am. Chem. Soc.*, 1993, **115**, 6636.
[249] B. M. Trost, S. Sharma, and T. Schmidt, *Tetrahedron Lett.*, 1993, **34**, 7183.
[250] M. Janson, I. Kvarnström, S. C. T. Svensson, B. Classon, and B. Samuelsson, *Synthesis*, 1993, 129.
[251] A. Ohsuka, T. Fujimori, T. Hirao, H. Kurosawa, and I. Ikeda, *J. Chem. Soc., Chem. Commun.*, 1993, 1039.
[252] F. Garro-Helion, A. Merzouk, and F. Guibé, *J. Org. Chem.*, 1993, **58**, 6109.
[253] J. P. Genêt, E. Blart, M. Savignac, S. Lemeune, S. Lemaire-Audoire, and J. M. Bernard, *SYNLETT*, 1993, 680.
[254] J. P. Genêt, E. Blart, M. Savignac, S. Lemeune, and J.-M. Paris, *Tetrahedron Lett.*, 1993, **34**, 4189.
[255] Y. Uozumi, A. Tanahashi, and T. Hayashi, *J. Org. Chem.*, 1993, **58**, 6826.
[256] A. Yamazaki, T. Morimoto, and K. Achiwa, *Tetrahedron: Asymmetry*, 1993, **4**, 2287.
[257] H. Kubota, M. Nakajima, and K. Koga, *Tetrahedron Lett.*, 1993, **34**, 8135.
[258] M. Bovens, A. Togni, and L. M. Venanzi, *J. Organomet. Chem.*, 1993, **451**, C28.

Reiser[259] has been using the ligand (69), which combines two nitrogen atoms with a pair of pendent oxygen donors.

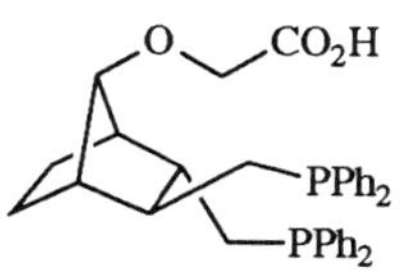

(66)

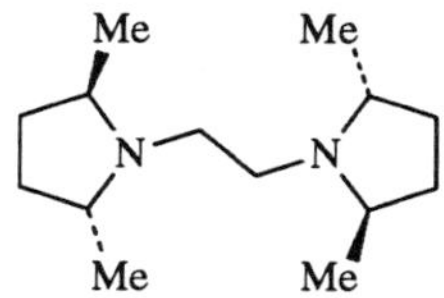

(67)

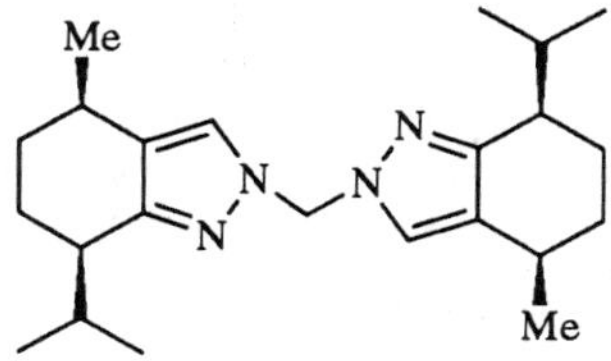

(68)

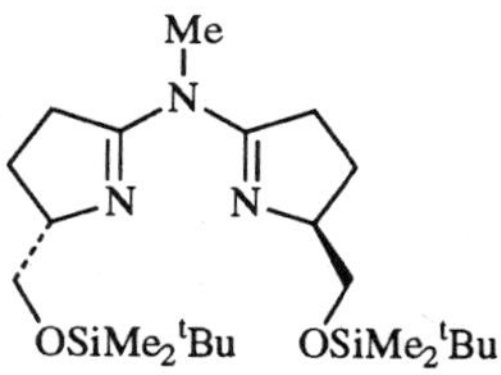

(69)

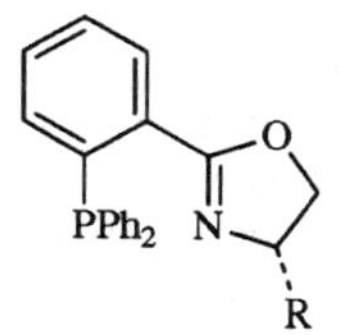

(70)

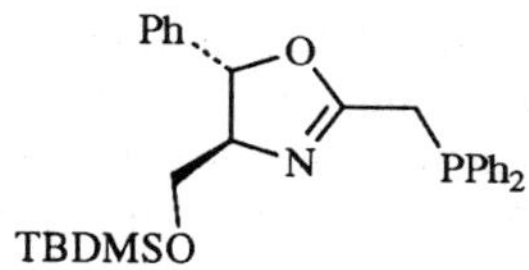

(71)

Work continues on ligands in which phosphorus and nitrogen donors are combined with a view to distorting the coordination sphere of palladium metal. The groups of Helmchen[260] and Williams[261] have evaluated (70), while Pfaltz[262] has also worked with (71). Optical yields greater than 97% enantiomeric excess are now obtained in these reactions. Sulfur and nitrogen centres are combined in the thiophene derivative (72)[263] and in the thiophenol derivative (73) (X = SMe)[264] and the corresponding phenol (X = OH)[265] and diarylthioether (73) (X = SPh).[266] In (74) the spacing

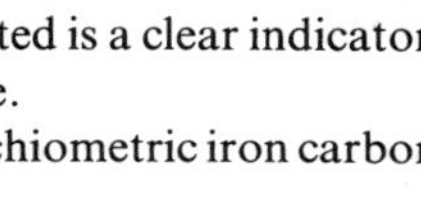

(72)

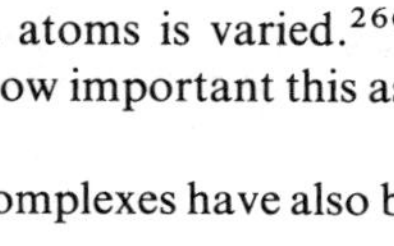

(73)

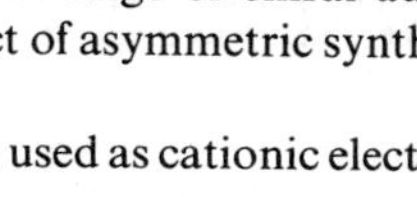

(74)

between the sulfur and nitrogen atoms is varied.[266] The range of chiral auxiliaries illustrated is a clear indicator of how important this aspect of asymmetric synthesis has become.

Stoichiometric iron carbonyl complexes have also been used as cationic electrophiles

[259] O. Reiser, *Angew. Chem., Int. Ed. Engl.*, 1993, **32**, 547.
[260] J. Sprinz and G. Helmchen, *Tetrahedron Lett.*, 1993, **34**, 1769.
[261] G. J. Dawson, C. G. Frost, J. M. J. Williams, and S. J. Coote, *Tetrahedron Lett.*, 1993, **34**, 3149.
[262] P. von Matt and A. Pfaltz, *Angew. Chem., Int. Ed. Engl.*, 1993, **32**, 566.
[263] C. G. Frost and J. M. J. Williams, *Tetrahedron Lett.*, 1993, **34**, 2015.
[264] C. G. Frost and J. M. J. Williams, *Tetrahedron: Asymmetry*, 1993, **4**, 1785.
[265] H. Yang, M. A. Khan, and K. M. Nicholas, *Organometallics*, 1993, **12**, 3485.
[266] G. J. Dawson, C. G. Frost, C. J. Martin, J. M. J. Williams, and S. J. Coote, *Tetrahedron Lett.*, 1993, **34**, 7793.

in the η^3-series. Acyl directing groups direct enol ether nucleophiles to the far end of the allyl system.[267] Optically active acrylates[268] and pyrrolin-2-ones[269] have been employed in enantioselective allylic functionalization steps. Amines and allylsilanes, respectively, were used as nucleophiles, and products had enantiomeric excesses greater than 95%. η^3-Allyl(dicarbonyl)cyclopentadienylmolybdenum complexes have been prepared from allylic diphenylphosphinates;[270] ruthenium complexes have been examined as catalysts for allylic displacement;[271] rhodium complexes to carbonylate allyl phosphate esters;[272] and an η^3-propargyl complex of platinum has been studied.[273] The bent geometry of the η^3-1-ene-2-yne unit corresponds to that of the cationic rhenium example discussed last year.

Anionic η^3-complexes have also received attention. Functional and crystallographic studies have been made of the tricarbonyl(cycloheptatrienyl)iron complex,[274] and, with a new twist to palladium allyl chemistry, an η^3 intermediate has been converted (by use of a dialkylzinc reagent) into an allylzinc nucleophile, effectively providing the ümpolung of the electrophilic allyl complex.[275] A neutral cyclopentadienylmolybdenum complex with an η^3-ligand and chirality at the metal atom has been resolved.[276] This neutral reagent serves as a nucleophile in a stereocontrolled reaction with benzaldehyde. In another organomolybdenum example, a pendent enone was used in a stereoselective Michael reaction.[277] Enolates next to dicarbonyl(η^3-cyclopentadienyl)molybdenum complexes have been used in aldol reactions. Subsequent CO replacement by NO^+ prepares for a cyclization forming isoxazoles and conjugated dienones.[278]

η^4-Complexes.—New structural studies of a well-known organometallic complex is a rare event. Tricarbonyl(butadiene)iron was one of the earliest of the π-complexes to be studied; in 1993 microwave spectroscopy has provided an exceptionally detailed structural characterization of this material.[279] A new heterocyclic tricarbonyliron complex containing a silacyclopentadiene ligand has been prepared.[280] Ruthenium dicarbonylphosphine complexes of η^4-enone ligands can be made by photolysis of tetracarbonylrutheniumphosphine complexes.[281]

When nucleophiles are added to η^4-ligands, cationic molybdenum complexes are often used to promote electrophilicity. Indenyl(dicarbonyl)molybdenum complexes help functionalize spirocyclopentadiene by reaction with dimethyl cuprate.[282]

267 T. Zhou and J. R. Green, *Tetrahedron Lett.*, 1993, **34**, 4497.
268 D. Enders and M. Finkam, *SYNLETT*, 1993, 401.
269 W.-J. Koot, H. Hiemstra, and W. N. Speckamp, *J. Chem. Soc., Chem. Commun.*, 1993, 156.
270 J. S. McCallum, J. T. Sterbenz, and L. S. Liebeskind, *Organometallics*, 1993, **12**, 927.
271 S.-W. Zhand, T.-a. Mitsudo, T. Kondo, and Y. Watanabe, *J. Organomet. Chem.*, 1993, **450**, 197.
272 Y. Imada, O. Shibata, and S.-I. Murahashi, *J. Organomet. Chem.*, 1993, **451**, 183.
273 P. W. Blosser, D. G. Schimpff, J. C. Gallucci, and A. Wojcicki, *Organometallics*, 1993, **12**, 1993.
274 M. Airoldi, G. Deganello, G. Gennaro, M. Moret, and A. Sironi, *Organometallics*, 1993, **12**, 3964.
275 K. Yasui, Y. Goto, T. Yajima, Y. Taniseki, K. Fugami, A. Tanaka, and Y. Tamaru, *Tetrahedron Lett.*, 1993, **34**, 7619.
276 J. W. Faller, J. T. Nguyen, W. Ellis, M. R. Mazzieri, *Organometallics*, 1993, **12**, 1434.
277 S.-H. Lin, W.-J. Cheng, Y.-L. Liao, S.-L. Wang, G.-H. Lee, S.-M. Peng, and R. S. Liu, *J. Chem. Soc., Chem. Commun.*, 1993, 1391.
278 S.-H. Lin, G.-H. Lee, S.-M. Peng, and R.-S. Liu, *Organometallics*, 1993, **12**, 2591.
279 S. G. Kukolich, M. A. Roehrig, D. W. Wallace, and G. L. Henderson, *J. Am. Chem. Soc.*, 1993, **115**, 2021.
280 M. Kako, S. Oba, and Y. Nakadaira, *J. Organometal. Chem.*, 1993, **461**, 173.
281 A. Marcuzzi, A. Linden, and W. Von Philipsborn, *Helv. Chim. Acta*, 1993, **76**, 976.
282 D. J. Norris, J. F. Corrigan, Y. Sun, N. J. Taylor, and S. Collins, *Can. J. Chem.*, 1993, **71**, 1029.

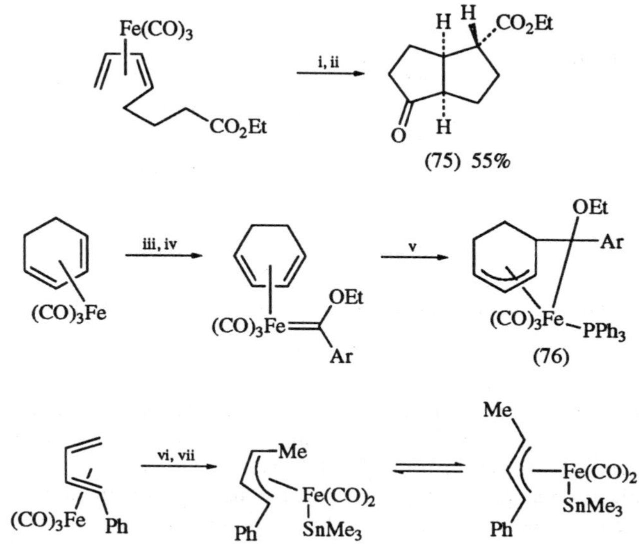

(75) 55%

(76)

(77)

(78)

Reagents: i, LDA, CO; ii, CF_3CO_2H; iii, ArLi, Et_2O, $-50\,°C$; iv, Et_3OBF_4; v, PPh_3; vi, $[HBEt_3]^-$; vii, $SnMe_3Cl$

Scheme 38

Neutral tricarbonyliron complexes can also react with nucleophiles (Scheme 38). An intramolecular example involving enolate addition to an inner position on the diene afforded the bicyclic product (75). Carbonyl insertion, followed by protonation at iron, and reductive elimination completes the cyclopentenone ring.[283] With aryllithium reagents, nucleophile addition occurs at a carbonyl group. The product can be trapped with Meerwein's reagent to form a carbene which, in the presence of an incoming ligand, adds to the η^4-ligand to produce (76).[284] Hydride addition to neutral η^4-tricarbonyliron complexes affords η^3-anion intermediates and can be trapped with trimethyltin chloride.[285] The *anti,syn* product (77) interconverts into the preferred *syn,syn* structure (78).

Examples of highly stereoselective nucleophile addition at ketone groups adjacent to neutral η^4-tricarbonyliron diene complexes have been reported.[286] Methyllithium and triethylaluminium gave complimentary stereoselectivity. Similar reactions at imines have also been described.[287] Stereoselective synthesis of diols from alkenes adjacent to the tricarbonyliron diene unit is now well known, but an unusual epimerization at the

[283] M.-C. P. Yeh, B.-A. Sheu, H.-W. Fu, S.-I. Tau, and L.-W. Chuang, *J. Am. Chem. Soc.*, 1993, **115**, 5941.
[284] J. Chen, Y. Yu, L. Hu, and Z. Jin, *J. Organomet. Chem.*, 1993, **447**, 113.
[285] S. Chang, P. S. White, and M. Brookhart, *Organometallics*, 1993, **12**, 3636.
[286] Y. Takemoto, J. Takeuchi, and C. Iwata, *Tetrahedron Lett.*, 1993, **34**, 6067.
[287] Y. Takemoto, J. Takeuchi, and C. Iwata, *Tetrahedron Lett.*, 1993, **34**, 6069.

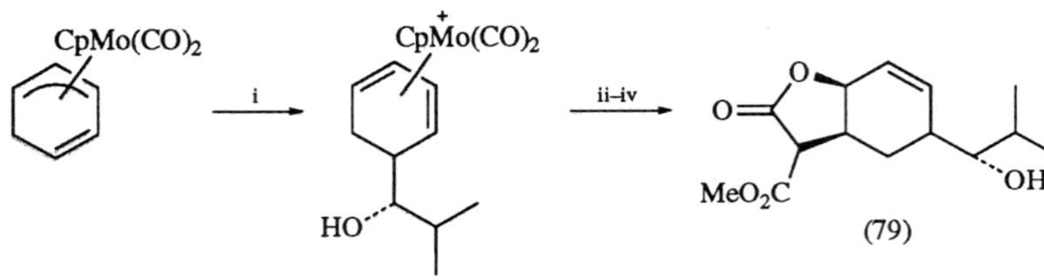

(79)

Reagents: i, 2-methylpropanal, $BF_3 \cdot OEt_2$; ii, dimethyl malonate enolate; iii, NO^+; iv, $NaCO_3$

Scheme 39

α-carbon has been reported through the action of tetrabutylammonium fluoride at −20 °C.[288] Friedel–Crafts acylation of 1,3-diene tricarbonyliron complexes has been extended to alkoxyalkylation using $EtOCH_2Cl$ with aluminium trichloride.[289] In cycloheptadienone chemistry, enolates developed towards an allylic position between the ketone group and the alkene can be alkylated stereoselectively with methyl iodide.[290] Nitrile oxides and nitrile imides have been used in the synthesis of bicyclic derivatives of cycloheptadiene by dipolar cycloaddition reactions with tricarbonyl(η^4-cycloheptatriene)iron.[291] Cationic η^4-diene complexes are obtained by the reaction of electrophiles (formed from aldehydes and boron trifluoride etherate) with (η^3-cyclohexadienyl)dicarbonyl(cyclopentadienyl)molybdenum complexes. The η^4 products can be taken on in reactions with nucleophiles, following a linear sequence of steps to form the bicyclic product (79) with efficient stereocontrol (Scheme 39).[292] The product from the initial Lewis-acid-catalysed reaction with the aldehyde is an η^4-diene complex, formed with complete stereoselectivity.

η^5-Complexes.—Tricarbonyl(pentadienyl)iron complexes often show different properties to their cyclic counterparts. In particular, since they are not locked in a ring, switching between cisoid and transoid products complicates work in the acylic series. A series of disubstituted examples have now been studied in detail (Scheme 40). In the case of (80)[293] and (81) (X = Me[293] and OMe[294]), addition of triphenylphosphine affords products which retain their cisoid geometry. Carbon–carbon bond formation from (81) (X = OMe) with allyltrimethylsilane affords a transoid product with a stereodefined methyl substituent originating from the methyl group at the terminus of the dienyl system. Triphenylphosphine addition to the indenyl tricarbonyliron complex (82), however, proceeds with displacement of a carbonyl group from the metal.[295]

288 J.-P. Lellouche, A. Gibou-Barbedette, and R. Grée, *J. Organomet. Chem.*, 1993, **461**, 167.
289 M. Frank-Neumann, P. Bissinger, and P. Geoffroy, *Tetrahedron Lett.*, 1993, **34**, 4643.
290 A. J. Pearson and K. Chang, *J. Org. Chem.*, 1993, **58**, 1228.
291 A. Gambe, R. Gandolfi, and P. Grunanger, *Gazz. Chim. Ital.*, 1993, **123**, 209.
292 S.-H. Wang, Y.-C. Cheng, G.-H. Lee, S.-M. Peng, and R.-S. Liu, *Organometallics*, 1993, **12**, 3282.
293 W. A. Donaldson, M.-J. Jin, and P. T. Bell, *Organometallics*, 1993, **12**, 1174.
294 W. A. Donaldson and M.-J. Jin, *Bull. Soc. Chim. Belg.*, 1993, **102**, 297.
295 D. A. Brown, N. J. Fitzpatrick, W. K. Glass, H. A. Ahmed, D. Cunningham, and P. McArdle, *J. Organomet. Chem.*, 1993, **455**, 157.

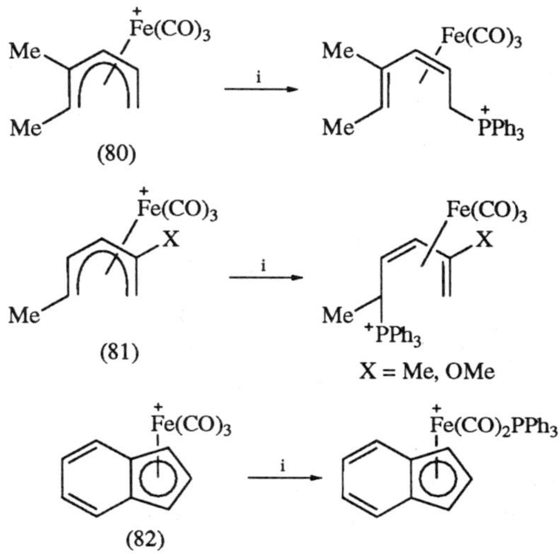

Reagents: i, PPh_3

Scheme 40

Substituted cyclohexadienylones react with hydroxide (sodium carbonate in water) to introduce an OH group at the far end of the η^5 unit. Oxidation affords a diene–dione complex. Substituents include methyl, phenyl, and a chlorine atom, the last of which provides a rare example of a 1-chloro substituted dienyl complex of tricarbonyliron.[296]

Double functionalization of cyclooctadienyl systems extends methods developed in the heptadienyl series.[297] The two nucleophile addition steps follow an iterative sequence, alternating between cationic dienyl complexes and neutral η^4-addition products. With smaller ring sizes, the reformation of the dienyl complex following nucleophilic addition is blocked. A strategy for reactivation that overcomes this disadvantage has been reported.[298] Protonation of a side-chain alkene followed by a hydrogen shift reforms the dienyl salt. A kinetic study of the electrophilicity of cationic η^5-tricarbonyliron complexes has been published.[299]

Several natural product syntheses starting from tricarbonyliron–dienyl complexes have been described during the year (Scheme 40). Knölker has utilized in addition the substituted aniline (83) in a short synthesis of carbazomycins A and B.[300] A variety of oxidative cyclizations of this type have been reported.[301]

The Donaldson group has completed an enantioselective lipoxygenase-inhibitor synthesis in which the alkylation of a pentadienyliron complex provides a key step.[302]

[296] N. Morita, S. Ito, and T. Asao, *J. Organomet. Chem.*, 1993, **460**, 67.
[297] A. J. Pearson, S. Balasubramanian, and K. Srinivasan, *Tetrahedron*, 1993, **49**, 5663.
[298] I. J. Alexander, N. J. Hales, and G. R. Stephenson, *J. Organomet. Chem.*, 1993, **453**, 111.
[299] H. Mayr, K.-H. Müller, and D. Rau, *Angew. Chem., Int. Ed. Engl.*, 1993, **32**, 1630.
[300] H.-J. Knölker and M. Bauermeister, *Helv. Chim. Acta*, 1993, **76**, 2500.
[301] H.-J. Knölker, M. Bauermeister, and J. B. Pannek, *Tetrahedron*, 1993, **49**, 841.
[302] C. Tao and W. A. Donaldson, *J. Org. Chem.*, 1993, **58**, 2134.

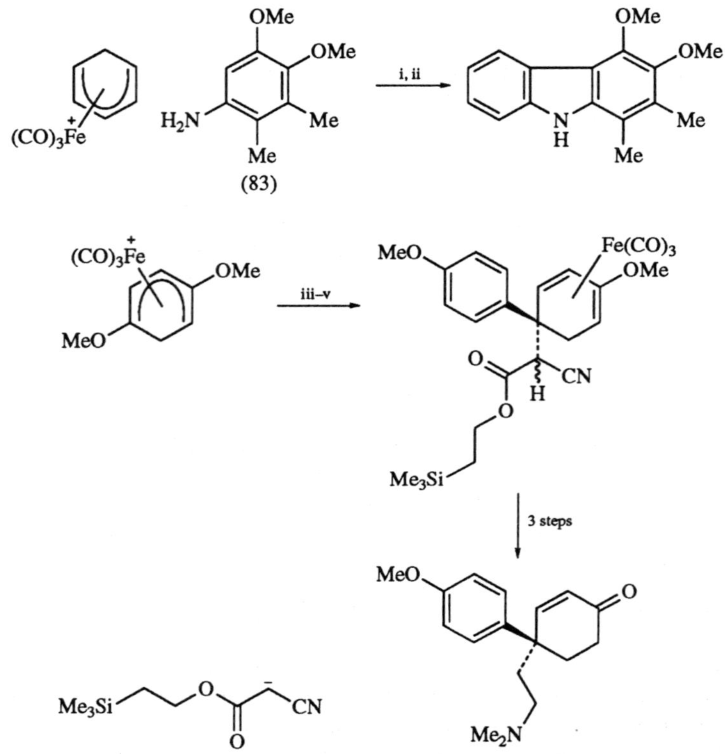

Reagents: i, MeCN; ii, MnO_2; iii, 4-methoxyphenylithium; iv; HPF_6; v, (84)

Scheme 41

The sequence of two nucleophile additions to a cyclohexadienyl complex completes a synthesis of (±)-*O*-methyljoubertiamine (85). Again, an iterative reaction sequence was used. In this case, the reactivation step employed a leaving group in place of hydrogen abstraction.[303]

Tricarbonyl(cyclohexadienyl)iron complexes have been used as selective labels in reactions with peptides, and even with the protein lysozyme,[304] and as a recoverable protecting group in peptide synthesis.[305] New examples of spirolactone formation based on cyanide addition to η^5-tricarbonyliron complexes employ trimethylsilyl cyanide to introduce the nitrile.[306]

[303] G. R. Stephenson, H. Finch, D. A. Owen, and S. Swanson, *Tetrahedron*, 1993, **49**, 5649.
[304] J. A. Carver, B. Fates, and L. A. P. Kane-Maguire, *J. Chem. Soc., Chem. Commun.*, 1993, 928.
[305] S. Fu, J. A. Carver, and L. A. P. Kane-Maguire, *J. Organomet. Chem.*, 1993, **454**, C11.
[306] C. W. Ong and C. J. Chien, *Organometallics*, 1993, **12**, 241.

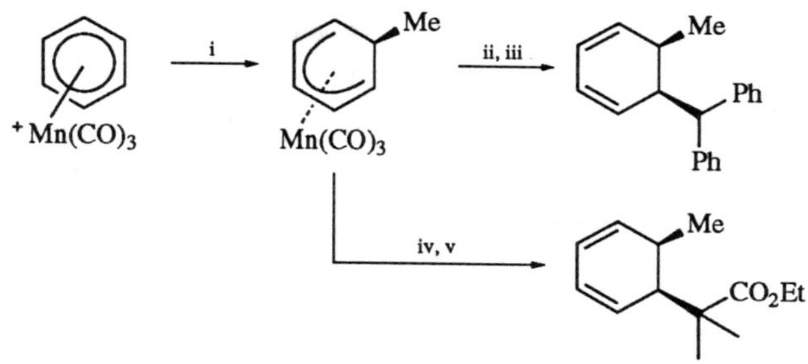

Reagents: i, MeLi; ii, Ph_2CHLi; iii, O_2; iv, $Me_2C(CO_2Et)Li$; v, O_2

Scheme 42

Progress with nucleophilic addition to neutral η^5-tricarbonylmanganese complexes has opened the way for a linear double functionalization leading to 5,6-disubstituted cyclohexadienes. A range of nucleophilic additions to the η^5 intermediates have been examined. When the starting material for the reaction sequence is a cationic η^6-arene complex, a sequence of two nucleophilic additions (first to form a neutral η^5 intermediate and then to an anionic η^4 adduct with two substituents) can provide a successful synthetic route if the final anionic intermediate is oxidized *in situ*.[307] Two examples are illustrated in Scheme 42. This reaction sequence represents a considerable step forward, since previous attempts to promote electrophilicity for the second nucleophile addition by replacement of carbon monoxide by NO^+, while affording cationic η^5 intermediates, had not proved successful in the identification of general nucleophile addition reactions to introduce the second substituent.

Among the unusual η^5-complexes reported this year, two heterocyclic examples, tricarbonyl(pentamethylpyrrole)chromium[308] and pentacarbonyl(2,5-dimethylbismolyl)manganese,[309] are notable.

η^6-Complexes.—By attachment of a transition metal to an aromatic ring, methods are now well established to manipulate the reactivity, not only of the metal-bound ring itself, but also at positions more remote from the site of metal attachment. Some of the developments highlighted in the preceding sections with small metal complexes advance these methodologies in this direction, but it is in the chemistry of η^6-tricarbonylchromium complexes, in particular, that the most highly developed general methods that utilize the metal in a variety of ways have become established as procedures for organic synthesis. Natural product syntheses reported during 1993 illustrate the power of the methodology. The Schmalz group have completed a total synthesis of (1S,4S)-7,8-dihydoxycalamenene (Scheme 43).[310] The ability of tricarbonylchromium complexes to facilitate metallation of the arene, and to stabilize benzylanions and benzyl cations are all brought into play in this synthesis. The optical

[307] B. C. Roell, Jr., K. F. McDaniel, W. S. Vaughan, and T. S. Macy, *Organometallics*, 1993, **12**, 224.
[308] N. Kuhn, J. Kreutzberg, E.-M. Lampe, D. Bläser, and R. Boese, *J. Organomet. Chem.*, 1993, **458**, 125.
[309] A. J. Ashe, III, J. W. Kampf, and D. B. Puranik, *J. Organomet. Chem.*, 1993, **447**, 197.
[310] H.-G. Schmalz, J. Hollander, M. Arnold, and G. Dürner, *Tetrahedron Lett.*, 1993, **34**, 6259.

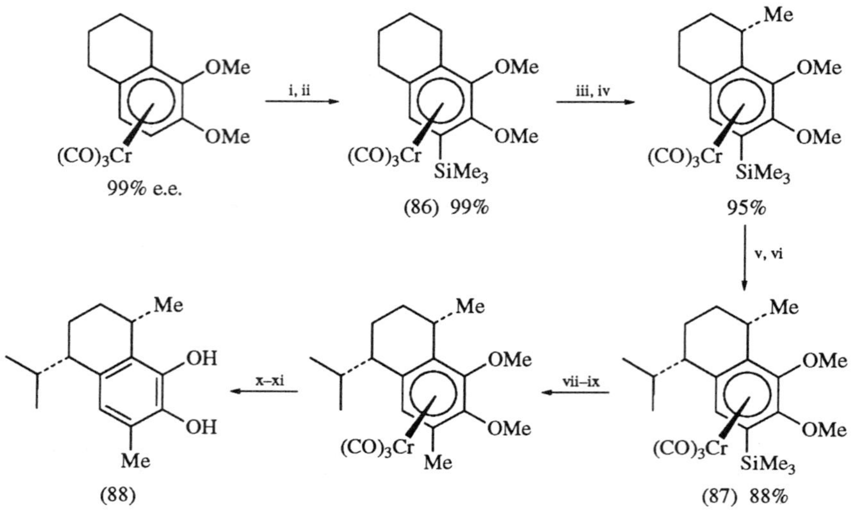

Reagents: i, BuLi; ii, $SiMe_3Cl$; iii, BuLi, THF, HMPA; iv, MeI; v, Bu^sLi, THF; vi, Pr^iI, HMPA; vii, TBAF; viii, BuLi; ix, MeI; x, I_2; xi, BBr_3

Scheme 43

purity of the final product was checked at the metal-free diether stage and was found to be greater than 99%. The key to the synthesis is the control of sites of metallation. First a silyl group was introduced on the aromatic ring to block the most acidic site. The product (86) was then elaborated through a sequence of two benzyl anion complexes with complete regio- and stereocontrol. Very high yields are reported for this route to (87). The trimethylsilyl blocking group was then removed (TBAF), and a repeat of the initial metallation followed by reaction with methyl iodide completes the substitution pattern on the aromatic ring. All that remains is removal of the chromium with iodine, and ether cleavage to complete the target molecule (88). Enantioselective modification of this route employed a dimethoxytetralone as a prochiral starting material. Enantioselective oxazaborolidine-catalysed boron hydride reduction was followed by diastereoselective complexation. It is here that the benzyl cation complexes come into play. The unnecessary OH group was removed by silane reduction ($HSiEt_3$) in trifluroacetic acid. In this step, protonation of the OH group followed by loss of water created a chromium-stabilized benzyl cation which was reduced by the silane. This enantioselective synthesis of (88) uses established bond-forming methods in an elegant and efficient way, and provides an excellent demonstration of the superb control of diastereoselectivity available from the use of the tricarbonylchromium group.

Planar chirality of tricarbonylchromium complexes is exploited in an initial aldol reaction to control stereocentres α and β to the metal-bound ring at the start of a synthesis of (+)-goniofufurone (91).[311] In this synthetic route (Scheme 44), the silyl

[311] C. Mukai, I. J. Kim, and M. Hanaoka, *Tetrahedron Lett.*, 1993, **34**, 6081.

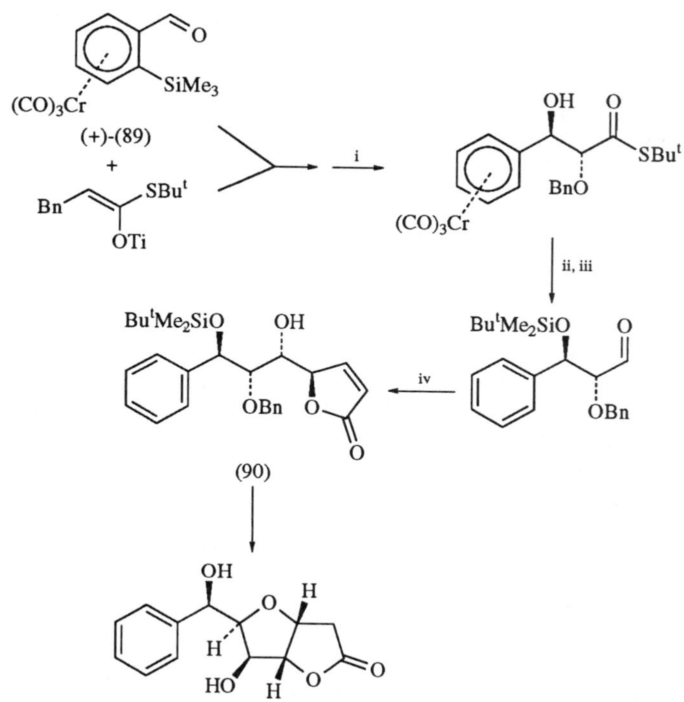

Reagents: i, desilylation; ii, metal removal; iii, steps; iv, 2-(trimethylsilyloxy)furan

Scheme 44

group on the aromatic ring of (89) is necessary to introduce chirality. The later steps, after removal of the chromium, utilize conventional chemistry to elaborate a further chiral centre and complete the carbon chain in (90). The relative stereochemistry in the intermediate (90), obtained in a racemic series of experiments, was proved by *X*-ray crystallography.

Organochromium routes towards (−)-indolactam V[312] and teleocidin A[313] have been investigated. 4-Fluoroindole complexes react with amines replacing the fluorine atom, but extension of this procedure to the ester amino acids for a route to indolactam V was unsuccessful. The target molecule, however, was completed by an alternative route. The preparation of the fluoroindole complex used an interesting trick. The ethylzinc adduct (at nitrogen) of the fluoroindole was formed prior to complexation, adding electron density to the ring. Only ethylzinc was suitable for this purpose. Copper, sodium, magnesium, and even zinc bromide were unsuccessful. Nonetheless, the yield of the indole complex was low. In the case of teleocidin (93), an

[312] M. F. Semmelhack and H. Rhee, *Tetrahedron Lett.*, 1993, **34**, 1395.
[313] M. F. Semmelhack and H. Rhee, *Tetrahedron Lett.*, 1993, **34**, 1399.

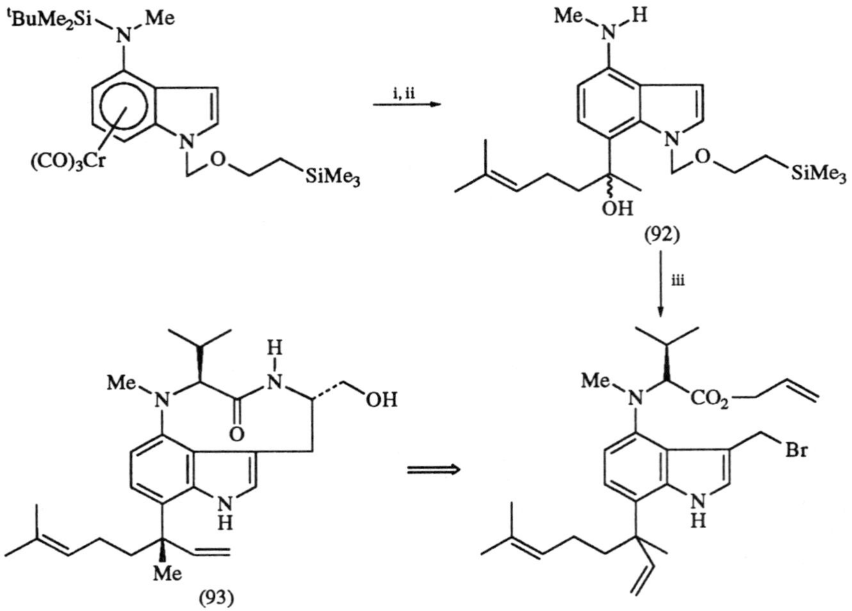

Reagents: i, 2-methyl-6-cyanohept-2-en-6-yllithium; ii, I_2; iii, steps

Scheme 45

organochromium-based route (Scheme 45) was successful. The tricarbonylchromium group was needed to promote electrophilicity on the indole ring. Nucleophilic addition followed by oxidation of the resulting metal anion afforded (92), which was elaborated to complete a formal total synthesis of the target molecule. Without a nitrogen substituent at C4 of the indole, nucleophilic addition occurs preferentially there. This has been used (Scheme 46) in a synthesis of (94), a key intermediate in a route to clavicipitic acid.[314]

An unusual cycloaddition reaction has been reported involving an aniline-derived imine, a pendent alkene, and one carbon–carbon double bond of a chromium-bound aromatic ring. Lewis acids are employed to promote this hetero cycloaddition process and the final product is a tricyclic chromium arene complex.[315] Electrocyclic addition at pendent groups on tricarbonylchromium complexes are more common. Stereoselective nitrile oxide addition[316] and the reverse process,[317] in which a metal-free alkene is combined with a nitrile oxide unit attached to a tricarbonylchromium complex, illustrate new developments. Under photolytic conditions, isocyanate adds to the

[314] M. F. Semmelhack, P. Knochel, and T. Singleton, *Tetrahedron Lett.*, 1993, **34**, 5051.
[315] S. Laschat, R. Noe, M. Riedel, and C. Krüger, *Organometallics*, 1993, **12**, 3738.
[316] C. Baldoli, P. Del Buttero, S. Maiorana, G. Zecchi, and M. Moret, *Tetrahedron Lett.*, 1993, **34**, 2529.
[317] C. Mukai, I. J. Kim, W. J. Cho, M. Kido, and M. Hanaoka, *J. Chem. Soc., Perkin Trans.* 1, 1993, 2495.

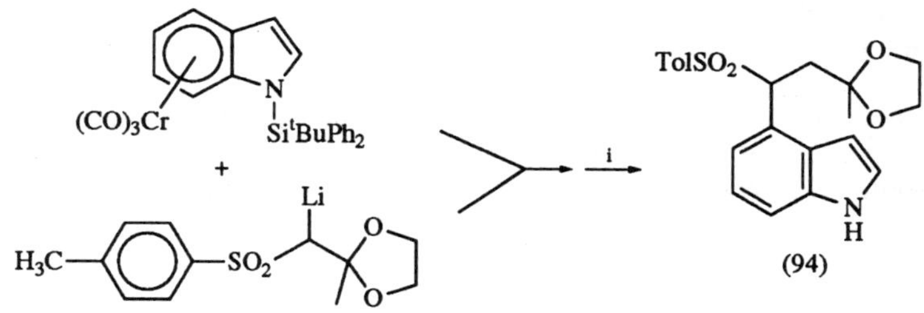

Reagents: i, Bu_4NH

Scheme 46

tricarbonylchromium complex of cycloheptatrienes.[318] Activated alkene dienophiles also react in a [6 + 2] cycloaddition to complexes of seven-membered ring trienes.[319] An intramolecular version has been reported. In this case, no electron withdrawing end-group was present on the dienophile and photochemical conditions were employed. A [6 + 4] cycloaddition is also described. This reaction involves a pendent diene.[320] There are examples of electrocyclic ring openings involving tricarbonyl(*o*-quinodimethane)chromium[321] obtained from an alkoxybenzocyclobutane complex, and related reactions employing dihydroxybenzocyclobutanes.[322] Stereocontrolled nucleophile addition adjacent to a chromium arene complex continues to be developed in work on acetylide addition,[323] and also in carbanion addition to aldehydes and imines.[324] A more unusual reaction involves cyclopropanation in the presence of the tricarbonylchromium complex.[325] An asymmetric oxidation of a chromium-bound arylthioether affords a sulfinyl substituted complex with high diastereoselectivity.[326] In another oxidation, a 1,2-dione replaces the carbon–carbon double bond in a bis(tricarbonyl)chromium complex of a trimethoxystilbene.[327]

Conformational studies of the $Cr(CO)_3$ tripod units in phenanthrene complexes have been made.[328] Conformational preferences in fluorobiphenyl complexes containing tricarbonylchromium units have been studied by ^{13}C and ^{19}F NMR spectroscopy.[329] Restricted rotation in polysubstituted aromatic complexes has been inves-

318 J. H. Rigby, G. Ahmed, and M. D. Ferguson, *Tetrahedron Lett.*, 1993, **34**, 5397.
319 J. H. Rigby, H. S. Ateeq, N. R. Charles, J. A. Henshilwood, K. M. Short, and P. M. Sugathapala, *Tetrahedron*, 1993, **49**, 5495.
320 J. H. Rigby and V. P. Sandanayaka, *Tetrahedron Lett.*, 1993, **34**, 935.
321 E. P. Kündig and J. Leresche, *Tetrahedron*, 1993, **49**, 5599.
322 M. Brands, R. Goddard, H. G. Wey, and H. Butenschön, *Angew. Chem., Int. Ed. Engl.*, 1993, **32**, 267.
323 C. Baldoli, P. Del Buttero, E. Licandro, S. Maiorana, A. Papagni, and M. Torchio, *Tetrahedron Lett.*, 1993, **34**, 7943.
324 E. P. Kündig, L. H. Xu, P. Romanens, and G. Bernardinelli, *Tetrahedron Lett.*, 1993, **34**, 7049.
325 S. Ganesh, K. M. Sathe, M. Nandi, P. Chakrabarti, and A. Sarkar, *J. Chem. Soc., Chem. Commun.*, 1993, 224.
326 A. Pérez-Encabo, S. Perrio, A. M. Z. Slawin, S. E. Thomas, A. T. Wierzchleyski, and D. J. Williams, *J. Chem. Soc., Chem. Commun.*, 1993, 1059.
327 Z. V. Todres and E. A. Ionina, *J. Organomet. Chem.*, 1993, **453**, 193.
328 D. J. Peitz, R. T. Palmer, L. J. Radonovich, and N. F. Woolsey, *Organometallics*, 1993, **12**, 4580.
329 P. Szczeciński and J. Zachara, *J. Organomet. Chem.*, 1993, **447**, 241.

tigated by variable temperature NMR spectroscopy.[330]

FTIR spectroscopy is useful for the trace measurement of the presence of organometalcarbonyl tags on biologically active substances. A new contribution to this area, which previously had been almost the exclusive preserve of the Jaouen group in Paris, has involved the preparation of tricarbonylchromium and dicarbonylchromiumphosphite derivatives of aminoglycoside antibiotics, such as (95). The complexes will be used as metallotracers for immunoassay work.[331]

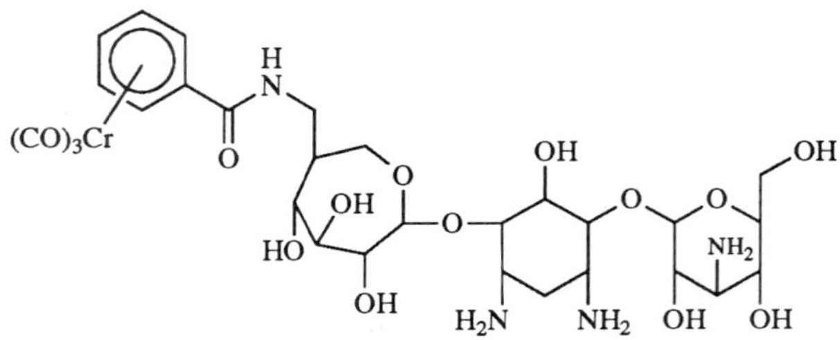

(95)

Access methods to optically pure tricarbonylchromium units include the use of a microorganism to effect an asymmetric reduction of an acetophenone,[332] and direct metallation of a prochiral chromium-bound aromatic ring in the presence of a chiral auxiliary that promotes orthometallation.[333] Organochromium-based auxiliaries are effective as asymmetric catalysts for the addition of an ethyl group from diethylzinc to aldehydes. Examples include chiral auxiliaries with two chiral centres in a side chain attached to an achiral tricarbonyl(arene)chromium complex,[334] and a case where the two coordinating groups that bind the zinc are in separate substituents in a chromium complex of a 1,2-disubstituted benzene ring.[335] In this latter case, planar chirality is present on the aromatic ring. From the point of view of asymmetric synthesis, the main thrust this year has been the use of tricarbonylchromium units as chiral auxiliaries, for example, in a catalytic enantioselective synthesis of (*R*)-(+)-lasiodiplodin,[336] and asymmetric Diels–Alder reactions.[337]

The chemistry of other metal complexes of aromatic rings develops in a similar way. Cyclopentadienyliron cations promote nucleophilic replacement of halides from aryl halide complexes from mono- and dichloroarenes.[338,339] Nucleophile addition to tricarbonylmanganese complexes of indoles has been examined,[340] as has the regiocontrol of nucleophilic addition to silyl-substituted manganese-bound aromatic

330 J. A. S. Howell, M. G. Palin, P. McArdle, D. Cunningham, Z. Goldschmidt, H. E. Gottlieb, and D. Hezroni-Langerman, *Organometallics*, 1993, **12**, 1694.
331 J. Szymoniak, B. El Mouatassim, J. Besançon, C. Moïse, and P. Brossier, *Tetrahedron*, 1993, **49**, 3109.
332 Y. Yamazaki and H. Kobayashi, *Tetrahedron*: *Asymmetry*, 1993, **4**, 1287.
333 Y. Kondo, J. R. Green, and J. Ho, *J. Org. Chem.*, 1993, **58**, 6182.
334 G. B. Jones and S. B. Heaton, *Tetrahedron*: *Asymmetry*, 1993, **4**, 261.
335 M. Uemura, R. Miyake, K. Nakayama, M. Shiro, and Y. Hayashi, *J. Org. Chem.*, 1993, **58**, 1238.
336 G. B. Jones and R. S. Huber, *SYNLETT*, 1993, 367.
337 M. Uemura, Y. Hayashi, and Y. Hayashi, *Tetrahedron*: *Asymmetry*, 1993, **4**, 2291.
338 A. S. Abd-El-Aziz and C. R. de Denus, *J. Chem. Soc.*, *Perkin Trans.* 1, 1993, 293.
339 S. Motallebi and P. Müller, *Organometallics*, 1993, **12**, 4668.
340 W. J. Ryan, P. E. Peterson, Y. Cao, P. G. Williard, D. A. Sweigart, C. D. Baer, C. F. Thompson, Y. K. Chung, and T. M. Chung, *Inorg. Chim. Acta*, 1993, 211.

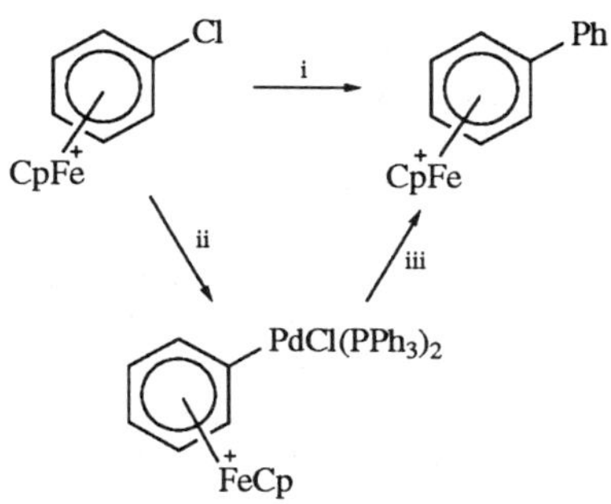

Reagents: i, palladium(0) (cat.), $SnPhBu_3$; ii, palladium(0) (1 eq.); iii, $SnPhBu_3$

Scheme 47

rings,[341] the use of Wittig reagents as nucleophiles with manganese arene complexes,[342] and electrochemical reduction in the presence of phosphines (which converts tricarbonyl(arene)manganese complexes into the dicarbonylphosphine adducts, but does not proceed with the corresponding rhenium complexes).[343] The formation of ruthenium arene complexes by demethylation of steroids[344] has received further attention, and bimetallic cyclopentadienylruthenium complexes of dibenzo-*p*-quinodimethane ligands have been prepared.[345]

Two unusual reports conclude this section. Trifluorophosphine is often used in place of carbonyl groups because of its roughly comparable electron-withdrawing effect. Hydride reduction of tris(trifluorophosphine)chromium complexes has produced an intermediate with agostic hydrogens shared between the aromatic ligand and the metal.[346]

In the cyclopentadienyliron complexes, an unusual coupling reaction has been performed. Palladium-catalysed cross-coupling (Scheme 47) replaced a chlorine atom from the cationic chlorobenzene complex with a substituent transferred from a tributyltin reagent. When one equivalent of a palladium complex was used, a stoichiometric bimetallic product could be isolated in 65% yield. This is the cyclopentadienyliron complex of a palladated aromatic ring. This intermediate reacts with tributyltin reagents to afford the same product as those arising from the cross-coupling reaction.[347]

7 Other Aspects

Not all areas of organometallic chemistry can be covered in detail every year in an

341 S. S. Lee, J.-S. Lee, and Y. K. Chung, *Organometallics*, 1993, **12**, 4640.
342 S. Lee, Y. K. Chung, T.-S. Yoon, and W. Shin, *Organometallics*, 1993, **12**, 2873.
343 C. C. Neto, C. D. Baer, Y. K. Chung, and D. A. Sweigart, *J. Chem. Soc., Chem. Commun.*, 1993, 816.
344 F. Urbanos, M. A. Halcrow, J. Fernandez-Baeza, F. Dahan, D. Labroue, and B. Chaudret, *J. Am. Chem. Soc.*, 1993, **115**, 3484; M. A. Halcrow, F. Urbanos, and B. Chaudret, *Organometallics*, 1993, **12**, 955.
345 D. T. Glatzhofer, Y. Liang, and M. A. Khan, *Organometallics*, 1993, **12**, 624.
346 E. P. Kündig, D. Amurrio, G. Bernardinelli, and R. Chowdhury, *Organometallics*, 1993, **12**, 4275.
347 N. Jevnaker, T. Benneche, and K. Undheim, *Acta Chem. Scand.*, 1993, **47**, 406.

Annual Report. The main theme this year has been a discussion of alkyne chemistry and coupling reactions. Nonetheless, in each year there are developments in all aspects of organometallic chemistry. During 1993, there has been considerable progress with organometallic dendrimers containing iron[348,349] and ruthenium.[349] In one case, iron complexes have been placed at the centre and the periphery of the tentacled structures.[350]

In the field of ferrocene chemistry, much synthetic work has been performed. The most notable products are designed for special purposes, such as thermotropic crystals;[351–353] as end groups on a structure along which a cyclophane moiety migrates;[354] as polydecker pentalene complexes linked in stacks at opposite faces of the back-to-back five-membered rings,[355] or even larger stacked structures (formed by palladium-catalysed cross-coupling to ferrocenylzinc derivatives to produce a range of ferrocene units linked by naphthalenes[356] or by binaphthyl[357] units), or dimeric metallocenes connected by a (dicyclopentadienyl)-methylene unit.[358] Ferrocenes span macrocycles[359,360] and, in a smaller ring system, the iron participates as a fourth ligand in an interaction with a palladium acetonitrile complex.[361] Tetrathiofulvalene,[362] fulgides,[363] and tetrazole hydrazone[364] systems have been prepared. Boronic acid substituents on ferrocene units can participate in cross-coupling in the presence of the metal attached to the cyclopentadienyl ligand,[365] but in (arene)tricarbonylchromium complexes, coupling is prevented by the presence of the metal.[366] Chiral centres[367,368] including an example at sulfur,[369] have been formed, and in one case, asymmetry was induced by microbial reduction.[370] Ferrocene-based chiral auxiliaries are useful in palladium[371] and Grignard[372] cross-coupling reactions.

Work on acyliron complexes has continued, with their utility being illustrated by

[348] F. Moulines, L. Djakovitch, R. Boese, B. Gloaguen, W. Thiel, J.-L. Fillaut, M.-H. Delville, and D. Astruc, *Angew. Chem., Int. Ed. Engl.*, 1993, **32**, 1075.
[349] Y.-H. Liao and J. R. Moss, *J. Chem. Soc., Chem. Commun.*, 1993, 1774.
[350] J.-L. Fillaut and D. Astruc, *J. Chem. Soc., Chem. Commun.*, 1993, 1320.
[351] R. Deschenaux, M. Rama, and J. Santiago, *Tetrahedron Lett.*, 1993, **34**, 3293.
[352] C. Loubser, C. Imrie, and P. H. Van Rooyen, *Adv. Mater.*, 1993, **5**, 45.
[353] R. Deschenaux and J. Santiago, *J. Mater. Chem.*, 1993, **3**, 219.
[354] A. C. Benniston and A. Harriman, *Angew. Chem., Int. Ed. Engl.*, 1993, **32**, 1459.
[355] B. Oelckers, I. Chávez, J. M. Manríquez, and E. Román, *Organometallics*, 1993, **12**, 3396.
[356] H. M. Nugent and M. Rosenblum, *J. Am. Chem. Soc.*, 1993, **115**, 3848.
[357] B. M. Foxman, M. Rosenblum, V. Sokolov, and N. Khrushchova, *Organometallics*, 1993, **12**, 4805.
[358] M. Watanabe, S. Iwamoto, S. Nakashima, H. Sakai, I. Motoyama, and H. Sano, *J. Organomet. Chem.*, 1993, **448**, 167.
[359] P. D. Beer, Z. Chen, M. G. B. Drew, J. Kingston, M. Ogden, and P. Spencer, *J. Chem. Soc., Chem. Commun.*, 1993, **13**, 1046.
[360] R. A. Holwerda, J. S. Kim, T. W. Robison, R. A. Bartsch, and R. P. Czech, *J. Organomet. Chem.*, 1993, **443**, 123.
[361] M. Sato, H. Asano, and S. Akabori, *J. Organomet. Chem.*, 1993, **452**, 105.
[362] A. J. Moore, P. J. Skabara, M. R. Bryce, A. S. Batsanov, J. A. K. Howard, and S. T. A. K. Daley, *J. Chem. Soc., Chem. Commun.*, 1993, 417.
[363] R. W. McCabe, D. E. Parry, and S. P. Saberi, *J. Chem. Soc., Perkin Trans. 1*, 1993, 1023.
[364] W. Shaozu, L. Benyan, Z. Yulan, and L. Zenglu, *Synth. React. Inorg. Met. Org. Chem.*, 1993, **23**, 1139.
[365] R. Knapp and M. Rehahn, *J. Org. Chem.*, 1993, **452,** 235.
[366] M. A. Beckett, R. J. Gilmore, and K. Idrees, *J. Organomet. Chem.*, 1993, **455**, 47.
[367] H. Wally, C. Kratky, W. Weissensteiner, M. Widhalm, and K. Schögl, *J. Organomet. Chem.*, 1993, **450**, 185.
[368] O. Riant, O. Samuel, and H. B. Kagan, *J. Am. Chem. Soc.*, 1993, **115**, 5835.
[369] F. Rebière, O. Riant, L. Ricard, and H. B. Kagan, *Angew. Chem., Int. Ed. Engl.*, 1993, **32**, 568.
[370] Y. Yamazaki, A. Maruyama, K. Hosono, T. Higashihara, H. Kobayashi, *Z. Naturforsch., BioSci.*, 1993, **48**, 451.
[371] B. Jedicka, C. Kratky, W. Weissensteiner, and M. Widhalm, *J. Chem. Soc., Chem. Commun.*, 1993, 1329.
[372] Z.-X. Wang and Z.-Y. Cheng, *Youji Huaxue*, 1993, **13**, 496.

their role as intermediates in syntheses of (−)-actinonin,[373] (*S*)-(−)-methyl tropinate,[374] and a precursor for verapamil.[375] The triphenylphosphine(cyclopentadienyl)iron complex has been used as a chiral auxiliary to induce stereoselective reactions in dihydronicotinoyl systems.[376] Two resolution procedures for this iron-based auxiliary have been reported,[377] and pentamethylcyclopentadienyl[378] and diphenylalkylphosphine[379] analogues have been studied.

Acknowledgement. I am grateful to Professor S. Maiorana and his coworkers, and to the Library of the Dipartimento di Chimica Organica e Industriale, for assistance with bibliographic work during my visit to the Università degli Studi di Milano in 1993.

373 G. Bashiardes, G. J. Bodwell, and S. G. Davies, *J. Chem. Soc.*, *Perkin Trans.* 1, 1993, 459.
374 T. M. Baker, G. J. Bodwell, S. G. Davies, A. J., Edwards, and M. R. Metzlev, *Tetrahedron*, 1993, **49**, 5635.
375 H. Brunner, S. Forster, and B. Nuber, *Organometallics*, 1993, **12**, 3819.
376 R. P. Beckett, V. A. Burgess, S. G. Davies, and M. Whittaker, *Tetrahedron Lett.*, 1993, **34**, 3617.
377 R. W. Baker and S. G. Davies, *Tetrahedron*: *Asymmetry*, 1993, **4**, 1479; S. C. Case-Green, J. F. Costello, S. G. Davies, N. Heaton, C. J. R. Hedgecock, and J. C. Prime, *J. Chem. Soc., Chem. Commun.*, 1993, 1621.
378 J.-P. Barras, S. G. Davies, M. R. Metzler, A. J. Edwards, V. M. Humphreys, and K. Prout, *J. Organomet. Chem.*, 1993, **461**, 157.
379 H. Stępowska and A. Zamojski, *J. Organomet. Chem.*, 1993, **456**, 221.

9 Synthetic Methods

By N. J. LAWRENCE
Department of Chemistry, UMIST, PO Box 88, Manchester, M60 1QD, UK

1 Introduction

One of the year's more interesting accounts of organic synthesis may be found in John Cornforth's Robert Price lecture entitled 'The Trouble with Synthesis'.[1] He discusses the changing reasons for undertaking a synthesis and the development of organic synthesis. The recurrent theme of most papers describing synthetic methods this year has undoubtedly been catalytic asymmetric synthesis. The need for efficient asymmetric synthetic methods is driven, in part, by the pharmaceutical industry's need for enantiopure intermediates; several reviews describing chiral drugs[2,3] have appeared this year. Among the most notable total syntheses of the year are those of the immunosuppressant rapamycin achieved by the groups of Nicolaou,[4] Schrieber,[5] and Danishefsky,[6] and the synthesis of strychnine by Overman and coworkers.[7] An excellent behind-the-scenes account of Nicolaou's[8] synthesis of calicheamicin γ_1^I has also been published this year. Specialist reference works that have been published this year include Volumes 43[9] and 44[10] of *Organic Reactions*, which review carbonyl alkylidination using titanium reagents,[9a] anion-assisted sigmatropic rearrangements,[9b] the Baeyer–Villager oxidation,[9c] selenoxide elimination,[10a] and enone–olefin [2 + 2] photochemical cyclization.[10b] Two *Tetrahedron* Symposia-in-print describe synthetic methods; 'synthesis of optically active compounds – prospects for the 21st century'[11] and 'transition metal organometallics in organic synthesis'.[12] Books published this year include works describing asymmetric catalysis in organic syn-

[1] J. W. Cornforth, *Aus. J. Chem.*, 1993, **46**, 157.
[2] S. C. Stinson, *Chem. Eng. News*, September 27, 1993, 38.
[3] H.-J. Federsel, *CHEMTECH*, December 1993, 24.
[4] K. C. Nicolaou, T. K. Chakraborty, A. D. Piscopio, N. Minowa, and P. Bertinato, *J. Am. Chem. Soc.*, 1993, **115**, 4419.
[5] D. Romom, S. D. Meyer, D. D. Johnson, and S. L. Schrieber, *J. Am. Chem. Soc.*, 1993, **115**, 7906.
[6] C. M. Hayward, D. Yohannes, and S. J. Danishefsky, *J. Am. Chem. Soc.*, 1993, **115**, 9345.
[7] S. D. Knight, L. E. Overman, and G. Paraudeau, *J. Am. Chem. Soc.*, 1993, **115**, 9293.
[8] K. C. Nicolaou, *Angew. Chem., Int. Ed. Engl.*, 1993, **32**, 1377.
[9] (a) S. H. Pine, *Org. React. (N.Y.)*, 1993, **43**, 1; (b) S. R. Wilson, *ibid.*, 1993, **43**, 93; (c) G. R. Crow, *ibid.*, 1993, **43**, 251.
[10] (a) H. J. Reich and S. Wollowitz, *Org. React. (N.Y.)*, 1993, **44**, 1; (b) M. T. Crimmins and T. L. Reingold, *ibid.*, 1993, **44**, 297.
[11] *Tetrahedron*, 1993, **49**, 1711.
[12] *Tetrahedron*, 1993, **49**, 5415.

thesis;[13] stereoselective synthesis;[14,15] catalytic asymmetric synthesis;[16,17] chiral auxiliaries;[18] oxidative enzymes;[19] industrial chirality;[20] natural products;[21] and stereochemistry.[22] The use of enzymes in organic synthesis[23] continues to be popular but is not reviewed here since it is well covered in chapter 10.

2 Carbon–Carbon Bond Formation

The asymmetric allylation of aldehydes has been achieved in several ways.[24] Keck *et al.* describe, in a series of papers,[25] the remarkably efficient allylation of aldehydes using a catalyst derived from binaphthol and titanium tetraisopropoxide (Scheme 1a).

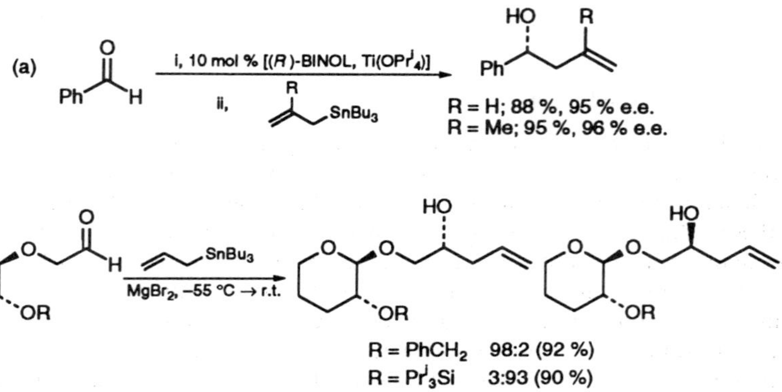

Scheme 1

Tagliavani, Umani-Ronchi, and coworkers report equally impressive enantioselectivity when the reaction is catalysed by the BINOL–$TiCl_2$ complex.[26] Tetraallyltin reacts selectively with aldehydes in the presence of ketones in aqueous acidic media (THF–aq. HCl);[27] the extremely high chemoselectivity (>98.98 : 0.02) is not observed with Lewis acids such as BF_3·OEt_2. Allyltrichlorosilanes add efficiently to aldehydes in DMF without catalyst under neutral conditions; DMF is thought to coordinate to silicon to

[13] R. Noyori, 'Asymmetric Catalysis in Organic Synthesis', Wiley, Chichester, 1993.
[14] Atta-ur-Rahman and Z. Shah, 'Stereoselective Synthesis in Organic Chemistry', Springer-Verlag, Berlin, 1993.
[15] E. Ottow and K. Schoellkopf, 'Stereoselective Synthesis', Springer-Verlag, Berlin, 1993.
[16] 'Catalytic Asymmetric Synthesis', ed. I. Ojima, VCH, New York, 1993.
[17] H. Brunner and W. Zettlmeier, 'Handbook of Enantioselective Catalysis', VCH, New York, 1993.
[18] J. Mulzer and E. Berger, 'Chiral Auxiliaries; Applications in Organic Synthesis', Ballen Vieveg and Sohn Int., Germany 1993.
[19] H. L. Holland, 'Organic Synthesis with Oxidative Enzymes', Verlag, Berlin, 1993.
[20] 'Chirality in Industry', ed. A. N. Collins, Wiley, Chichester, 1993.
[21] 'Dictionary of Natural Products', Chapman and Hall, London, 1993.
[22] E. Eliel and S. H. Wilen, 'Stereochemistry of Organic Chemistry', Wiley, Chichester, 1993.
[23] For an illustration of current research into the use of enzymes in organic synthesis see *Tetrahedron Asymmetry*, 1993, **4**, Nos. 4 and 5.
[24] Y. Nishigaichi, A. Takuwa, Y. Naruta, and K. Maruyama, *Tetrahedron*, 1993, **49**, 7395.
[25] G. E. Keck, K. H. Tarbet, and L. S. Geraci, *J. Am. Chem. Soc.*, 1993, **115**, 8467; G. E. Keck and L. S. Geraci, *Tetrahedron Lett.*, 1993, **34**, 7827; G. E. Keck, D. Krishnamurthy, and M. C. Grier, *J. Org. Chem.*, 1993, **58**, 6543.
[26] A. L. Costa, M. G. Piazza, E. Tagliavini, C. Trombini, and A. Umani-Ronchi, *J. Am. Chem. Soc.*, 1993, **115**, 7001.
[27] A. Yanagisawa, H. Inoue, M. Morodome, and H. Yamamoto, *J. Am. Chem. Soc.*, 1993, **115**, 10356.

produce a hypervalent allylsilicate.[28] Allylation of the chiral glyoxalate (1) is highly diastereoselective and very sensitive to the nature of the protecting group;[29] changing from benzyl to triisopropylsilyl reverses the selectivity (Scheme 1b).

Catalytic asymmetric cyanohydrin synthesis has received considerable attention this year.[30] Corey and Whang describe the use of a pair of chiral catalysts, (2) to activate the aldehyde and (3) to provide an equivalent of chiral cyanide ion (Scheme 2a), to achieve the asymmetric silylcyanation of aldehydes.[31] The chiral cyanide donor is thought to be a complex formed between (3) and trace HCN. A related synthesis of cyano ethers involves the ring opening of the oxazolidinium salt (4) with sodium cyanide to give the α-hydroxy acid (5) after quaternization and treatment with hydrochloric acid (Scheme 2b).[32] A more conventional synthesis of cyanohydrin derivatives has been described by de Vries and coworkers. They use the pantolactone-derived titanium complex (6) to catalyse the asymmetric silylcyanation of aldehydes with modest selectivity (Scheme 2c).[33]

The asymmetric catalytic addition of organozinc reagents[34] to aldehydes has proved as popular as ever this year (Figure 1). Some of the catalysts used include the azetidine (7),[35] the thiophosphoramidate (8),[36] the β-hydroxysulfide (9),[37] the hydroxymethyl oxazoline (10),[38] the diethanolamine (11),[39] and the chromium arene complex (12).[40] The study of organolanthanoid complexes has also seen great activity this year. Organocerium reagents add selectively to SAMP hydrazones (13), thereby providing an afficient route to β-amino aldehydes and acids (Scheme 3).[41] Similarly, α-amino aldehydes are synthesized from the SAMP monohydrazone of protected glyoxal.[42] A new method for the preparation of anhydrous lanthanoid chlorides from the metal and hexachloroethane should prove useful for the synthesis of the corresponding organolanthanoid reagents.[43]

Several groups have reported the use of chiral ligands to control the stereochemistry of palladium-catalysed allylic substitution (Scheme 4). The groups of Helmchen, Pfaltz, and Williams have each used similar bidentate ligands, phosphines (14),[44–46] sulfides (15),[47] and thiophenes (16),[48] which all incorporate a chirality-controlling

[28] S. Kobayashi and K. Nishio, *Tetrahedron Lett.*, 1993, **34**, 3453.
[29] A.B. Charette, C. Mellon, L. Rouillard, and E. Malenfant, *SYNLETT*, 1993, 81.
[30] M. North, *SYNLETT*, 1993, 807.
[31] E.J. Corey and Z. Whang, *Tetrahedron Lett.*, 1993, **34**, 4001.
[32] C. Andrés, M. Delgado, R. Pedrosa, and R. Rodríguez, *Tetrahedron Lett.*, 1993, **34**, 8325.
[33] D. Callant, D. Stanssens, and J.G. de Vries, *Tetrahedron: Asymmetry*, 1993, **4**, 185.
[34] P. Knochel and R.D. Singer, *Chem. Rev.*, 1993, **93**, 2117.
[35] W. Behnen, T. Mehler, and J. Martens, *Tetrahedron: Asymmetry*, 1993, **4**, 1413.
[36] K. Soai, Y. Hirose, and Y. Ohno, *Tetrahedron: Asymmetry*, 1993, **4**, 1473.
[37] M.C. Carreño, J.L. García Ruano, M.C. Maestro, and L.M. Martín Cabrejas, *Tetrahedron: Asymmetry*, 1993, **4**, 727.
[38] J.V. Allen, C.G. Frost, and J.M.J. Williams, *Tetrahedron: Asymmetry*, 1993, **4**, 649.
[39] E.F.J. de Vries, J. Brussee, C.G. Kruse, and A. van der Gen, *Tetrahedron: Asymmetry*, 1993, **4**, 1987.
[40] G.B. Jones and S.B. Heaton, *Tetrahedron: Asymmetry*, 1993, **4**, 261.
[41] D. Enders, M. Klatt, and R. Funk, *SYNLETT*, 1993, 226.
[42] D. Enders, R. Funk, M. Klatt, G. raabe, and E.R. Hovestreydt, *Angew. Chem., Int. Ed. Engl.*, 1993, **32**, 418.
[43] G.B. Deacon, T. Feng, S. Nickel, B.W. Skelton, and A.H. White, *J. Chem. Soc., Chem. Commun.*, 1993, 1328.
[44] P. von Matt and A. Pfaltz, *Angew. Chem., Int. Ed. Engl.*, 1993, **32**, 566.
[45] J. Sprinz and G. Helmchen, *Tetrahedron Lett.*, 1993, **34**, 1769.
[46] G.J. Dawson, C.G. Frost, J.M.J. Williams, and S.J. Coote, *Tetrahedron Lett.*, 1993, **34**, 3149.
[47] G.J. Dawson, C.G. Frost, C.J. Martin, J.M.J. Williams, and S.J. Coote, *Tetrahedron Lett.*, 1993, **34**, 7793.
[48] C.G. Frost and J.M.J. Williams, *Tetrahedron Lett.*, 1993, **34**, 2015.

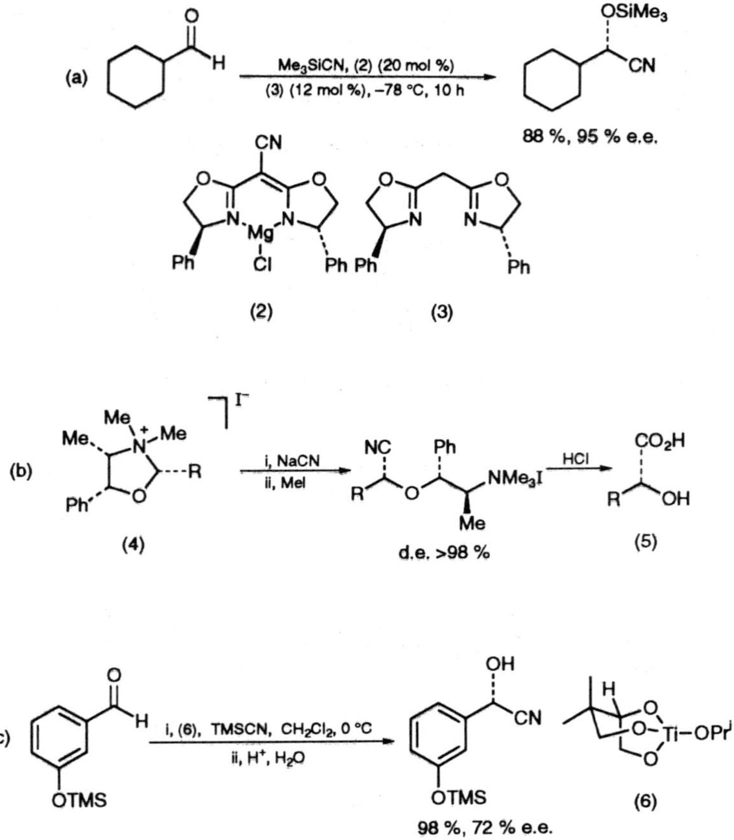

Scheme 2

oxazoline group. Substitution of the acetate in (17) with dimethyl malonate is impressively enantioselective with (14) and (15) (e.e. > 95%). Allylic alcohols may be used in this type of reaction by prior reaction of the allylic alkoxide with triphenylboron.[49] Zhou and Pfaltz[50] have also used the mercaptoaryl-oxazoline (18) as a ligand for the asymmetric copper-catalysed conjugate addition of Grignard reagents to α,β-unsaturated ketones (e.e. > 50%).

Several new Lewis-acid catalysts for aldol reactions have been described. Tris(pentafluorophenyl)boron[51] and lanthanum triflate[52] are excellent air-stable, water-tolerant Lewis-acid catalysts for the aldol and Michael reactions of silyl enol ethers. The methods can be used for the direct reaction of aldehydes supplied commercially as aqueous solutions (*e.g.* formaldehyde). On the other hand, Kobayashi

[49] I. Stary, I. G. Stará, and P. Kocovsky, *Tetrahedron Lett.*, 1993, **34**, 179.
[50] Q.-L. Zhou and A. Pfaltz, *Tetrahedron Lett.*, 1993, **34**, 7725.
[51] K. Ishihara, N. Hananki, and H. Yamamoto, *SYNLETT*, 1993, 577.
[52] S. Kobayashi, I. Hachiya, and T. Takahori, *Synthesis*, 1993, 371.

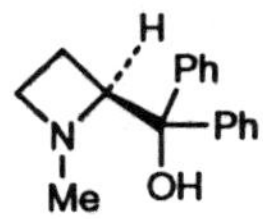

(7) (*S*); 5 mol %; 98 % e.e.

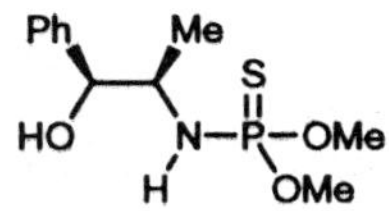

(8) (*S*); 5 mol %; 95 % e.e.

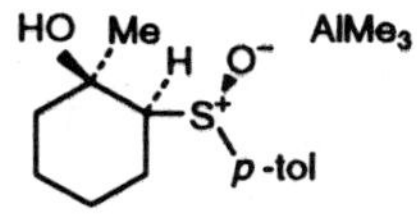

(9) (*S*); 2 mol %; 55 % e.e.

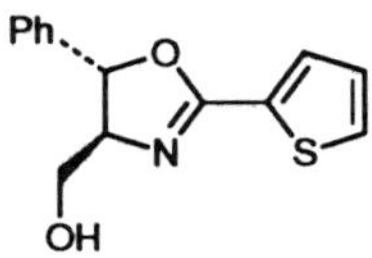

(10) (*S*); 5 mol %; 57 % e.e.

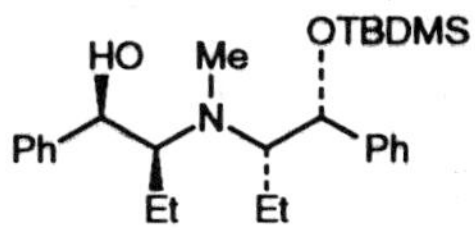

(11) (*R*); 5 mol %; 85 % e.e.

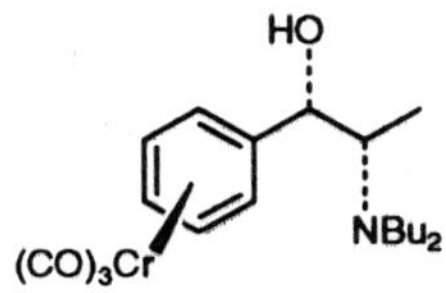

(12) (*R*); 10 mol %; 99 % e.e.

Figure 1 Selectivity in the addition of Et_2Zn to PhCHO

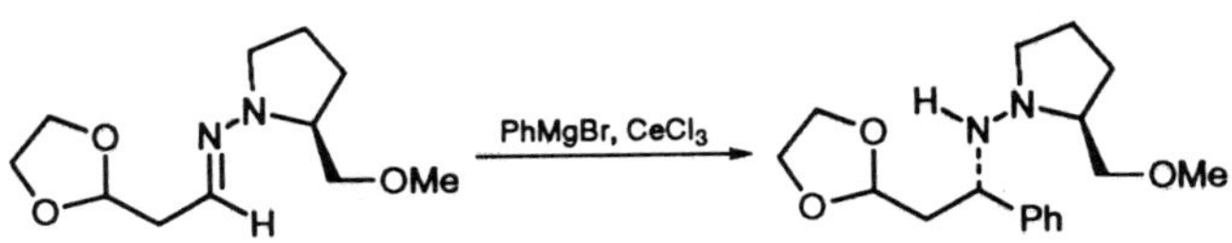

(13) 97 %, 99 % d.e.

Scheme 3

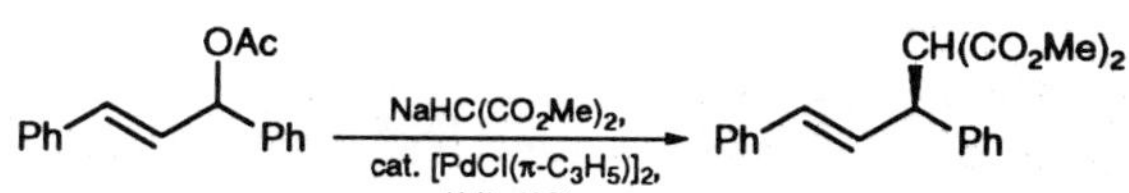

(17)

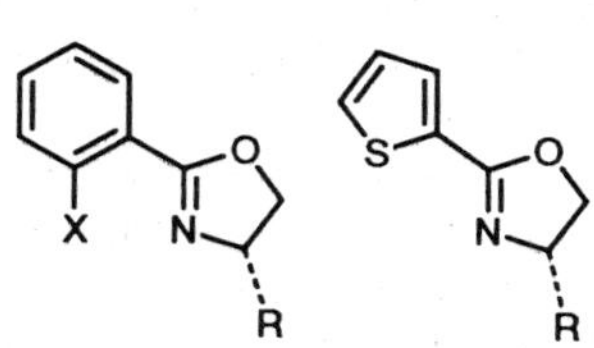

(14) X = PPh_2 (16)
(15) X = SPh
(18) X = SH

Scheme 4

and Nishio report the use of dimethyl(triflyl)silyl enol ethers as enolate equivalents for aldol and Michael reactions without Lewis-acid catalysis.[53] The asymmetric nitro-aldol reaction[54,55] (Henry reaction) has been used by Shibasaki and coworkers to synthesize the β-blocker (*S*)-propranolol (19) (Scheme 5).[56]

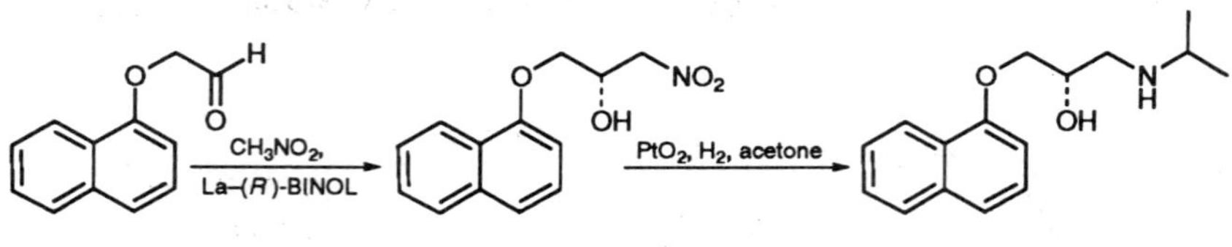

80 %, 92 % e.e.   (19) (*S*)-(-)-propranolol

Scheme 5

The disclosure of new chiral auxiliaries always generates great interest. Such is the case of a new paper from Yamamoto and coworkers. They have found that 2,2,6,6-tetramethyl-3,5-heptanediol (TMHDiol) derivatives, such as the monobenzoate (20), are excellent conformationally rigid acylic chiral auxiliaries.[57] The α,β-unsaturated ester (21) is highly selective in the conjugate addition of lithium amides [(21) → (22)] and Diels–Alder reactions. The high selectivity is associated with the rigid structure (21), as evidenced from *X*-ray and NMR analysis (Scheme 6a). Vedejs *et al.*[58] report an interesting and highly efficient method for the alkylation of chiral α-amino acid enolate equivalents, involving transfer of chirality from starting material to product *via* the transient boron complex (25) (Scheme 6b). The oxazaborolidinone (24), prepared selectiviely from the α-amidino carboxylate (23) and $PhBF_2$ *in situ*, is deprotonated and alkylated [(24) → (25)] with, in most cases, exceptional stereoselectivity.

Symmetrical α-diones have been prepared in excellent yield by conventional reaction of organometallic reagents with the C-2 units 1,1′-oxalylimidazole[59] and *N*,*N*-dimethylpiperazine-2,3-dione.[60]

Alkene Synthesis.—Many important improvements to the Wittig reaction have been made this year. Vedejs *et al.* have shown that the phosphole-derived ethylide (26)[61] is extremely *E* selective in the reaction with both aldehydes and ketones.[62] In comparison, the reaction of $Ph_3P{=}CHCH_3$ exhibits *Z* selectivity, whilst the reaction with ketones is very much substrate dependent (Scheme 7a). Patil and Schlosser have introduced a new class of phosphorane (27), derived from tris(2-methoxymethoxy)-phosphine, with enhanced *cis* selectivity (Scheme 8b). Stabilized ylides ($X = CO_2Me$)

[53] S. Kobayashi and K. Nishio, *J. Org. Chem.*, 1993, **58**, 2647.
[54] H. Sasai, T. Suzuki, N. Itoh, K. Tanaka, T. Date, K. Okamura, and M. Shibasaki, *J. Am. Chem. Soc.*, 1993, **115**, 10372.
[55] H. Sasai, T. Suzuki, N. Itoh, S. Arai, and M. Shibasaki, *Tetrahedron Lett.*, 1993, **34**, 2657.
[56] H. Sasai, N. Itoh, T. Suzuki, and M. Shibasaki, *Tetrahedron Lett.*, 1993, **34**, 855.
[57] I. Suzuki, H. Kin, and Y. Yamamoto, *J. Am. Chem. Soc.*, 1993, **115**, 10139.
[58] E. Vedejs, S.C. Fields, and M.R. Schrimpf, *J. Am. Chem. Soc.*, 1993, **115**, 11612.
[59] R.H. Mitchell and V.S. Iyer, *Tetrahedron Lett.*, 1993, **34**, 3683.
[60] U.T. Mueller-Westerhoff and M. Zhou, *Tetrahedron Lett.*, 1993, **34**, 571.
[61] E. Vedejs and M.J. Peterson, *J. Org. Chem.*, 1993, **58**, 1985.
[62] E. Vedejs, J. Cabaj, and M.J. Peterson, *J. Org. Chem.*, 1993, **58**, 6509.

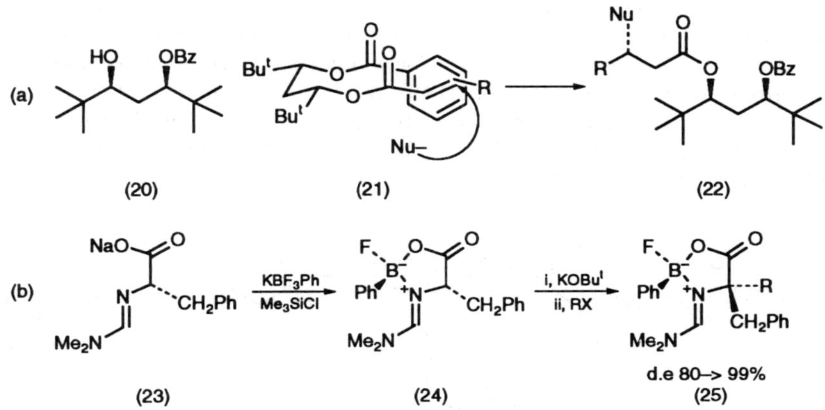

Scheme 6

show a curious solvent effect; in methanol the reaction is *cis* selective, whilst in hexane it is *trans* selective.[63] When the phosphorane bears an α-heteroatom the reaction is again impressively *cis* selective – much more so than when a triphenylphosphine derived reagent is used.[64]

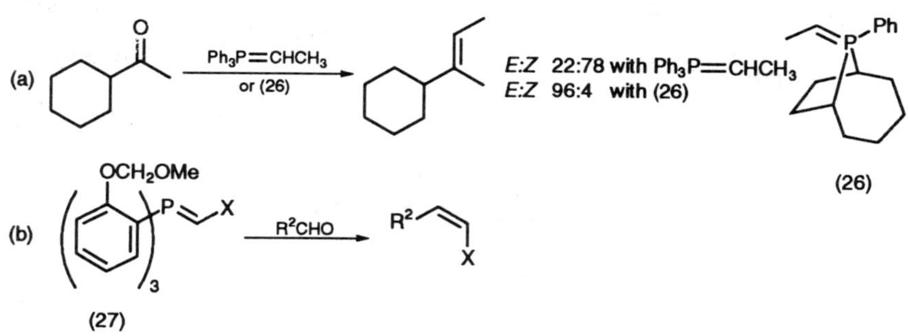

Scheme 7

Asymmetric Wittig reactions are proving to be popular. The chiral phosphonate (28) derived from 8-phenylmenthol reacts selectively with the *meso* dialdehyde (29)[65] to give the α,β-unsaturated ester (30) with the expected *E* selectivity (Scheme 8a). The chiral phosphonate (31) also shows high selectivity in the Horner–Wadsworth–Emmons (HWE) reaction with the *meso* 1,2-diketone (32) (Scheme 8b). The reaction is highly enantioselective (e.e. $\approx$ 100%) and somewhat surprisingly, for an HWE reaction, *Z*

[63] V. Patil and M. Schlosser, *SYNLETT*, 1993, 125.
[64] X. P. Zhang and M. Schlosser, *Tetrahedron Lett.*, 1993, **34**, 1925.
[65] N. Kann and T. Rein, *J. Org. Chem.*, 1993, **58**, 3802.

selective.[66] Chiral ketones have been kinetically resolved by HWE reaction with the phosphonate (33) derived from D-mannitol.[67] The groups of Paterson[68] and Koskinen[69] report the use of barium hydroxide and potassium carbonate in acetonitrile respectively as mild bases for the HWE reaction of β-ketophosphonates. The modifications are particularly advantageous when base-sensitive aldehydes are used. Baldwin *et al.* have described an interesting Wittig reaction of β-lactams; stabilized ylides react with *N*-Boc protected monocyclic β-lactams.[70]

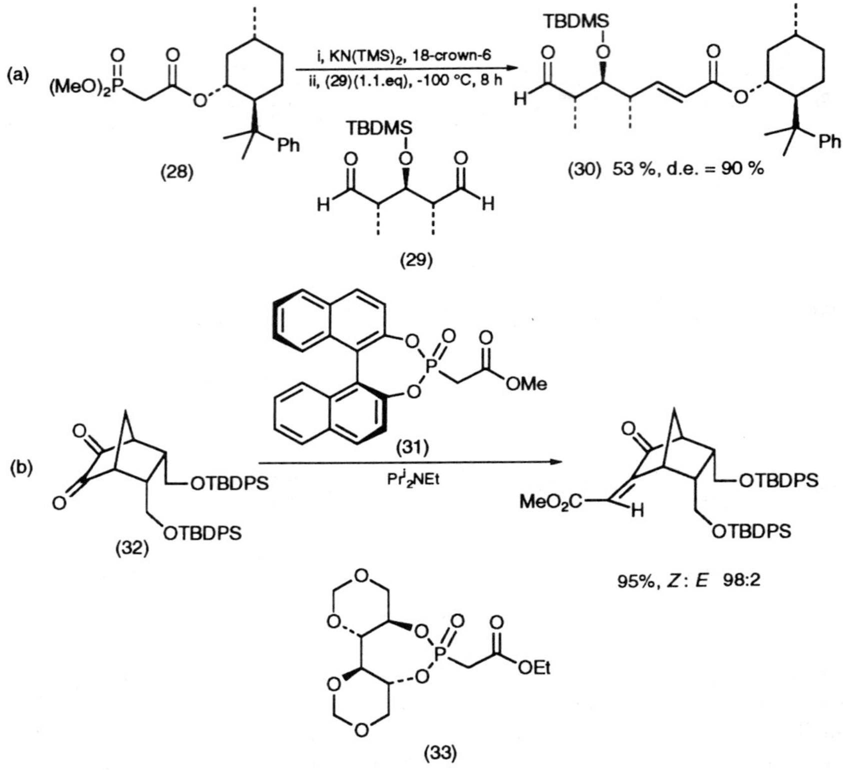

Scheme 8

Denmark and Amburgey have reported a general stereoselective method for the synthesis of trisubstituted alkenes (Scheme 9a).[71] They found that both the alkylation of the racemic β-keto phosphonamidate (34) and subsequent reduction of the carbonyl group are highly diastereoselective. The customary base-induced HWE elimination of (35) proved problematic; however, simple thermal cycloelimination proceeded cleanly

[66] K. Tanaka, Y. Ohta, K. Fuji, and T. Taga, *Tetrahedron Lett.*, 1993, **34**, 4071.
[67] K. Narasaka, E. Hidai, Y. Hayashi, and J.-L. Gras, *J. Chem. Soc., Chem. Commun.*, 1993, 102.
[68] I. Paterson, K.-S. Yeung, and J. B. Smaill, *SYNLETT*, 1993, 774.
[69] A. M. P. Koskinen and P. M. Koskinen, *SYNLETT*, 1993, 501.
[70] J. E. Baldwin, A. J. Edwards, C. N. Farthing, and A. T. Russell, *SYNLETT*, 1993, 49.
[71] S. E. Denmark and J. Amburgey, *J. Am. Chem. Soc.*, 1993, **115**, 10 386.

to give the trisubstituted alkene. Le Corre and coworkers[72] have shown that α-metallo phosphine–boranes[73] react with aldehydes in a Horner–Wittig-like manner to give *E* alkenes; when butyllithium is used as the base, the intermediate hydroxy phosphine-borane can be isolated after work-up (Scheme 9b). Most Wittig-like reactions proceed by *syn* elimination. However, Lawrence, and Muhammad have shown that Horner–Wittig intermediates (36) undergo *anti* elimination by reduction ($LiAlH_4$–$CeCl_3$) and phosphorus trichloride mediated elimination [(37) → (38)] (Scheme 9c).[74]

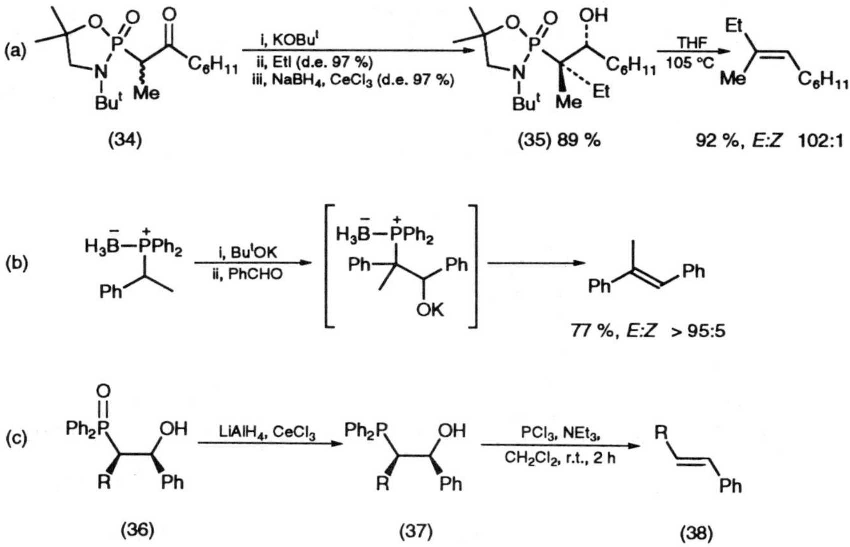

Scheme 9

Chiral alkenes are produced by enantioselective dehydrohalogenation of prochiral alkylbromides by chiral alkoxides derived from *N*-methylephedrine (Scheme 10a).[75] Similarly, asymmetric oxidation of the prochiral selenide (39) using Davis' chiral oxaziridine gives a reactive selenoxide (40) of undefined configuration, which eliminates to the chiral alkene (41) (Scheme 10b).[76]

Several examples of silicon-mediated alkene synthesis have appeared. The dibromovinylsilane (42) provides a useful synthon for the construction of a variety of geometrically pure alkenes. For example, both bromides react with higher order cuprates to give a vinylsilane (43) that is easily desilylated with iodine in wet benzene (Scheme 11).[77] Chauret and Chong[78] have shown that α-epoxy(triethylsilane) serves as

72 Y. Gourdel, A. Ghanimi, P. Pellon, and M. Le Corre, *Tetrahedron Lett.*, 1993, **34**, 1011.
73 T. Imamoto, *Pure and Applied*, 1993, **65**, 655.
74 N. J. Lawrence and F. Muhammad, *J. Chem. Soc., Chem. Commun.*, 1993, 1187.
75 J. Vadecard, J.-C. Plaquevent, L. Duhamel, and P. Duhamel, *J. Chem. Soc., Chem. Commun.*, 1993, 116.
76 N. Komatsu, S. Matsunaga, T. Sugita, and S. Uemura, *J. Am. Chem. Soc.*, 1993, **115**, 5847.
77 R. Angell, P. J. Parsons, and A. Naylor, *SYNLETT*, 1993, 189.
78 D. C. Chauret and J. M. Chong, *Tetrahedron Lett.*, 1993, **34**, 3695.

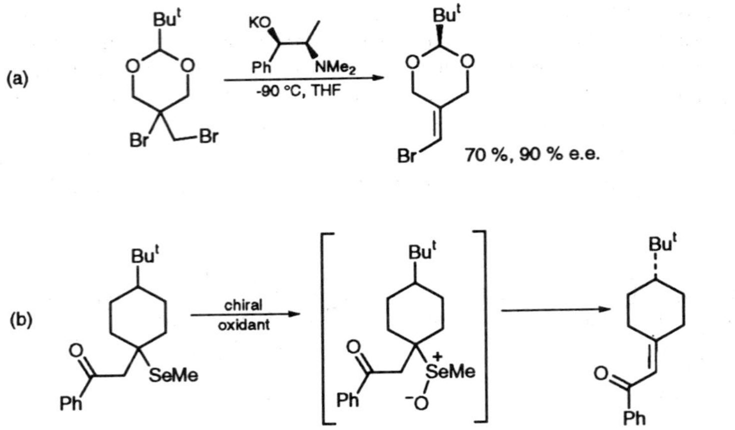

(39) (40) (41) 96 %, 83% e.e.

Scheme 10

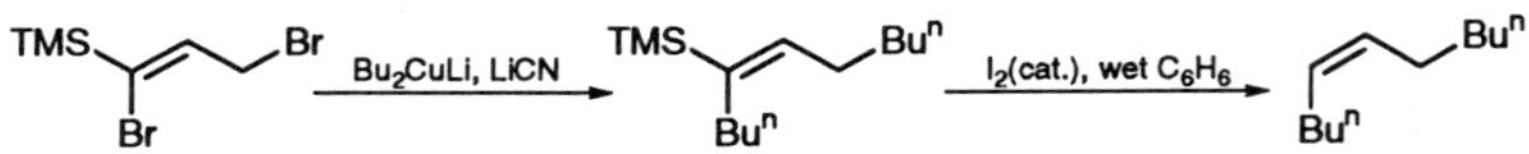

(42) (43)

Scheme 11

an excellent precursor to *E* and *Z* disubstituted alkenes by Hudrlik–Peterson chemistry.

Finally, Fry *et al.* have shown that benzal chloride ($PhCHCl_2$) is converted to stilbene (87%, *E*:*Z* 1:1.1) by the action of electrochemically generated cobalt(I) salen.[79]

Cycloadditions.—Kobayashi has shown that scandium(III) triflate is an excellent catalyst for the Diels–Alder reaction.[80] The reaction is *endo* selective and can be performed in both organic and aqueous media. In addition, the catalyst is easily recovered and may be used at low concentrations (1 mol%). Chiral Lewis acids that have been used as catalysts for the asymmetric Diels–Alder reaction[81] (Figure 2) include the copper(II) complexes of the bis(imine) (44)[82] and bis(oxazoline) (45);[83] the substituted biaryls (46)[84] and (47)[85] (which forms an interesting titanium(IV) helical

79 A. J. Fry, U. N. Sirisoma, and A. S. Lee, *Tetrahedron Lett.*, 1993, **34**, 809.
80 S. Kobayashi, I. Hachiya, M. Araki, and H. Ishitani, *Tetrahedron Lett.*, 1993, **34**, 3755.
81 U. Pindur, G. Lutz, and C. Otto, *Chem. Rev.*, 1993, **93**, 741.
82 D. A. Evans, T. Lectka, and S. J. Miller, *Tetrahedron Lett.*, 1993, **34**, 7027.
83 D. A. Evans, S. J. Miller, and T. Lectka, *J. Am. Chem. Soc.*, 1993, **115**, 6460.
84 J. Bao, W. D. Wulff, and A. L. Rheingold, *J. Am. Chem. Soc.*, 1993, **115**, 3814.
85 K. Maruoka, N. Murase, and H. Yamamoto, *J. Org. Chem.*, 1993, **58**, 2938.

complex); and the 2-amino-1-indanol (48).[86] Ito and Katsuki[87] have used the copper(I) complex of the chiral bipyridine (49) to catalyse the asymmetric cyclopropanation of alkenes (Scheme 12). The reaction proceeds with good *trans*: *cis* selectivity and excellent enantioselectivity.

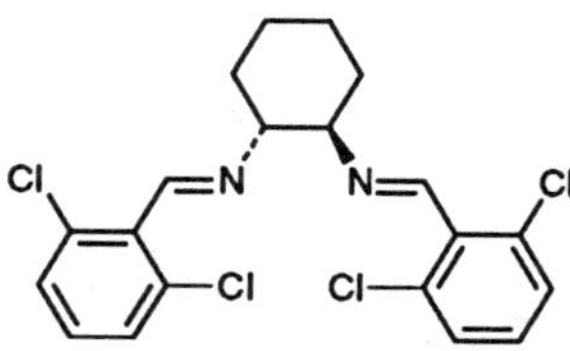

(44)

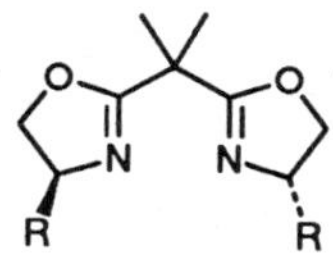

(45)

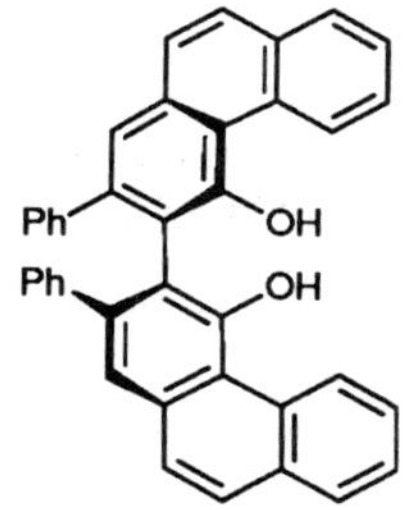

(46)

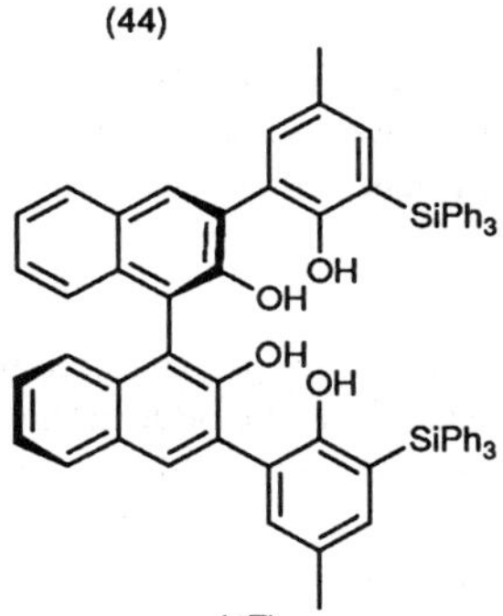

(47)

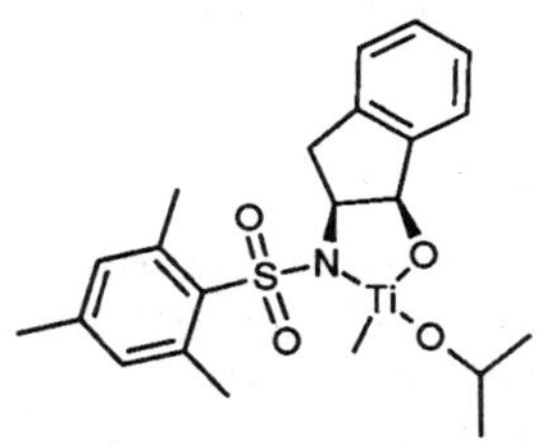

(48)

Figure 2

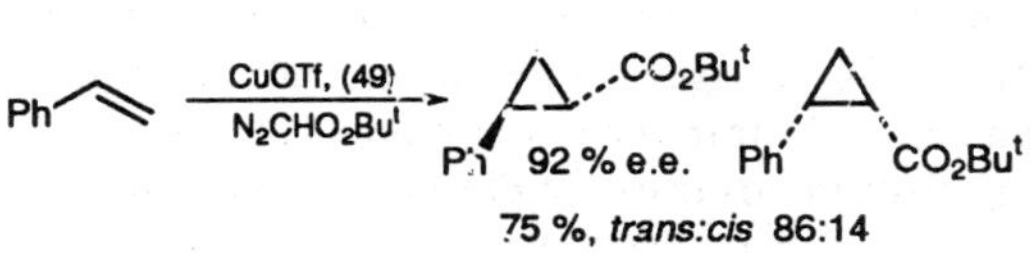

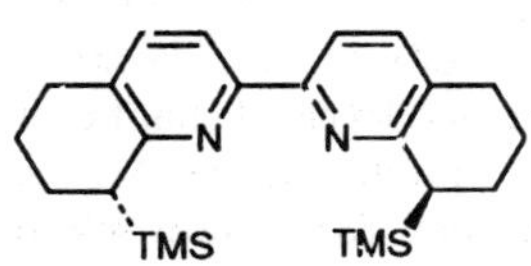

(49)

Scheme 12

Radical-based Methods.—Many advances have been made this year in the use of synthetic methods that involve radical intermediates.[88] Ryu, Sonoda, and co-workers[89] have shown that tris(trimethylsilyl)silane (TTMSS) is an excellent reagent

[86] E. J. Corey, T. D. Roper, K. Ishihara, and G. Sarakinos, *Tetrahedron Lett.*, 1993, **34**, 8399.
[87] K. Ito and T. Katsuki, *Tetrahedron Lett.*, 1993, **34**, 2661.
[88] P. Dowd and W. Zhang, *Chem. Rev.*, 1993, **93**, 2091.
[89] I. Ryu, M. Hasegawa, A. Kurihara, A. Ogawa, S. Tsunoi, and N. Sonoda, *SYNLETT*, 1993, 143.

for the free-radical formylation of alkyl halides (Scheme 13). Additionaly, when an electron-deficient alkene is present, unsymmetrical ketones (50) are produced. TTMSS is superior to tin hydrides for this purpose due to its modest ability to donate a hydrogen atom to both the intermediate alkyl and acyl radical. TTMSS also effects the isomerization of some *Z* alkenes to their *E* isomers by an addition–elimination pathway.[90]

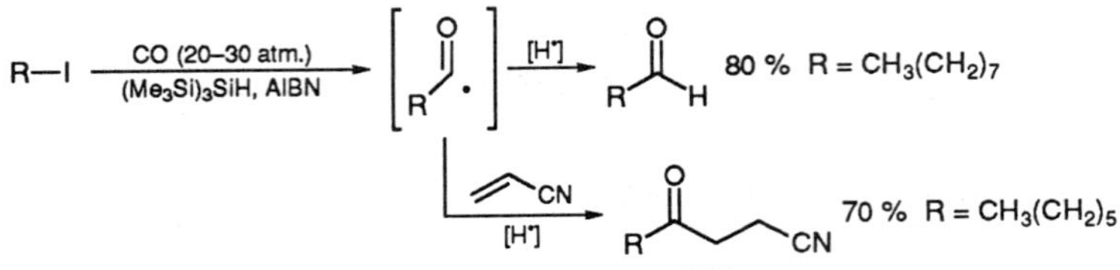

Scheme 13

Finally, tributylstannyl radicals may be generated, in the absence of tin hydride species, by mild photochemical methods employing acetone as a triplet sensitizer,[91] or by thermal decomposition of *O,O'*-bistributyltin benzopinacolate.[92]

3 Reduction

Oxazaborolidine catalysts have seen widespread use for the reduction of ketones [(51) → (52)] (Scheme 14). New oxazaborolidine catalysts introduced this year include the phenylglycine-derived (53),[93] the hydroxysulfoximine (54),[94] and the *erythro* 2-amino-1,2-diphenylethanol derived (55).[95] Interestingly, the complex (56) derived *in situ* from the corresponding amino alcohol behaves in the same way as the alkylated complex (55);[96] the method therefore avoids the potentially troublesome isolation of the oxazaborolidine. Liotta and coworkers have published the results of a theroretical study of such oxazaborolidine-catalysed ketone reductions and suggest that hydride transfer occurs *via* a chair-like transition state (58) in contrast to the previously proposed boat transition state (57) (Scheme 15).[97] It has also been shown that the use of oxazaborolidine–borane complexes in combination with triethylamine leads to increased levels of enantioselectivity. The origins of this curious effect are not entirely clear although it appears the amine might trap the initial monoalkoxyborane, which itself may show lower selectivity than the oxazaborolidinone–borane complex.[98]

Evans *et al.* report a highly efficient asymmetric version of the Meerwein–Pondorf–Verley (MPV) reduction [(59) → (60)] (Scheme 16). They found that the samarium complex (61), obtained from samarium(III) iodide and a styrene-oxide-derived tridentate ligand catalyses the hydrogen transfer from isopropanol to a range

[90] C. Ferreri, M. Ballestri, and C. Chatgilialoglu, *Tetrahedron Lett.*, 1993, **34**, 5147.
[91] M. Harendza, J. Junggebauer, K. Lessmann, W.P. Neumann, and H. Tews, *SYNLETT*, 1993, 286.
[92] M.J. Tomaszewski and J. Warkentin, *J. Chem. Soc., Chem. Commun.*, 1993, 1407.
[93] C. Dauelsberg and J. Martens, *Synth. Commun.*, 1993, **23**, 2091.
[94] C. Bolm and M. Felder, *Tetrahedron Lett.*, 1993, **34**, 6041.
[95] G.J. Quallich and T.M. Woodall, *Tetrahedron Lett.*, 1993, **34**, 4145.
[96] G.J. Quallich and T.M. Woodall, *SYNLETT*, 1993, 929.
[97] D.K. Jones, D.C. Liotta, I. Shinkai, and D.J. Mathre, *J. Org. Chem.*, 1993, **58**, 799.
[98] D. Cai, D. Tschaen, Y.-J. Shi, T.R. Verhoeven, R.A. Reamer, and A.W. Douglas, *Tetrahedron Lett.*, 1993, **34**, 3243.

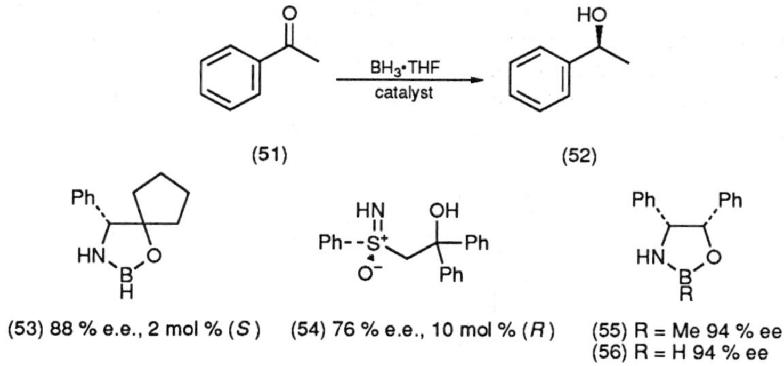

Scheme 14

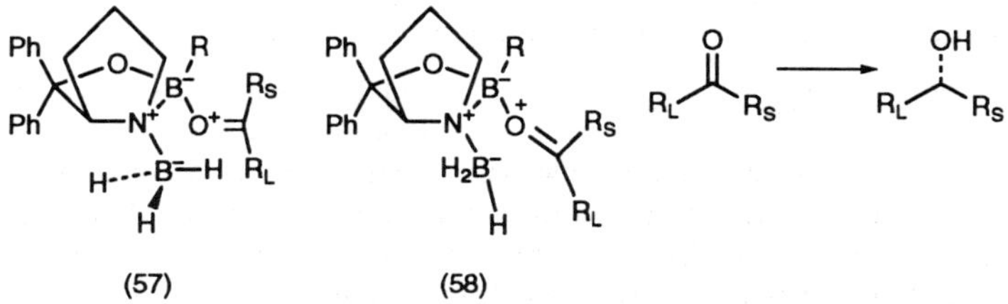

Scheme 15

of aromatic ketones.[99] The chiral ruthenium and rhodium complexes (62)[100] and (63)[101] have also been used in a similar manner to obtain moderate selectivity. Molander and McKie report the highly stereoselective samarium(III)-catalysed intramolecular MPV reaction.[102] Other modified MPV catalysts include silica gel-supported zirconium(IV)[103] and zirconia/titania[104] for the reduction of carboxylic acids and sterically hindered ketones.

The year has seen significant developments in the area of asymmetric hydrogenation reactions. Buchwald and coworkers have used the titanocene catalyst (64) to achieve the asymmetric hydrogenation of unfunctionalized trisubstituted alkenes (Scheme 17a).[105] Prior to this disclosure only alkenes that possessed a chelating substituent could successfully be reduced stereoselectively. The same catalyst has been used to reduce cyclic imines to the corresponding cyclic amine.[106] Faller and Parr have

99 D. A. Evans, S. G. Nelson, M. R. Gagné, and A. R. Muci, *J. Am. Chem. Soc.*, 1993, **115**, 9800.
100 J.-P. Genêt, V. Ratovelomanana-Vidal, and C. Pinel, *SYNLETT*, 1993, **34**, 478.
101 P. Gamez, F. Fache, P. Mageney, and M. Lemaire, *Tetrahedron Lett.*, 1993, **34**, 6897.
102 G. A. Molander and J. A. McKie, *J. Am. Chem. Soc.*, 1993, **115**, 5821.
103 K. Inada, M. Shibagaki, Y. Nakanishi, and H. Matsushita, *Chem. Lett.*, 1993, 1795.
104 K. Takahasi, M. Shibagaki, H. Kuno, and H. Matsushita, *Chem. Lett.*, 1993, 839.
105 R. D. Broene and S. L. Buchwald, *J. Am. Chem. Soc.*, 1993, **115**, 12 569.
106 C. A. Willoughby and S. L. Buchwald, *J. Org. Chem.*, 1993, **58**, 7627.

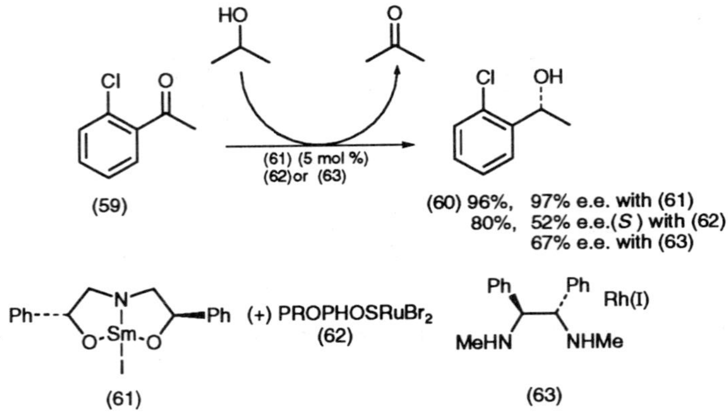

Scheme 16

introduced the concept of chiral poisoning as a new strategy for asymmetric catalysis (Scheme 17b).[107] This is especially useful when the catalyst is expensive [such as chiraphos (65)] and the poison cheap. The principal is illustrated by the reduction of alkenes with the racemic chiraphos rhodium complex (66). Addition of the *S* isomer of the thiophosphinite (67) results in selective poisoning of the (*S*,*S*)-chiraphos species. Faller has also applied the same principle to the kinetic resolution of allylic alcohols, by selectively reducing one alkene with a poisoned ruthenium catalyst.[108]

Cycloalkanones are reduced with impressive enantioselectivity (>90% e.e.) by hydrogenation using iridium(I) BINAP catalysts in the presence of bis(*o*-*N*,*N*-dimethylaminophenyl)phenylphosphine.[109] The (*R*)-BICHEP–Ru(II) complex (68) is also an excellent catalyst for the enantioselective hydrogenation of α-ketoamides (69) (Scheme 18a).[110] Asymmetric hydrosilylation of ketones [(70) → (71)] is achieved with high enantioselectivity with the new chiral bipyridine Bipymox (72) (Scheme 18b),[111] and TADDOL-derived cyclic phosphonites and phosphites.[112]

Reduction with complex hydride reagents is the subject of many reports. Carboxylic acids are conveniently reduced to aldehydes in a one-pot procedure with piperidine and sodium diethyldihydroaluminate.[113] Singaram's newly reported lithium aluminium hydride equivalent, lithium pyrrolidinoborohydride, is highly regiospecific in a 1,2 sense in the reduction of α,β-unsaturated ketones.[114] The same reagent reduces tertiary amides to the corresponding alcohol.[115] Reduction of esters[116] and acid chlorides[117] is

[107] J. W. Faller and J. Parr, *J. Am. Chem. Soc.*, 1993, **115**, 804.
[108] J. W. Faller and M. Tokunaga, *Tetrahedron Lett.*, 1993, **34**, 7359.
[109] X. Zhang, T. Taketomi, T. Yoshizumi, H. Kumobayashi, S. Akutagawa, K. Mashima, and H. Takaya, *J. Am. Chem. Soc.*, 1993, **115**, 3318.
[110] T. Chiba, A. Miyashita, H. Nohira, and H. Takaya, *Tetrahedron Lett.*, 1993, **34**, 2351.
[111] H. Nishiyama, S. Yamaguchi, S.-B. Park, and K. Itoh, *Tetrahedron: Asymmetry*, 1993, **4**, 143.
[112] J.-i. Sakaki, W. B. Schweizer, and D. Seebach, *Helv. Chim. Acta*, 1993, **76**, 2654.
[113] N. M. Yoon, K. I. Choi, Y. S. Gyoung, and W. S. Jun, *Synth. Commun.*, 1993, **23**, 1775.
[114] J. C. Fuller, E. L. Stangeland, C. T. Goralski, and B. Singaram, *Tetrahedron Lett.*, 1993, **34**, 257.
[115] G. B. Fisher, J. C. Fuller, J. Harrison, C. T. Goralski, and B. Singaram, *Tetrahedron Lett.*, 1993, **34**, 1091.
[116] N. M. Yoon, J. H. Ahn, D. K. An, and Y. S. Shon, *J. Org. Chem.*, 1993, **58**, 1941.
[117] J. S. Cha and H. C. Brown, *J. Org. Chem.*, 1993, **58**, 4732.

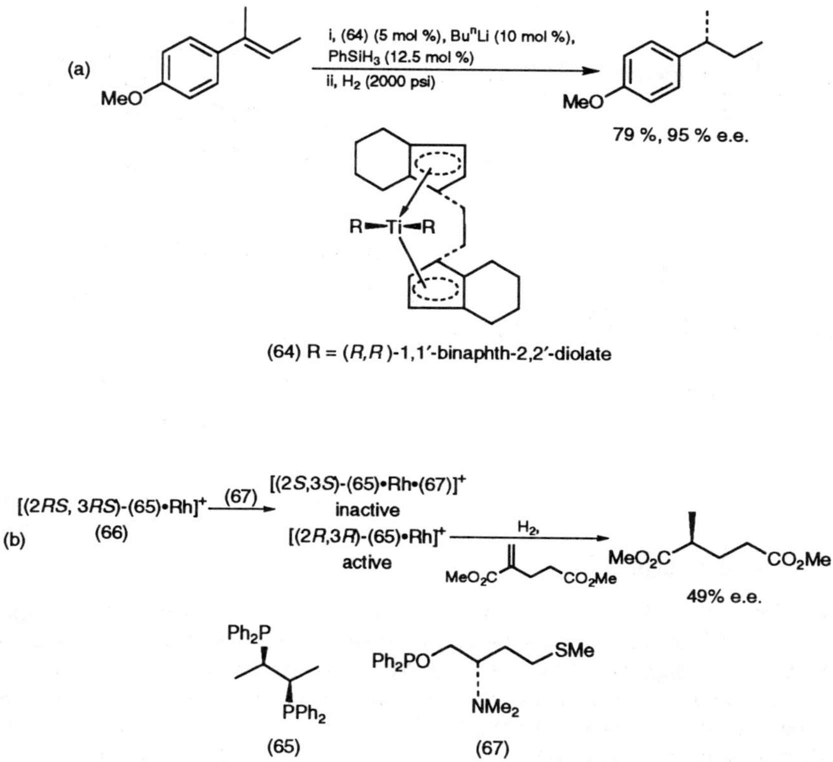

Scheme 17

achieved with sodium diethylpiperidinylaluminohydride and sodium tris-t-butoxyaluminohydride respectively.

Esters, carboxylic acids, amides, and nitriles are efficiently reduced with samarium diiodide in the presence of water.[118] Secondary amides are reductively deoxygenated by Schwartz's reagent (Cp_2ZrHCl) to the corresponding imine.[119] Alcohols are reduced to alkanes[120] by the action of sodium borohydride and the phosphonium anhydride reagent $(Ph_3P^+)_2O(CF_3SO_2^-)_2$. α-Amino acids are conveniently reduced, without racemization, to the corresponding amino alcohol by sodium borohydride and iodine.[121] Miscellaneous reagents for the reduction of ketones include copper(II)-exchanged-cation resin–$NaBH_4$ and lithium tris[(3-t-butyl-3-pentyl)oxy] aluminium hydride.[122,123]

[118] Y. Kamochi and T. Kudo, *Chem. Lett.*, 1993, 1495.
[119] D. J. A. Schedler, A. G. Godfrey, and B. Ganem, *Tetrahedron Lett.*, 1993, **115**, 5035.
[120] J. B. Hendrickson, M. Singer, and M. S. Hussoin, *J. Org. Chem.*, 1993, **58**, 6913.
[121] M. J. McKennon, A. I. Meyers, K. Drauz, and M. Schwarm, *J. Org. Chem.*, 1993, **58**, 3568.
[122] A. Sarkar, B. R. Rao, and B. Ram, *Synth. Commun.*, 1993, **23**, 291.
[123] G. Boireau, A. Deberly, and R. Toneva, *SYNLETT*, 1993, 585.

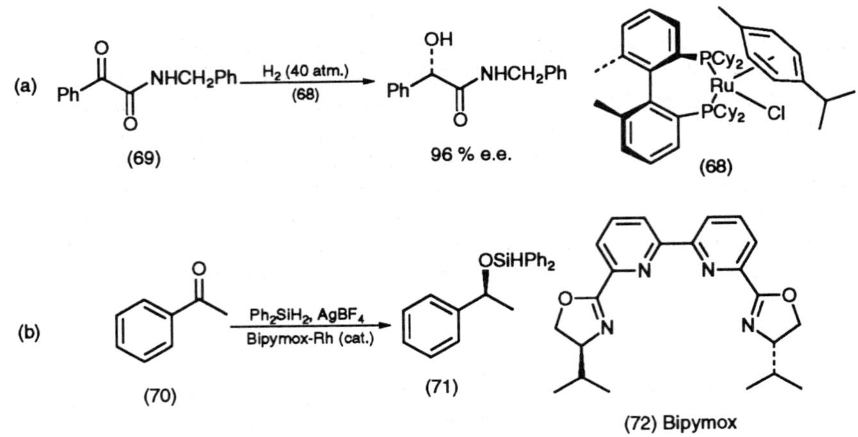

Scheme 18

4 Oxidation

Oxidation is the subject of issue 164 of *Topics in Current Chemistry*;[124] peroxygen reagents,[124a] dioxiranes,[124b] enantioselective epoxidation,[124c] and catalytic oxidation with peroxide reagents[124d] are reviewed. Sharpless and coworkers continue their pioneering studies of asymmetric dihydroxylation of alkenes. The osmium-catalysed asymmetric dihydroxylation of terminal alkenes is efficiently achieved by use of the pyrimidine-linked ligand $(DHQD)_2$–PYR (73),[125] or $(DHQD')_2$–PHAL (74);[126] for this type of alkene the new ligands are superior to the previously described phthalazine catalysts, whose preparation has been recently described.[127] The asymmetric dihydroxylation of tetrasubstituted double bonds occurs with low to very high enantioselectivity with both phthalazine and pyrimidine ligand classes.[128] Among other substrates treated with AD-mixes are α-substituted styrenes and α-phenylacrylates,[129] α,β-unsaturated ketones,[130] α,β- and β,γ-unsaturated amides,[131] aryl allyl ethers,[132] allylsilanes,[133,134] and tertiary allylic alcohols.[135] The procedure has seen several uses

[124] (a) H. Heaney, *Top. Curr. Chem.*, 1993, **164**, 1; (b) W. Adam and L. Hadjiarapoglou, *ibid.*, 1993, **164**, 45; (c) E. Höft, *ibid.*, 1993, **164**, 63; (d) R. Sheldon, *ibid.*, 1993, **164**, 21.

[125] G. A. Crispino, K.-S. Jeong, H. C. Kolb, Z.-M. Wang, D. Xu, and K. B. Sharpless, *J. Org. Chem.*, 1993, **58**, 3785.

[126] M. P. Arrington, Y. L. Bennani, T. Göbel, P. Walsh, S.-H. Zhao, and K. B. Sharpless, *Tetrahedron Lett.*, 1993, **34**, 7375.

[127] W. Amberg, Y. L. Bennani, R. K. Chadha, G. A. Crispino, W. D. Davis, J. Hartung, K.-S. Jeong, Y. Ogino, T. Shibata, and K. B. Sharpless, *J. Org. Chem.*, 1993, **58**, 844.

[128] K. Morikawa, J. Park, P. G. Andersson, T. Hashiyama, and K. B. Sharpless, *J. Am. Chem. Soc.*, 1993, **115**, 8463.

[129] Z.-M. Wang and K. B. Sharpless, *SYNLETT*, 1993, 603.

[130] P. J. Walsh and K. B. Sharpless, *SYNLETT*, 1993, 605.

[131] Y. L. Bennani and K. B. Sharpless, *Tetrahedron Lett.*, 1993, **34**, 2079.

[132] Z.-M. Wang, X.-L. Zhang, and K. B. Sharpless, *Tetrahedron Lett.*, 1993, **34**, 2267.

[133] S. Okamoto, K. Tani, F. Sato, K. B. Sharpless, and D. Zargarian, *Tetrahedron Lett.*, 1993, **34**, 2509.

[134] J. A. Soderquist, A. M. Rane, and C. J. López, *Tetrahedron Lett.*, 1993, **34**, 1893.

[135] Z.-M. Wang and K. B. Sharpless, *Tetrahedron Lett.*, 1993, **34**, 8225.

in natural product synthesis.[136–139] Not surprisingly, asymmetric dihydroxylation may be used to resolve racemic olefins kinetically; there being a 30-fold difference in the rate of reaction of each enantiomer with AD-mix α [containing $(DHQ)_2$–PHAL] and AD-mix β [containing $(DHQD)_2$–PHAL] (Scheme 19).[140] Similarly, secondary allylic acetates may be kinetically resolved using the teraphthalyl linked ligand (75).[141] Corey's and Sharpless' groups have sought to explain the origin of the high enantioselectivity. Corey argues that the reaction proceeds *via* a μ-oxo-bridged bisOs(VIII) species, $QO_3Os[O_2]OsO_3Q$.[142] However, the conclusion from the Sharpless group is different; they believe that the two quinuclidine units do not act in concert (*i.e.* the active species is a 1 : 1 OsO_4–phthalazine ligand complex).[143]

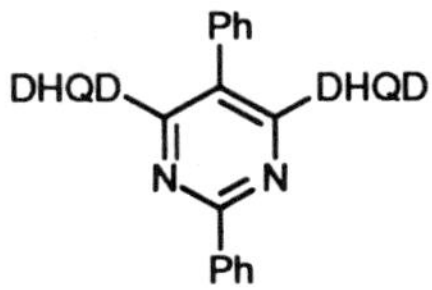

(73) $(DHQD)_2$-PYR

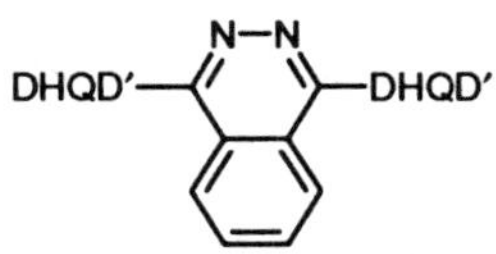

(74)

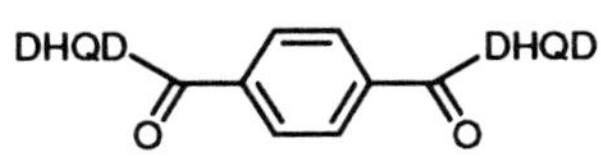

(75)

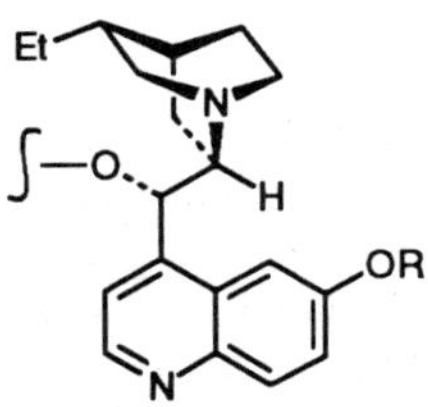

DHQD (R = Me)
DHQD′ (R = ^{i}pentyl)

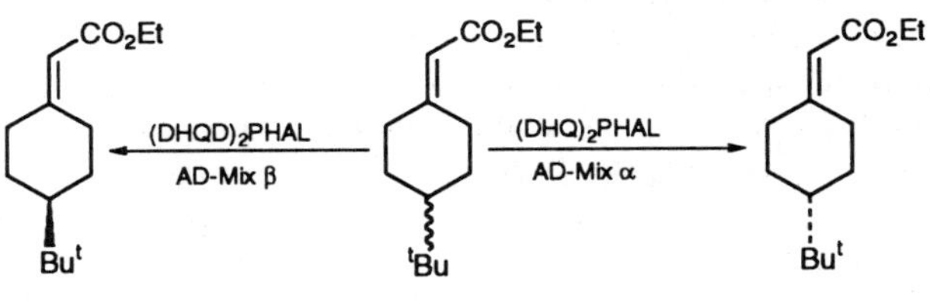

Scheme 19

136 G. A. Crispino and K. B. Sharpless, *SYNLETT*, 1993, 47.
137 G. A. Crispino and K. B. Sharpless, *Synthesis*, 1993, 777.
138 H. C. Kolb, Y. L. Bennani, and K. B. Sharpless, *Tetrahedron: Asymmetry*, 1993, **4**, 133.
139 Y. L. Bennani and K. B. Sharpless, *Tetrahedron Lett.*, 1993, **34**, 2083.
140 M. S. VanNieuwenhze and K. B. Sharpless, *J. Am. Chem. Soc.*, 1993, **115**, 7864.
141 B. B. Lohray and V. Bhushan, *Tetrahedron Lett.*, 1993, **34**, 3911.
142 E. J. Corey, M. C. Noe, and S. Sarshar, *J. Am. Chem. Soc.*, 1993, **115**, 3828; E. J. Corey and M. C. Noe, *J. Am. Chem. Soc.*, 1993, **115**, 12 579.
143 H. C. Kolb, P. G. Andersson, Y. L. Bennani, G. A. Crispino, K.-S. Jeong, H.-L. Kwong, and K. B. Sharpless, *J. Am. Chem. Soc.*, 1993, **115**, 12 226; T. Göbel and K. B. Sharpless, *Angew. Chem., Int. Ed. Engl.*, 1993, **32**, 1329.

Hanessian *et al.* have used the readily available C_2 symmetric 1,2-diamine (76) to control the enantioselectivity of osmium tetroxide mediated dihydroxylation of a wide range of alkenes (Scheme 20a). Although the enantioselectivities are impressive, at present stoichiometric amounts of osmium and amine are required.[144] Henry and Weinreb report a useful method for the oxidative cleavage of alkenes [(77) → (78)] with catalytic osmium tetroxide and excess Jones' reagent (Scheme 20b).[145] Similarly, alkenes are oxidatively cleaved to aldehydes by the action of potassium permanganate supported on alumina – yields are good and the conditions mild.[146]

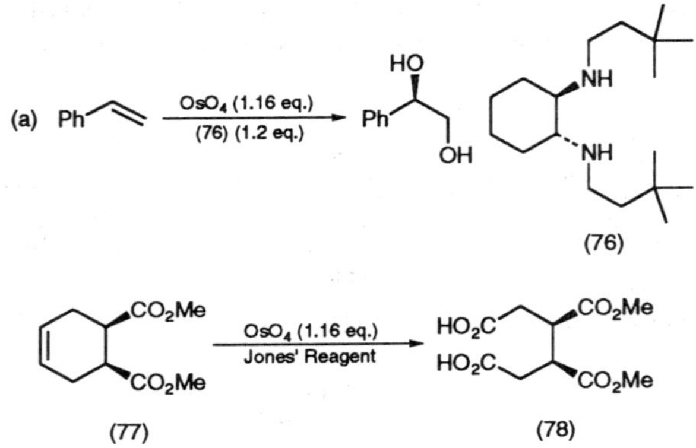

Scheme 20

Many new reagents have been developed for the oxidation of alcohols. Allylic alcohols are selectively oxidized, in the presence of primary alcohols, with stoichiometric palladium(II) salts;[147] this may prove more useful when a catalytic version is developed. Aerobic oxidation with metal catalysts is attractive for economic and environmental reasons. For example, simple secondary alkyl alcohols are oxidized by molecular oxygen with a ruthium(III)–cobalt(II) bimetallic catalyst.[148] Molecular oxygen has also been used in the Baeyer–Villiger oxidation of ketones to lactones in the presence of benzaldehyde, with or without metal [Cu(II) or Ni(II)] catalysis.[149] Other reagents for the oxidation of alcohols include MnO_2–bentonite–microwave;[150] trichloromelamine;[151] hypochlorous acid in methanol (to yield esters directly from primary alcohols);[152] CrO_3 (cat.)–70% aq. ButOOH;[153] chromium-substituted aluminophosphate-5 zeolite–Bu^{t}OOH;[154] and the complex HOF·CH_3CN.[155] The

144 S. Hanessian, P. Meffre, M. Girard, S. Beaudoin, J.-Y. Sancéau, and Y. Bennani, *J. Org. Chem.*, 1993, **58**, 1991.
145 J. R. Henry and S. M. Weinreb, *J. Org. Chem.*, 1993, **58**, 4745.
146 D. G. Lee, T. Chen, and Z. Wang, *J. Org. Chem.*, 1993, **58**, 2918.
147 V. Bellosta, R. Benhaddou, and S. Czernecki, *SYNLETT*, 1993, 861.
148 S.-I. Murahashi, T. Naota, and N. Hirai, *J. Org. Chem.*, 1993, **58**, 7318.
149 C. Bolm, G. Schlingloff, and K. Weickhardt, *Tetrahedron Lett.*, 1993, **34**, 3405.
150 L. A. Martinez, O. Garcia, F. Delgado, C. Alvarez, and R. Patiño, *Tetrahedron Lett.*, 1993, **34**, 5293.
151 S. Kondo, M. Ohira, S. Kawasoe, H. Kunisada, and Y. Yuki, *J. Org. Chem.*, 1993, **58**, 5003.
152 C. E. McDonald, L. E. Nice, A. W. Shaw, and N. B. Nestor, *Tetrahedron Lett.*, 1993, **34**, 2741.
153 J. Muzart and A. N'Ait Ajjou, *Synthesis*, 1993, 785.
154 J. D. Chen, J. Dakka, E. Neeleman, and R. A. Sheldon, *J. Chem. Soc., Chem. Commun.*, 1993, 1379.
155 S. Rozen, Y. Bareket, and M. Kol, *Tetrahedron*, 1993, **49**, 8169.

Dess–Martin periodinane, a useful reagent for the oxidation of alcohols, has seen an improved and large-scale synthesis.[156] Aromatic aldehydes are oxidized efficiently to the corresponding benzoic acid with hydrogen peroxide in formic acid.[157]

Epoxidation continues to draw much attention. Iqbal and coworkers have found that the epoxidation of alkenes with molecular oxygen is achieved using the hydroperoxide (79), derived by *in situ* cobalt(II)-catalysed autoxidation of (80) (Scheme 21a).[158] Schwenkreis and Berkessel have used the manganese(III) complex (81) of the dihydrosalen ligand bearing a pendant imidazole group as a biomimetic catalyst for epoxidation of alkenes.[159] Collman *et al.*[160] have similarly used a threitol-strapped manganese porphyrin as an epoxidation catalyst; oxidation of 1,2-dihydronaphthalene (82) is highly enantioselective (88% e.e.). Katsuki and coworkers have used the dihydrosalen catalyst (83) bearing axially chiral binaphthyl groups to oxidize the same alkene (Scheme 21b).[161] Komiya and coworkers also report the epoxidation of alkenes with molecular oxygen and catalytic copper(II) salts; however, the reaction is not stereospecific.[162]

Other reagents for epoxidation include bis(ephedrine)cobalt(II)–O_2;[163,164] clay-impregnated-with-Ni(II)–O_2 and isobutyraldehyde;[165] and polymer-supported Mn(III)–salen complexes and iodosylbenzene.[166] Mukaiyama *et al.* have reported the use of propionaldehyde diethyl acetal as a reductant in the cobalt(II)-catalysed aerobic epoxidation of alkenes.[167] The method is neutral and can be used to synthesize acid-sensitive epoxides.[168] A different approach to catalytic epoxidation has been taken by Manoury *et al.*;[169] the tartrate-derived borate (84) catalyses oxygen transfer from t-butylhydroperoxide to prochiral alkenes with moderate enantioselectivity (Scheme 22a). Chiral epoxides are obtained in moderate enantiomeric excess from racemates by chiral Lewis-acid catalysed ring-opening with diethylamine (Scheme 22b).[170] Terminal epoxides are efficiently constructed in a two-step process from chiral trichloromethylcarbinols (accessible by oxazaborolidine reduction of the ketone) by selective reduction and base-catalysed ring closure of the resulting chlorohydrin.[171] Hindered alkenes are efficiently epoxidized with nitrogen dioxide.[172] Epoxides are also synthesized by a modification of Corey's reaction of carbonyl compounds and sulfonium ylides by generation of the sulfonium ylide *in situ* from dimethyl sulfide, methanol, and sulfuric acid.[173] Epoxides are converted directly into carboxylic acids

[156] R. E. Ireland and L. Liu, *J. Org. Chem.*, 1993, **58**, 2899.
[157] R. H. Dodd and M. Le Hyaric, *Synthesis*, 1993, 295.
[158] T. Punniyamurthy, B. Bhatia, and J. Iqbal, *Tetrahedron Lett.*, 1993, **34**, 4657.
[159] T. Schwenkreis and A. Berkessel, *Tetrahedron Lett.*, 1993, **34**, 4785.
[160] J. P. Collman, V. J. Lee, X. Zhang, J. A. Ibers, and J. I. Brauman, *J. Am. Chem. Soc.*, 1993, **115**, 3834.
[161] H. Sasaki, R. Irie, and T. Katsuki, *SYNLETT*, 1993, 300; for other complexes see N. Hosoya, R. Irie, and T. Katsuki, *SYNLETT*, 1993, 261.
[162] S.-I. Murahashi, Y. Oda, T. Naota, and N. Komiya, *J. Chem. Soc., Chem. Commun.*, 1993, 139.
[163] J. Iqbal, S. Bhatia, and M. M. Reddy, *Synth. Commun.*, 1993, **23**, 2285.
[164] S. Bhatia, T. Punniyamurthy, B. Bhatia, and J. Iqbal, *Tetrahedron*, 1993, **49**, 6101.
[165] E. Bouhlel, P. Laszlo, M. Levart, M.-T. Montaufier, and G. P. Singh, *Tetrahedron Lett.*, 1993, **34**, 1123; P. Laszlo and M. Levart, *Tetrahedron Lett.*, 1993, **34**, 1127.
[166] B. B. De, B. B. Lohray, and P. K. Dhal, *Tetrahedron Lett.*, 1993, **34**, 2371.
[167] T. Mukaiyama, K. Yorozu, T. Takai, and T. Yamada, *Chem. Lett.*, 1993, 439.
[168] K. Yorozu, T. Takai, T. Yamada, and T. Mukaiyama, *Chem. Lett.*, 1993, 1579.
[169] E. Manoury, H. A. H. Mouloud, and G. G. A. Balavoine, *Tetrahedron: Asymmetry*, 1993, **4**, 2339.
[170] M. Brunner, L. Mussmann, and D. Vogt, *SYNLETT*, 1993, 893.
[171] E. J. Corey and C. J. Helal, *Tetrahedron Lett.*, 1993, **34**, 5227.
[172] E. Bosch and J. K. Kochi, *J. Chem. Soc., Chem. Commun.*, 1993, 667.
[173] J. Forrester, R. V. H. Jones, P. N. Preston, and E. S. C. Simpson, *J. Chem. Soc., Perkin Trans. 1*, 1993, 1937.

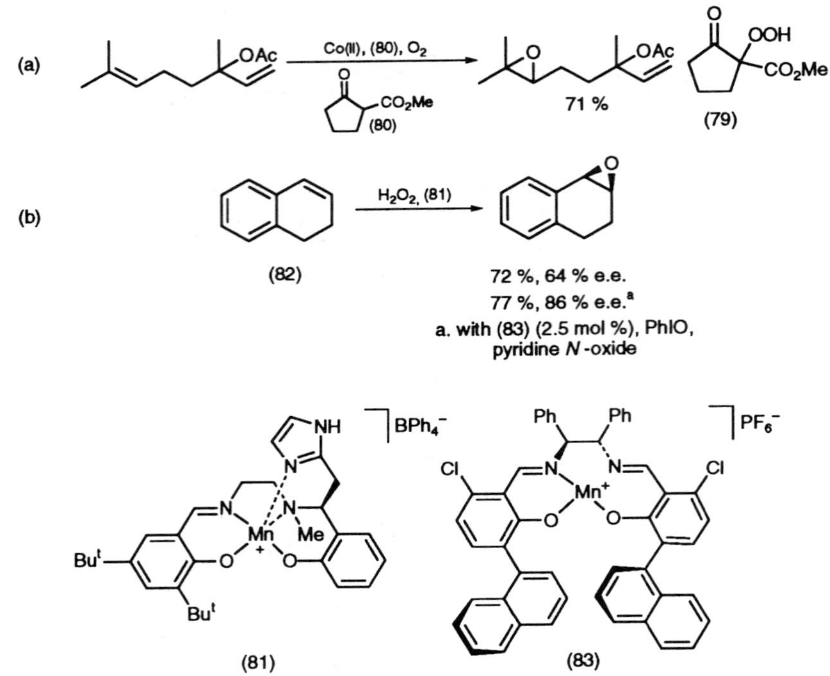

Scheme 21

by the action of bismuth(III) mandelate; olefins and alcohols are tolerated by the reagent (Scheme 23).[174]

New methods for the asymmetric aziridination of alkenes continue to be developed. Jacobsen and coworkers report the use of the copper(I) complexes of the chiral Schiff base (85) to catalyse the aziridination of alkenes with the iodinane (86) (Scheme 24a);[175] aziridination of chromene derivatives is particularly efficient. Katsuki and coworkers have also used the manganese–salen catalyst in a similar manner to achieve asymmetric aziridination with moderate selectivity in the best case.[176] Similarly, Evans *et al.* report the chiral aziridination of phenyl cinnamate with the copper complex of the bis(oxazoline) (87) (Scheme 24b).[177] *N*-hydroxy-*N*-pivaloylaniline has also been used for the aziridination of alkenes.[178]

Sulfides are oxidized selectively to sulfoxides with the peracetoxyimidic acid

[174] T. Zevaco, E. Duñach, and M. Postel, *Tetrahedron Lett.*, 1993, **34**, 2601.
[175] Z. Li, K. R. Conser, and E. N. Jacobsen, *J. Am. Chem. Soc.*, 1993, **115**, 5326.
[176] K. Noda, N. Hosoya, R. Irie, Y. Ito, and T. Katsuki, *SYNLETT*, 1993, 469.
[177] D. A. Evans, M. M. Faul, M. T. Bilodeau, B. A. Anderson, and D. M. Barnes, *J. Am. Chem. Soc.*, 1993, **115**, 5328.
[178] M. M. Pereira, P. P. O. Santos, L. V. Reis, A. M. Lobo, and S. Prabhakar, *J. Chem. Soc., Chem. Commun.*, 1993, 38.

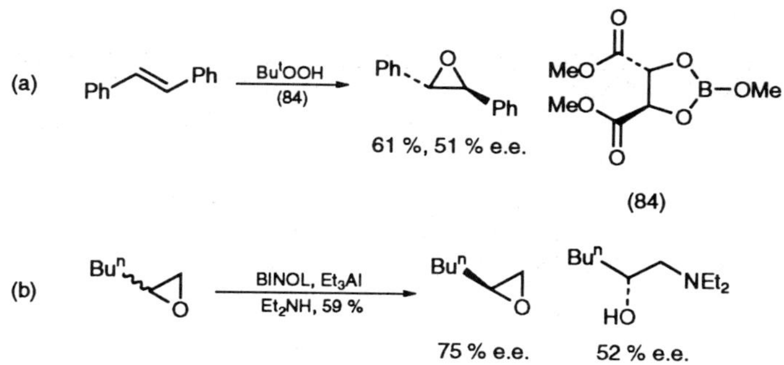

Scheme 22

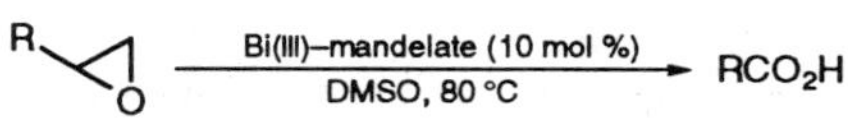

Scheme 23

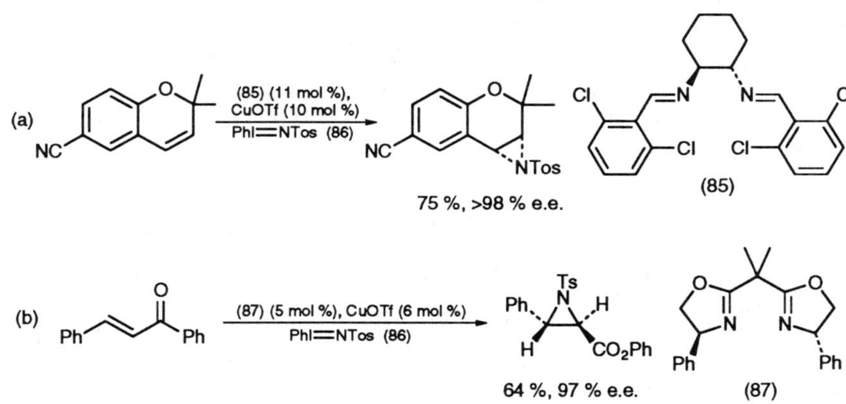

Scheme 24

generated *in situ* from acetonitrile and hydrogen peroxide;[179] a similar transformation is affected by photolysis in the presence of tetranitromethane.[180] Tetrapropylammonium perruthenate is an efficient catalyst for the oxidation of sulfides to sulfones.[181] Sulfoxides are kinetically resolved by titanium-binaphthol-mediated oxidation of one enantiomer to the sulfone.[182]

[179] P.C. Bulman Page, A.E. Graham, D. Bethell, and K.B. Park, *Synth. Commun.*, 1993, **23**, 1507.
[180] D. Ramkumar and S. Sankararaman, *Synthesis*, 1993, 1057.
[181] K.R. Guertin and A.S. Kende, *Tetrahedron Lett.*, 1993, **34**, 5369.
[182] N. Komatsu, M. Hashizume, T. Sugita, and S. Uemura, *J. Org. Chem.*, 1993, **58**, 7624.

5 Protection

Protecting group methodology has seen much study this year. The use of silyl ethers as protecting groups, especially in conjunction with oxidation reactions, has been reviewed.[183] Primary and secondary alcohols are rapidly protected as their t-butyldiphenylsilyl ethers by the action of t-butyldiphenylsilyl chloride and ammonium nitrate or perchlorate,[184] and as their trimethylsilyl ethers with hexamethyldisilazane and zinc chloride.[185] Lipshutz *et al.*[186] report the use of trimethylsilylfluorosulfonate (TMSOFS) as a cheap alternative to trimethylsilyl triflate for the protection of sterically hindered alcohols. The TMSOFS is simply prepared *in situ* by protodesilylation of allyltrimethylsilane with fluorosulfonic acid.

Several methods for the deprotection of allyl-protected amines and alcohols have appeared. Allyl glycosides are deprotected by conversion into the 2-oxopropyl glycoside by Wacker oxidation [Pd(II), Cu(I), O_2], followed by photolysis.[187] Allyl and 1-propenyl ethers are deprotected in a similar fashion [Pd(II), Cu(I), H_2O_2], but without recourse to photolysis.[188] Allyl ethers and amines are deprotected by *in situ* generated dicyclopentadienyl zirconium (Cp_2Zr);[189] tetrahydropyranyl ethers, diisopropylidene acetals, and esters (partially) are tolerated by this reagent. Allylamines are also rapidly deprotected by palladium(0) allylic substitution with *N,N'*-dimethylbarbituric acid.[190] Propargylic ethers are selectively cleaved to the corresponding alcohol with low-valent titanium ($TiCl_3$/Mg) in the presence of silyl, benzyl, and methyl ethers.[191] Benzyl-protected alcohols are selectively deprotected in the presence of t-butyldiphenylsilyl ethers by boron-trichloride–dimethyl-sulfide complex.[192] Tetrahydropyranyl ethers are efficiently made from dihydropyran with catalytic $[Ru(CH_3CN)_3(triphos)](OTf)_2$ [triphos = $CH_3C(CH_2PPh_2)_3$].[193] Tetrahydropyranyl ethers are also formed by molybdenum(VI) acetylacetonate,[194] ceric ammonium nitrate,[195] or H-Y zeolite[196] and 3,4-dihydro-2*H*-pyran. Alcohols are protected as their 2-tetrahydrofuranyl ethers by n-tetrabutylammonium-peroxydisulfate-mediated radical coupling with tetrahydrofuran.[197] Dimethoxytrityl-protected alcohols and thiols are selectively deprotected, in the same molecule, by 80% acetic acid (aq.) and silver nitrate–dithioerythritol, respectively.[198] Both 1,2 and 1,3-diols are rapidly and efficiently protected as their base-labile cyclic carbonates by reaction with triphosgene.[199] BOC-protected amides are selectively deprotected with catalytic

[183] J. Muzart, *Chem. Rev.*, 1993, **93**, 11.
[184] S. A. Hardinger and N. Wijaya, *Tetrahedron Lett.*, 1993, **34**, 3821.
[185] H. Firouzabadi and B. Karimi, *Synth. Commun.*, 1993, **23**, 1633.
[186] B. H. Lipshutz, J. Burgess-Henry, and G. P. Roth, *Tetrahedron Lett.*, 1993, **34**, 995.
[187] J. Lünning, U. Möller, N. Debski, and P. Welzel, *Tetrahedron Lett.*, 1993, **34**, 5871.
[188] H. B. Mereyala and S. Guntha, *Tetrahedron Lett.*, 1993, **34**, 6929.
[189] H. Ito, T. Taguchi, and Y. Hanzawa, *J. Org. Chem.*, 1993, **58**, 774.
[190] F. Garro-Helion, A. Merzouk, and F. Guibé, *J. Org. Chem.*, 1993, **58**, 6109.
[191] S.K. Nayak, S.M. Kadam, and A. Banerji, *SYNLETT*, 1993, 581.
[192] M.S. Congreve, E.C. Davison, M.A.M. Fuhry, A.B. Holmes, A.N. Payne, R.A. Robinson, and S.E. Ward, *SYNLETT*, 1993, 663.
[193] S. Ma and L. M. Venanzi, *Tetrahedron Lett.*, 1993, **34**, 5269.
[194] M.L. Kantam and P.L. Santhi, *Synth. Commun.*, 1993, **23**, 2225.
[195] G. Maity and S.C. Roy, *Synth. Commun.*, 1993, **23**, 1667.
[196] P. Kumar, C.U. Dinesh, R.S. Reddy, and B. Pandey, *Synthesis*, 1993, 1069.
[197] J.C. Jung, H.C. Choi, and Y.H. Kim, *Tetrahedron Lett.*, 1993, **34**, 3581.
[198] Z. Huang and S.A. Benner, *SYNLETT*, 1993, 83.
[199] R.M. Burk and M.B. Roof, *Tetrahedron Lett.*, 1993, **34**, 395.

magnesium(II) perchlorate in acetonitrile even in the presence of BOC-protected amines.[200]

Esters are often used to protect alcohols and several new methods for their synthesis and hydrolysis[201] have been described. Vedejs *et al.* report the use of tributylphosphine as a remarkable catalyst for the acetylation and benzoylation of alcohols.[202] In the latter case tributylphosphine is a more efficient catalyst than the well-documented acylation catalyst DMAP. The use of bulky amines (Pr^i_2EtN or 1,2,2,6,6-pentamethylpiperidine) with acetyl chloride ($-78\,°C \rightarrow$ r.t.) results in the selective acylation of primary alcohols in the presence of secondary alcohols (>99.6:0.4).[203] Folmer and Weinreb[204] report a potentially useful and mild one-pot method for ester synthesis involving activation of carboxylic acids with Appel's salt (88) to acyl-1,2,3-dithiazoles (89) (Scheme 25). Formate esters are efficiently made from alcohols with thionyl

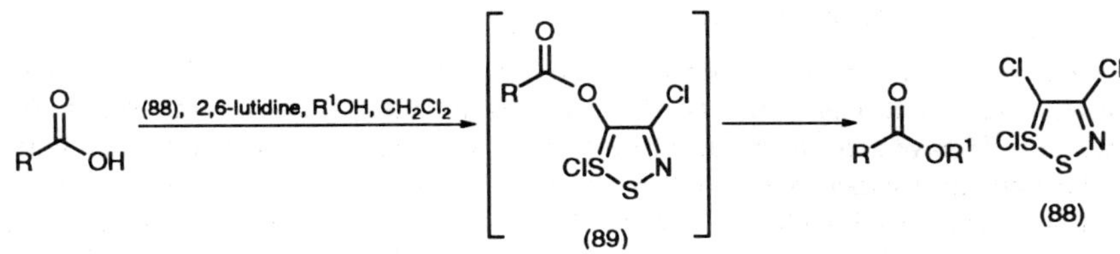

Scheme 25

chloride–DMF–lithium iodide; when potassium iodide is used the corresponding alkyl iodide is isolated.[205] Another simple procedure for the formylation of alcohols and amines by reaction with cyanomethylformate is described.[206] The deprotection of acetate and pivaloyl esters is achieved rapidly (in just a few minutes) by microwave promoted hydrolysis on alumina.[207] Similarly, t-butyldimethylsilyl ethers,[208] benzaldehyde diacetates,[209] acetates,[210] and benzyl ethers[211] are deprotected by microwave promoted hydrolysis on alumina. The procedures involve no reaction solvent and minimal work up.

Several new methods for the removal of acetal protecting groups have been reported. Thioacetals and thioketals are deprotected photochemically ($\lambda > 350$ nm) in the presence of 2,3-dichloro-5,6-dicyano-*p*-benzoquinone (DDQ); alternatively, the same transformation is achieved thermally by heating in acetonitrile under reflux.[212] Node

200 J. A. Stafford, M. F. Brackeen, D. S. Karanewsky, and N. L. Valvano, *Tetrahedron Lett.*, 1993, **34**, 7873.
201 C. J. Salomon, E. G. Mata, and O. A. Mascaretti, *Tetrahedron*, 1993, **49**, 3691.
202 E. Vedejs and S. T. Diver, *J. Am. Chem. Soc.*, 1993, **115**, 3358; E. Vedejs, N. S. Bennet, L. M. Conn, S. T. Diver, M. Gingras, S. Lin, P. A. Oliver, and M. J. Peterson, *J. Org. Chem.*, 1993, **58**, 7286.
203 K. Ishihara, H. Kurihara, and H. Yamamoto, *J. Org. Chem.*, 1993, **58**, 3791.
204 J. J. Folmer and S. M. Weinreb, *Tetrahedron Lett.*, 1993, **34**, 2737.
205 I. Fernández, B. Garcia, S. Muñoz, J. R. Pedro, and R. de la Salud, *SYNLETT*, 1993, 489.
206 J. Deutsch and H.-J. Niclas, *Synth. Commun.*, 1993, **23**, 1561.
207 S. V. Ley and D. M. Mynett, *SYNLETT*, 1993, 793.
208 R. S. Varma, J. B. Lamture, and M. Varma, *Tetrahedron Lett.*, 1993, **34**, 3029.
209 R. S. Varma, A. K. Chatterjee, and M. Varma, *Tetrahedron Lett.*, 1993, **34**, 3207.
210 R. S. Varma, M. Varma, and A. K. Chatterjee, *J. Chem. Soc., Perkin Trans. 1*, 1993, 999.
211 R. S. Varma, A. K. Chatterjee, and M. Varma, *Tetrahedron Lett.*, 1993, **34**, 4603.
212 L. Mathew and S. Sankararaman, *J. Org. Chem.*, 1993, **58**, 7576.

et al.[213] have developed a reagent system (silver nitrate–iodine) for the rapid and mild deprotection of both dithioacetal and monothioacetal groups. Dithioacetals are efficiently hydrolysed by the electrophilic fluorinating reagent *N*-fluoro-2,4,6-trimethylpyridinium triflate in aqueous THF.[214] Roskamp and Ford report an improvement for the deprotection of acetals with tin(II) chloride; the rate of hydrolysis is greatly increased by simply adding catalytic naphthalene or, remarkably, C_{60}.[215]

On occasions in a synthesis it is necessary to change protecting groups to circumvent incompatibility problems. Efficient one-pot procedures are desirable for this purpose. To this end, benzyl-protected alcohols are efficiently converted into their corresponding acetates by the action of catalytic tin(II) bromide and acetyl bromide.[216] Similarly, amine protecting groups may be changed; N^{α}-fluorenylmethoxycarbonyl (Fmoc) groups are converted into an N^{α}-benzyloxycarbonyl (Z) group by potassium fluoride and *N*-benzyloxycarbonyl-5-norbornene-2,3-dicarboximide (BCN).[217]

6 Miscellaneous Preparations

Chiral acylations are normally best carried out by the use of enzymes. However, many workers are developing reagents and catalysts to effect this process. Evans and coworkers[218] have used the imide (90) as a chiral acylating agent to resolve secondary alcohols kinetically (Scheme 26a). Kinetic selectivity of 20–30 : 1 for the *R* enantiomer of the magnesium alkoxide is observed. Katsuki and co-workers[219] have used the chiral benzimidazole (91) for the chiral acylation of lithium amide–enolates (Scheme 26b).

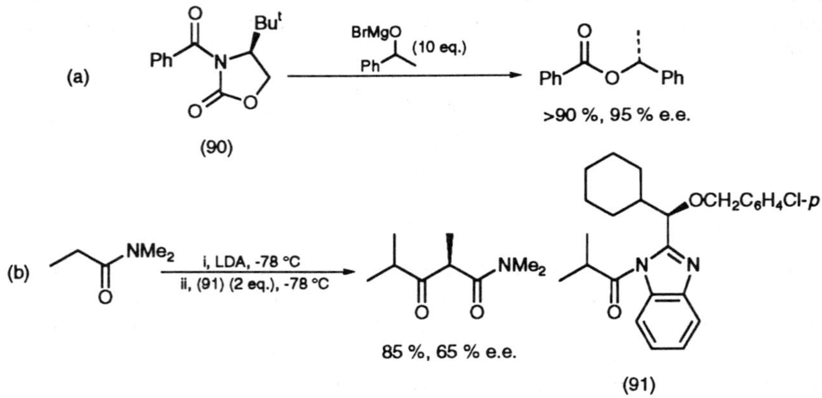

Scheme 26

[213] K. Nishide, K. Yokota, D. Nakamura, T. Sumiya, M. Node, M. Ueda, and K. Fuji, *Tetrahedron Lett.*, 1993, **34**, 3425.
[214] A. S. Kiselyov, L. Strekowski, and V. V. Semenov, *Tetrahedron Lett.*, 1993, **49**, 2151.
[215] K. L. Ford and E. J. Roskamp, *J. Org. Chem.*, 1993, **58**, 4142.
[216] T. Oriyama, M. Kimura, M. Oda, and G. Koga, *SYNLETT*, 1993, 437.
[217] W.-R. Li, J. Jiang, and M. M. Joullié, *SYNLETT*, 1993, 362.
[218] D. A. Evans, J. C. Anderson, and M. K. Taylor, *Tetrahedron Lett.*, 1993, **34**, 5563.
[219] M. Ogata, T. Yoshimura, H. Fujii, Y. Ito, and T. Katsuki, *SYNLETT*, 1993, 728.

Several reports describe the development of reagents that are chiral proton equivalents for the protonation of prochiral nucleophiles. Amines have rarely been used as the source of chiral protons for the enantioselective protonation of prochiral enolates [(94) → (95)] (Scheme 27). However, the groups of Fuji[220] and Koga[221] have reported the use of the piperazine (92) and the triamine (93) respectively as reagents for the stereoselective protonation of lithium enolates.

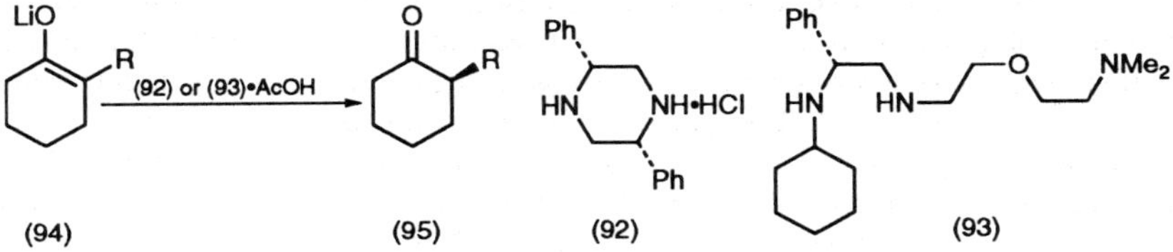

(94) (95) (92) (93)

Scheme 27

Davies *et al.* have continued to illustrate the usefulness of the enantioselective conjugate addition of (α-methylbenzyl)benzylamine to t-butyl cinnamate by application to the syntheses of α-methyl-β-phenylalanines,[222] the taxol-side chain,[223] and *cis*-pentacin.[224]

Perhaps just as important as asymmetric synthetic methods are the methods for the exact determination of enantiomeric excess. Accordingly, Feringa and coworkers have developed the phosphoric acid chloride (96) for the determination of the optical purity of chiral alcohols and amines by ^{31}P NMR spectroscopy.[225] An interesting account of the accurate determination of extremely high enantiomeric purity appears in a paper by Rautenstrauch *et al.*[226] The principles are applied to the determination of the enantiomeric purity (>99.5% e.e.) of several commercially available samples of camphor.

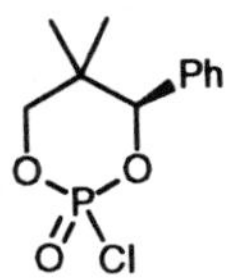

(96)

The year has seen several reports of novel procedures for the hydroboration of alkenes. Evans *et al.* report the use of achiral lanthanoid complexes as catalysts for the hydroboration of olefins with catecholborane.[227] The Lewis acidity of the boron atom is thought to be dramatically increased by coordination of the lanthanide (*e.g.* samarium) to a catechol oxygen atom. The chiral rhodium complex (97) is also an excellent catalyst for the hydroboration of substituted styrenes [(98) → (99)] (Scheme

220 K. Fuji, K. Tanaka, and H. Miyamoto, *Tetrahedron: Asymmetry*, 1993, **4**, 247.
221 T. Yasukata and K. Koga, *Tetrahedron: Asymmetry*, 1993, **4**, 35.
222 S. G. Davies, N. M. Garrido, O. Ichihara, and I. A. S. Walters, *J. Chem. Soc., Chem. Commun.*, 1993, 1153.
223 M. E. Bunnage, S. G. Davies, and C. J. Goodwin, *J. Chem. Soc., Perkin Trans. 1*, 1993, 1375.
224 S. G. Davies, O. Ichihara, and I. A. S. Walters, *SYNLETT*, 1993, 461.
225 R. Hulst, R. W. J. Zijlstra, B. L. Feringa, N. K. de Vries, W. ten Hoeve, and H. Wynberg, *Tetrahedron Lett.*, 1993, **34**, 1339.
226 V. Rautenstrauch, M. Lindström, B. Bourdin, J. Currie, and E. Oliveros, *Helv. Chim. Acta*, 1993, **76**, 607.
227 D. A. Evans, A. R. Muci, and R. Stürmer, *J. Org. Chem.*, 1993, **58**, 5307.

28a).[228] A process that is equivalent to the hydroboration of alkenes involves selenoalkoxylation and subsequent removal of the seleno group. Déziel *et al.*[229] have introduced a chiral version of this procedure. The C_2 symmetric phenylselenyl triflate (100) reacts with alkenes in the presence of an alcohol to give *anti* selenyl ethers (101), which can be reduced to give chiral protected alcohols (Scheme 28b). Alkenes are hydrated in an anti-Markovnikov fashion by titanium(III) borohydride species or zinc borohydride;[230] the lack of both stereospecificity and need of oxidant seems to rule out a hydroboration mechanism.[231] The hydroboration–oxidation of enamines by Ipc_2BH yields β-amino alcohols with high enantioselectivity (e.e. 50–86%).[232]

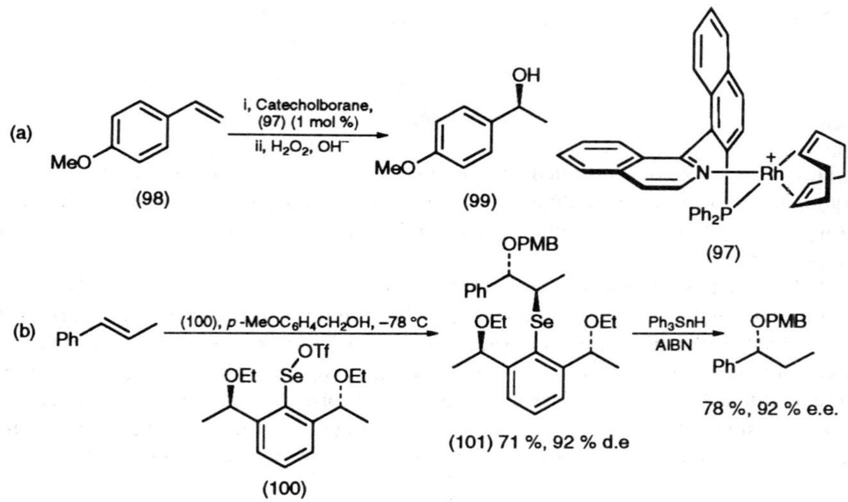

Scheme 28

Microwave-mediated synthesis has received much attention this year (see also section 5). Amides are prepared simply by heating (5 min) an equimolar mixture of a secondary amine and carboxylic acid in a microwave oven.[233] Conventional heating (160–180 °C) of a similar mixture requires somewhat longer reaction times.[234] Similarly, esters are formed from carboxylic acids, alcohols, and an acid catalyst under microwave irradiation.[235] Anhydrides are synthesized from carboxylic acids, isopropenyl acetate, and an acid catalyst.[236] β-Aminoesters and β-lactams are also prepared by montmorillonite-catalysed coupling of silyl ketene acetals and imines by the action of microwave radiation.[237]

228 J. M. Brown, D. I. Hulmes, and T. P. Layzell, *J. Chem. Soc., Chem. Commun.*, 1993, 1673.
229 R. Déziel, S. Goulet, L. Grenier, J. Bordeleau, and J. Bernier, *J. Org. Chem.*, 1993, **58**, 3619.
230 B. C. Ranu, R. Chakraborty, and M. Saha, *Tetrahedron Lett.*, 1993, **34**, 4659.
231 K. S. Ravi Kumar, S. Baskaran, and S. Chandrasekaran, *Tetrahedron Lett.*, 1993, **34**, 171.
232 G. B. Fisher, C. T. Goralski, L. W. Nicholson, and B. Singaram, *Tetrahedron Lett.*, 1993, **34**, 7693.
233 M. P. Vázquez-Tato, *SYNLETT*, 1993, 506.
234 B. S. Jursic and Z. Zdravkovski, *Synth Commun.*, 1993, **23**, 2761.
235 A. Loupy, A. Petit, M. Ramdani, C. Yvanaeff, M. Majdoub, B. Labiad, and D. Villemin, *Can. J. Chem.*, 1993, **71**, 90.
236 D. Villemin, B. Labiad, and A. Loupy, *Synth. Commun.*, 1993, **23**, 419.
237 F. Texier-Boullet, R. Latouche, and J. Hamelin, *Tetrahedron Lett.*, 1993, **34**, 2123.

Some interesting functional group transformations described this year include the conversion of aldehydes into nitriles by the one-pot reaction of the corresponding *N,N*-dimethylhydrazone with magnesium monoperoxyphthalate.[238] Triphosgene in combination with triphenylphosphine provides a mild reagent for the conversion of primary and secondary alcohols into alkyl chlorides.[239] Miller *et al.* report a convenient conversion of carboxylic acids into alk-1-enes of one less carbon atom by a palladium-catalysed decarbonylation–dehydration of the mixed-acid–acetic-acid anhydride.[240] Hexamethylene tetraamine tribromide provides a new mild reagent for electrophilic bromination of aromatic compounds.[241]

The synthesis of chiral organofluorine compounds[242] is currently receiving considerable attention. Davis' group has used the N–F camphorsultam (102) as a chiral electrophilic fluorinating agent to synthesize α-fluoroketones from sodium enolates [(103) → (104)] (Scheme 29).[243] The new electrophilic fluorinating agent (105) has seen several uses this year, including the fluoro-destannylation of vinylstannanes,[244] and the fluorination of alkenes, arenes, and carbanions.[245] The chiral quaternary ammonium fluoride (106) derived from quinine has been used to catalyse the aldol reaction of silyl enol ethers and aldehydes with high enantioselectivity.[246] *N,N,N*-trimethyl-1-adamantylammonium fluoride[247] and proton-sponge fluoride[248] provide completely anhydrous sources of fluoride ion.

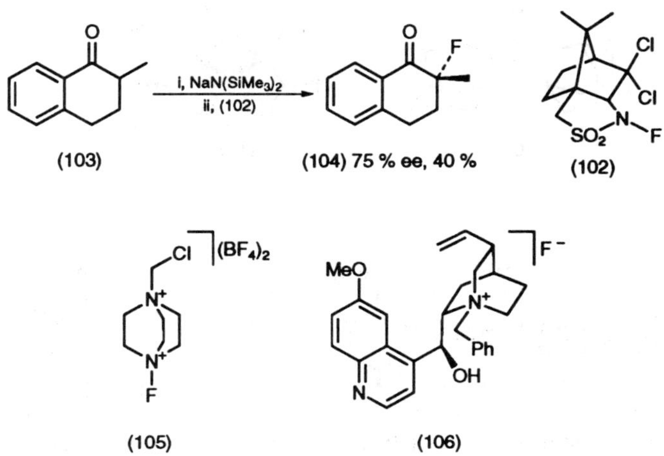

Scheme 29

[238] R. Fernández, C. Gasch, J.-M. Lassaletta, J.-M. Llera, and J. Vázquez, *Tetrahedron Lett.*, 1993, **34**, 141.
[239] I. A. Rivero, R. Somanathan, and L. H. Hellberg, *Synth. Commun.*, 1993, **23**, 711.
[240] J. A. Miller, J. A. Nelson, and M. P. Byrne, *J. Org. Chem.*, 1993, **58**, 18.
[241] S. C. Bisarya and R. Rao, *Synth. Commun.*, 1993, **23**, 779.
[242] G. Resnati, *Tetrahedron*, 1993, **49**, 9385.
[243] F. A. Davis, P. Zhou, and C. K. Murphy, *Tetrahedron Lett.*, 1993, **34**, 3971.
[244] D. P. Matthews, S. C. Miller, E. T. Jarvi, J. S. Sabol, and J. R. McCarthy, *Tetrahedron Lett.*, 1993, **34**, 3057.
[245] G. S. Lal, *J. Org. Chem.*, 1993, **58**, 2791.
[246] A. Ando, T. Miura, T. Tatematsu, and T. Shioiri, *Tetrahedron Lett.*, 1993, **34**, 1507.
[247] K. M. Harmon, B. A. Southworth, K. E. Wilson, and P. K. Keefer, *J. Org. Chem.*, 1993, **58**, 7294.
[248] R. D. Chambers, E. F. Holmes, S. R. Korn, and G. Sandford, *J. Chem. Soc., Chem. Commun.*, 1993, 855.

Amide bases continue to be the topic of many important papers. Collum and coworkers have found that the extremely hindered base lithium bis(2-adamantyl)amide (107) is an excellent reagent for the generation of *E* enolates from ketones (*E*: *Z* 50:1)(*c.f.*, LDA, *E*: *Z B* 2.5:1).[249] Williard has published the *X*-ray structure of an LDA–THF complex.[250] The crystalline aggregate, prepared from LDA (from diisopropylamine, lithium metal, and styrene in ether) and THF, is a bis-solvated dimer. Such studies are undoubtedly of help to those researchers designing homochiral lithium amide bases. Indeed the *X*-ray structure of the chiral lithium amide (108) has recently been reported.[251] Koga and coworkers have found that the chiral lithium amides (109) and (110), the latter containing a fluorinated alkyl group, induce high enantioselectivity in kinetic deprotonation of 4-substituted cyclohexanones (Scheme 30).[252] They have also shown that in certain cases the amide bases can further control the alkylation of chiral enolates.[253] Bunn and Simpkins report that the enantioselectivity of chiral-amide-base-mediated deprotonation under external quench conditions is improved by the addition of lithium chloride; this now allows direct reaction of the chiral enolate with electrophiles other than silyl chlorides, in high enantioselectivity.[254] Milne and Murphy have used the dilithium salt of norephedrine to effect enantioselective deprotonation of prochiral epoxides, thereby providing allylic alcohols.[255]

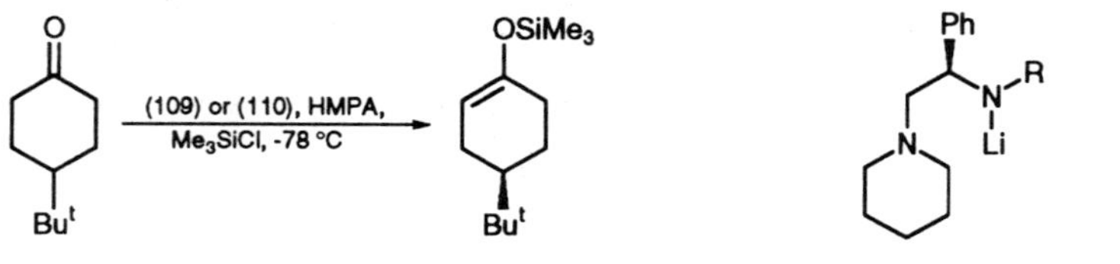

Scheme 30

[249] K. Sakuma, J. H. Gilchrist, F. E. Romesberg, C. E. Cajthami, and D. B. Collum, *Tetrahedron Lett.*, 1993, **34**, 5213.
[250] P. G. Williard and J. M. Salvino, *J. Org. Chem.*, 1993, **58**, 1.
[251] A. J. Edwards, S. Hockey, F. S. Mair, P. R. Raithby, R. Snaith, and N. S. Simpkins, *J. Org. Chem.*, 1993, **58**, 6942.
[252] K. Aoki, H. Noguchi, K. Tomioka, and K. Koga, *Tetrahedron Lett.*, 1993, **34**, 5105.
[253] Y. Hasegawa, H. Kawasaki, and K. Koga, *Tetrahedron Lett.*, 1993, **34**, 1963.
[254] B. J. Bunn and N. S. Simpkins, *J. Org. Chem.*, 1993, **58**, 533.
[255] D. Milne and P. J. Murphy, *J. Chem. Soc., Chem. Commun.*, 1993, 884.

Beak and Du[256] report the stereoselective alkylation and silylation [(111) → (112)] of carbanions in the presence of (−)-sparteine (Scheme 31). This method for asymmetric carbon–carbon bond formation is unlike that involving chiral lithium amide bases since the chiral catalyst is introduced only after deprotonation has occurred. The selectivity is probably a result of interconversion of the diastereoisomeric substrate–sparteine complexes since the organolithium is configurationally unstable.

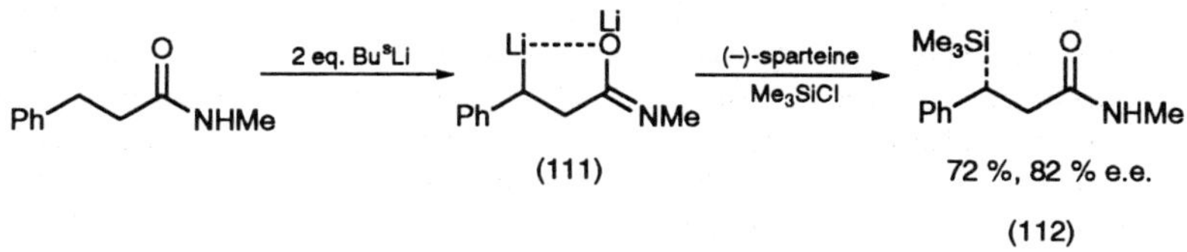

Scheme 31

Several reports have appeared describing efficient procedures for the resolution of the often-used auxiliary 1,1′-binaphthalene-2,2′-diol, *via* its *N*-[(*S*)-α-methylbenzylamine]thiophosphoramidate[257] and *O*-(−)-menthyl phosphate.[258] Finally, Fritz-Langhals[259] reports a useful way to separate diastereoisomers. Conventionally this is carried out by fractional recrystallization or chromatography but in this study the racemic acid (113) is resolved by fractional distillation of the chiral amides (114) and (115), using a spinning-band column (Scheme 32). The boiling points of the amides (114) and (115) differ by 7 K which allows for efficient separation of optically pure material. It is highly likely that this technique will find many applications.

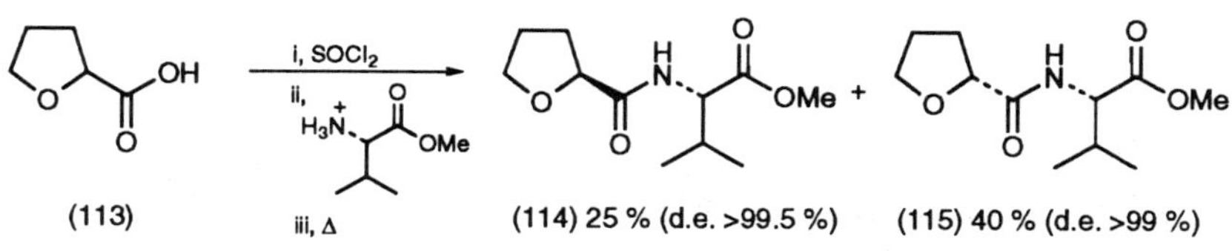

Scheme 32

[256] P. Beak and H. Du, *J. Am. Chem. Soc.*, 1993, **115**, 2516.
[257] D. Fabbri, G. Delogu, and O. De Lucchi, *J. Org. Chem.*, 1993, **58**, 1748.
[258] J.-M. Brunel and G. Buono, *J. Org. Chem.*, 1993, **58**, 7313.
[259] E. Fritz-Langhals, *Angew. Chem., Int. Ed. Engl.*, 1993, **32**, 753.

10 Enzyme Chemistry

By A. G. SUTHERLAND

School of Applied Chemistry, University of North London, London, N7 8DB, UK

1 Introduction

As some biotransformation methods, particularly the kinetic resolution of carboxylate esters by hydrolysis or condensation, are increasingly thought of as a standard weapon in a synthetic chemist's armoury, there has been a marked tendency for specialists in the enzyme chemistry field to move on to new areas. Thus the past year has seen an increase in the investigation of areas such as amide and epoxide hydrolysis and the use of oxidoreductase systems in resolution processes. Accordingly, in line with the trend of last year,[1] this report will deal in greater detail with these emerging methods rather than attempting to reflect the relative volume of papers published on each of the topics below.

The rate of publication on enzymes in synthesis has now made the publication of a full review of the subject a virtual impossibility, but a large symposium-in-print goes much of the way to delineating the state of the art.[2] A review of the use of enzymes in the synthesis of chiral drugs provides a different viewpoint.[3] Two doyens of the field have written perspectives on their recent work: thus Jones has discussed his seminal research on the prediction and control of enzyme specificity,[4] while Wong has summarized his syntheses of aza sugars using several enzyme systems.[5]

2 Hydrolysis and Condensation Reactions

Alcohols, Carboxylic Acids, and Esters.—The resolution of racemic, typically secondary, alcohols (or their esters) by acylation (or by hydrolysis or transesterification) forms the bulk of the current literature on biotransformations and, as a relatively routine technique, will only be covered briefly here. The transesterification methods have been reviewed[6] and compared with similar chemical methods.[7]

Secondary alcohol resolutions are at their most powerful when used to resolve small, highly functionalized, and readily available starting materials such as the butenol

1 A. G. Sutherland, *Ann. Rep. Prog. Chem., Sect. B, Org. Chem.*, 1992, **89**, 281.
2 *Tetrahedron: Asymmetry*, 1993, **4**, 757–1396.
3 A. L. Margolin, *Enzyme Microb. Technol.*, 1993, **15**, 266.
4 J. B. Jones, *Can. J. Chem.*, 1993, **71**, 1273.
5 G. C. Look, C. H. Fotsch, and C.-H. Wong, *Acc. Chem. Res.*, 1993, **26**, 182.
6 E. Santaniello, P. Ferraboschi, and P. Grisenti, *Enzyme Microb. Technol.*, 1993, **15**, 367.
7 J. Otera, *Chem. Rev.*, 1993, **93**, 1449.

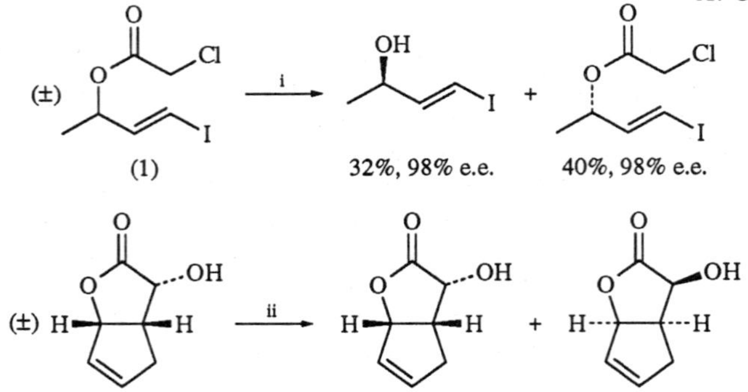

(1) 32%, 98% e.e. 40%, 98% e.e.

(2) > 95% e.e. (40% conversion) > 95% e.e. (60% conversion)

Reagents: i, *Pseudomonas* sp. lipase, toluene, H_2O; ii, *Pseuodomonas fluorescens* lipase, vinyl acetate

Scheme 1

derivative (1)[8] or the lactone (2)[9] (Scheme 1), the last of which has been used to synthesize the anti-HIV carbocyclic nucleoside precursor carbovir[9] and also the key hydroxy-lactone moiety of the HMG CoA reductase inhibitors based on compactin.[10] Other notable and extremely efficient resolutions include that of a general inositol precursor[11] and the chiral auxiliary *trans*-2-(α-cumyl)cyclohexanol.[12] Dasrhadi and O'Hagan have explored the resolution of diarylmethanols and found that the extent of discrimination between relatively similar rings (*e.g.* 2,5-difluorophenyl *versus* phenyl) can be surprisingly high.[13]

The past year has seen more examples of successful resolutions of tertiary alcohols coming to light. Thus, hydrolysis at the more hindered carbonyl of the oxalate (3) gave the resulting norbornanol in good enantiomeric excess (Scheme 2),[14] while the enantiomers of bridgehead alcohols in a series of [4.1.0]heptanes have been resolved by hydrolysis of the corresponding chloroacetates.[15]

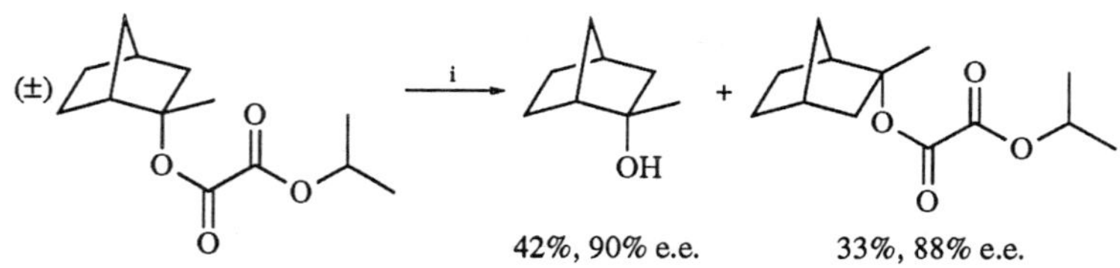

(3) 42%, 90% e.e. 33%, 88% e.e.

Reagents: i, porcine pancreatic lipase, Bu^tOH, H_2O

Scheme 2

[8] M. Bazinger, G. J. Griffiths, and J. F. McGarrity, *Tetrahedron: Asymmetry*, 1993, **4**, 723.
[9] R. A. MacKeith, R. McCague, H. F. Olivo, C. F. Palmer, and S. M. Roberts, *J. Chem. Soc., Perkin Trans. 1*, 1993, 313.
[10] R. McCague, H. F. Olivo, and S. M. Roberts, *Tetrahedron Lett.*, 1993, **34**, 3785.
[11] L. Ling and S. Ozaki, *Tetrahedron Lett.*, 1993, **34**, 2501.
[12] D. L. Comins and J. M. Salvador, *Tetrahedron Lett.*, 1993, **34**, 801.
[13] L. Dasradhi and D. O'Hagan, *Bioorg. Med. Chem. Lett.*, 1993, **3**, 1655.
[14] I. Brackenridge, R. McCague, S. M. Roberts, and N. J. Turner, *J. Chem. Soc., Perkin Trans. 1*, 1993, 1093.
[15] J. P. Barnier, L. Blanco, G. Rousseau, E. Guibe-Jampel, and I. Fresse, *J. Org. Chem.*, 1993, **58**, 1570.

The desymmetrization of *meso* diols (the '*meso* trick') by lipase-catalyzed acylation remains a popular approach.[16,17] This technique has been extended by Uguen and coworkers to a tetraol (4) with two-fold *meso* symmetry.[18] Acylation gave a mixture of C_2-symmetrical and *meso* diacetates (Scheme 3), which, after conversion into the corresponding bis(phenylsulfides), were separated to give the chiral product (5) as one enantiomer. An elegant exploitation of the *meso* trick in combination with a kinetic resolution has been used to deracemize and dediastereoisomerize the mixture of isomers of hexan-2,4-diol (Scheme 3).[19] It was shown that acylation of these substrates only occurs at the *S* alcohol positions so that Mitsonobu inversion of the crude mixture, followed by hydrolysis, gives only the *S*,*S* enantiomer.

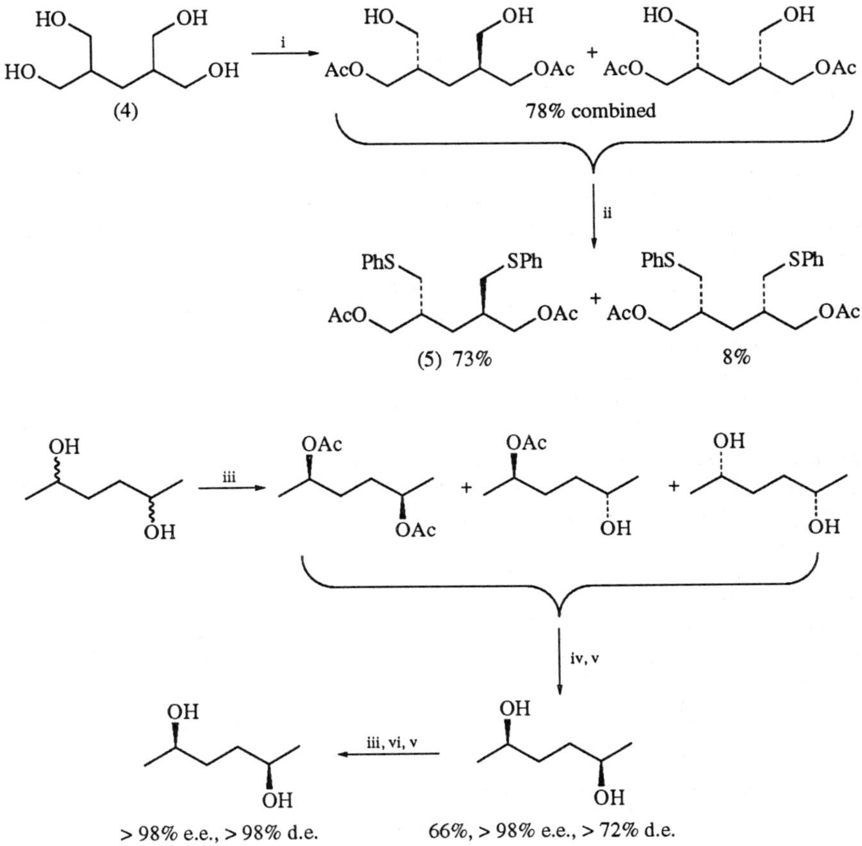

Reagents: i, *Pseudomonas fluorescens* lipase, vinyl acetate; ii, PhSSPh, PBu_3, CH_2Cl_2; iii, *Pseudomonas* sp. lipase, vinyl acetate, Bu^{t}OMe; iv, DEAD, *p*-nitrobenzoic acid, PPh_3; v, LiOH; vi, separation

Scheme 3

16 S. Takano, M. Moriya, Y. Higashi, and K. Ogasawara, *J. Chem. Soc., Chem. Commun.*, 1993, 177.
17 N. Toyooka, A. Nishino, and T. Momose, *Tetrahedron Lett.*, 1993, **34**, 4539.
18 P. Breuilles, T. Schmittberger, and D. Uguen, *Tetrahedron Lett.*, 1993, **34**, 4205.
19 M.-J. Kim and I. S. Lee, *SYNLETT*, 1993, 767.

Over recent years lipases have become popular as catalysts for regioselective ester hydrolysis (or condensation) and further examples of this use in the carbohydrate[20,21] and phospholipid[22,23] fields continue to be reported. This methodology has now been extended to polyhydroxybenzopyran,[24] benzopyranone,[25] and other aromatic[26] systems with considerable success.

The kinetic resolution of racemic carboxylic acids, largely by hydrolysis of a simple alkyl ester, also maintains its popularity – notably in the resolution of non-natural amino acids.[27,28] A particularly topical resolution is the transesterification of methyl *trans*-β-phenylglycidate under *Mucor miehei* lipase catalysis. Both enantiomerically pure products can be converted enantioconvergently into the taxol C-13 side chain (6) (Scheme 4).[29]

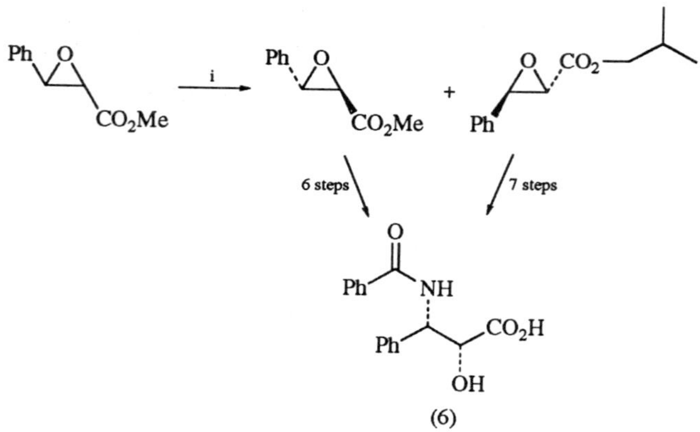

Reagents: i, *Mucor miehei* lipase, Bu^{i}OH, hexane

Scheme 4

A striking example of the use of the *meso* trick in the context of complex carboxylic acids has been provided by the direct, enantioselective synthesis of the calcium antagonist (7) (Scheme 5).[30]

Pig liver esterase has been shown to perform diastereoselective hydrolyses of arylidene and dialkylidene malonate diesters.[31,32] In the examples studied, reaction tended to occur at the *Z* position ester with moderate selectivity, *e.g.* (8) (R^1 = n-

[20] G. B. Oguntimein, H. Erdmann, and R. D. Schmid, *Biotechnol. Lett.*, 1993, **15**, 175.
[21] R. Lopez, C. Perez, A. Fernandez-Mayoralas, and S. Conde, *J. Carbohydr. Chem.*, 1993, **12**, 165.
[22] G. Lin, F.-C. Wu, and S.-H. Liu, *Tetrahedron Lett.*, 1993, **34**, 1959.
[23] T. Morimoto, N. Murakami, A. Nagatsu, and J. Sakakibara, *Tetrahedron Lett.*, 1993, **34**, 2487.
[24] D. Lambusta, G. Nicolosi, A. Patti, and M. Piattelli, *Synthesis*, 1993, 1155.
[25] V. S. Parmar, A. K. Prasad, N. K. Sharma, A. Vardhan, H. N. Pati, S. K. Sharma, and K. S. Bisht, *J. Chem. Soc., Chem. Commun.*, 1993, 27.
[26] G. Nicolosi, M. Piattelli, and C. Sanfilippo, *Tetrahedron*, 1993, **49**, 3143.
[27] B. Nielsen, H. Fisker, B. Bjarke, U. Madsen, D. R. Curtis, P. Krogsgaard-Larsen, and J. J. Hansen, *Bioorg. Med. Chem. Lett.*, 1993, **3**, 107.
[28] B. Imperiali, T. J. Prins, and S. L. Fisher, *J. Org. Chem.*, 1993, **58**, 1613.
[29] D.-M. Gou, Y.-C. Liu, and C. S. Chen, *J. Org. Chem.*, 1993, **58**, 1287.
[30] T. Adachi, M. Ishii, Y. Ohta, T. Ota, T. Ogawa, and K. Hanada, *Tetrahedron: Asymmetry*, 1993, **4**, 2061.
[31] T. Schirmeister and H.-H. Otto, *J. Org. Chem.*, 1993, **58**, 4819.
[32] T. Schirmeister and H.-H. Otto, *Angew. Chem., Int. Ed. Engl.*, 1993, **32**, 572.

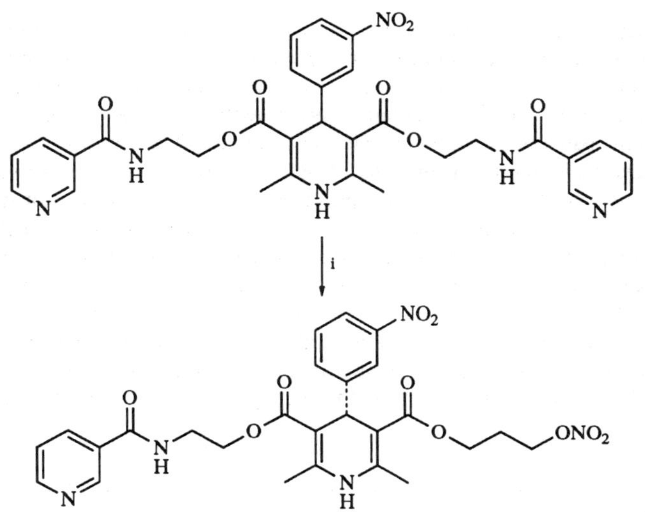

(7) 77%, > 99% e.e.

Reagents: i, *Aspergillus melleus* protease, 3-nitrooxypropan-1-ol, H_2O

Scheme 5

C_5H_{11}, $R^2 = CH_3$), 36% de,[31] to complete selectivity, *e.g.* (8) (R^1 = Ph, R^2 = H) and (9).[31,32]

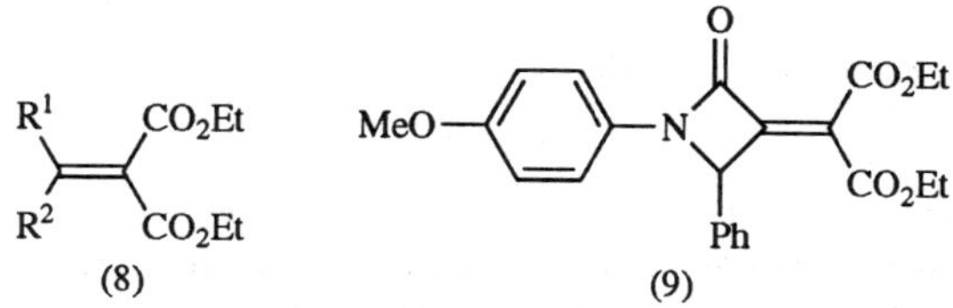

(8) (9)

Amines, Amides, and Lactams.—The past year has seen a considerable upsurge in interest in the enzyme-catalysed hydrolysis and condensation reactions of amides. An example of this, which could prove to be of some significance, is the report by Sih and coworkers of the lipase-catalysed methanolysis of the β-lactam (10) (Scheme 6), which provides direct access to the C-13 side chain of taxol (6).[33]

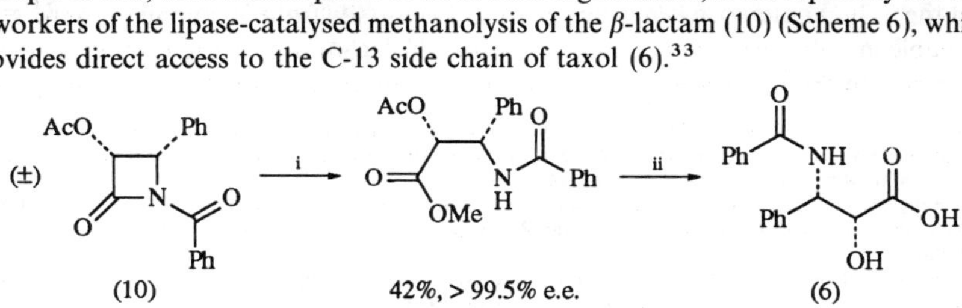

(10) 42%, > 99.5% e.e. (6)

Reagents: i, *Pseudomonas* sp. lipase, Bu^{t}OMe, MeOH; ii, NaOH(aq.)

Scheme 6

[33] R. Brieva, J. Z. Crich, and C. J. Sih, *J. Org. Chem.*, 1993, **58**, 1068.

The penicillin acylase from *Escherichia coli* has emerged in the important new application of the kinetic resolution of racemic amines *via* the hydrolysis of the corresponding phenylacetamides. This protocol has been used to synthesize both β-[34] and γ-amino acids[35] in high enantiomeric purity. The refinement of using *p*-hydroxyphenylacetamides in these reactions, which often provides an increase in enantioselectivity, also looks promising.[36] In the opposite reaction direction, Sheldon and coworkers have demonstrated that lipases can catalyse the conversion of carboxylic esters into the corresponding primary amides (in the presence of ammonia),[37] and have further observed that this process can display a higher enantioselectivity than the analogous aqueous hydrolysis (Scheme 7).[37] Gotor and colleagues have shown that this type of procedure can be extended to more functionalized amine systems,[38] while it has also been demonstrated that the protease 'alcalase' can catalyse the conversion of a peptide C-terminal ester into the analogous primary amide in a similar process.[39]

Proteases have seen further use in the incorporation of (protected) amino acid

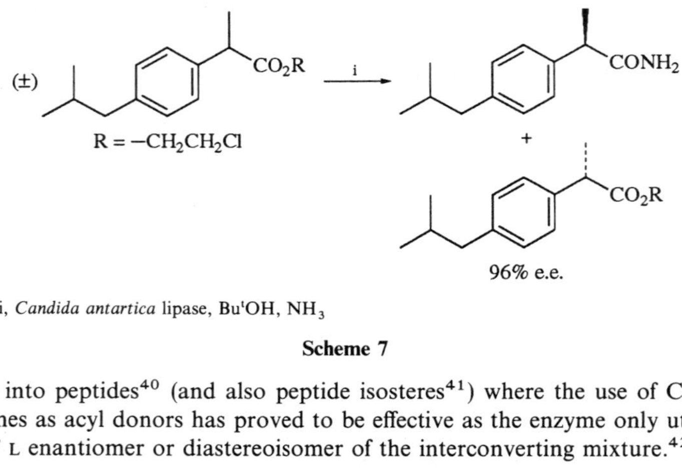

Reagents: i, *Candida antartica* lipase, Bu^{t}OH, NH_3

Scheme 7

residues into peptides[40] (and also peptide isosteres[41]) where the use of C-terminal oxazolones as acyl donors has proved to be effective as the enzyme only utilizes the 'natural' L enantiomer or diastereoisomer of the interconverting mixture.[42]

Epoxides.—The use of epoxide hydrolases to provide enantiomerically pure epoxides or *vicinal* diols has always seemed attractive. This approach has been limited by the fact that the well-characterized enzymes in this class, usually from mammalian sources, are available in only very small quantities (although elegant uses of these systems are still appearing in the literature).[43,44] However this situation is now likely to change as a

34 V. A. Soloshonok, Y. K. Svedas, V. P. Kukhar, A. G. Kirilenko, A. V. Rybakova, V. A. Soldenko, N. A. Fokina, O. V. Kogut, I. Yu. Galaev, E. V. Kozlova, I. P. Shishkina, and S. V. Galushko, *SYNLETT*, 1993, 339.
35 A. L. Margolin, *Tetrahedron Lett.*, 1993, **34**, 1239.
36 A. Guy, A. Dumant, and P. Sziraky, *Bioorg. Med. Chem. Lett.*, 1993, **3**, 1041.
37 M. C. de Zoete, A. C. Kock-van Dalen, F. van Rantwijk, and R. A. Sheldon, *J. Chem. Soc., Chem. Commun.*, 1993, 1831.
38 S. Puertas, R. Brieva, F. Rebolledo, and Y. Gotor, *Tetrahedron*, 1993, **49**, 4007.
39 S.-T. Chen, M.-K. Jang, and K.-T. Wang, *Synthesis*, 1993, 858.
40 S.-T. Chen, C.-C. Tu, and K.-T. Wang, *Bioorg. Med. Chem. Lett.*, 1993, **3**, 539.
41 M. Schuster, B. Munoz, W. Yuan, and C.-H. Wong, *Tetrahedron Lett.*, 1993, **34**, 1247.
42 B. K. Hwang, Q.-M. Gu, and C. J. Sih, *J. Am. Chem. Soc.*, 1993, **115**, 7912.
43 B. Borhan, J. Nourooz-Zadeh, T. Uemetsu, B. D. Hammock, and M. J. Kurth, *Tetrahedron*, 1993, **49**, 2601.
44 P. Barili, G. Berti, and E. Mastrorilli, *Tetrahedron*, 1993, **49**, 6263.

number of microbial sources of these enzymes have come to light in the past year. While some of these systems, such as those from *Saccharomyces cerevisiae*[45] and *Rhodococcus* sp.,[46] have yet to be shown to have potential in enantioselective reactions others, from *Aspergillus niger*[47] and *Beauveria sulfurescens*,[48] have considerable promise in this context. The reaction of the last two systems with racemic styrene oxide is of particular interest (Scheme 8).[48] Thus *A. niger* hydrolyses the *R* epoxide selectively to give the *R* diol, while in contrast *B. sulfurescens* hydrolyses the *S* epoxide, but with inversion of configuration at the benzylic position again to give the *R* diol. Therefore, when this sequence is taken to its logical conclusion and the racemic epoxide is treated with a mixture of the two organisms, both epoxide enantiomers are converted into the *R* diol in high yield and enantioselectivity (Scheme 8).

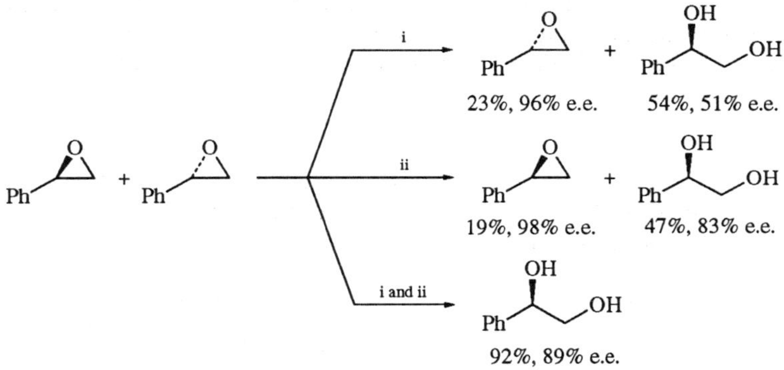

Reagents: i, *Aspergillus niger*; ii, *Beauvaria sulfurescens*

Scheme 8

3 Oxidation Reactions

Boyd *et al.* have studied the *cis* dihydroxylation of benzo-fused unsaturated heterocycles by the dioxygenase of *Pseudomonas putida* (in whole cell form).[49] A clear pattern emerged (Scheme 9) where non-aromatic heterocyclic alkenes, such as the chromene (11), were oxidized to give an *S*-configuration alcohol at the benzylic position, while heteroarenes, *e.g.* benzothiophene, were oxidized to give an *R* configuration. Other products were obtained from the heteroarenes through mutarotation and through oxidation of the benzenoid ring. The products of this type of dihydroxylation reaction have seen many applications in asymmetric synthesis in recent years and this trend continues, for example in the Hudlicky group's versatile synthesis of a range of biologically active aza sugars.[50]

[45] G. Fouche, R. M. Horak, and O. Meth-Cohn, *J. Chem. Soc., Chem. Commun.*, 1993, 119.
[46] P. Hechtberger, G. Wirnberger, M. Mischitz, N. Klempier, and K. Faber, *Tetrahedron: Asymmetry*, 1993, **4**, 1161.
[47] X.-J. Chen, A. Archelas, and R. Furstoss, *J. Org. Chem.*, 1993, **58**, 5528.
[48] S. Pedragosa-Moreau, A. Archelas, and R. Furstoss, *J. Org. Chem.*, 1993, **58**, 5533.
[49] D. R. Boyd, N. D. Sharma, R. Boyle, B. T. McMurray, T. A. Evans, J. F. Malone, H. Dalton, J. Chima, and G. N. Sheldrake, *J. Chem. Soc., Chem. Commun.*, 1993, 49.
[50] T. Hudlicky, J. Rouden, and H. Luna, *J. Org. Chem.*, 1993, **58**, 985.

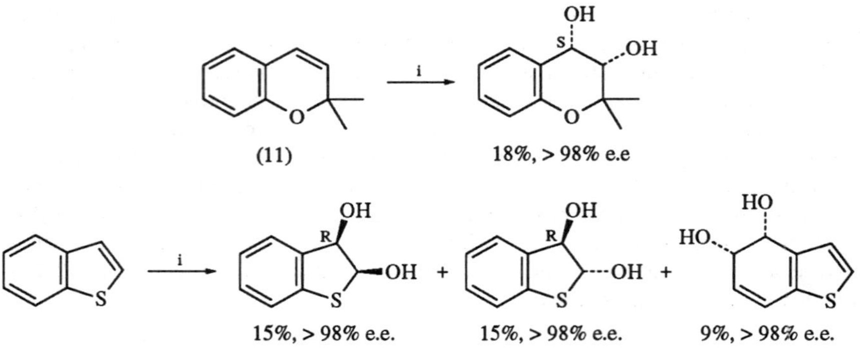

Reagents: i, *Pseudomonas putida* UV4

Scheme 9

The enzymatic Baeyer–Villiger reaction is an attractive way to introduce both extra functionality and optical activity into a molecule. To date, attempts to use the requisite monooxygenase, obtained from *Acinetobacter calcoaceticus*, in isolated form have been hindered by the difficulty in recycling the cofactor required, NADPH. However, Roberts and coworkers have obtained a new monooxygenase from *Pseudomonas putida* that requires the easily recyclable NADH cofactor and have demonstrated that

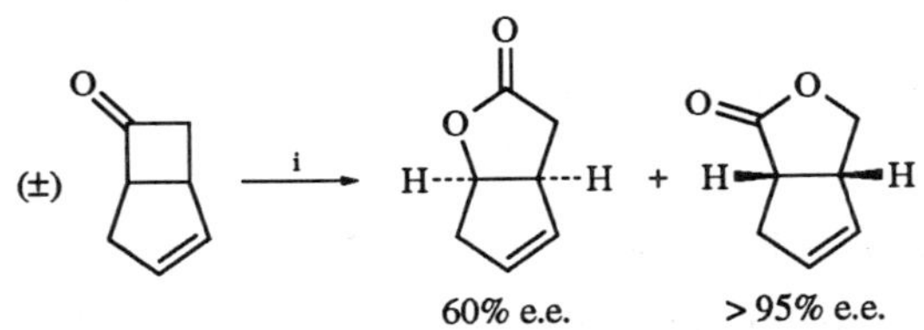

Reagents: i, *Pseudomonas putida* NCIMB 10007 monooxygenase, air, NADH, formate, *Candida boidinnii* formate dehydrogenase

Scheme 10

this enzyme is of considerable potential in these oxidations (Scheme 10).[51]

The commercially available enzyme *Caldariomyces fumago* chloroperoxidase has been shown to catalyse the epoxidation of alkenes by hydrogen peroxide. Although the reaction is generally restricted to the oxidation of *cis*-disubstituted alkenes, the high enantiomeric excesses available make this a very promising discovery (Scheme 11).[52]

An interesting development in the past year has been the use of enzyme systems in 'enantioselective destruction' processes, in which one enantiomer of a racemate is removed by oxidation to an achiral product. Using such a procedure, a number of D-phenylalanine analogues have been prepared by conversion of the L enantiomer into the keto acid using an L-amino acid oxidase.[53] Similarly, a number of enantiomerically

[51] G. Grogan, S. M. Roberts, and A. J. Willetts, *J. Chem. Soc., Chem. Commun.*, 1993, 699.
[52] E. J. Allain, L. P. Hager, L. Deng, and E. N. Jacobsen, *J. Am. Chem. Soc.*, 1993, **115**, 4415.
[53] M. C. Pirrung and N. Krishnamurthy, *J. Org. Chem.*, 1993, **58**, 957.

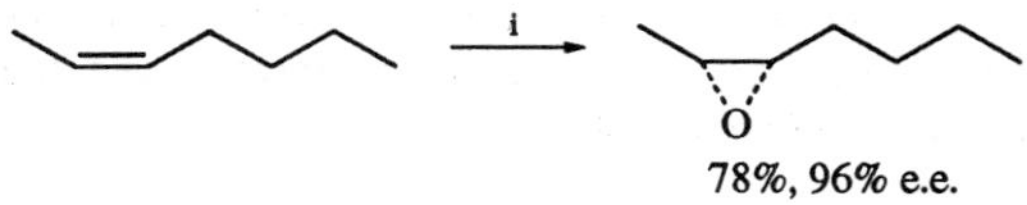

Reagents: i, chloroperoxidase, H_2O_2 (added slowly), pH 5, acetone

Scheme 11

pure 1-arylethanols have been prepared by oxidation of the other enantiomer using a range of whole-cell systems, which have included bakers' yeast (*Saccharomyces cerevisiae*),[54] although other microorganisms have proved more efficient (Scheme 12).[55]

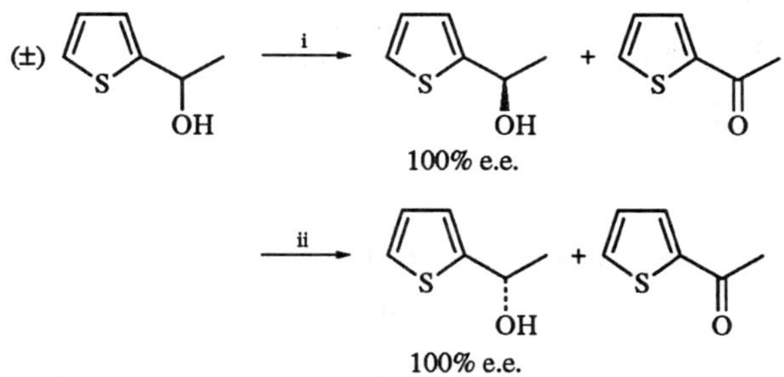

Reagents: i, *Bacillus stearothermophilus*; ii, *Acinetobacter calcoaceticus*

Scheme 12

4 Reduction Reactions

The area of enzyme-mediated reductions continues to be dominated by whole cell transformations of β-ketoesters. Bakers' yeast has enduring popularity in this area and continues to prove highly effective, for providing either simple chirons,[56] or more sophisticated targets (Scheme 13).[57]

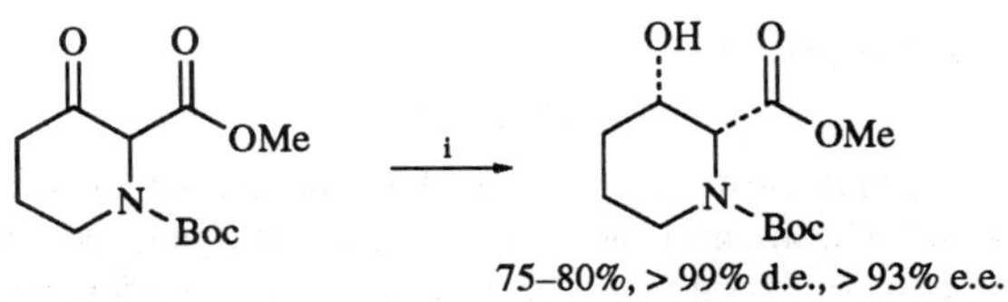

Reagents: i, *Saccharomyces cerevisiae*, sucrose, 30 °C

Scheme 13

Fuganti *et al.* have looked at the reduction of β-ketoesters with a range of microorganisms as a potential route to the β-lactam antibiotic precursor (12). While

[54] G. Fantin, M. Fogagnolo, A. Medici, P. Pedrini, S. Poli, and M. Sinigaglia, *Tetrahedron Lett.*, 1993, **34**, 883.
[55] G. Fantin, M. Fogagnolo, A. Medici, P. Pedrini, S. Poli, and F. Gardini, *Tetrahedron: Asymmetry*, 1993, **4**, 1607.
[56] H. Akita, R. Todoroki, H. Endo, Y. Ikari, and T. Oishi, *Synthesis*, 1993, 513.
[57] D. W. Knight, N. Lewis, A. C. Share, and D. Haigh, *Tetrahedron: Asymmetry*, 1993, **4**, 625.

the obvious precursor (13) failed to give useful diastereoselectivity, the use of the sulfide-substituted analogue (14) gave the required intermediate (after desulfurization) as a single diastereoisomer (Scheme 14).[58]

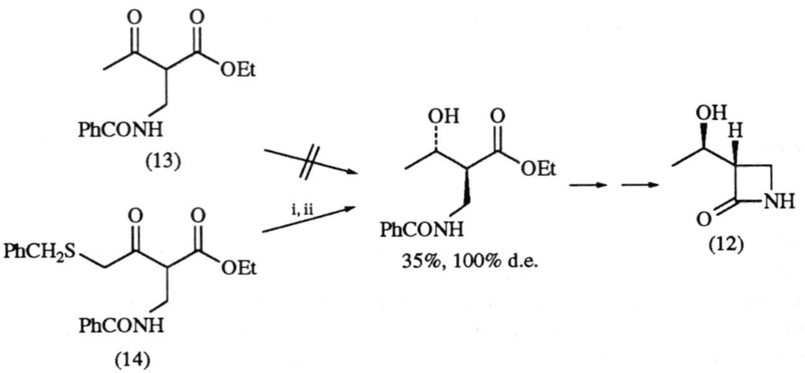

Reagents: i, *Candida guilliermondii*, glucose; ii, Raney nickel, EtOH

Scheme 14

Aside from β-ketoesters, remarkable selectivity has been reported in the reduction of a benzophenone, where the only differentiation between the two aromatic rings is at the *para* position and hence remote from the ketone (Scheme 15).[59]

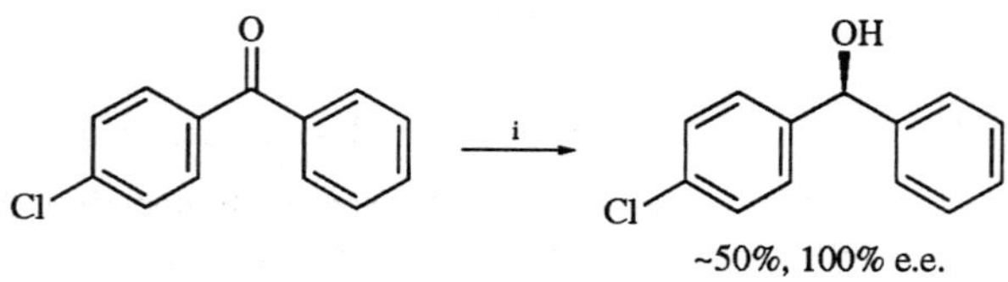

Reagents: i, *Debaryomyces marama*, 30 °C

Scheme 15

Considerable attention continues to be paid to the reaction conditions of bakers' yeast reductions, with the intention of optimizing not only yield and enantioselectivity but also ease of operation. It is now well established that immobilizing yeast in or on a solid support (*e.g.* calcium alginate beads), a relatively simple procedure, has considerable benefits with respect to the latter objective. Bhalerao *et al.* have now shown that, using such immobilized systems, exquisite and concomitant control of enantioselectivity and yield can be obtained by careful and continuous control of the reaction pH (Scheme 16), a factor largely ignored by others.[60] In a remarkably different approach to the same end, it has been shown that ethyl acetoacetate can be reduced

[58] C. Fuganti, S. Lanati, S. Servi, A. Tagliani, A. Bedeschi, and G. Francheschi, *J. Chem. Soc., Perkin Trans. 1*, 1993, 2247.
[59] G. Spassov, V. Pramatorova, R. Vlahov, and G. Snatzke, *Tetrahedron: Asymmetry*, 1993, **4**, 301.
[60] U. T. Bhalerao, Y. Chandraprakash, R. L. Babu, and N. W. Fadnavis, *Synth. Commun.*, 1993, **23**, 1201.

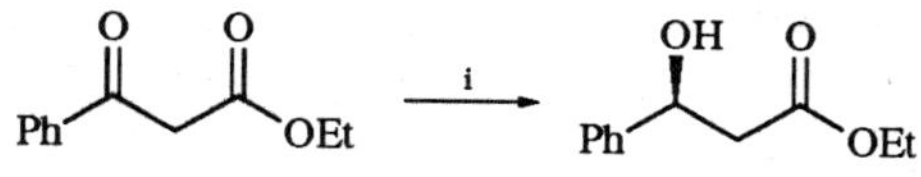

pH	product % e.e.
3	82
4.5	99
6	80
8	35

Reagents: i, *Saccharomyces cerevisiae*

Scheme 16

with high enantioselectivity and in good yield by the simple expedient of stirring the substrate with commercial, freeze dried yeast in wet light petroleum.[61]

Adam *et al.* have reported an unusual reductive kinetic resolution. Treatment of racemic hydroperoxides with horseradish peroxidase in the presence of an oxidation substrate leads to enantioselective reduction of the *R* hydroperoxide (Scheme 17).[62]

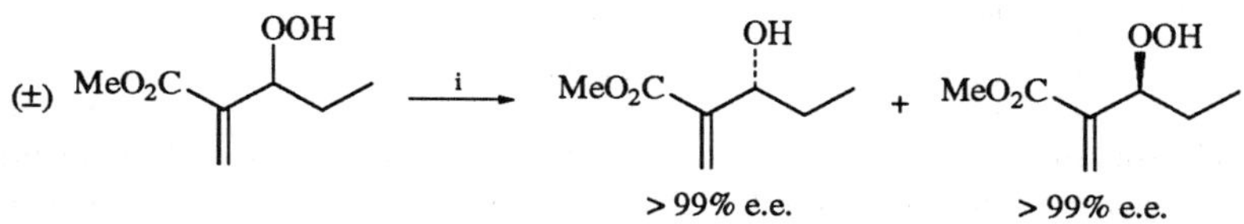

Reagents: i, Horseradish peroxidase, 2-methoxyphenol

Scheme 17

5 Carbon–Carbon Bond-forming and Cleaving Reactions

Although there is interest in the use of many different carbon–carbon bond-forming enzymes, such as oxynitrilase,[63] transketolase,[64] and pyruvate decarboxylase,[65] this area is dominated in the literature by reports of aldolase-catalysed reactions.

Aldolase research seems evenly divided between the discovery and characterization of new enzymes (usually from a microorganism), as typified by Wong's study of the 3-deoxy-D-manno-2-octulosonic acid aldolase from *Aureobacterium barkerei*,[66] and the detailed investigation of the extremes of substrate specificity of more well-characterized catalysts, such as rabbit muscle aldolase.[67]

Whitesides and coworkers have found that *N*-acetyl neuraminic acid aldolase accepts *N*-carboxybenzyl mannosamine (15) as a substrate (Scheme 18). This allows a subsequent, simple nitrogen deprotection providing, ultimately, access to the glycoside

61 L. Y. Jayasinghe, A. J. Smallridge, and M. A. Trewhella, *Tetrahedron Lett.*, 1993, **34**, 3949.
62 W. Adam, U. Hoch, C. R. Saha-Moller, and P. Schreier, *Angew. Chem., Int. Ed. Engl.*, 1993, **32**, 1737.
63 M. North, *SYNLETT*, 1993, 807.
64 G. R. Hobbs, M. D. Lilly, N. J. Turner, J. M. Ward, A. J. Willetts, and J. M. Woodley, *J. Chem. Soc., Perkin Trans. 1*, 1993, 165.
65 V. Kren, D. H. G. Crout, H. Dalton, D. W. Hutchinson, W. Konig, M. M. Turner, G. Dean, and N. Thomson, *J. Chem. Soc., Chem. Commun.*, 1993, 341.
66 T. Sugai, G. J. Shen, Y. Ichikawa, and C.-H. Wong, *J. Am. Chem. Soc.*, 1993, **115**, 413.
67 W. J. Lees and G. M. Whitesides, *J. Org. Chem.*, 1993, **58**, 1887.

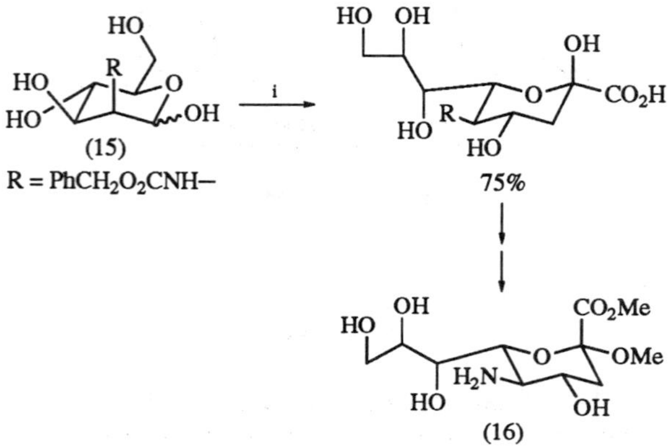

Reagents: i, NeuAc aldolase, sodium pyruvate

Scheme 18

(16), which is a general precursor to a range of *N*-acyl neuraminic acid analogues of potential as inhibitors of virus-cell adhesion.[68]

Herbert *et al.* have discovered a β-hydroxy-α-amino acid aldolase in *Streptomyces amakusaensis* that is highly selective for the cleavage of the 2*S*, 3*R* enantiomer (and diastereoisomer) and hence could have considerable potential in the synthesis of the 2*R*, 3*S* systems by selective degradation of the racemate.[69]

[68] M. A. Sparks, K. W. Williams, C. Lukacs, A. Schrell, G. Priebe, A. Spaltenstein, and G. M. Whitesides, *Tetrahedron*, 1993, **49**, 1.

[69] R. B. Herbert, B. Wilkinson, G. J. Ellames, and E. K. Kunec, *J. Chem. Soc., Chem. Commun.*, 1993, 205.

11 Biosynthesis

By R. A. HILL
Department of Chemistry, University of Glasgow, Glasgow, G12 8QQ, UK

1 Introduction

The previous Report[1] on the biosynthetic studies of secondary metabolites and related work covered the period 1989–91. This report covers the period 1992–93. The study of biosynthetic pathways continues to be a highly active area. This report is a personal view of some of the highlights in the area of biosynthesis.

The production of new microbial metabolites using unnatural precursors, a technique known as precursor-directed biosynthesis,[2] the use of plant cell cultures,[3] and the biosynthesis of marine natural products[4,5] and isocyanides and cyanides[6] have been reviewed.

2 Fatty Acid and Polyketide Biosynthesis

Reviews have appeared on the biosynthesis of fatty acid and polyketide metabolites[7,8] and the angucucline antibiotics.[9]

Deuterium labelling studies were used to show that the conversion of palmitic acid (1) into the sex pheromone (2) of the moth *Mamestra brassicae* occurs by a formal *syn* elimination of the 11-(*pro-R*)- and 12-(*pro-R*)-hydrogen atoms.[10] This result indicates that the insect Δ^{11} desaturase is similar to the Δ^{9} desaturase found in mammals, plants, and bacteria. Labelling studies have shown that palmitic acid (1) is converted into diabolic acid (3) by *Butyrivibrio fibrisolvens* without the loss of label from either C-14 or C-16 of palmitic acid (1).[11] This rules out the possibility of the intermediacy of Δ^{14}- or Δ^{15}-palmitic acids in the biosynthesis of diabolic acid (3). The degradation of the keto acid (4) to produce δ-decalone (5) in *Sporobolomyces odorus* involves inversion of the stereochemistry. This has been shown to be due to the intermediacy of a ketone, which is later reduced to give the opposite stereochemistry.[12] Degradation of α-linolenic and

[1] R. A. Hill, *Annu. Rep. Prog. Chem., Sect. B*, 1991, **88**, 283.
[2] R. Thiericke and J. Rohr, *Nat. Prod. Rep.*, 1993, **10**, 265.
[3] J. P. Kutney, *Acc. Chem. Res.*, 1993, **26**, 559.
[4] M. J. Garson, *Chem. Rev.*, 1993, **93**, 1699.
[5] G. Cimino and G. Sodan, *Topics Current Chem.*, 1993, **167**, 77.
[6] P. J. Scheuer, *Acc. Chem. Res.*, 1992, **25**, 433.
[7] D. O'Hagan, *Nat. Prod. Rep.*, 1992, **9**, 447.
[8] D. O'Hagan, *Nat. Prod. Rep.*, 1993, **10**, 593.
[9] J. Rohr and R. Thiericke, *Nat. Prod. Rep.*, 1992, **9**, 103.
[10] W. Boland, C. Frössl, M. Schlöttler, and M. Tóth, *J. Chem. Soc., Chem. Commun.*, 1993, 1155.
[11] W. Fitz and D. Arigoni, *J. Chem. Soc., Chem. Commun.*, 1992, 1533.
[12] W. Albrecht, M. Schwarz, J. Heidlas, and R. Tressl, *J. Org. Chem.*, 1992, **57**, 1954.

linoleic acids in green leaves produces a 'green odour'. This has been characterized as a mixture of C_6 aldehydes and alcohols. The biogeneration of the 'green odour' has been reviewed.[13] Marine brown algae (Phaeophyceae) produce pheromones by metabolism of unsaturated icosanoic acids to give a series of $C_{11}H_{16}$ hydrocarbons. Deuterium labelling studies have been used to investigate their biosynthesis.[14] For example it has been shown that all-*cis*-icosapenta-5,8,11,14,17-enoic acid (6) is metabolized to giffordene (7) by *Giffordia mitchellae*. The labelling studies are consistent with a thermally-allowed antarafacial [1,7] hydrogen shift during the biosynthesis (Scheme 1).

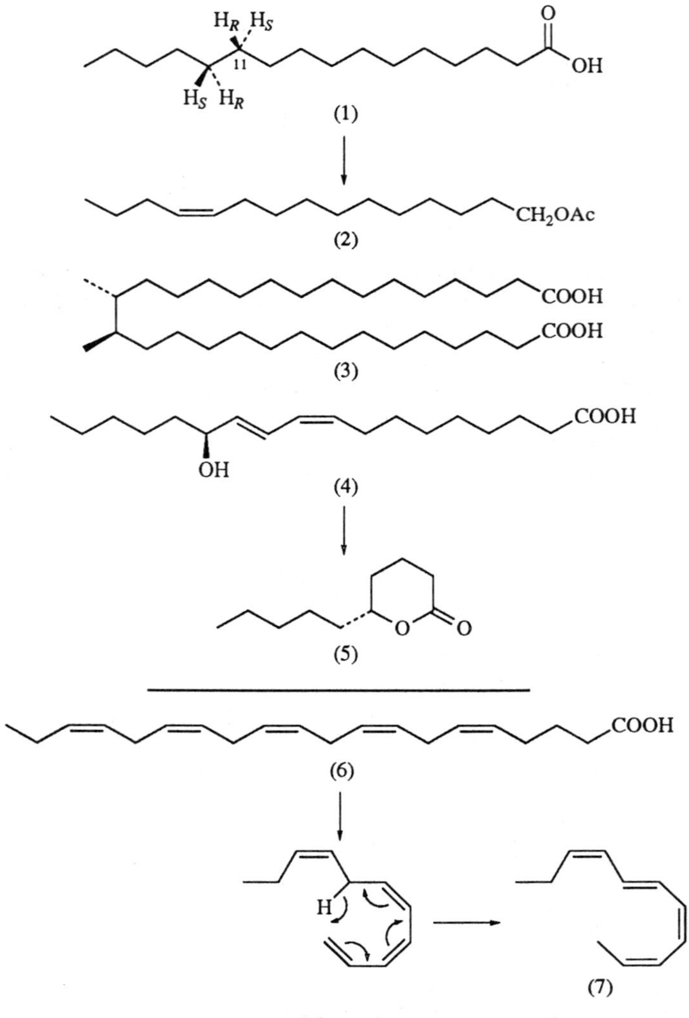

Scheme 1

[13] A. Hatanaka, *Phytochemistry*, 1993, **34**, 1201.
[14] K. Stratmann, W. Boland, and D.G. Müller, *Tetrahedron*, 1993, **49**, 3755.

Studies with single and double-labelled ^{13}C acetates have shown that thiarubine A (8) from *Ambrosia artemisiifolia* is formed from seven acetate units with the terminal acetate being cleaved.[15] All the carbons of oncorhyncolide (9) from the marine bacterial isolate MK157 have been shown to be derived from acetate.[16] The methyl branches arise from C-2 of acetate. This type of branching has only been observed in a few biosynthetic pathways.

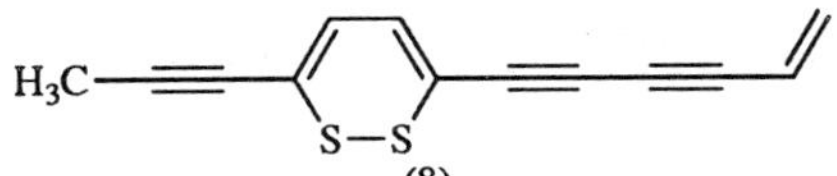

(8)

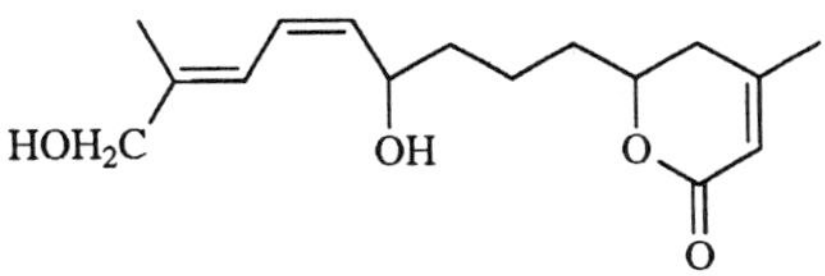

(9)

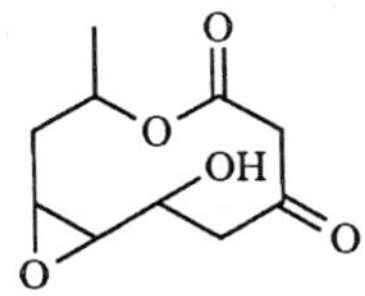

(10)

Extensive labelling studies have shown that decarestricine B (10) from *Penicillium simplicissimum* is a pentaketide. The epoxide oxygen is derived from molecular oxygen and the hydroxyl oxygen is derived from the medium. The biosynthetic relationships between the various members of the decarestricine family have been presented.[17]

There is more evidence for the processive mode of polyketide biosynthesis. The evidence has been extended to the polyether antibiotics with the incorporation of the *N*-acetylcysteamyl thioester of 5-oxo-2,4-dimethylhexanoic acid (11) labelled with deuterium into monensin A (12) in *Streptomyces cinnamonensis*.[18] During this study it was found that the presence of 2,6-*O*-dimethyl-*β*-cyclodextrin and a *β*-oxidation

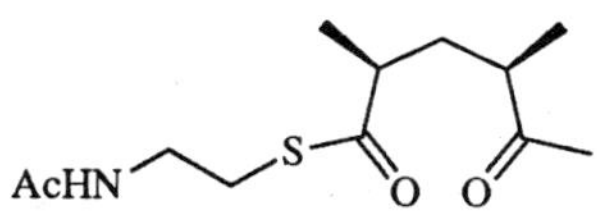

(11)

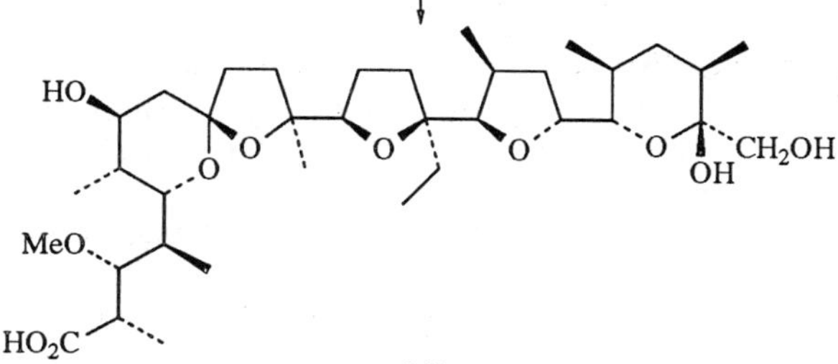

(12)

[15] M. L. Gomez-Barrios, F. J. Parodi, D. Vargas, L. Quijano, M. A. Hjortso, H. E. Flores, and N. H. Fischer, *Phytochemistry*, 1992, **31**, 2703.
[16] J. Needham, R. J. Andersen, and M. T. Kelly, *J. Chem. Soc., Chem. Commun.*, 1992, 1367.
[17] M. Mayer and R. Thiericke, *J. Chem. Soc., Perkin Trans. 1*, 1993, 495.
[18] H. Patzelt and J. A. Robinson, *J. Chem. Soc., Chem. Commun.*, 1993, 1258.

inhibitor during the fermentation was essential for successful incorporation. Further compelling evidence for the processive mode has also been provided with studies on methymycin (13) from *Streptomyces venezuelae*[19] and nargenicin (14) from *Streptomyces aureus*.[20] In the latter study the proposed (4 + 2) cycloaddition to form the octalin ring of nargenicin (14) is also supported.

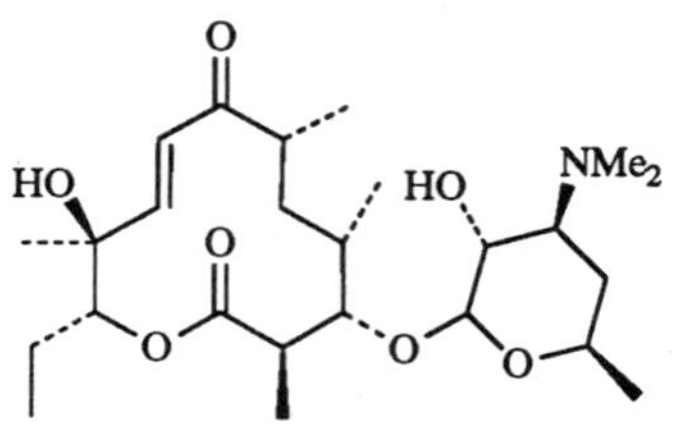

(13)

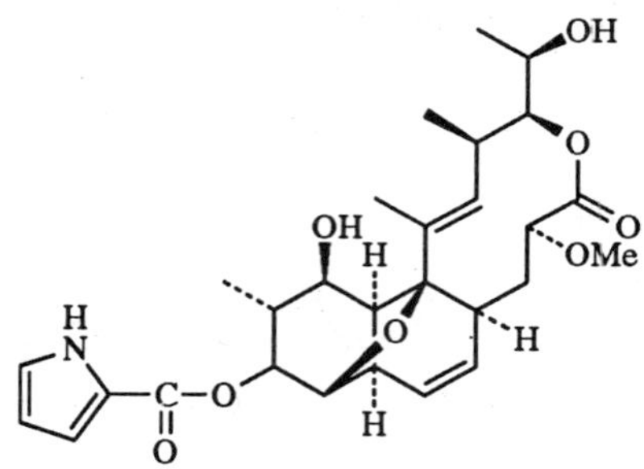

(14)

Incorporation of labelled acetate, propionate, butyrate, and glucose into elaiophylin (15) from a *Streptomyces* species has established the biosynthesis of this macrolide. A detailed analysis of the fermentation time course of *Streptoverticillium baldacii* subsp. *netropse* that produces the desertomycin family of macrolide antibiotics has provided details of the relationships between the various members of the family.[21] Extensive labelling studies have established the polyketide origin of the [7.7]paracyclophanes

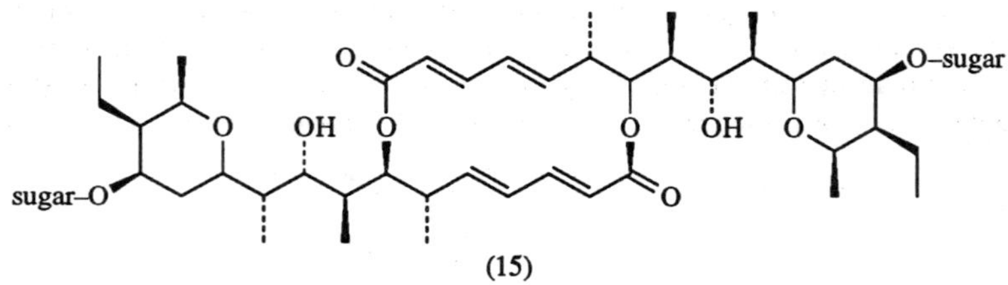

(15)

such as nostocyclophane D (16) from the cyanobacteria *Nostoc linckia* and *Cylindrospermum licheniforme*.[22] The experiments indicate that these [7.7]paracyclophanes arise by dimerization of nonaketide intermediates. The fungus *Artropsis truncata* produces a range of interesting metabolites. Incorporation studies on one of these metabolites, arthropsadiol A (17), have indicated that one of the terminal methyl groups (C-12) arises from the methyl group of methionine.[23] Methylation of a hexaketide intermediate is proposed to explain the results.

Labelled acetate and methionine were used to establish that spiciferone A (18) and spiciferin (19) from *Cochliobolus spicifer* are derived from a hexaketide with two C_1 units.[24] They probably both arise by oxidative cleavage of a bicyclic system (Scheme 2).

[19] D. E. Cane, R. H. Lambalot, P. C. Prabhakaran, and W. R. Ott, *J. Am. Chem. Soc.*, 1993, **115**, 522.
[20] D. E. Cane, W. Tan, and W. R. Ott, *J. Am. Chem. Soc.*, 1993, **115**, 527.
[21] M. Mayer and R. Thiericke, *J. Chem. Soc., Perkin Trans. 1*, 1993, 2525.
[22] S. C. Bobzin and R. E. Moore, *Tetrahedron*, 1993, **49**, 7615.
[23] W. A. Ayer and P. A. Craw, *Can. J. Chem.*, 1992, **70**, 1348.
[24] H. Nakajima, R. Matsumoto, Y. Kimura, and T. Hamasaki, *J. Chem. Soc., Chem. Commun.*, 1992, 1654; H. Nakajima, H. Fujimoto, R. Matsumoto, and T. Hamasaki, *J. Org. Chem.*, 1993, **58**, 4526.

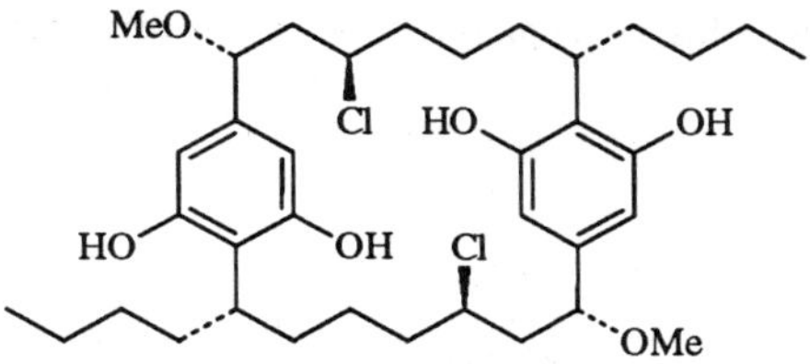

(16)

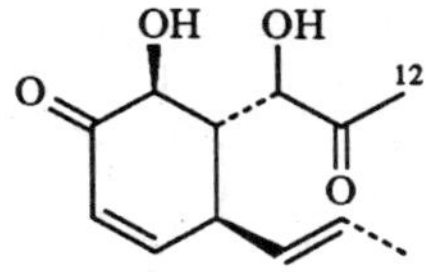

(17)

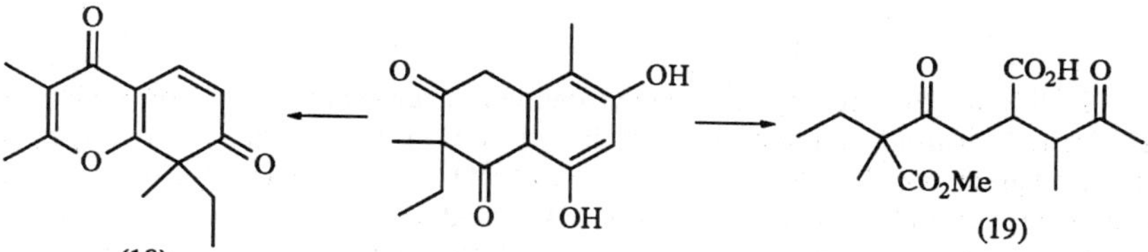

(18) (19)

Scheme 2

Similar studies were used to establish the biosynthetic origin of zaragozic acid A (20), a squalene synthase inhibitor from an unidentified fungus (MF5443).[25] The remaining carbons arise from benzoic acid and succinic acid.

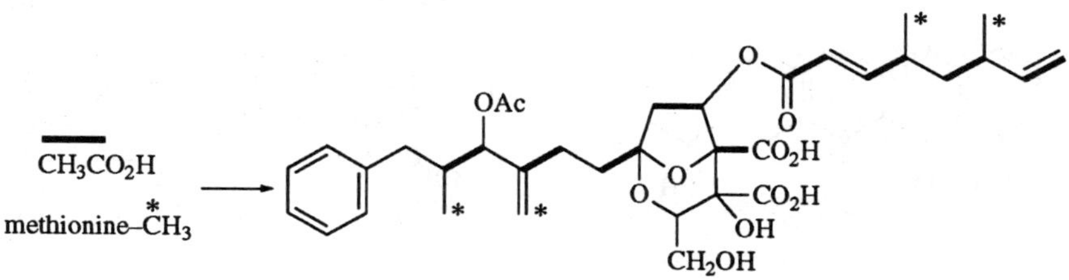

(20)

Chirally labelled malonyl CoA has been incorporated into orsellinic acid using orsellinic acid synthase from *Penicillium cyclopium*.[26] The results are similar to those obtained earlier for 6-methylsalicyclic acid.[1] The hydrogen atoms removed from the two methylene groups at the 2- and 4-positions of the putative polyketide intermediate have opposite stereochemistry. Doubly labelled acetate was used to show that mellein and 2,4-dihydroxyacetophenone are biosynthesized in the ant *Rhytidoponera chalybaea* in the same way as in fungi.[27]

Detailed labelling studies using ^{13}C, ^{2}H, and ^{18}O labelled acetates and $^{18}O_2$ gas on LL-D253α (21), a metabolite of *Phoma pigmentivora*, have shown that the metabolite is derived from two preformed polyketide chains.[28] The label observed in the hydroxy-

[25] K. M. Bryne, B. H. Arison, M. Nallin-Omstead, and L. Kaplan, *J. Org. Chem.*, 1993, **58**, 1019.
[26] J. B. Spencer and P. M. Jordan, *J. Chem. Soc., Chem. Commun.*, 1992, 646.
[27] C.-M. Sun and R. F. Toia, *J. Nat. Prod.*, 1993, **56**, 953.
[28] I. M. Chandler, C. R. McIntyre, and T. J. Simpson, *J. Chem. Soc., Perkin Trans. 1*, 1992, 2285.

methyl side chain is randomized. This is explained by the involvement of a spirocyclopropyl intermediate.

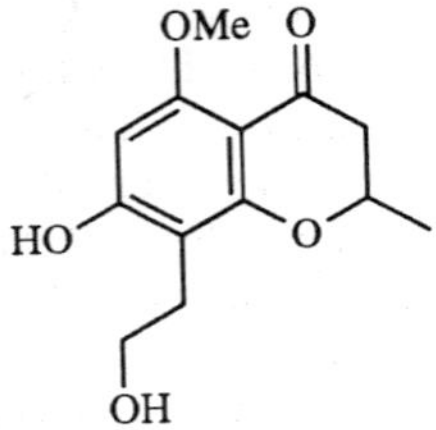

(21)

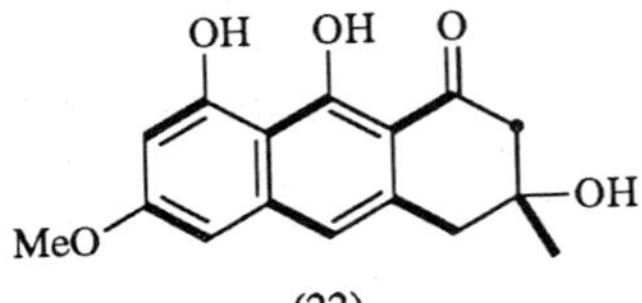

(22)

[1,2-$^{13}C_2$]Acetate feeding results have defined the chain folding of torosachrysone (22) from *Dermatocybe* mushrooms.[29] Incorporation experiments with simple and advanced precursors labelled with ^{13}C, ^{2}H, and ^{18}O have shown that tajixanthone (23) is derived through cleavage of an octaketide-derived anthraquinone with the introduction of two dimethylallyl moieties.[30] Purpactin A (24) from *Penicillium purpurogenum* has also been shown to be derived by oxidative cleavage of an octaketide-derived anthraquinone.[31] Xanthoquinodin A (25) from *Humicola* sp. FO-88 has been shown to be produced from two octaketide chains, one having been oxidatively cleaved.[32] A similar biosynthesis has been shown for *Cercospora beticola* toxin (CBT) (26).[33]

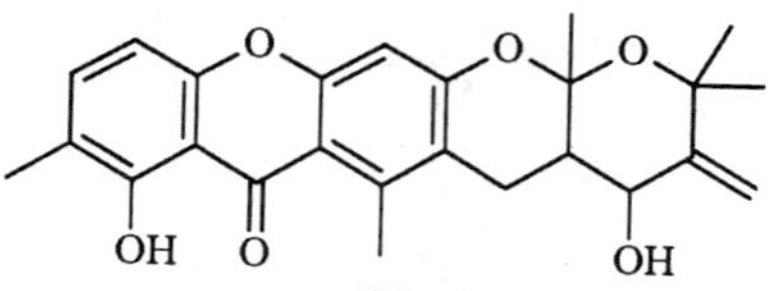

(23)

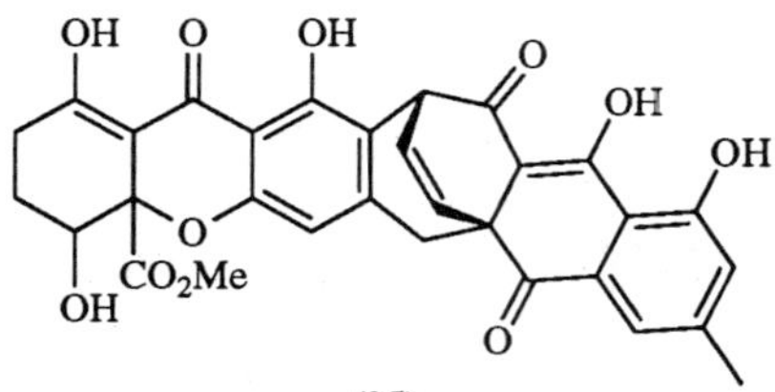

(25)

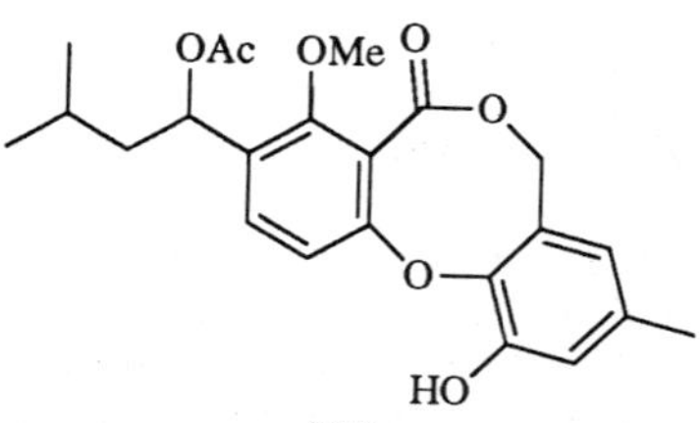

(24)

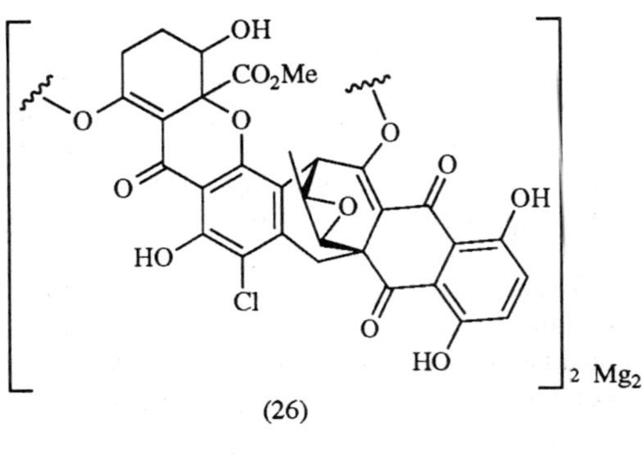

(26)

[29] M. Gill, A. Giménez, and R. Watling, *J. Nat. Prod.*, 1992, **55**, 372.
[30] S. A. Ahmed, E. Bardshiri, C. R. McIntyre, and T. J. Simpson, *Aust. J. Chem.*, 1992, **45**, 249.
[31] H. Nishida, H. Tomoda, S. Okuda, and S. Omura, *J. Org. Chem.*, 1993, **57**, 1271.
[32] H. Tabata, H. Tomoda, K. Matsuzaki, and S. Omura, *J. Am. Chem. Soc.*, 1993, **115**, 8558.
[33] A. Arnone, G. Nasini, L. Merlini, E. Ragg, and G. Assante, *J. Chem. Soc., Perkin Trans. 1*, 1993, 145.

A useful folding code using *E*/*Z* terminology has been devised for the comparison of multicyclic polyketides and this has been used to compare their biosynthetic origins.[34]

The genetic understanding of polyketide synthesis continues to improve. This is aptly demonstrated by a study in which genes from the polyketide synthase (PKS) from *Streptomyces roseofulvus*, which produces both frenolicin B (27) and nanaomycin A (28), were expressed in *Streptomyces colelicolor* A3 with components of the actinorhodin (29) gene cluster. The primary products of the recombinant strain are the new metabolites RM18 (30), RM18b (31), and the anthraquinone (32).[35] Analysis of the results confirms earlier conclusions regarding the protein determinants of chain length, keto reduction, and cyclization specificities. The biosynthetic model in which keto reduction occurs only after the complete polyketide chain has been synthesized is supported.

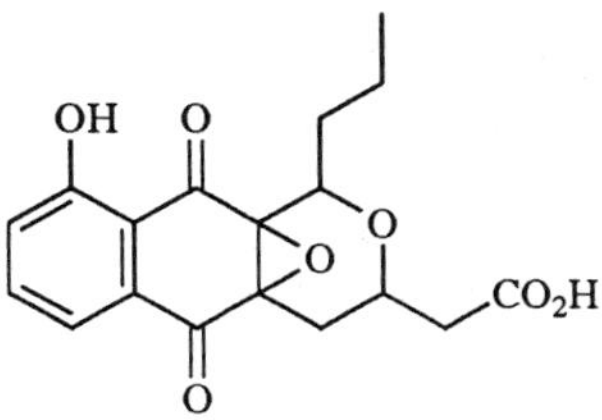

(27)

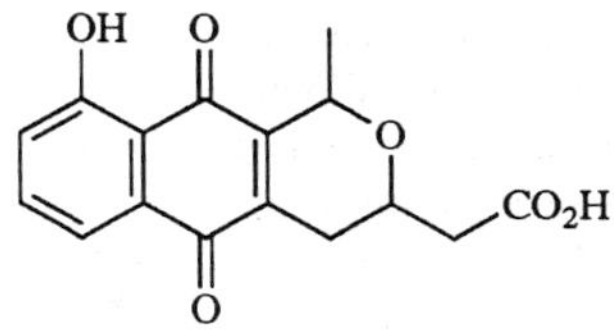

(28)

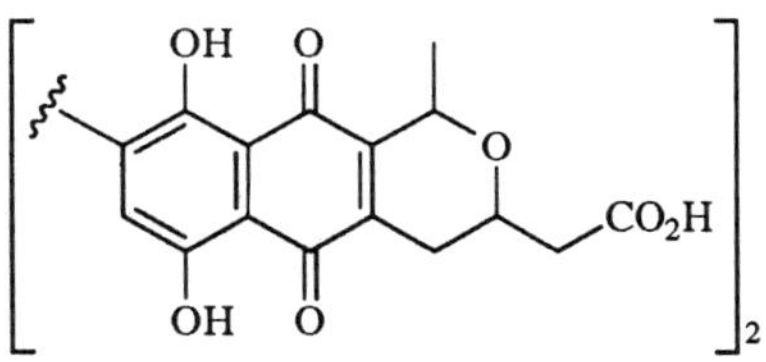

(29)

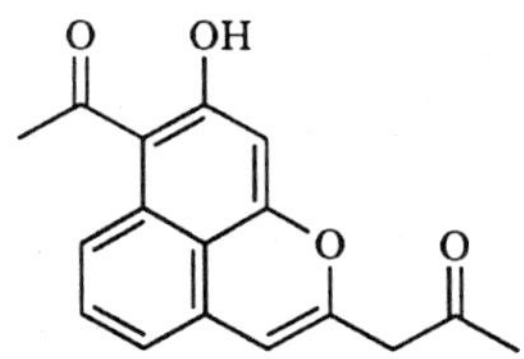

(30)

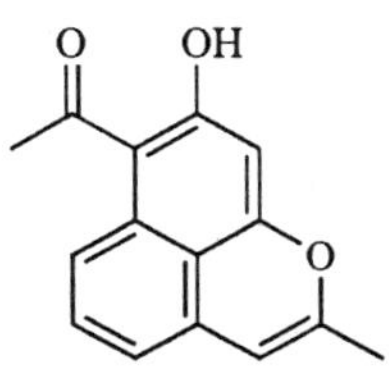

(31)

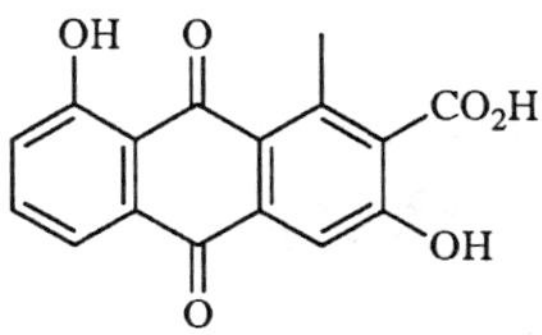

(32)

The angucyclin group of antibiotics continues to provide interesting results. Blocked mutants of *Streptomyces fradiae* have produced five new products. Their relevance to postulated biosynthetic pathways is discussed.[36] The origins of the oxygen atoms in aquamycin (33) from *Streptomyces fradiae*[37] and antibiotic PD 116198 (34)[38] from

[34] J. Rohr, *J. Org. Chem.*, 1992, **57**, 5217.

[35] R. McDaniel, S. Ebert-Khosla, D. A. Hopwood, and C. Khosla, *J. Am. Chem. Soc.*, 1993, **115**, 11 671.

[36] J. Rohr, M. Schönewolf, G. Udvarnoki, K. Eckardt, G. Schumann, C. Wagner, J. M. Beale, and S. D. Sorey, *J. Org. Chem.*, 1993, **58**, 2547.

[37] G. Udvarnoki, T. Henkel, R. Machinek, and J. Rohr, *J. Org. Chem.*, 1992, **57**, 1274.

[38] S. J. Gould and X.-C. Cheng, *Tetrahedron*, 1993, **48**, 11 135.

Streptomyces phaeochrogenes WP 3688 have been studied (Scheme 3). Most of the angucyclins appear to be derived from a decaketide intermediate folding into an angular conformation; however labelling studies on antibiotic PD 116198 (34) are consistent with rearrangement of an initially-formed linear tetracyclic intermediate. Feeding experiments using *Streptomyces murayamensis* that produces dehydroabelomycin (35) and tetrangulol (36) indicate that deoxygenation occurs at a prearomatic stage.[39]

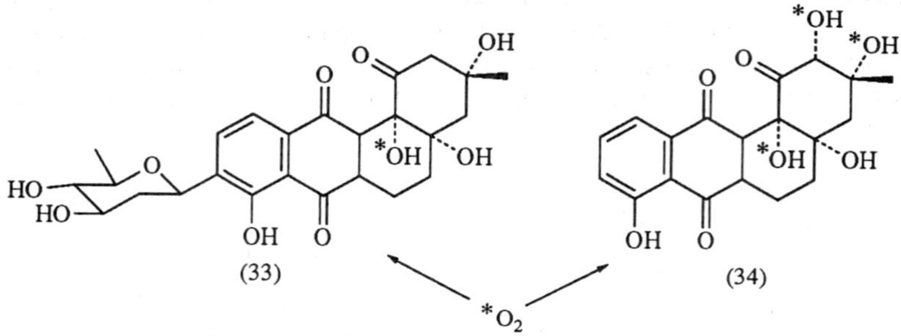

Scheme 3

Early intermediates on the biosynthetic pathway of tetracenomycin C from *Streptomyces glaucescens* have been identified in blocked mutants.[40] Labelling studies on esperamicin A_1 from *Actinomadura verrucosospora* have indicated that the enediyne core (37) is derived from an octaketide intermediate with loss of the terminal carboxyl group.[41]

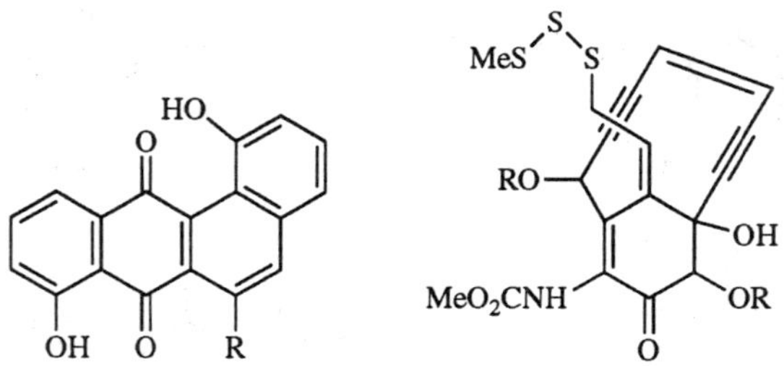

(35) R = OH
(36) R = H

(37)

3 Terpenoids

Reviews have appeared on HMG CoA reductase inhibitors,[42] the folding of squalene,[43] sterol metabolism in insects,[44] the biosynthesis of marine sterols,[45,46] and

[39] S. J. Gould, X.-C. Cheng, and K. A. Halley, *J. Am. Chem. Soc.*, 1992, **114**, 10066.
[40] B. Shen, H. Nakayama, and C. R. Hutchinson, *J. Nat. Prod.*, 1993, **56**, 1288.
[41] K. S. Lam, J. A. Veitch, J. Golik, B. Krishnan, S. E. Klohr, K. J. Volk, S. Forenza, and T. W. Doyle, *J. Am. Chem. Soc.*, 1993, **115**, 12340.
[42] A. Endo and K. Hasumi, *Nat. Prod. Rep.*, 1993, **10**, 541.
[43] R. Bohlmann, *Angew. Chem., Int. Ed. Engl.*, 1992, **31**, 582.
[44] N. Ikekawa, M. Morisaki, and Y. Fujimoto, *Acc. Chem. Res.*, 1993, **26**, 139.
[45] B. J. Baker and R. G. Russel, *Topics Current Chem.*, 1993, **167**, 1.
[46] J.-L. Giner, *Chem. Rev.*, 1993, **93**, 1735.

the bacteriohopanoids.[47,48] A new pathway for the biosynthesis of isopentenyl diphosphate in bacteria has been identified.[49] Geranyl diphosphate synthase has been used to study the stereochemistry of the formation of geranyl diphosphate from dimethylallyl diphosphate and isopentenyl diphosphate.[50] Detailed labelling studies have been conducted to investigate the biosynthesis of limonene (38) from geranyl diphosphate.[51] Deuterium labelled geranyl diphosphate has been used in biosynthetic studies of monoterpenes from *Salvia officinalis*.[52] These results cast doubt on the use of natural abundance deuterium NMR for biosynthetic studies.

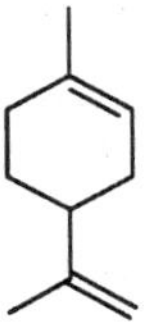

(38)

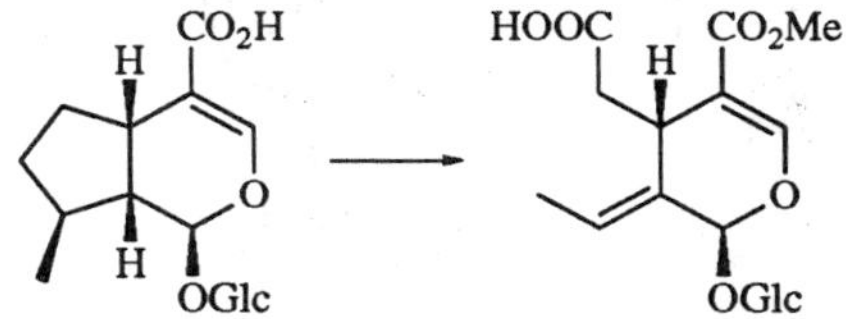

(39) (40)

Deoxyloganic acid (39) has been efficiently converted into oleoside methyl ester (40) using *Fraxinus excelsior* and *Syringa josikaea*.[53] Analysis of the results of feeding deuteriated mevalonates to *Perilla frutescens* callus revealed that the biosynthesis of α-curcumene (41) involves a 1,2-hydride shift and the biosynthesis of cuparene (42) involves a 1,3-hydride shift.[54] Deuteriated farnesols were incorporated into dehydrogeosmin (43) using flower heads of the cactus *Rebutia marsoneri*.[55] The results are consistent with dehydrogeosmin (43) being formed by loss of the three-carbon side chain of a eudesmane intermediate.

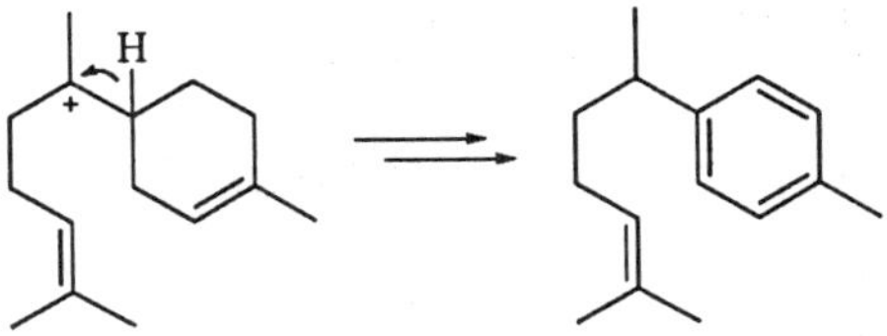

(41)

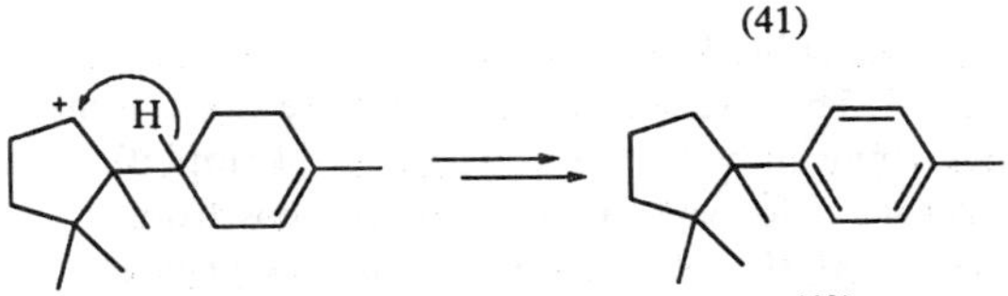

(42)

Competitive feeding experiments indicated that isotrichodiol (44) and isotrichotriol

[47] G. Ourison and M. Rohmer, *Acc. Chem. Res.*, 1992, **25**, 403.
[48] M. Rohmer, *Pure Appl. Chem.*, 1993, **65**, 1293.
[49] M. Rohmer, M. Knani, P. Simonin, B. Sutter, and H. Sahm, *Biochem. J.*, 1993, **295**, 517.
[50] T. Endo and T. Suga, *Phytochemistry*, 1992, **31**, 1565.
[51] H.-J. Pyun, R. M. Coates, K. C. Wagschal, P. McGeady, and R. B. Croteau, *J. Org. Chem.*, 1993, **58**, 3998.
[52] D. Arigoni, D. E. Cane, J. H. Shim, R. Croteau, and K. Wagschal, *Phytochemistry*, 1993, **32**, 623.
[53] S. Damtoft, H. Franzyk, and S. R. Jensen, *Phytochemistry*, 1993, **34**, 1291.
[54] K. Nabela, K. Kawakita, Y. Yada, and H. Okuyama, *Biosci. Biotech. Biochem.*, 1992, **57**, 792.
[55] Z. Feng, U. Huber, and W. Boland, *Helv. Chim. Acta*, 1993, **76**, 2547.

(45), rather than trichodiol (46) and trichotriol (47), are intermediates in trichothecene biosynthesis.[56] Labelled velutinal (48) is incorporated into the lactaranes vellerol (49) and velleral (50) by bruised fruiting bodies of the mushroom *Lactarius vellereus*.[57] [^{14}C]Arteannuic acid (51) has been incorporated into both arteannuin **B** (52) and artemisinin (qinghaosu) (53) using *Artemisia annua*, suggesting that arteannuic acid (51) might be a common intermediate in the biosynthesis of (52) and (53).[58]

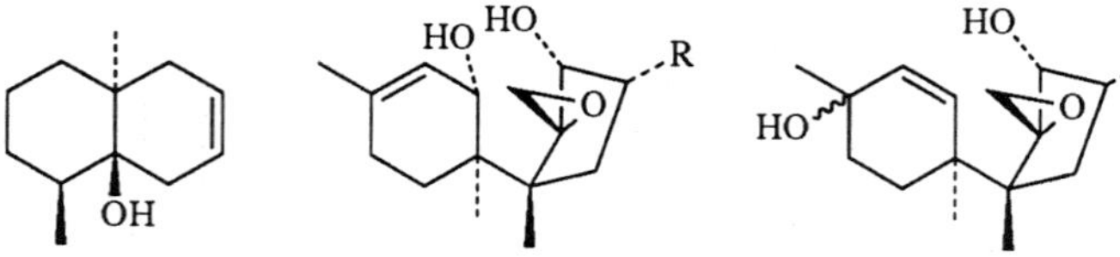

(43) (44) R = H (45) R = OH (46) R = H (47) R = OH

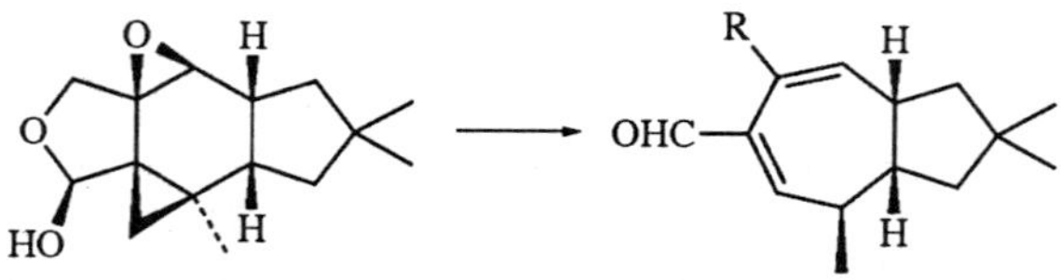

(48) (49) R = CH_2OH (50) R = CHO

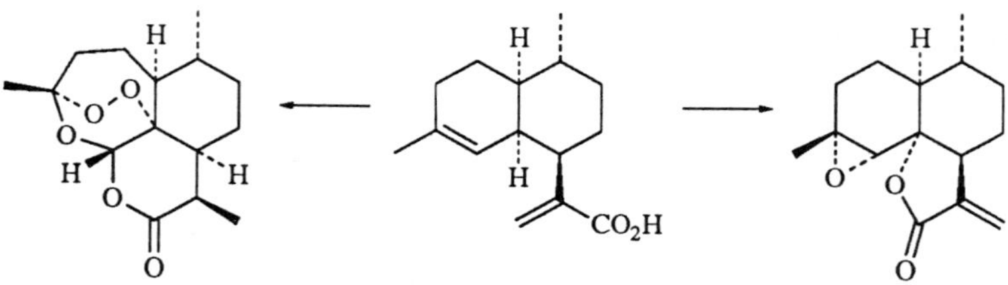

(53) (51) (52)

The biosynthesis of ether lipids such as di-*O*-phytanylglycerol (54) has been studied in Archaebacteria.[59,60] The results indicate that there is a common pathway for the biosynthesis of these lipids in both methanogenic and halophilic bacteria and that di-*O*-phytanylglycerol (54) is formed in two distinct steps from glycerol and geranylgeranyl diphosphate in *Methanobacterium thermoautotrophicum*.

Feeding experiments with [1,2-$^{13}C_2$]acetate and [2-^{13}C,$^2H_3$]acetate in *Amsonia elliptica* have provided details of the biosynthesis of the side chains of campesterol (55) and dihydrobrassicasterol (56).[61] The results are consistent with the biosynthesis

[56] A. R. Hesketh, B. W. Bycroft, P. M. Dewick, and J. Gilbert, *Phytochemistry*, 1993, **32**, 105.
[57] T. Hansson, Z. Pang, and O. Sterner, *Acta Chem. Scand.*, 1993, **47**, 403.
[58] R. S. Sangwan, K. Agarwal, R. Luthra, R. S. Thakur, and N. Singh-Sangwan, *Phytochemistry*, 1993, **34**, 1301.
[59] D. Zhang and C. D. Poulter, *J. Org. Chem.*, 1993, **58**, 3919.
[60] D. Zhang and C. D. Poulter, *J. Am. Chem. Soc.*, 1993, **115**, 1270.
[61] S. Seo, A. Uomori, Y. Yoshimura, K. Takeda, H. Noguchi, Y. Ebizuka, U. Sankawa, and H. Seto, *J. Chem. Soc., Perkin Trans. 1*, 1992, 569.

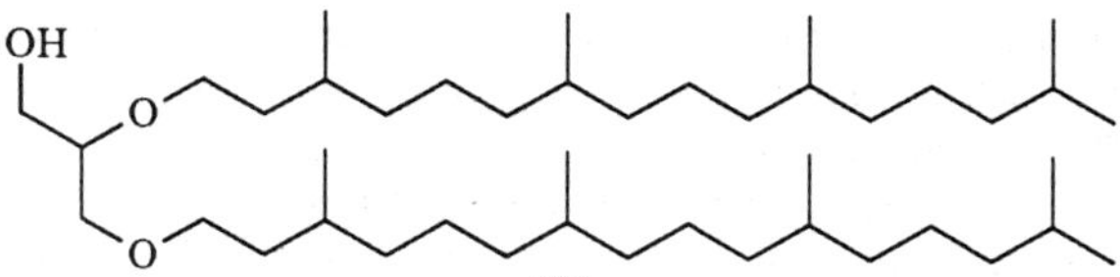

(54)

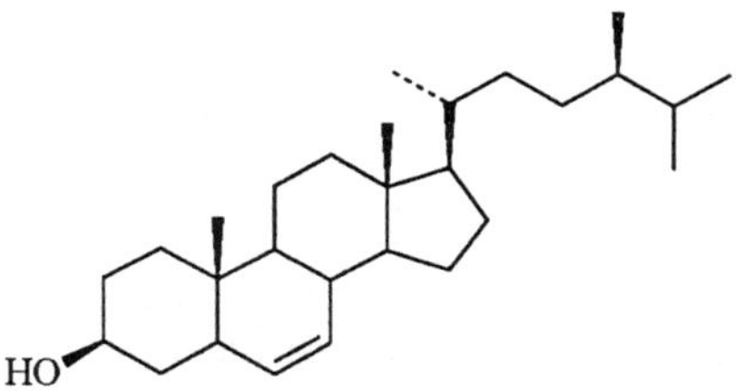

(55)

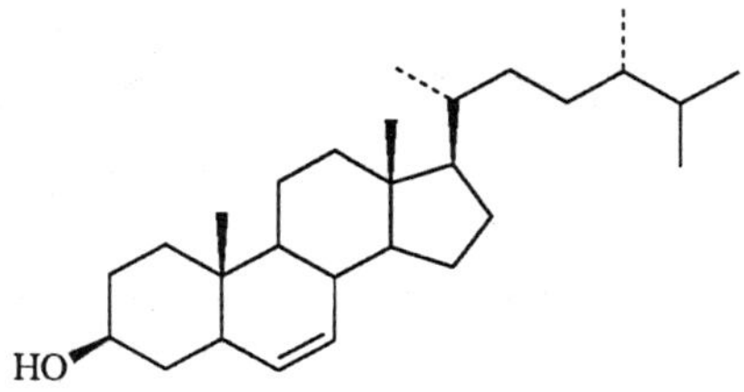

(56)

involving regiofacial specific reduction of a C-24 double bond which is formed by double bond isomerization of a $\Delta^{24(28)}$-sterol (Scheme 4). Detailed studies on the biosynthesis of estrogens by human microsomal placental aromatase have shown that

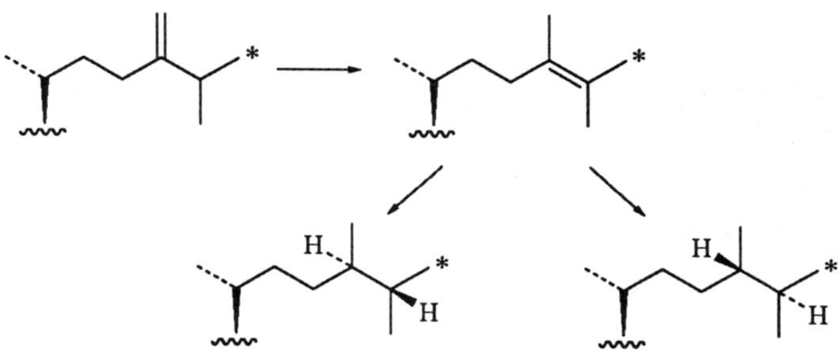

Scheme 4

the biosynthesis proceeds through several different pathways.[62,63]

Incorporation of ^{13}C-labelled acetate and formate into citreohybridone B (57) by a hybrid strain derived from *Penicillium citreo-viride* indicated that citreohybridone B (57) is formed by a meroterpenoid pathway similar to that for terretonin.[64]

4 Shikimate and Related Metabolites

Reviews have appeared on the biosynthesis of shikimate metabolites,[65,66] phenyl-

62 E. Caspi, H. R. W. Dharmaratne, E. Roitman, and C. Shackleton, *J. Chem. Soc., Perkin Trans. 1*, 1993, 1191.
63 H. R. W. Dharmaratne, J. L. Kilgore, E. Roitman, C. Shackleton, and E. Caspi, *J. Chem. Soc., Perkin Trans. 1*, 1993, 1529.
64 S. Kosemura, H. Miyata, K. Matsunaga, and S. Yamamura, *Tetrahedron Lett.*, 1992, **33**, 3882.
65 P. Dewick, *Nat. Prod. Rep.*, 1992, **9**, 153.

propanoids and isoflavonoids in alfalfa,[67] isoflavone and pterocarpan phytoalexins,[68] and phenylpropanoids.[69]

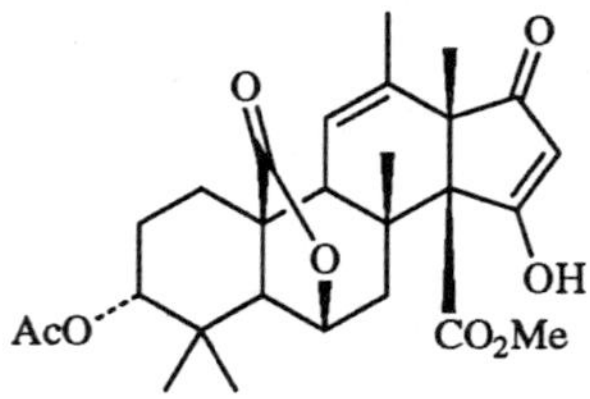

(57)

The enzymes of the shikimate pathway have continued to come under extensive scrutiny and have been reviewed.[70] The dehydration of 3-dehydroquinate (58) to 3-dehydroshikimate (59) is catalysed by 3-dehydroquinase. There are two distinct classes of 3-dehydroquinase, which are quite different in physical and biochemical properties.[71] The active site lysine that is known to form a Schiff's base with the ketone carbonyl of 3-dehydroquinate (58) is found in Type I enzymes but not in Type II enzymes. The properties of isochorismate hydroxymutase, which catalyses the interconversion of chorismic acid (60) and isochorismic acid (61), have been described.[72,73]

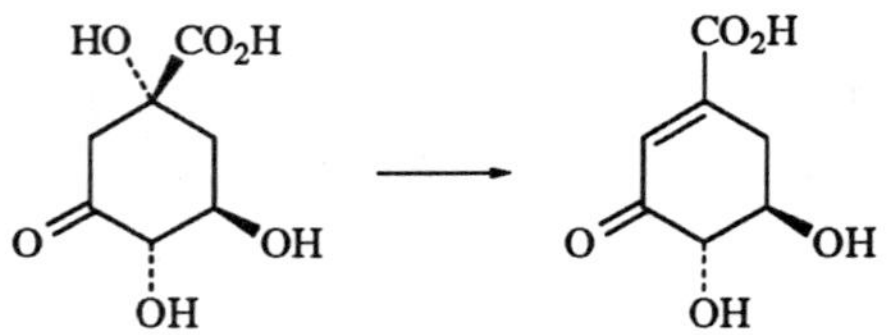

(58) (59)

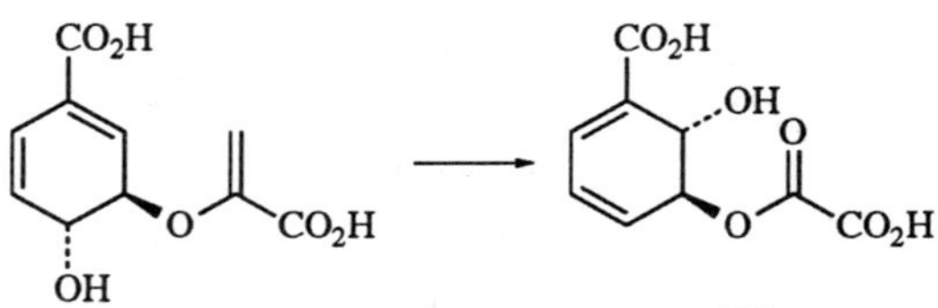

(60) (61)

[1-^{13}C]Phenylalanine was incorporated into the phenylpropanoid (62) without

[66] P. Dewick, *Nat. Prod. Rep.*, 1993, **10**, 233.
[67] R.A. Dixon, A.D. Choudhary, K. Dalkin, R. Edwards, T. Fahrendorf, G. Gowri, M.J. Harrison, C.J. Lamb, G.L. Loake, C.A. Maxwell, J. Orr, and N.L. Paiva, *Recent Adv. Phytochem.*, 1992, **26**, 91.
[68] W. Barz and R. Welle, *Recent Adv. Phytochem.*, 1992, **26**, 139.
[69] C.J. Douglas, M. Ellard, K.D. Hauffe, E. Molitor, M. Moniz de Sá, S. Reinold, R. Subramaniam, and F. Williams, *Recent Adv. Phytochem.*, 1992, **26**, 63.
[70] B.K. Singh, D.L. Siehl, and J.A. Connelly, *Oxford Surv. Plant Mol. Cell Biol.*, 1991, **7**, 143.
[71] C. Kleanthous, R. Deka, K. Davis, S.M. Kelly, A. Cooper, S.E. Harding, N.C. Price, A.R. Hawkins, and J.R. Coggins, *Biochem. J.*, 1992, **282**, 687.
[72] P.M.M. Schaaf, L.E. Heide, E.W. Leistner, Y. Tani, M. Karas, and R. Deutzmann, *J. Nat. Prod.*, 1993, **56**, 1294.
[73] P.M.M. Schaaf, L.E. Heide, E.W. Leistner, Y. Tani, and M.M. El-Olemy, *J. Nat. Prod.*, 1993, **56**, 1304.

rearrangement of the side chain.[74] Labelled 1-(4-methoxyphenyl)prop-1-ene (63) was incorporated into epoxypseudoisoeugenol 2-methylbutyrate (64) in tissue cultures of *Pimpinella anisum*; the labelling of the metabolite indicated that an NIH shift of the side chain had occurred during the introduction of the aromatic hydroxyl group.[75] A similar migration is probably involved in the biosynthesis of the tyrosine kinase inhibitor erbstatin (65) from a *Streptomyces* species, as labelled shikimic acid, phenylalanine, and tyrosine are incorporated efficiently into erbstatin (65).[76] Biosynthetic studies on axenomycin D (66) show that the aromatic moiety is derived from a metabolite of the shikimate pathway and that the biosynthesis is different to that of the menaquinones.[77]

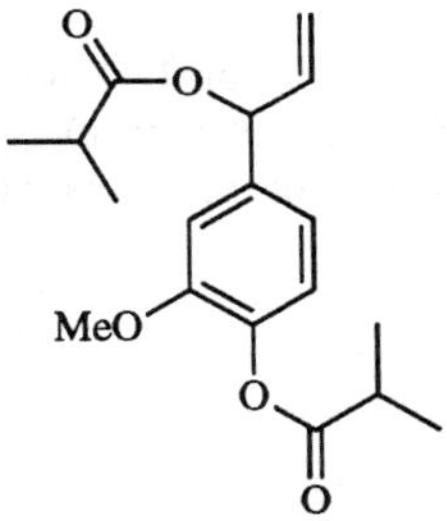

(62)

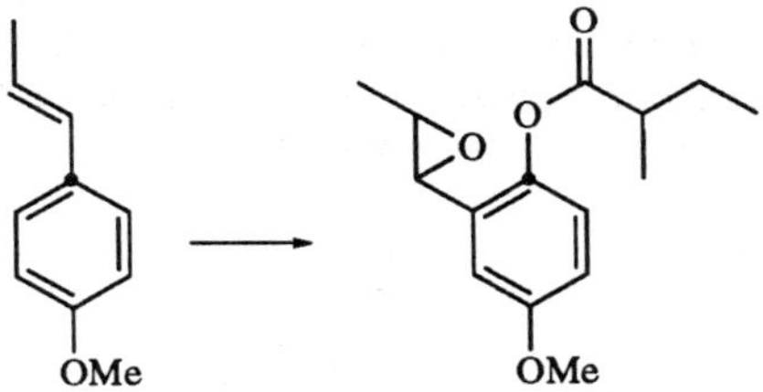

(63) (64)

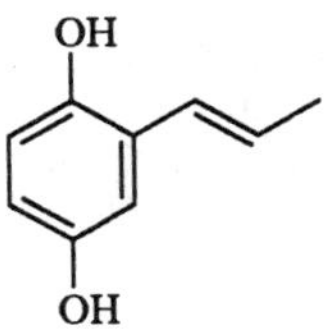

(65)

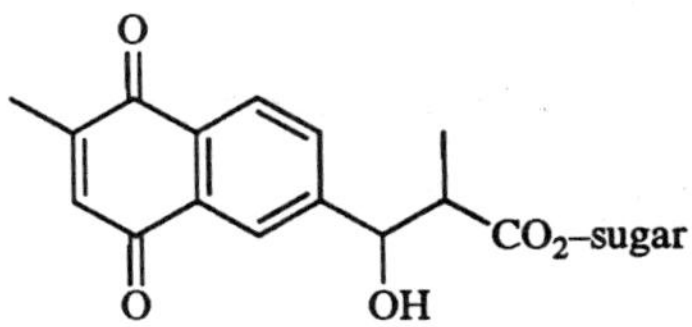

(66)

The role of salicyclic acid as a plant hormone has been reviewed.[78] The biosynthetic pathway of salicyclic acid from cinnamic acid is not clear at present. Levels of salicyclic acid in plants appear to be controlled by conjugation with glucose.[79]

Sporopellenin from *Curcurbita maxima* is a biopolymer that is resistant to chemical, physical, and biological degradation. Although its structure is not known, [U-^{14}C]-phenylalanine has been shown to be efficiently incorporated into sporopollenin.[80–82]

Labelling studies using [U-^{13}C]glucose, [3-^{13}C]phenylalanine, and [1,2-$^{13}C_2$]-phenylacetic acid (67) into thiotropocin (68) from *Pseudomonas* CB-104 are consistent with a shikimate origin involving symmetrical phenylpyruvate and phenylacetate

[74] K. H. Horz and J. Reichling, *Phytochemistry*, 1993, **33**, 349.
[75] R. Martin and J. Reichling, *Phytochemistry*, 1992, **31**, 511.
[76] Y. Tabata, M. Imoto, and K. Umezawa, *J. Antibiot.*, 1992, **45**, 1382.
[77] V. Firese, A. Boos, H.-J. Bauch, and E. Leistner, *Phytochemistry*, 1993, **32**, 613.
[78] I. Raskin, *Annu. Rev. Plant Physiol. Plant Mol. Biol.*, 1992, **43**, 439.
[79] N. Yalpani, N. E. Blake, and M. Schultz, *Plant Physiol.*, 1992, **100**, 1114.
[80] S. Gubatz and R. Wiermann, *Bot. Acta*, 1992, **105**, 407.
[81] S. Gubatz and R. Wiermann, *Z. Naturforsch., C Biosci.*, 1993, **48**, 10.
[82] S. Wilmesmeier, S. Steuernagel, and R. Wiermann, *Z. Naturforsch., C Biosci.*, 1993, **48**, 697.

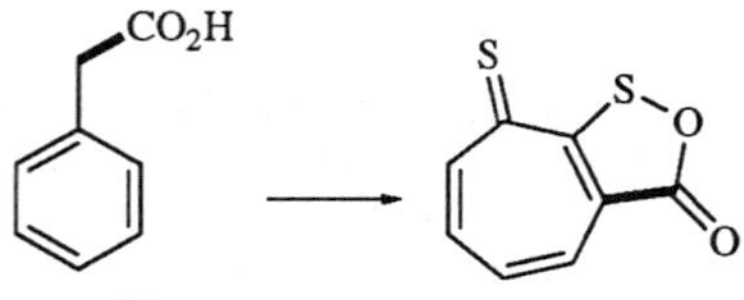

(67) (68)

intermediates.[83] Phenylacetic acid (67) presumably undergoes oxidative ring expansion followed by further oxidation and incorporation of sulfur.

The biosynthesis of ansatriene A (69) from *Streptomyces collinus* has been extensively studied.[84–87] The results indicate that the cyclohexanecarboxylic acid portion is derived from shikimate.

Full details of the biosynthetic studies on rotenone (70) and amorphigenin (71) by *Amorpha fruticosa* seedlings have been published.[88–90] A mechanism involving a spirodienone intermediate has been proposed for the migration of the phenyl group in the biosynthesis of isoflavonoids.[91]

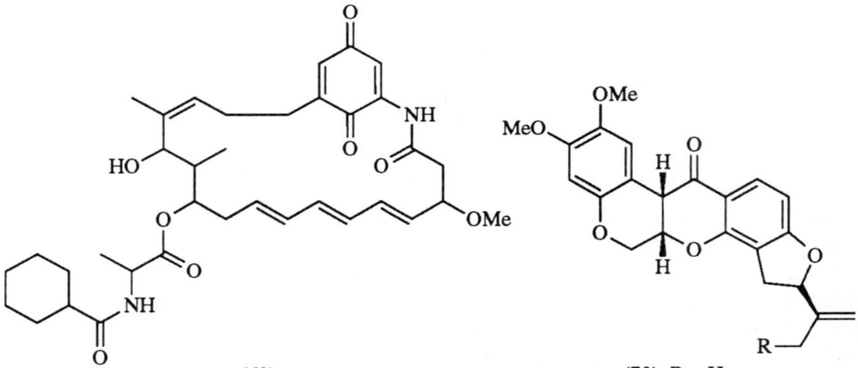

(69)

(70) R = H
(71) R = OH

The biosynthesis of the aromatic side chains of taxol (74) has been studied.[92] The results indicate that phenylalanine is incorporated *via* β-phenylalanine (72) and phenylisoserine (73). The benzoyl moieties are also derived from β-phenylalanine (72) either directly or *via* phenylisoserine (73) (Scheme 5).

83 D. E. Cane, Z. Wu, and J. E. Van Epp, *J. Am. Chem. Soc.*, 1992, **114**, 8479.
84 K. A. Reynolds, P. Wang, K. M. Fox, and H. G. Floss, *J. Antibiot.*, 1992, **45**, 411.
85 K. A. Reynolds, *J. Nat. Prod.*, 1993, **56**, 175.
86 K. A. Reynolds, N. Seaton, K. M. Fox, K. Warner, and P. Wang, *J. Nat. Prod.*, 1993, **56**, 825.
87 B. S. Moore, H. Cho, R. Casati, E. Kenedy, K. A. Reynolds, U. Mocek, J. M. Beale, and H. G. Floss, *J. Am. Chem. Soc.*, 1993, **115**, 5254.
88 P. Bhandari, L. Crombie, P. Daniels, I. Holden, N. Van Bruggen, and D. A. Whiting, *J. Chem. Soc., Perkin Trans. 1*, 1992, 839.
89 P. Bhandari, L. Crombie, G. W. Kilbee, S. J. Pegg, G. Proudfoot, J. Rossiter, M. Sanders, and D. A. Whiting, *J. Chem. Soc., Perkin Trans. 1*, 1992, 851.
90 P. Bhandari, L. Crombie, M. F. Harper, J. Rossiter, M. Sanders, and D. A. Whiting, *J. Chem. Soc., Perkin Trans. 1*, 1992, 1685.
91 L. Crombie and D. A. Whiting, *Tetrahedron Lett.*, 1992, **33**, 3663.
92 P. E. Flemming, U. Mocek, and H. G. Floss, *J. Am. Chem. Soc.*, 1993, **115**, 805.

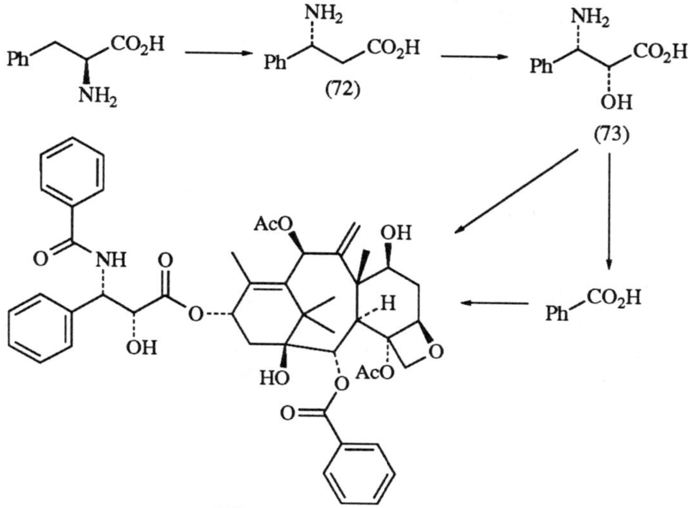

(74)

Scheme 5

5 Alkaloids and Other Amino-acid-derived Metabolites

Reviews have appeared on the biosynthesis of plant alkaloids and nitrogenous microbial metabolites.[93,94]

Labelled serine is a better precursor than alanine for the carbons and nitrogen of the dehydroalanyl portion of valanimycin (75) from *Streptomyces viridifaciens*. Isobutylamine and isobutylhydroxylamine are incorporated into the remaining carbons and nitrogen, the latter very efficiently (Scheme 6).[95] It is not known how the N–N

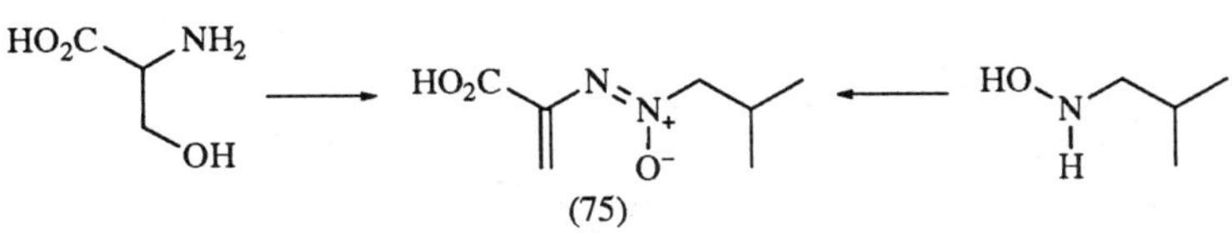

(75)

Scheme 6

bond is formed in valanimycin (75). Labelling studies have established that ornithine is the precursor of avicin (76) from *Streptomyces aviceus*[96] and that domoic acid (77) from *Nitzscia pungens* is derived from a geranyl precursor and a metabolite of the citric acid pathway.[97] The hydroxycyclopentenone portion of reductiomycin (78) from *Streptomyces xanthochromogenus* has been shown to be derived from 5-aminolevulinic acid

[93] R. B. Herbert, *Nat. Prod. Rep.*, 1992, **9**, 507.
[94] R. B. Herbert, *Nat. Prod. Rep.*, 1993, **10**, 575.
[95] R. J. Parry, Y. Li, and F.-L. Lii, *J. Am. Chem. Soc.*, 1992, **114**, 10 062.
[96] S. J. Gould and S. Ju, *J. Am. Chem. Soc.*, 1992, **114**, 10 166.
[97] D. J. Douglas, U. P. Ramsey, J. A. Walker, and J. L. C. Wright, *J. Chem. Soc., Chem. Commun.*, 1992, 714.

whereas the right-hand portion is derived by ring cleavage of a symmetrical product of the shikimic acid pathway.[98]

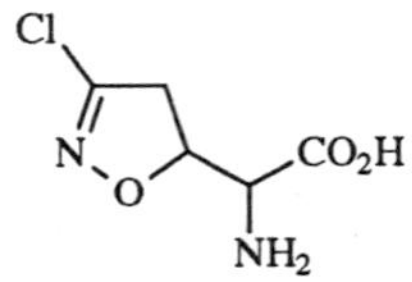

(76)

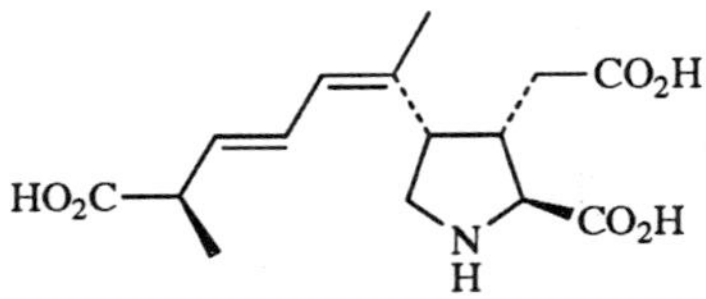

(77)

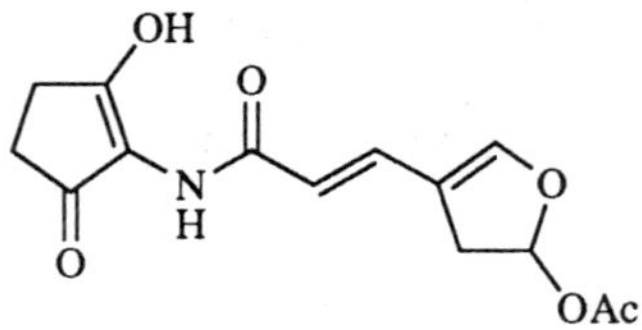

(78)

Several groups have been involved in biosynthetic studies of the tropane alkaloids. [1-^{13}C]Acetate has been efficiently incorporated into tropane alkaloids using hairy root cultures of *Hyoscyamus albus*.[99] Intermediate trapping and competitive feeding experiments have established that phenyllactic acid (79) is a precursor of the tropic acid moiety of hyoscyamine and scopolamine (hyoscine) (81) from *Datura stramonium*.[100–102] Transformed root cultures of a *Datura* hybrid were used to demonstrate that the 6α and 7α hydrogens of hyoscyamine (80) are retained in the formation of scopolamine (81).[103]

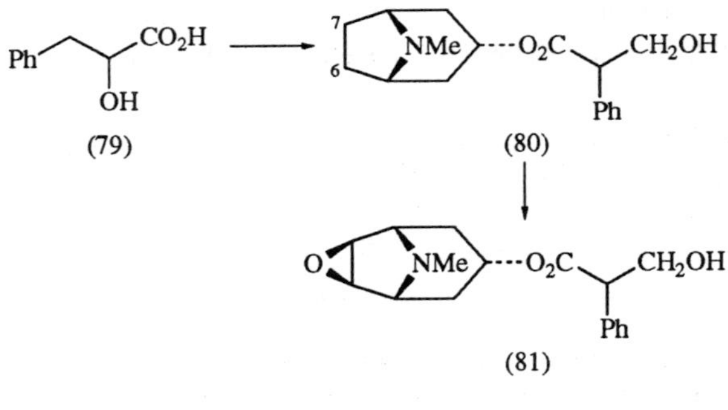

(79) (80) (81)

Transformed root cultures of *Nicotiana rustica* have been used to convert the unnatural precursor *N*-ethylputrescine (82) into (*S*)-*N*-ethylnornicotine (83), with an

[98] H. Cho, J. M. Beale, C. Graff, U. Mocek, A. Nakagawa, S. Omura, and H. G. Floss, *J. Am. Chem. Soc.*, 1993, **115**, 12296.
[99] M. Sauerwein, K. Shimomura, and M. Wink, *Phytochemistry*, 1993, **32**, 905.
[100] M. Ansarin and J. G. Wooley, *Phytochemistry*, 1993, **32**, 1183.
[101] M. Ansarin and J. G. Wooley, *J. Nat. Prod.*, 1993, **56**, 1211.
[102] Y. Kitamura, S. Nishima, H. Miura, and T. Kinoshita, *Phytochemistry*, 1993, **34**, 425.
[103] A. B. Watson, I. K. A. Freer, D. J. Robins, and N. J. Walton, *J. Nat. Prod.*, 1993, **56**, 1234.

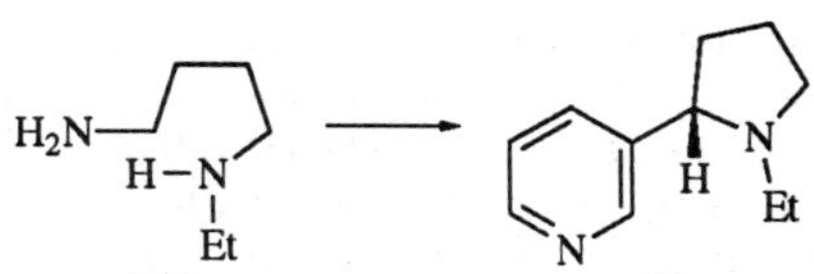

(82) (83)

efficiency similar to that of the natural system, which uses *N*-methylputrescine.[104]

Detailed labelling experiments using benzoic acid, benzaldehyde, 1-phenylpropane-1,2-dione (84), and cathinone (85) to investigate the biosynthesis of the *Ephedra* alkaloids such as ephedrine (86) have been reported.[105] The results indicated that benzoic acid rather than benzaldehyde is involved in the biosynthetic pathway and that two carbons are added from pyruvate to produce 1-phenylpropane-1,2-dione (84) and hence by transamination into cathinone (85) (Scheme 7).

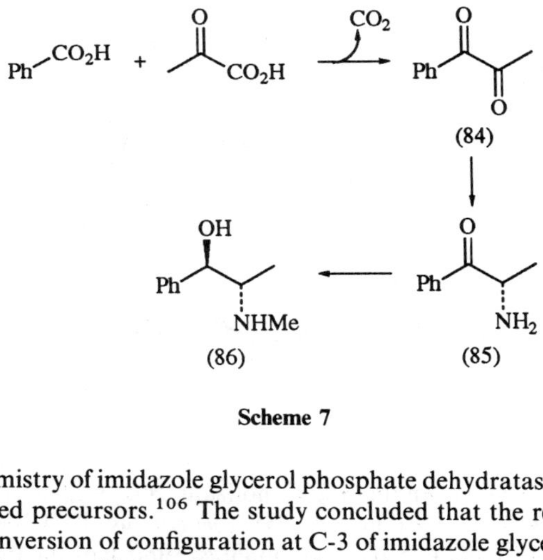

Scheme 7

The stereochemistry of imidazole glycerol phosphate dehydratase was studied using deuterium labelled precursors.[106] The study concluded that the reaction (Scheme 8) proceeded with inversion of configuration at C-3 of imidazole glycerol phosphate (87) and that the hydrogen added to C-2 comes from the medium.

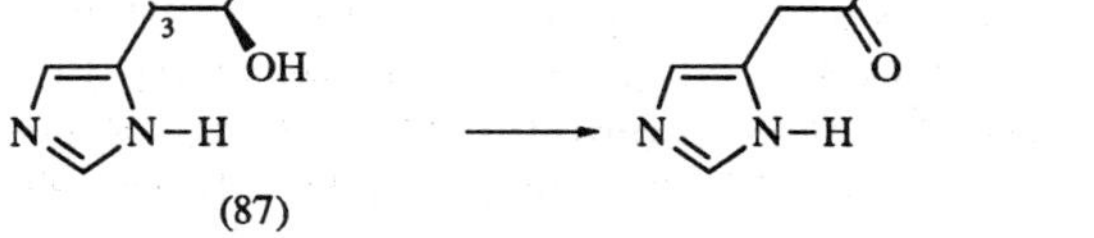

Scheme 8

Anatoxin a(s) (88) is a potent anticholinesterase produced by *Anabaena flosaquae*. Feeding experiments using ^{14}C-labelled precursors demonstrated that the carbon

[104] H. D. Boswell, A. B. Watson, N. J. Walton, and D. J. Robins, *Phytochemistry*, 1993, **34**, 153.
[105] G. Grue-Sørensen and I. D. Spenser, *J. Am. Chem. Soc.*, 1993, **115**, 2052.
[106] J. A. Moore, A. R. Parker, V. J. Davisson, and J. M. Schwab, *J. Am. Chem. Soc.*, 1993, **115**, 3338.

atoms of the main skeleton are derived from arginine and the three methyl groups come from the C_1 pool.[107] The labelled thiol (89) is incorporated into biotin (90) without loss of label; this finding gives some insight into the introduction of sulfur into biotin.[108]

Full details of the biosynthetic studies on obafluorin (91) from *Pseudomonas fluorescens* have been published.[109,110] The labelling studies indicated that 4-aminophenylalanine, 2,3-dihydroxybenzoic acid, and glycine (which supplies carbons 1 and 2) are precursors.

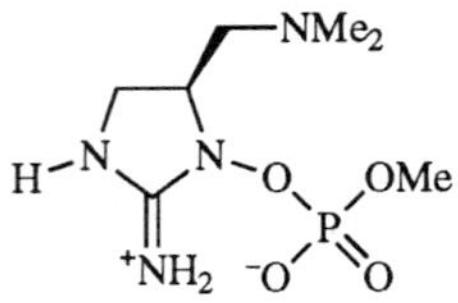

(88)

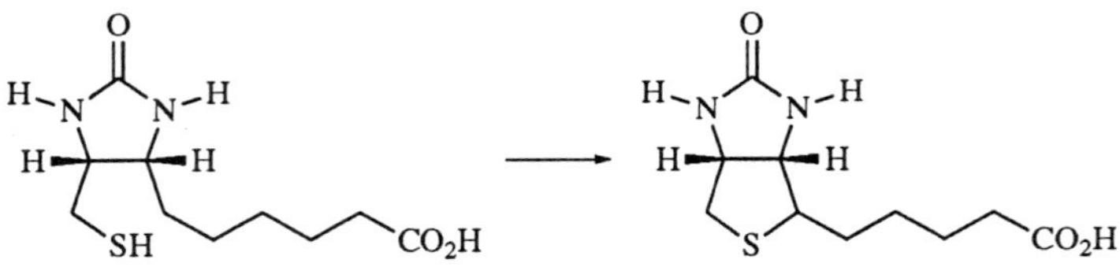

(89) (90)

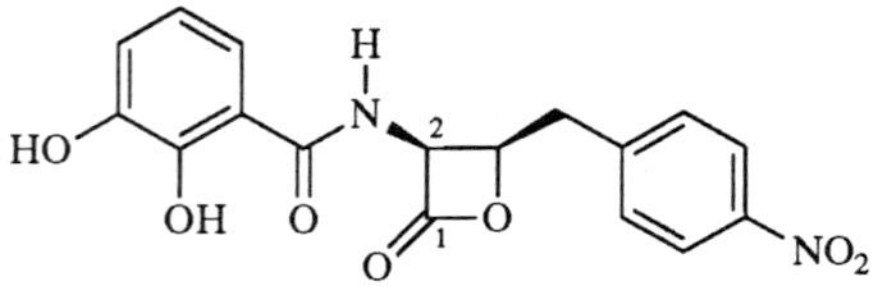

(91)

Labelling studies have shown that the quinolizidine (92) is an effective precursor of vertine (93) and lythrine (94) in *Decodon verticillatus*.[111] Labelled lysine was incorporated into ring A of these alkaloids and the results indicate that a symmetrical intermediate is involved. Full details of the biosynthetic studies on anosmine (95) from *Dendrobium anosmum* have appeared.[112]

A detailed study on the biosynthesis of streptazolin (96) from a *Streptomyces* species has revealed that it has a polyketide origin with the carbonyl carbon being derived from the C_1 pool.[113] The hydroxyl oxygen comes from molecular oxygen, which suggests an epoxide intermediate. Feeding experiments using [1,2-$^{13}C_2$]acetate to *Lycopodium tristachyum* indicated that the acetate derived C_3 units of lycopodine (97) are formed

107 B. S. Moore, I. Ohtani, C. B. de Konig, R. E. Moore, and W. W. Carmichael, *Tetrahedron Lett.*, 1992, **33**, 6595.

108 A. Marguet, F. Frappier, G. Guillerm, M. Azaulay, D. Florentin, and J.-C. Tabet, *J. Am. Chem. Soc.*, 1993, **115**, 2139.

109 R. B. Herbert and A. R. Knaggs, *J. Chem. Soc., Perkin Trans. 1*, 1992, 103.

110 R. B. Herbert and A. R. Knaggs, *J. Chem. Soc., Perkin Trans. 1*, 1992, 109.

111 S. H. Hedges, R. B. Herbert, and P. C. Wormald, *J. Nat. Prod.*, 1993, **56**, 1259.

112 T. Hemscheidt and I. D. Spencer, *J. Nat. Prod.*, 1993, **56**, 1281.

113 M. Mayer and R. Thiericke, *J. Org. Chem.*, 1993, **58**, 3486.

though a symmetrical intermediate.[114] The phenazine antibiotic saphenamycine (98) from *Streptomyces antibioticus* has been shown to be derived from metabolites of the shikimic pathway.[115]

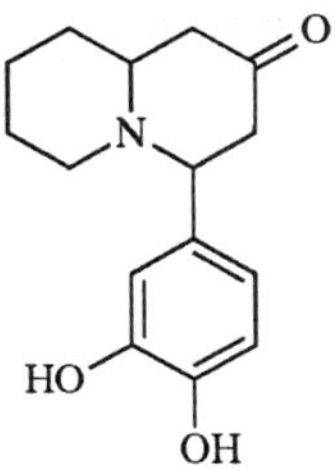

(92)

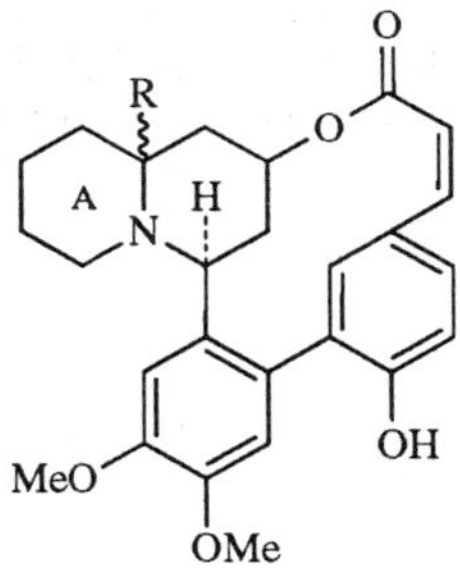

(93) R = αH
(94) R = βH

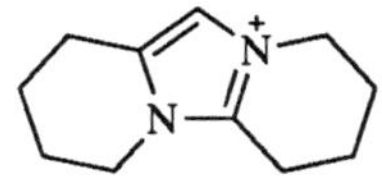

(95)

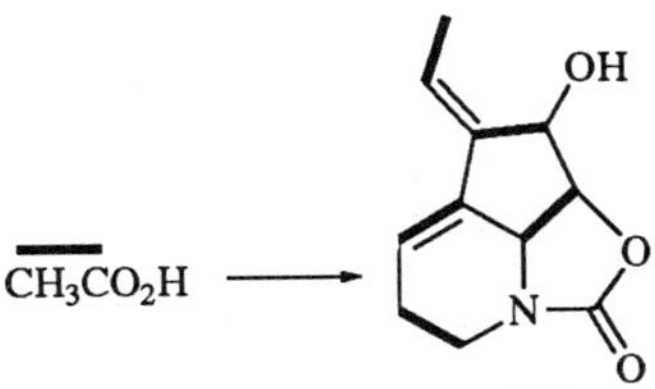

(96)

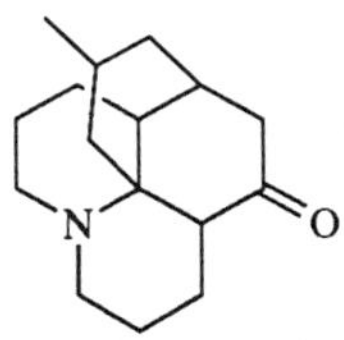

(97)

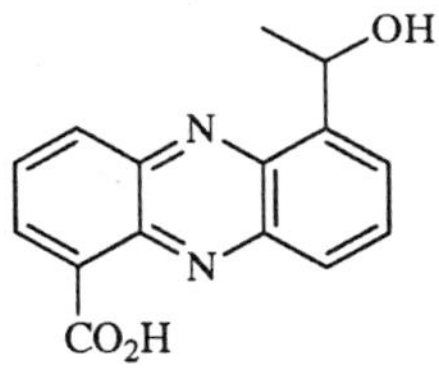

(98)

Oxazolomycin (99) has been shown to have a very interesting biosynthesis.[116] Glycine is incorporated as the starter unit of two separate polyketide chains. Propionate is not incorporated; all the methyl groups, including the *gem*-dimethyl groups, are derived from methionine. The biosynthetic origin of the remaining carbons is still to be established.

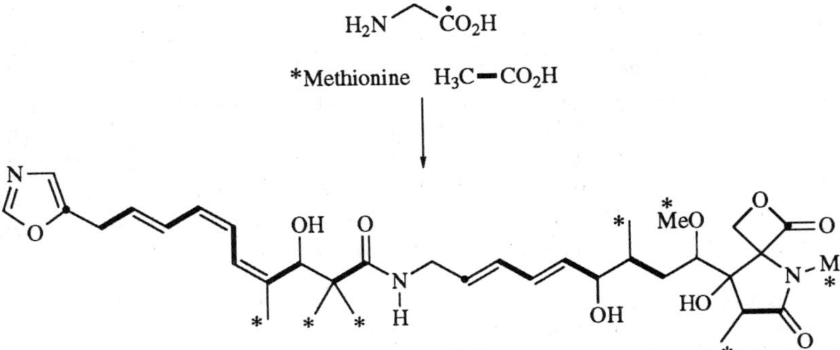

(99)

[114] T. Hemscheidt and I. D. Spencer, *J. Am. Chem. Soc.*, 1993, **115**, 3020.
[115] C. W. Van't Land, U. Mocek, and H. G. Floss, *J. Org. Chem.*, 1993, **58**, 6576.
[116] U. Gräfe, H. Kluge, and R. Thiericke, *Leibigs Ann. Chem.*, 1992, 429.

Further details of the incorporation of glycerol into the *m*-C_7N unit of asukamycin (100) from *Streptomyces nodosus* ssp. *asukaensis* have been published.[117] Biosynthetic studies on the aporphine and protoberberine alkaloids produced by cell cultures of *Peumus boldus* and *Berberis stolonifera* indicate demethylation and remethylation occur readily *via* the C_1 pool.[118] The biosynthesis of the cularine alkaloids in *Corydalis claviculata* and *Sarcapnos crassifolia* plants has been studied using ^{13}C and ^{18}O-labelled intermediates.[119] Precursor-directed biosynthesis of unnatural ergot alkaloids in *Claviceps purpurea* has been reported.[120]

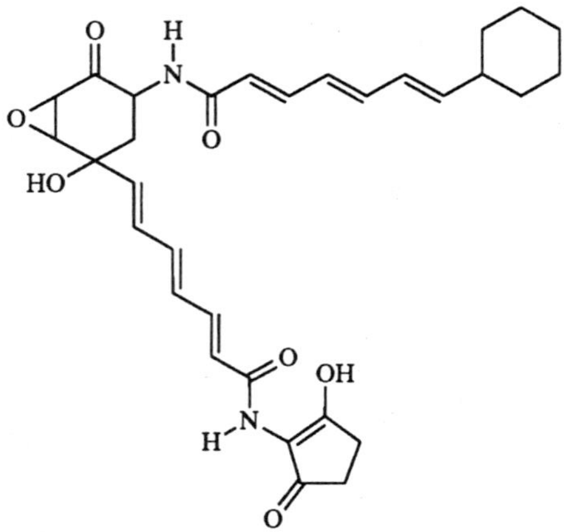

(100)

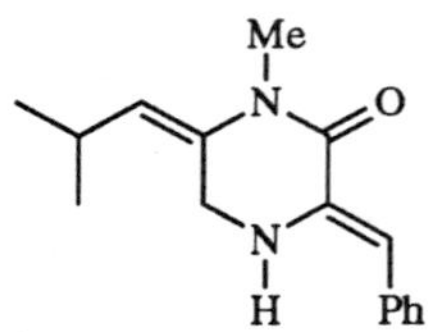

(101)

The biosynthetic pathways of a range of diketopiperazine alkaloids have been studied. *N*-Methylalbonoursin (101) from a *Streptomyces* species has been shown to be derived from leucine and phenylalanine.[121] Labelling studies using acetate and tyrosine have shown that all nine carbons of tyrosine are incorporated into sirodesmins A (102), B (103), and C (104) in *Sirodesmium diversum*.[122] The studies indicate that Claisen-type rearrangement of an *O*-(3,3-dimethylallyl)tyrosine derivative is involved in the biosynthesis. [2,4,5,6,7-$^2H_5$]Tryptophan was incorporated into roquefortine (105) by *Penicillium roqueforti* and into aszonalenin (106) by *Aspergillus zonatus* with retention in each case of all five deuterium atoms.[123] The 5a-hydrogen of both metabolites is retained; this precludes the involvement of any 2-substituted indole intermediates in the biosynthesis. Incorporation studies using labelled acetate and mevalonolactones have established that the C-22 methyl groups become scrambled during the biosynthesis of fumitremorgin B (107) and verruculogen (108) by *Penicillium verruculosum*.[124] It is proposed that the scrambling of the methyl groups occurs during

117 H. Cho, I. Sattlet, J. M. Beale, A. Zeech, and H. G. Floss, *J. Org. Chem.*, 1993, **58**, 7925.
118 B. Schneider and M. H. Zenk, *Phytochemistry*, 1993, **32**, 897.
119 M. J. Mueller and M. H. Zenk, *Leibigs Ann. Chem.*, 1993, 557.
120 N. C. Perellino, J. Malyszko, M. Ballabio, B. Gioia, and A. Minghetti, *J. Nat. Prod.*, 1992, **55**, 424.
121 K. A. Gurney and P. G. Mantle, *J. Nat. Prod.*, 1993, **56**, 1194.
122 J. D. Bu'Lock and L. E. Clough, *Aust. J. Chem.*, 1992, **45**, 39.
123 B. Bhat, D. M. Harrison, and H. M. Lamont, *Tetrahedron*, 1993, **49**, 10 663.
124 R. Veggaar, R. M. Horak, and V. J. Maharaj, *J. Chem. Soc., Chem. Commun.*, 1993, 274.

the formation of ring C. Evidence from feeding studies indicated that brevianamide A (109) from *Penicillium brevicompactum* is probably not formed *via* an indole derivative by a (4 + 2) cycloaddition.[125,126]

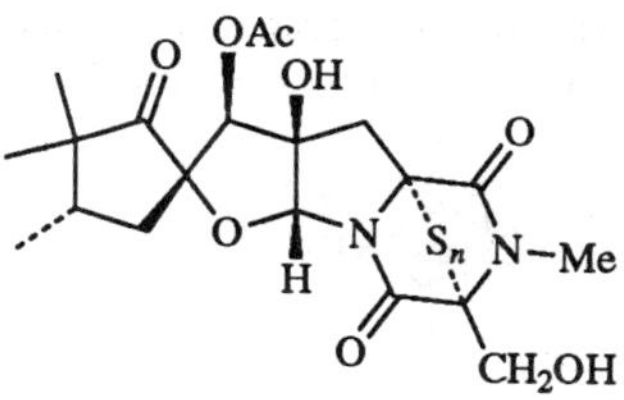

(102) $n = 2$
(103) $n = 3$
(104) $n = 4$

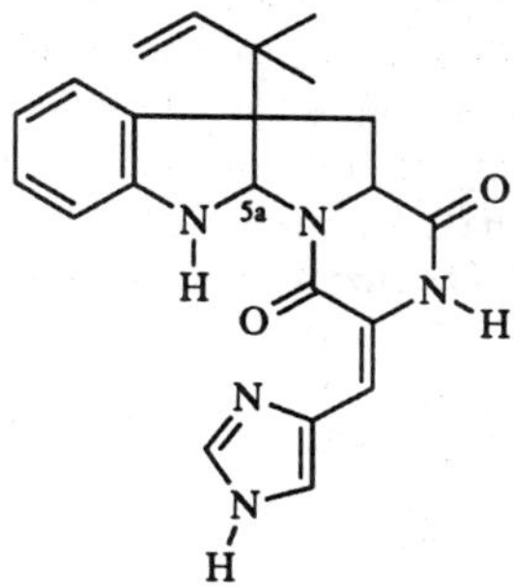

(105)

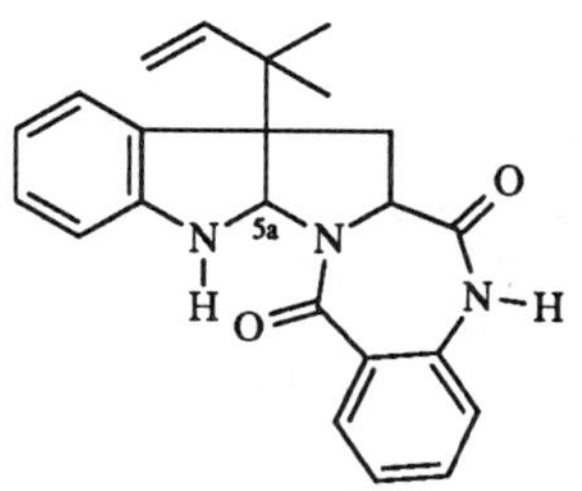

(106)

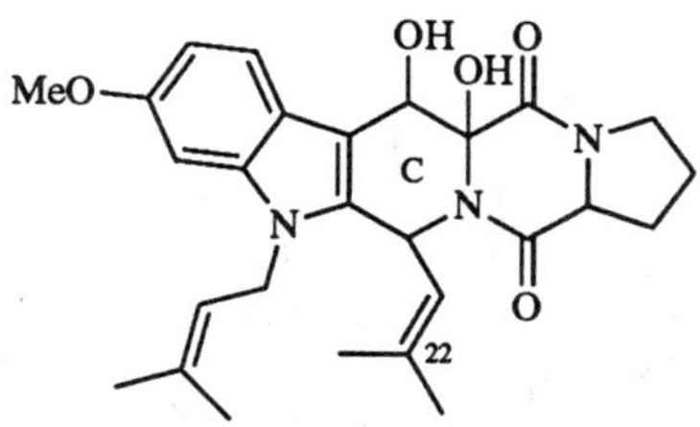

(107)

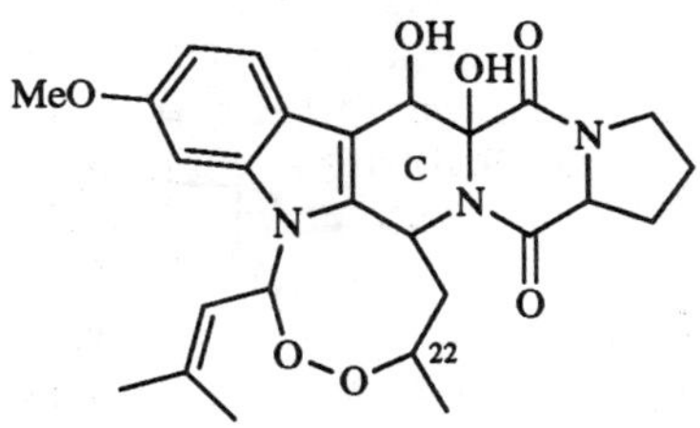

(108)

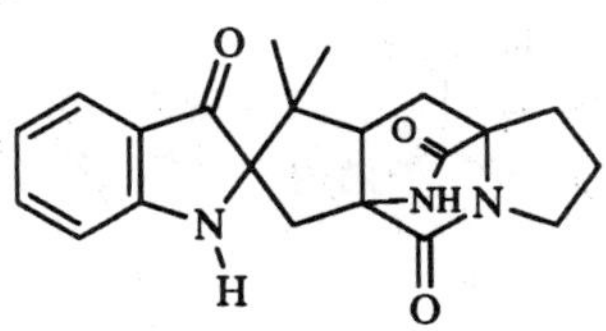

(109)

The biosyntheses of the modified peptides thiostreptin from *Streptomyces azureus* and *Streptomyces laurentii*,[127] nosiheptide from *Streptomyces actuosus*, [128] and the destruxins from *Metarhizum anisopliae*[129] have been investigated.

[125] J. F. Sanz-Cervera, T. Gliaka, and R. M. Williams, *J. Am. Chem. Soc.*, 1993, **115**, 347.
[126] J. F. Sanz-Cervera, T. Gliaka, and R. M. Williams, *Tetrahedron*, 1993, **49**, 8471.
[127] U. Mocek, Z. Zeng, D. O'Hagan, P. Zhou, L.-D. G. Fan, J. M. Beale, and H. G. Floss, *J. Am. Chem. Soc.*, 1993, **115**, 7992.
[128] U. Mocek, A. R. Knaggs, R. Tsuchuja, T. Nguyen, J. M. Beale, and H. G. Floss, *J. Am. Chem. Soc.*, 1993, **115**, 7557.
[129] A. Jegorov, P. Sedmera, and V. Matha, *Phytochemistry*, 1993, **33**, 1403.

More results have been published in the area of β-lactam antibiotic biosynthesis. Full details of the studies using labelled valine in the formation of the ACV tripeptide by ACV synthase have been reported.[130,131] The incorporation of ^{18}O from both molecular oxygen and water into β-lactams using deacetoxy and deacetylcephalosporin C synthase is interpreted as the first example of water exchange of intermediates in a non-haem iron, α-ketoglutarate dependent dioxygenase.[132]

Illuminating studies on the biosynthesis of clavulanic acid (10) in *Streptomyces clavuligerus* have been reported.[133–135] The pathway has been elucidated and is shown in Scheme 9. The enzyme that catalyses the conversion of proclavaminic acid (111) into

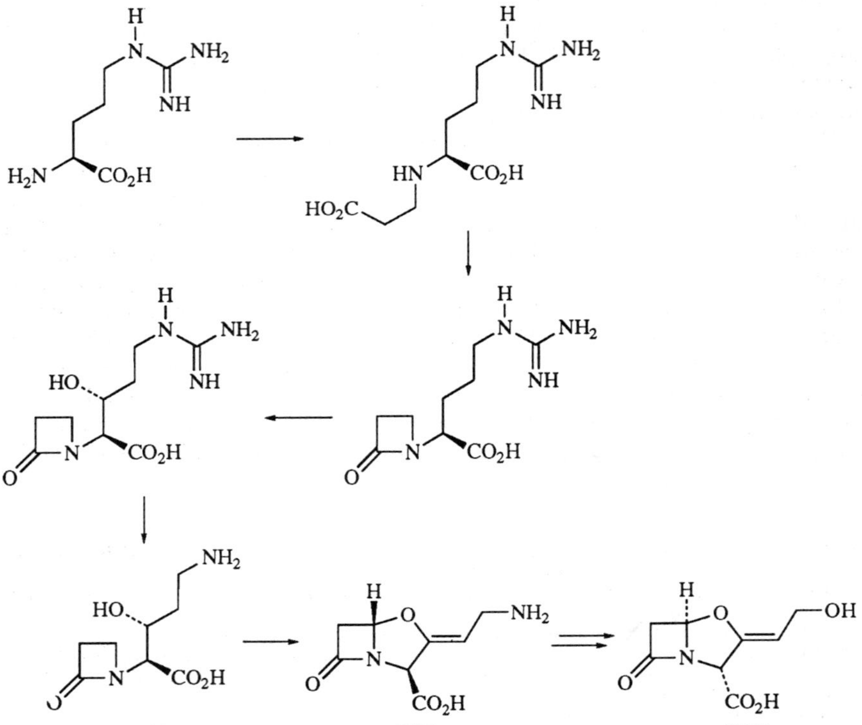

Scheme 9

[130] J.E. Baldwin, R.M. Adlington, J.W. Bird, R.A. Field, N.M. O'Callaghan, and C.J. Schofield, *Tetrahedron.*, 1992, **48**, 1099.

[131] J.E. Baldwin, M.F. Byford, R.A. Field, C.-Y. Shiau, W.J. Sobey, and C.J. Schofield, *Tetrahedron.*, 1993, **49**, 3221.

[132] J.E. Baldwin, R.M. Adlington, N.P. Crouch, I.A.C. Pereira, R.T. Aplin, and C. Robinson, *J. Chem. Soc., Chem. Commun.*, 1993, 105.

[133] B.P. Valentine, C.R. Bailey, A. Doherty, J. Morris, S.W. Elson, K.H. Baggaley, and N.H. Nicholson, *J. Chem. Soc., Chem. Commun.*, 1993, 1210.

[134] S.W. Elson, K.H. Baggaley, M. Fulston, N.H. Nicholson, J.W. Tyler, J. Edwards, H. Holms, I. Hamilton, and D. Mousdale, *J. Chem. Soc., Chem. Commun.*, 1993, 1211.

[135] S.W. Elson, K.H. Baggaley, M. Davison, M. Fulston, N.H. Nicholson, G.D. Risbridger, and J.W. Tyler, *J. Chem. Soc., Chem. Commun.*, 1993, 1212.

clavaminic acid (112) has been well studied.[136–140] Unnatural precursors have been used to probe the mechanism of this enzyme.

6 Porphyrins

A book covering the biosynthesis of both linear and cyclic tetrapyrroles has been published.[141] Phycobilins are open-chain tetrapyrroles that function as chromophores of light-harvesting chromoproteins in some photosynthetic organisms. The biosynthesis of these phycobilins has been reviewed.[142] Genetic studies have had a great impact on studies of the biosynthesis of the tetrapyrroles.[143]

Further evidence for the involvement of a spiro intermediate in the formation of uroporphyrinogen III from a linear hydroxymethylbilane has been presented.[144] A synthetic spirolactam related to the proposed spiro intermediate was shown to be an inhibitor of cosynthetase (uroporphyrinogen III synthase).

The studies of the biosynthesis of vitamin B_{12} have been progressing and the pathway is now very well understood (Scheme 10).[145] The fate of the oxygen atoms in precorin-2 (112) when it is converted enzymatically into cobyrinic acid has been studied.[146] The results indicated that the carboxyl group of the C-2 acetate group (*) in precorrin-2 (113) loses one labelled oxygen atom. This is consistent with the involvement of this carboxyl group in lactone formation. The structure of precorrin-3A (114) has been confirmed.[147] Precorrin-3A (114) is converted into the hydroxy γ-lactone, precorrin-3B (3x) (115) by the enzyme encoded by gene CobG from *Pseudomonas denitrificans*.[148,149] This enzyme contains iron and incorporates molecular oxygen into precorrin-3B (115).[150] The ring contraction of precorrin-3B (115) is catalysed by the enzyme corresponding to gene CobJ. The product of this transformation was identified as precorrin-4 (116).[151] Other workers identified factor IV, the

136 J. E. Baldwin, R. M. Adlington, J. S. Bryans, M. D. Lloyd, T. J. Sewell, C. J. Schofield, K. H. Baggaley, and R. Cassels, *J. Chem. Soc., Chem. Commun.*, 1992, 877.
137 J. E. Baldwin, K. D. Merritt, C. J. Schofield, S. W. Elson, and K. H. Baggaley, *J. Chem. Soc., Chem. Commun.*, 1993, 1301.
138 D. B. McIlwaini and C. A. Townsend, *J. Chem. Soc., Chem. Commun.*, 1993, 1346.
139 D. Iwata-Reuyl, A. Basak, L. S. Silverman, C. A. Engle, and C. A. Townsend, *J. Nat. Prod.*, 1993, **56**, 1373.
140 J. E. Baldwin, V. Lee, M. D. Lloyd, C. J. Schofield, S. W. Elson, and K. H. Baggaley, *J. Chem. Soc., Chem. Commun.*, 1993, 1694.
141 'Biosynthesis of Tetrapyrroles', ed. P. M. Jordan, Elsevier, Amsterdam, 1991.
142 S. I. Beale, *Chem. Rev.*, 1993, **93**, 785.
143 A. I. Scott, *Tetrahedron.*, 1992, **48**, 2559.
144 W. M. Stark, C. J. Hawker, G. J. Hart, A. Philippides, P. M. Petersen, J. D. Lewis, F. J. Leaper, and A. R. Battersby, *J. Chem. Soc., Perkin Trans. 1*, 1993, 2875.
145 A. R. Battersby, *Acc. Chem. Res.*, 1993, **26**, 15.
146 R. A. Vishwakarma, S. Balachandran, A. I. D. Alanine, N. P. J. Stamford, F. Kiuchi, F. J. Leeper, and A. R. Battersby, *J. Chem. Soc., Perkin Trans. 1*, 1993, 2893.
147 M. J. Warren, C. A. Roessner, S.-I. Ozaki, N. J. Stolowich, P. J. Santander, and A. I. Scott, *Biochemistry*, 1992, **31**, 603.
148 A. I. Scott, C. A. Roessner, N. J. Stolowich, J. B. Spencer, C. Min, and S.-I. Ozaki, *FEBS Lett.*, 1993, **331**, 105.
149 L. Debussche, D. Thibaut, M. Danzer, F. Debu, D. Fréchet, F. Herman, F. Blanche, and M. Vuilhorgne, *J. Chem. Soc., Chem. Commun.*, 1993, 1100.
150 J. B. Spencer, N. J. Stolowich, C. A. Roessner, C. Min, and A. I. Scott, *J. Am. Chem. Soc.*, 1993, **115**, 11 610.
151 A. I. Scott, C. A. Roessner, N. J. Stolowich, J. B. Spencer, C. Min, and S.-I. Ozaki, *FEBS Lett.*, 1993, **331**, 105.

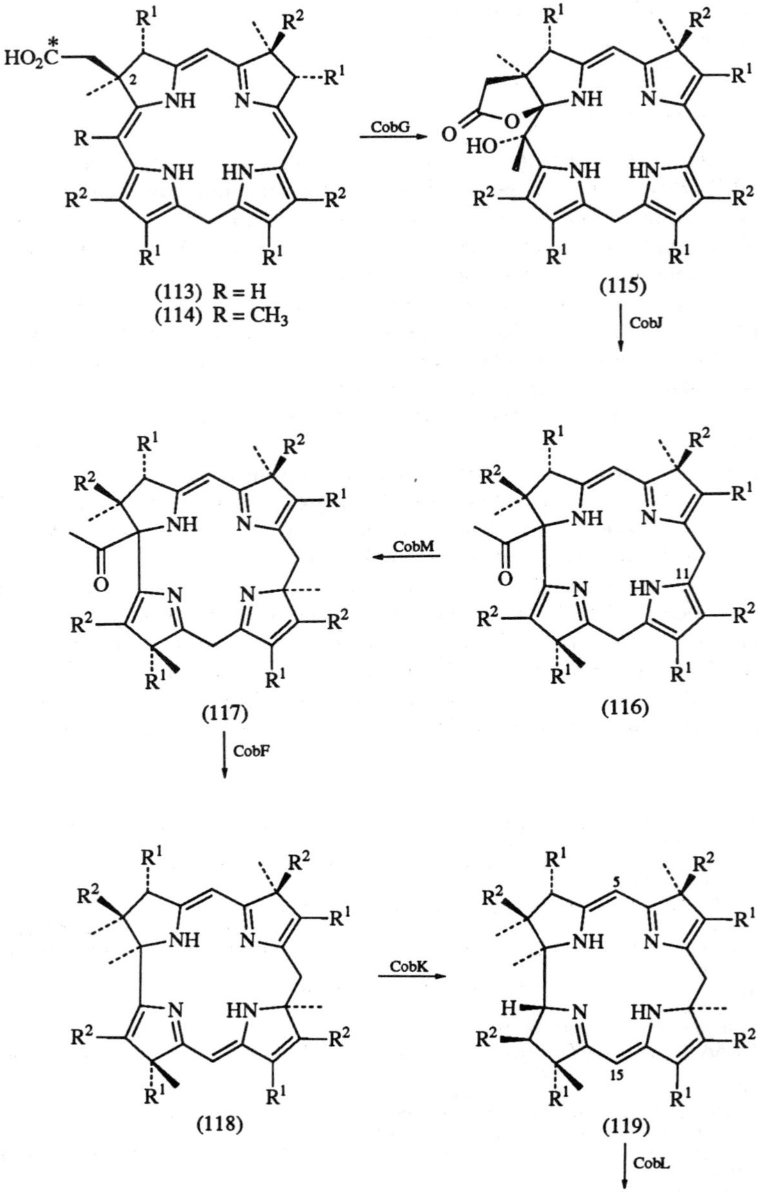

HO$_2$C*
2
R^1
R^2
R
NH
N
HN
CobG
O
HO
CobJ
CobM
11
CobF
CobK
H
5
15
CobL
(113) R = H
(114) R = CH$_3$
(115)
(116)
(117)
(118)
(119)

Scheme 10

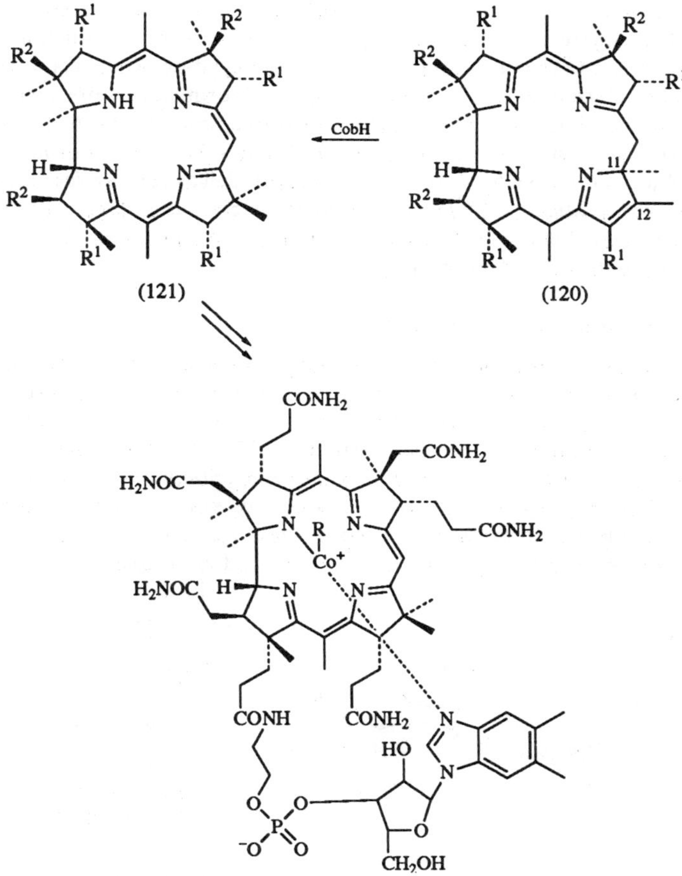

(121) (120)

$R^1 = CH_2CH_2CO_2H$; $R^2 = CH_2CO_2H$

Scheme 10

oxidation product of precorrin-4.[152,153] The next step in the biosynthetic pathway is the addition of a methyl to C-11 of precorrin-4 (116) to produce precorrin-5 (117).[154] The acetyl group of precorrin-5 (117) is replaced by a methyl group to produce precorrin-6A (6x) (118).[155] Precorrin-6A (6x) reductase[156,157] (encoded by the gene

[152] D. Thibaut, L. Debussche, D. Fréchet, F. Herman, M. Vuilhorgne, and F. Blanche, *J. Chem. Soc., Chem. Commun.*, 1993, 513.
[153] L. Debussche, D. Thibaut, M. Danzer, F. Debu, D. Fréchet, F. Herman, and M. Vuilhorgne, *J. Chem. Soc., Chem. Commun.*, 1993, 1100.
[154] C. Min, B. P. Atshaves, C. A. Roessner, N. J. Stolonwich, J. B. Spencer, and A. I. Scott, *J. Am. Chem. Soc.*, 1993, **115**, 10 380.
[155] F. Blanche, M. Kodera, M. Couder, F. J. Leper, D. Thibaut, and A. R. Battersby, *J. Chem. Soc., Chem. Commun.*, 1992, 138.
[156] F. Kiuchi, D. Thibaut, L. Debussche, F. J. Leeper, F. Blanche, and A. R. Battersby, *J. Chem. Soc., Chem. Commun.*, 1992, 306.
[157] K. Ichinose, M. Kodera, F. J. Leeper, and A. R. Battersby, *J. Chem. Soc., Chem. Commun.*, 1993, 515.

CobK[158]) catalyses the reduction of precorrin-6A (118) to form precorrin-6B (6y) (119).[159] The enzyme that is encoded by CobL[160] catalyses the introduction of methyl groups at C-5 and C-15 of precorrin-6B and decarboxylation of the acetate side chain on C-12 to produce precorrin-8x (120).[161] The transformation of precorrin-8x (120) into hydrogenobyrinic acid (121) is catalysed by the enzyme encoded by CobH[162] and involves the migration of the methyl group at C-11 to C-12 in a [1,5]-sigmatropic shift.

The pathway to vitamin B_{12} in anaerobic bacteria such as *Salmonella typhimurium* is parallel but nonidentical.[163] The methyl group of methionine has been shown to be the source of the methyl groups at C-2 and C-7 of the porphyrin of cytochrome c_3 in the anaerobic bacterium *Desulfovibrio vulgaris*.[164]

7 Miscellaneous Metabolites

Reviews on the biosynthesis of bacterial cell wall peptidoglycan[165] and the use of chiral methyl groups in biosynthetic studies[166] have been published. Further details of the biosynthesis of fosfomycin (122) from *Streptomyces fradiae* have been published.[167,168] The C-3 deoxygenation of pentopyranine C (123) from *Streptomyces griseochromogenes* has been studied.[169] Labelling studies using *Streptomyces subrutilus*, which produces 1-deoxynojirimycin (124) and 1-deoxymannonojirimycin (125), have shown that the biosynthetic pathway involves the isomerism of glucose to fructose followed by oxidation of C-6 to an aldehyde, transamination, and cyclization.[170]

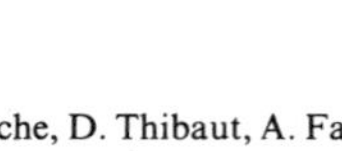

(122)

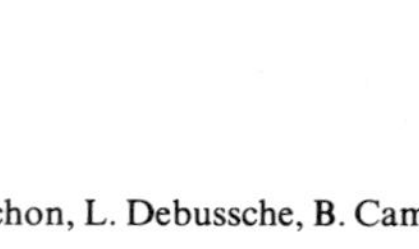

(123)

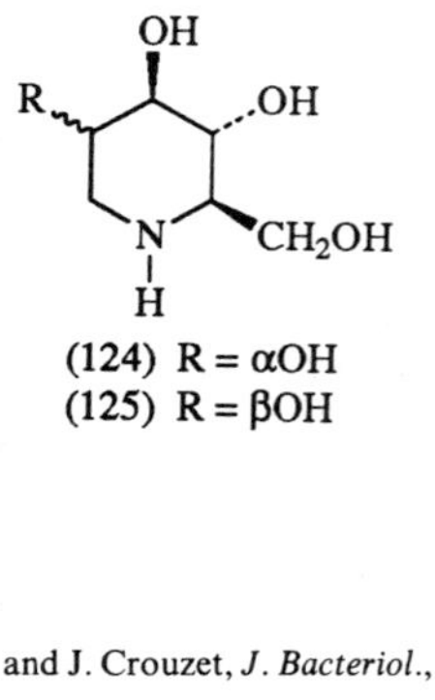

(124) R = αOH
(125) R = βOH

[158] F. Blanche, D. Thibaut, A. Famechon, L. Debussche, B. Cameron, and J. Crouzet, *J. Bacteriol.*, 1992, **174**, 1036.
[159] F. Blanche, F. Kuichi, L. Debussche, F. J. Leeper, F. Blanche, and A. R. Battersby, *J. Chem. Soc., Chem. Commun.*, 1992, 139.
[160] F. Blanche, A. Famechon, D. Thibaut, L. Debussche, B. Cameron, and J. Crouzet, *J. Bacteriol.*, 1992, **174**, 1050.
[161] D. Thibaut, F. Kiuchi, L. Debussche, F. Blanche, M. Kodera, F. J. Leeper, and A. R. Battersby, *J. Chem. Soc., Chem. Commun.*, 1992, 982.
[162] D. Thibaut, M. Couder, A. Famechon, L. Debussche, B. Cameron, J. Crouzet, and F. Blanche, *J. Bacteriol.*, 1992, **174**, 1050.
[163] J. R. Roth, J. G. Lawrence, M. Rubenfield, S. Kieffer-Higgins, and G. M. Church, *J. Bacteriol.*, 1993, **175**, 3303.
[164] H. Akutsu, J.-S. Park, and S. Sano, *J. Am. Chem. Soc.*, 1993, **115**, 12 185.
[165] T. D. H. Bugg and C. T. Walsh, *Nat. Prod. Rep.*, 1992, **9**, 199.
[166] H. G. Floss and S. Lee, *Acc. Chem. Res.*, 1993, **26**, 116.
[167] F. Hammerschmidt, *Leibigs Ann. Chem.*, 1992, 553.
[168] F. Hammerschmidt and H. Kählig, *Leibigs Ann. Chem.*, 1992, 1201.
[169] S. J. Gould and J. Guo, *J. Am. Chem. Soc.*, 1992, **114**, 10 176.
[170] D. J. Hardick, D. W. Hutchinson, S. J. Trew, and E. M. H. Wellington, *Tetrahedron*, 1992, **48**, 6258.

Author Index

Aase, K., 194
Abboud, K.A., 145
Abd-El-Aziz, A.S., 264
Abdourazak, A.H., 169
Abell, C., 29
Aboutayab, K., 86
Achiwa, K., 252
Adachi, T., 160, 302
Adam, W., 122, 156, 158, 182, 284, 309
Adams, J., 27
Adger, B.J., 181
Adlington, R.M., 186, 332, 333
Aebersold, R., 29
Afridi, A.S., 160
Agarwal, K., 320
Agarwal, R., 148, 171
Agati, V., 141
Agawa, A., 279
Agbossou, F., 248
Ager, D.J., 106
Agrios, K.A., 235
Aguilar, M.A., 60
Ahlberg, P., 148
Ahmed, G., 263
Ahmed, H.A., 256
Ahmed, S.A., 316
Ahn, J.H., 282
Ahuja, J.R., 107
Aihara, J., 148
Aimar, M.L., 162
Airoldi, M., 254
Akabori, S., 266
Åkermark, B., 198, 249
Akhrem, I., 159
Aki, S., 158
Akinyele, E.T., 162
Akita, H., 307
Akutagawa, S., 282
Akutsu, H., 336
Alagona, G., 58
Alami, M., 167, 232, 233, 235
Alanine, A.I.D., 333
Alberts, I.L., 80
Albini, A., 107, 168
Albizati, K.F., 112
Albrecht, M., 107
Albrecht, P., 159
Albrecht, W., 311
Alcaide, B., 186
Alcon, M., 132
Alexander, I.J., 257
Ali, S.A., 197
Aliev, A.E., 19
Allain, E.J., 121, 306
Allemand, P.M., 25
Allen, J.V., 271
Allen, M.H., 29
Aller, E., 196
Allinson, G., 148
Allison, C.E., 24
Allmaier, G., 27
Almena, J., 128
Almlof, J., 57
Almstead, N.G., 108
Alonso-Peramau, D., 75
Alpegiani, M., 185
Alper, H., 129
Alt, C., 248
Alvarez, C., 286
Alves, R.B., 138
Aly, A.A., 176
Amat, M., 108, 235
Amberg, W., 284
Amburgey, J., 115, 276
Amos, R.D., 53, 56, 57
Amurrio, D., 265
An, D.K., 282
Anaya, J., 183
Andemichael, Y.W., 93
Andersen, R.J., 313
Andersh, B., 86
Anderson, B.A., 123, 138, 150, 181, 288
Anderson, H.L., 105, 141
Anderson, J.C., 292
Anderson, S., 25, 105
Anderson, W.K., 159
Andersson, C.-M., 227
Andersson, P.G., 119, 122, 235, 247, 249, 284, 285
Ando, A., 295
Ando, C., 224
Ando, M., 128
Ando, Y., 26
Andrés, C., 271
Andrus, M.B., 105
Andzelm, J., 54
Angell, R., 277
Angle, S.R., 206, 208
Anglister, J., 6, 7
Anguyan, J.G., 60
Annan, R.S., 31
Annby, U., 234
Annunziata, R., 147
Ansarin, M., 326
Ansink, H.R.W., 162
Anthony, J., 174
Anzini, M., 210
Aoki, K., 296
Aoki, Y., 186
Aoyagi, K., 243, 244
Aoyama, T., 202
Aplin, R.T., 31, 332
Appella, E., 31
Aqvist, J., 58
Arai, I., 30
Arai, K., 233
Arai, S., 128, 273
Araki, M., 278
Arcadi, A., 226, 233, 237
Archelas, A., 305
Archer, D.B., 31
Arif, A.M., 141, 248
Arigoni, D., 311, 319
Arison, B.H., 315
Armentrout, P.B., 27
Arnold, D.R., 168
Arnold, M., 259
Arnone, A., 316
Arnswald, M., 159
Arredondo, Y., 199
Arrington, M.P., 284
Arseniyadis, S., 138
Arts, H.J., 204
Arumugam, S., 18
Asano, H., 266
Asano, K., 172
Asao, N., 106, 132
Asao, T., 257
Asayama, M., 235
Asensio, G., 182
Asfari, Z., 105
Ashby, E.C., 162
Ashby, G.A., 29
Ashe, A.J., III, 259
Ashimori, A., 243
Ashton, P.R., 147
Assante, G., 316
Astruc, D., 266

Ateeq, H.S., 74, 263
Atshaves, B.P., 335
Attardo, G., 202
Atta-ur-Rahman, 151
Audier, H.E., 22, 24
Audouin, M., 223
Aumann, R., 151
Aurrecoechea, J.M., 138
Austin, S.J., 148
Ay, M., 154, 219, 235
Ayer, S.W., 30
Ayer, W.A., 314
Azaulay, M., 328
Aznar, F., 141
Azuma, R., 102, 150

Baba, M.S., 26
Babin, P., 235
Babu, R.L., 308
Bach, T., 108
Bachi, M.D., 100
Bachrach, S., 79
Baciocchi, E., 95, 193
Badone, D., 108
Bäckvall, J.-E., 134, 247, 249, 252
Baer, C.D., 264, 265
Baer, T., 27
Baerends, E.J., 66
Bagatti, M., 75
Baggaley, K.H., 332, 333
Bahktiar, R., 27
Bahr, U., 28
Bailey, C.R., 332
Baker, B.J., 318
Baker, R.W., 267
Baker, T.M., 267
Balachandran, S., 333
Balasubramanian, R., 26
Balasubramanian, S., 257
Balavoine, G.G.A., 287
Baldenius, K.U., 149
Baldoli, C., 262, 263
Baldridge, K.K., 146, 149
Baldwin, J.E., 186, 276, 332, 333
Baldwin, K.P., 233
Ball, R.G., 125
Ballabio, M., 330
Ballestri, M., 82, 129, 280
Baltisberger, J.H., 16
Bando, T., 119
Banerjee, A.K., 141
Banerji, A., 290
Banfi, L., 106
Banister, J.A., 217
Bányai, I., 4
Bao, J., 138, 150, 278
Barbey, S., 200
Bardone, D., 227
Bardshiri, E., 316
Bareket, Y., 286
Barford, R.A., 59
Barili, P., 304
Barlow, S.J., 159
Barluenga, J., 141, 158
Barnes, C.J., 29
Barnes, D.M., 123, 181, 288
Barnier, J.P., 300
Barrall, G.A., 16
Barras, J.-P., 267
Barrett, A.G.M., 167, 244
Barrett, C., 181
Barton, D.H.R., 94, 183
Bartsch, R.A., 266
Barz, W., 322
Basak, A., 333
Bashiardes, G., 267
Basilevsky, M.V., 60
Baskaran, S., 294
Bastock, T.W., 159
Bates, R.W., 224
Bathnagar, S., 79
Bats, J.W., 108
Batsanov, A.S., 266
Batta, G., 4, 7
Battersby, A.R., 214, 216, 333, 335, 336
Bau, R., 57
Bauch, H.-J., 323
Bauermeister, M., 257
Bax, A., 6, 7, 9, 11
Bax, B.M., 151, 222
Bazinger, M., 300
Beak, P., 108, 166, 188, 297
Beale, J.M., 317, 324, 326, 330, 331
Beale, S.I., 333
Beaucage, S.L., 106
Beaudet, I., 240
Beaudoin, S., 121, 286
Beck, W., 29
Becke, A.D., 52, 57
Becker, D.P., 220
Becker, H., 108
Beckett, M.A., 266
Beckett, R.P., 267
Beckham, S., 149
Beckhaus, H.D., 78
Beckwith, A.L.J., 89, 106
Bedeschi, A., 237, 308
Beer, P.D., 266
Behar, V., 210
Behnen, W., 271
Beholz, L.G., 159
Beifuss, U., 107
Belenlkii, L.I., 160
Beley, M., 230
Belfoure, V., 160
Bell, P.T., 256
Bellina, F., 235
Bellosta, V., 286
Bellur, N.S., 107
Beltrane, P., 76
Belyk, K., 113
Benareb, A., 206
Ben-David, Y., 224, 227
Benhaddou, R., 286
Benincori, T., 196
Bennani, Y.L., 119, 121, 284, 285, 286
Benneche, T., 200, 250, 265
Benner, S.A., 290
Bennet, N.S., 291
Bennetau, B., 106, 158, 235
Benniston, A.C., 266
Bent, B.F., 167
Bentley, J.M., 77
Benyan, L., 266
Berces, A., 54, 57
Bergander, K., 4, 8
Bergdahl, M., 112
Berger, E., 270
Berger, S., 4, 8
Berinstain, A.B., 148
Berk, S.C., 220
Berk, S.E., 138
Berkessel, A., 287
Berkowitz, J., 23, 25
Bermel, W., 11
Bernard, J.M., 252
Bernardi, A., 66
Bernardi, F., 66, 67, 75, 145
Bernardinelli, G., 263, 265
Bernardini, C.B., 182
Bernier, J., 119, 294
Bernstein, M.A., 160
Bernstein, M.P., 108
Berrios-Peña, N.G., 249
Berson, J.A., 148
Bertha, F., 185
Berthomieu, D., 24
Berti, G., 304
Bertinato, P., 105, 231, 269
Bertrand, G., 187
Beruben, D., 135
Berven, H.M., 138
Besançon, J., 264
Beslim, P., 78
Bethell, D., 289
Beveridge, D.L., 58
Bhalerao, U.T., 308
Bhandari, P., 324
Bhat, B., 330
Bhatia, B., 287
Bhatia, S., 287
Bhatt, M.V., 156
Bhatt, R.K., 112
Bhattacharya, S., 166, 240
Bhushan, V., 285
Bicchierini, N., 155
Bichard, C.J.F., 191
Bickelhaupt, F.M., 66, 176
Biemann, K., 31
Biesemans, M., 3
Biggs, T.N., 155
Bigler, P., 8
Bilkis, I.I., 168
Billups, W.E., 145
Bilodeau, M.T., 123, 181, 288
Bird, J.W., 332
Birkett, M.A., 193

Bisarya, S.C., 158, 295
Bisht, K.S., 302
Bissinger, P., 256
Bissolino, P., 185
Bitsch, F., 30
Bjarke, B., 302
Björk, P., 229
Black, M., 86
Bläser, D., 145, 259
Blagg, J., 244
Blake, J.K., 71
Blake, N.E., 323
Blanche, F., 333, 335, 336
Blanco, L., 300
Blank, S., 188
Blart, E., 252
Blau, N., 31
Blechert, S., 117, 138, 219
Bley, J., 107
Blomberg, M., 249
Blosser, P.W., 254
Bobbitt, K.L., 128
Bobzin, S.C., 314
Bodenhausen, G., 10, 15
Bodwell, G.J., 267
Boere, B.B., 171
Boernsen, K.O., 23
Boese, R., 145, 259, 266
Bohlmann, R., 318
Bohm, S., 79
Bohme, D.K., 22
Boireau, G., 283
Boland, W., 311, 312, 319
Bolin, G., 227
Bolm, C., 106, 130, 280, 286
Bolton, G.L., 140
Bonjoch, J., 102
Bonner, M.P., 79, 80
Bonnert, R., 208
Booker-Milburn, K.I., 92
Boos, A., 323
Booze, J.A., 27
Bordeleau, J., 119, 294
Borden, W.T., 146
Borghi, D., 185
Borgis, D., 57
Borhan, B., 304
Bornmann, W.G., 230
Borzilleri, R.M., 192
Bosch, E., 100, 287
Bosch, J., 102, 108, 235
Boswell, H.D., 327
Bottoni, A., 75
Bouchoux, G., 24
Bouhlel, E., 287
Bourdin, B., 293
Bouzide, A., 138
Bovens, M., 253
Bovonsombat, P., 158
Bowen, R.D., 21, 24, 25
Bowie, J.H., 24
Boyd, D.R., 148, 305
Boyd, R.J., 57
Boyde, D.R., 171
Boylan, M.J., 192
Boyle, R., 232, 305
Boyle, T.J., 12
Brackeen, M.F., 291
Brackenridge, I., 300
Braddock, D.C., 141
Bradshaw, J.D., 233
Bradshaw, J.S., 105, 214
Brakta, M., 228
Braley, T.L., 171
Bramble, F.Q., Jr., 30
Brands, M., 263
Brandvold, T.A., 138, 150
Brauman, J.I., 287
Braun, M., 111
Bree, A.V., 57
Breen, A.P., 103, 104
Breitenbacher, J.G., 206
Brenna, E., 196
Brennan, J., 181
Breuilles, P., 301
Brezova, V., 168
Brieva, R., 303, 304
Brimble, M.A., 153
Brintzinger, H.H., 138, 236
Brisset, H., 131
Brodbelt, J., 22
Broene, R.D., 131, 281
Bronnimann, C.E., 13
Brook, A.G., 74
Brook, M.A., 23
Brookhart, M., 255
Brophy, J., 74
Brossier, P., 264
Brown, D., 191
Brown, D.A., 256
Brown, G.R., 147
Brown, H.C., 214, 282
Brown, J.M., 106, 131, 141, 294
Brown, R.F.C., 107, 146, 171
Brown, S., 237
Bruche, L., 213
Brückner, R., 111, 150, 233
Brummond, K.M., 140
Brumwell, J.E., 84, 138
Brunel, J.-M., 297
Brunner, H., 267, 270
Brunner, M., 287
Brussee, J., 271
Bryans, J.S., 333
Bryce, M.R., 266
Bssaibis, M., 77
Bubenitschek, P., 171
Buchanan, R., 30
Buchwald, S.L., 131, 138, 220, 244, 281
Bugg, T.D.H., 336
Bulman Page, P.C., 289
Bu'Lock, J.D., 330
Bunn, B.J., 296
Bunnage, M.E., 293
Bunt, R.C., 249
Bunz, U.H.F., 136, 241
Buono, G., 297
Bupp, J.E., 171
Burchat, A.F., 128
Burdeniuk, J., 224
Burdisso, M., 75
Bures, E.J., 29
Burgers, P.C., 21, 22, 23
Burgess, V.A., 267
Burgess-Henry, J., 290
Burk, M.J., 131
Burk, R.M., 290
Burlingame, A.L., 30
Burt, S.K., 57
Burton, D.J., 232
Burton, N.A., 59, 60
Busacca, C.A., 166, 225
Buser, H.R., 30
Bushby, R.J., 148
Busia, K., 156
Butenschön, H., 263
Buttery, C.D., 188
Bycroft, B.W., 320
Byford, M.F., 332
Byrne, K.M., 315
Byrne, M.P., 295

Cabaj, J., 273
Cabiddu, S., 165
Cabri, W., 237
Cacchi, S., 222, 226, 233, 237
Caddick, S., 86, 97
Cadogan, J.I.G., 86
Cadoni, E., 76
Caffyn, A.J.M., 245
Cai, D., 280
Cai, Z.-W., 227
Cajthami, C.E., 125, 296
Caldwell, K.A., 25
Calkins, T.L., 93, 138, 149
Callahan, J.H., 25
Calter, M.A., 126
Cameron, B., 336
Campbell, E.E.B., 26
Campbell, M.M., 107
Campbell, S.J., 211
Cámpora, J., 244
Campus, P.J., 158
Candiani, I., 237
Cane, D.E., 314, 319, 324
Cao, Y., 264
Cappelli, A., 210
Caputo, R., 164
Car, R., 56
Carless, H.A.J., 156
Carling, R.W., 190
Carlsen, P.H.J., 194
Carlson, H.A., 58
Carmichael, D., 208
Carmichael, W.W., 328
Carofiglio, T., 244
Carpenter, B.K., 78
Carpenter, K., 155
Carpita, A., 235
Carreño, M.C., 271

Carter, D.S., 165
Carter, J.T., 57
Cartmell, E., 211
Cartwright, G.A., 86
Carty, A.J., 14
Carvalho, W.A., 106
Carver, J.A., 258
Casarini, D., 147
Casati, R., 324
Case-Green, S.C., 267
Casida, K.C., 57
Casida, M.E., 57
Caspi, E., 321
Cassels, R., 333
Cassoschi, A., 72
Castedo, L., 138, 205
Castelijns, A.M., 204
Castellino, S., 108
Cavazza, M., 155
Cerfontain, H., 162
Cha, J.S., 282
Chadha, R.K., 284
Chakrabarti, P., 263
Chakraborty, R., 294
Chakraborty, T.K., 105, 231, 269
Challener, C.A., 138, 150
Challoner, R., 14
Chamberlain, S., 151, 222
Chamberlin, R., 162
Chambers, R.D., 295
Chan, C.-S., 230
Chan, J.B., 81
Chan, K.S., 230
Chandler, I.M., 315
Chandraprakash, Y., 308
Chandrasekaran, S., 294
Chang, C.J., 29
Chang, K., 256
Chang, S., 122, 255
Chang, S.-F., 108, 226
Chang, V.K., 219
Chapman, J.R., 28
Charette, A.B., 271
Charles, N.R., 74, 263
Chary, K.V.R., 3
Chatani, N., 153, 218
Chatgilialoglu, C., 82, 129, 280
Chatterjee, A.K., 291
Chatterton, W.J., 74
Chaudret, B., 265
Chauret, D.C., 277
Chávez, I., 266
Cheetham, A.K., 12
Chelain, E., 223
Chem, B.-L., 158
Chemla, F., 245
Chen, C., 84
Chen, C.-M., 138
Chen, C.S., 302
Chen, C.Z., 73
Chen, H., 57
Chen, J., 255
Chen, J.D., 286
Chen, L.Z., 25
Chen, Q.-Y., 95, 169
Chen, S., 30
Chen, S.-T., 304
Chen, T., 286
Chen, X.-J., 305
Chen, Y., 23, 155
Chen, Y.-J., 138
Chen, Y.T., 57
Chen, Z., 266
Cheng, C.-H., 236
Cheng, K., 158
Cheng, W.-J., 254
Cheng, X., 24
Cheng, X.-C., 317, 318
Cheng, Y.K., 58
Cheng, Z.-Y., 266
Chiacchio, U., 150
Chiarelli, R., 57
Chiba, T., 282
Chien, C.J., 258
Chignell, C.F., 168
Chima, J., 305
Chingas, G.C., 16
Chiu, R.-T., 138
Cho, B.P., 171
Cho, H., 324, 326, 330
Cho, I., 140
Cho, S.J., 76
Cho, W.J., 262
Chodorowski, S., 230
Choi, H.C., 290
Choi, K.I., 282
Chong, D.P., 57
Chong, J.M., 128, 277
Chou, C.-M., 169
Chou, T.S., 113
Choudhary, A.D., 322
Choueiry, D., 244
Chowdhury, C., 234
Chowdhury, R., 265
Chowdhury, S.K., 25
Christe, K.O., 57
Christensen, A.M., 18
Christian, J.F., 25
Christova, N.B., 75
Chuang, L.-W., 255
Chudinov, G.E., 60
Chung, D.S., 66
Chung, J., 11
Chung, T.M., 264
Chung, Y.K., 220, 221, 246, 264, 265
Church, G.M., 336
Cid, P., 75
Cimino, G., 311
Cinquini, M., 147
Cioslowski, J., 65, 72
Cladingboel, D.E., 93
Clairet, F., 22
Clark, J.H., 159
Clark, J.S., 203
Clark, T., 61
Clark, W.M., 155
Clarkson, S., 237
Clasby, M.C., 141
Classon, B., 252
Clausen, F.P., 200
Clayden, J., 121
Clayton, S.C., 197, 224
Clive, D.L.J., 138, 142
Clore, G.M., 11
Clough, L.E., 330
Coates, R.M., 319
Cobes, A., 205
Cochennac, C., 201
Coggins, J.R., 322
Cohen, L.A., 234
Cohen, T., 133
Colburn, A.W., 25
Collett, A., 182
Collin, J.-P., 230
Collington, E.W., 121
Collins, D.J., 148
Collins, J.J., 74
Collins, P.W., 106
Collins, S., 254
Collman, J.P., 287
Collum, D.B., 108, 125, 296
Colton, R., 28
Colwell, S.M., 57
Comins, D.L., 300
Compton, R.N., 26
Conçepcion, A.B., 132
Concha, M.C., 57
Conde, S., 302
Congreve, M.S., 290
Conn, L.M., 291
Connell, R.D., 198
Connelly, J.A., 322
Connolly, C.B., 247
Conser, K.R., 123, 288
Constantin, E., 28
Cooks, R.G., 22, 29
Cooper, A., 322
Coote, S.J., 114, 163, 198, 253, 271
Corey, E.J., 101, 119, 121, 136, 141, 271, 279, 285, 287
Cornelis, A., 159
Cornelisse, J., 171
Cornforth, J.W., 269
Corrigan, J.F., 14, 254
Corriu, R.J.P., 227
Cortez, C., 171
Cossi, P., 235
Cossio, F.P., 65, 75
Cossy, J., 138
Costa, A.L., 270
Costello, J.F., 267
Couder, M., 335, 336
Coughlan, S.M., 248
Courtemanche, G., 138
Couture, A., 205
Covey, T.C., 29
Cox, A.L., 31
Cox, D.B., 28
Cox, D.M., 25

Coxon, J.M., 108
Cozzi, F., 147
Cozzi, P.G., 244
Crabtree, G.R., 105
Craig, D., 141, 182
Craig, G.A., 27
Cramer, C.J., 57
Crampton, M.R., 162
Craw, P.A., 314
Cremer, D., 148
Crich, D., 84, 89
Crich, J.Z., 303
Crimmins, M.T., 106, 137, 269
Crisp, G.T., 232
Crispino, G.A., 119, 284, 285
Crist, D.F., 162
Crombie, L., 324
Cross, G.G., 105
Croteau, R.B., 161, 319
Crouch, N.P., 332
Crousse, B., 232
Crout, D.H.G., 105, 309
Crouzet, J., 336
Crow, G.R., 269
Crumrine, D.S., 72
Cucciolito, M.E., 249
Cuevas, C., 75
Cuisiat, S.V., 74
Cullen, S.R., 159
Cunningham, D., 256, 264
Curci, R., 132
Curran, D.P., 84, 86, 88, 93, 97, 99, 106, 129
Currie, J., 293
Curtis, D.R., 302
Curtis, J.M., 28
Czech, R.P., 266
Czernecki, S., 286

D'Accolti, L., 132
Dahan, F., 187, 265
Dai, W.M., 149
Dakka, J., 286
Dakternieks, D., 28
d'Alessandro, N., 168
Daley, S.T.A.K., 266
Dalkin, K., 322
Dalton, D.M., 248
Dalton, H., 305, 309
Dalvi, P.V., 142
Dambacher, G., 108
Damm, W., 97, 98, 99, 129
Damtoft, S., 319
Dance, I.G., 27
Daniell, G.J., 9
Daniels, P., 324
Danil de Namour, A.F., 105
Danis, P.O., 28
Danishefsky, S.J., 105, 108, 210, 227, 230, 269
Danzer, M., 333, 335
Daran, J.-C., 223
Dasgupta, S.K., 232
Da Silveira, E.F., 28
Dasradhi, L., 300
Dass, C.H., 24
Date, T., 108, 141, 273
Dauelsberg, C., 280
Daves, G.D., Jr., 228
Davies, H.M.L., 107
Davies, I.W., 247
Davies, S.G., 106, 168, 267, 293
Davis, F.A., 295
Davis, K., 322
Davis, M.F., 13
Davis, W.D., 284
Davison, E.C., 290
Davison, M., 332
Davisson, V.J., 327
Davister, M., 159
Dawson, G.J., 114, 163, 198, 253, 271
De, B.B., 287
Deacon, G.B., 271
Dean, G., 309
De Benedetto, L., 246
Deberly, A., 283
DeBernardinis, S., 237
Debrauwer, L., 24
Debski, N., 290
Debu, F., 333, 335
Debussche, L., 333, 335, 336
Dec, S.F., 13
DeCamp, A.E., 111, 232
De Clerq, E., 228
de Denus, C.R., 264
de Dios, A., 232
de Gala, S., 230
Deganello, G., 254
Deka, R., 322
de Koning, C.B., 328
Dekoning, L.J., 66
de la Salud, R., 291
Del Buttero, P., 262, 263
Delgado, F., 286
Delgado, M., 271
Del Grosso, M., 108
Dell, K.A., 155
Dell'Erba, C., 163
Delogu, G., 297
del Rio Portilla, F., 10
Deluca, M., 208
De Lucchi, O., 297
Delvalle, F.J.O., 60
Delville, M.-H., 266
de March, P., 75
de Meijere, A., 178
Demura, T., 162
Deng, L., 121, 306
Dengler, A., 107
Deniau, E., 205
DeNisco, M., 164
Denisov, V., 117, 244
Denmark, S.E., 108, 115, 276
Denney, B., 162
Denney, D.Z., 162
Deppe, A., 28
de Rossi, R.H., 162
Derrick, P.J., 24, 25, 27, 28, 30
Deschenaux, R., 266
Desimoni, G., 72, 78
Deslongchamps, P., 72
Desponds, O., 106
Destabel, C., 90
Detomaso, A., 132
Deutsch, J., 291
Deutzmann, R., 322
de Vargas, E.B., 162
de Vries, E.F.J., 271
de Vries, N.K., 293
de Vroom, E., 182
Dewey, M.A., 248
Dewick, P.M., 320, 321
de Wolf, W.H., 176
Déziel, R., 119, 294
de Zoete, M.C., 304
Dhal, P.K., 287
Dharmaratne, H.R.W., 321
DiCapua, F.M., 58
Dickhaut, J., 98, 129
Dickson, R.M., 57
Diederich, F., 174
Dietz, A., 107
Dillet, V., 60
DiMichele, L., 232
Dimitroff, M., 141
Dinesh, C.U., 290
Ding, D., 26
Ding, S., 227, 249
Disch, R.L., 146
Diver, S.T., 291
Dixon, D.A., 57
Dixon, R.A., 322
Djakovitch, L., 266
Djuric, S.W., 106
Do, J.Y., 90
Dobson, C.M., 12
Dodd, R.H., 287
Dörner, W., 138
Dötz, K.H., 223
Doherty, A., 79, 332
Doherty, S., 14
Doi, T., 150
Dol, P.P.M., 73
Dolbier, W.R., Jr., 78
Dolling, U.-H., 111, 232
Dominguez, D., 138
Domling, A., 199
Donaldson, W.A., 256, 257
Donchi, K.F., 24
Donogues, J., 158
Donohoe, T.J., 106, 168
Donovan, T.A., Jr., 22, 168
Dorado, M., 138
Dorsey, B.D., 125
Doty, M.J., 222
Doucet, J.P., 69
Douglas, A.W., 280
Douglas, C.J., 322
Douglas, D.J., 325
Dove, Y., 156

Dowd, P., 90, 106, 211, 279
Doyle, M.P., 135
Doyle, T.W., 318
Doyon, J., 138
Drader, J.J., 27
Dragovich, P.S., 149
Drauz, K., 283
Drew, M.G.B., 266
Drewello, T., 23
Drewes, S.E., 203
Drummond, M.H., 29
Du, H., 297
Dual, C., 56
DuBay, W.J., 190
Duchamp, J.C., 148
Duchêne, A., 240
Dürner, G., 108, 259
Duffy, P., 57
Duggan, P.J., 89
Duhamel, L., 117, 277
Duhamel, P., 117, 277
Dulcere, J.-P., 119, 141
Dumant, A., 304
Dumas, J., 238
Duñach, E., 288
Dunbar, R.C., 22
Duncan, S., Jr., 160
Dunn, M.J., 235
Dunn, W.J., 58
Dunogués, J., 106, 235
Durant, G.J., 58
Dutto, G., 246
Dvortsak, P., 12
Dwight, W.J., 108
Dwyer, T.J., 8
Dyker, G., 171
Dykstra, R.R., 128

Eades, R.A., 54, 57
Eady, R.R., 29
East, M.B., 106
Eastman, M.A., 16
Eastwood, F.W., 107, 146, 171
Eaton, P.E., 132
Eberbach, W., 214
Eberlein, T.H., 193
Eberlin, M., 22
Ebert, G.W., 168
Ebert-Khosla, S., 317
Ebizuka, Y., 320
Eckardt, K., 317
Eckert, H., 25
Edge, S.J., 179
Edwards, A.J., 125, 267, 276, 296
Edwards, J., 332
Edwards, J.P., 153, 241
Edwards, R., 322
Effenberger, F., 193
Egan, M., 138, 149
Eggenberger, U., 8
Eggers, A., 111, 233
Eggleston, D.S., 199
Eguchi, H., 158
Eichele, K., 14
Einaga, H., 57
Eisenberg, C., 113
Ekeberg, D., 27
Eksterowicz, J.E., 71, 98, 107
Elgendy, S.M.A., 155
El Hafa, H., 246
Eliel, E., 270
Ellames, G.J., 310
Ellard, M., 322
Ellinger, Y., 57
Ellis, D., 235
Ellis, W., 254
Ellstead, K.E., 140
El Mouatassim, B., 264
El-Olemy, M.M., 322
Elson, S.W., 332, 333
Emokpae, T.A., 162
Emsley, L., 8, 16
Emslie, N.D., 203
Enders, D., 141, 254, 271
Endo, A., 318
Endo, H., 307
Endo, T., 319
Engbersen, J.F.G., 190
Engel, G.T., 192
Engelhard, V.H., 31
Engle, C.A., 333
Enkelmann, V., 136, 241
Eppers, O., 4
Erdmann, H., 302
Eriksson, A., 234
Eriksson, L.A., 57
Eriksson, M., 112
Erker, G., 107
Eshghi, H., 159
Especl, P.H., 159
Essex, J.W., 58, 59
Etinger, A., 26
Evans, B., 108, 227
Evans, D.A., 106, 108, 123, 126, 130, 181, 278, 281, 288, 292, 293
Evans, D.H., 106
Evans, G.R., 141
Evans, S., 30
Evans, T.A., 305
Evans, W.J., 12
Evanseck, J.D., 65
Everberg, K.A., 137
Ezcurra, J.E., 153

Fabbri, D., 297
Faber, K., 305
Fabre, A., 88, 138
Fabrizi, G., 233
Fache, F., 281
Fadnavis, N.W., 308
Fahrendorf, T., 322
Fairfax, D.J., 150
Faita, G., 72, 78
Falck, J.R., 112
Falck-Pedersen, M.L., 250
Faller, J.W., 132, 254, 282
Fallis, A.G., 141
Famechon, A., 336
Fan, L.-D.G., 331
Fan, W.-Q., 160
Fang, D., 74
Fang, J., 30
Fang, Z., 164
Fantin, G., 307
Farata, T., 158
Farina, V., 167, 232
Farrant, R.D., 11, 12
Farrell, J., Jr., 21
Farthing, C.N., 276
Fates, B., 258
Fattuoni, C., 165
Faul, M.M., 123, 181, 288
Faulkner, D.J., 106
Faure, R., 141
Faust, R., 146
Fava, A., 107
Feaster, J.E., 131
Féaux de Lacroix, S., 16
Federsel, H.-J., 269
Fehr, C., 125
Feibelmann, S., 140, 220
Feigelson, G.B., 184
Feistner, G.J., 30
Felder, M., 130, 280
Feldman, K.S., 123, 138
Feltman, T.L., 160
Feng, T., 271
Feng, Z., 319
Fengl, R.W., 241
Fenn, J.B., 28
Fensterbank, L., 165
Ferguson, M.D., 74, 263
Feringa, B.L., 293
Fernández, I., 291
Fernández, R., 295
Fernandez-Acebes, A., 138
Fernandez-Baeza, J., 265
Fernandez-Mayoralas, A., 302
Ferraboschi, P., 299
Ferreri, C., 82, 129, 280
Ferri, F., 167, 233
Ferrier, R.J., 138
Fetter, J., 185
Feuilleaubois, E., 69
Fevig, J.M., 208
Fidalgo, J., 138
Fiedler, A., 27
Field, J.S., 203
Field, R.A., 332
Fields, S.C., 273
Fields, T.L., 101
Figatere, F., 107
Figuerdo, M., 75
Filipkowski, M.A., 77, 137
Filipp, N., 74
Filippini, L., 227
Fillaut, J.-L., 266
Finch, H., 108, 227, 258

Fink, D.M., 164
Finkam, M., 254
Firouzabadi, H., 290
Fischer, B., 187
Fischer, J.E., 25
Fischer, N.H., 313
Fischer, W., 200
Fishbein, K.W., 19
Fisher, E.R., 27
Fisher, G.B., 282, 294
Fisher, H., 124
Fisher, S.L., 302
Fisker, H., 302
Fitz, W., 311
Fitzgerald, G., 53, 57
Fitzgerald, J.J., 13
Fitzpatrick, N.J., 256
Fleet, G.W.J., 191
Fleming, I., 122
Flemming, P.E., 324
Fleukiger, P., 57
Flippin, L.A., 165
Florentin, D., 328
Flores, H.E., 313
Floriani, C., 244
Floris, B., 95
Floris, C., 76, 165
Floss, H.G., 108, 324, 326, 329, 330, 331, 336
Flynn, B.L., 232
Flynn, D.L., 220
Fogagnolo, M., 307
Fokina, N.A., 304
Fokkens, R.H., 24
Folga, E., 54, 57
Folkers, P.J.M., 4
Folmer, J.J., 291
Folmer, R.H.A., 4
Foltin, M., 27
Font, J., 75
Fontain, E., 107
Ford, K.L., 292
Forenza, S., 318
Formica, M., 249
Fornarini, S., 159
Forrester, J., 287
Forster, S., 267
Forsti, E., 147
Fothergill, M., 58
Fotsch, C.H., 105, 299
Foubelo, F., 128
Foubister, A.J., 147
Fouche, G., 305
Fournier, F., 24
Fournier, R., 57
Fowler, P.W., 148
Fox, D.N.A., 127
Fox, K.M., 324
Fox, T., 4, 61
Foxall, P.J.D., 12
Foxman, B.M., 266
Francheschi, G., 185, 308
Franck, B., 174
Franczyk, H., 319
Frank-Neumann, M., 256
Frappeir, F., 328
Fraser-Reid, B., 81
Freccero, M., 168
Fréchet, D., 333, 335
Freeman, R., 10
Freer, I.K.A., 326
Frenzen, G., 73
Fresse, I., 300
Fried, C.A., 249
Friese, V., 323
Frim, R., 178
Frish, M., 61
Fritch, P.C., 193
Fritz-Langhals, E., 297
Frössl, C., 311
Frost, C.G., 114, 163, 198, 253, 271
Frost, J.W., 155
Fry, A.J., 278
Frydman, L., 16
Fu, D.-K., 244
Fu, G.C., 106, 117
Fu, H.-W., 255
Fu, P.P., 156, 171
Fu, S., 258
Fu, X., 74, 76
Fuchs, P.L., 108
Fuest, M., 235
Fugami, K., 247, 254
Fuganti, C., 308
Fuhry, M.A.M., 290
Fuji, K., 106, 116, 132, 276, 292, 293
Fujihara, H., 153
Fujii, H., 292
Fujimori, T., 252
Fujimoto, H., 314
Fujimoto, K., 122, 235
Fujimoto, Y., 318
Fujisaka, S., 158
Fujisawa, M., 138
Fujita, M., 108
Fujiwara, K., 150
Fukuhara, T., 163
Fukumoto, K., 107, 136, 138, 200
Fukumoto, Y., 153, 218
Fukuno, A., 188
Fukushima, N., 57
Fukuyama, Y., 232
Fuller, D.J., 108
Fuller, J.C., 130, 282
Fulston, M., 332
Funakoshi, K., 138
Funk, R., 140, 271
Furstner, A., 106
Furstoss, R., 305
Furukawa, N., 153
Furukawa, S., 229
Furuta, T., 30
Fusco, C., 132
Futano, M., 249
Fyles, T.M., 105
Gabel, C.J., 224
Gable, K.P., 23
Gabrik, A.H., 30
Gagné, M.R., 130, 281
Galaev, I.Yu., 304
Galambos, G., 232
Galindo, J., 125
Gall, C., 162
Gallagher, D.J., 108, 166
Gallagher, R.T., 28
Gallagher, T., 247
Galli, C., 159
Gallucci, J.C., 254
Galushko, S.V., 304
Gambe, A., 256
Gamez, P., 281
Gandolfi, R., 75, 256
Ganem, B., 283
Ganesan, A., 208
Ganesh, P., 245
Ganesh, S., 263
Ganeskpure, P.A., 156
Ganghoff, A.R., 198
Ganguly, A.K., 201
Gann, S.L., 16
Gao, J.L., 58, 62
Gao, J.R., 22
Gao, O., 141
Garanti, L., 213
Garbow, J.R., 18
Garbutt, R., 27
Garcia, A.E., 58
Garcia, B., 291
Garcia, E., 167
Garcia, O., 286
Garcia-Granda, S., 141
Garcia-Martin, M.A., 158
Garcia Ruano, J.L., 271
Gardini, F., 307
Gardner, M., 66
Garner, P., 206
Garrett, B.C., 67, 68, 79
Garrett, D.S., 11
Garrido, N.M., 293
Garro-Helion, F., 252, 290
Garson, M.J., 311
Gasch, C., 295
Gasche, J., 108, 207, 226
Gasparrini, F., 147, 197
Gasteiger, J., 69
Gatti, R., 247, 252
Gauthier, A., 72
Gawley, R.E., 128, 201
Geib, S.V., 99
Gelli, G., 76, 165
Gelling, O.J., 218
Genêt, J.-P., 250, 252, 281
Geng, B., 227
Gennari, C., 66
Gennaro, G., 254
Gentili, P., 162
Geoffroy, P., 256
George, G.I., 183
George, M., 22

Geppert, T., 8
Geraci, L.S., 127, 270
Gere, R., 256
Gero, S.D., 183
Gerson, F., 178
Gervasio, G., 221
Ghanimi, A., 171, 277
Ghio, C., 58
Ghosh, A., 57
Ghosh, R., 135
Giannis, A., 105
Giblin, D.E., 25
Gibou-Barbedette, A., 256
Gielen, M., 3
Giese, B., 83, 97, 98, 99, 129
Giese, R.W., 156
Giesen, V., 235
Giguere, R.J., 141
Gilbert, A.M., 138, 157, 236
Gilbert, B.A., 118
Gilbert, J., 80, 320
Gilchrist, J.H., 108, 125, 296
Gilchrist, T.L., 235
Giles, M., 208
Gill, G.B., 77
Gill, M., 316
Gill, P.M.W., 53
Gillard, J.W., 202
Gillmann, T., 138
Gilmore, R.J., 266
Giménez, A., 316
Gimisis, T., 171
Gin, D.Y., 128
Giner, J.-L., 318
Gingras, M., 291
Gioia, B., 330
Giordano, F., 249
Giovannoli, M., 197
Girard, M., 121, 286
Girija, K., 165
Girijavallabhan, V.M., 201
Giuliano, R.M., 72
Gladstone, B.G., 152
Gladysz, J.A., 248
Glänzer, J., 223
Glaser, J., 4
Glass, W.K., 256
Glassl, B., 201
Glatzhofer, D.T., 265
Gleason, M.M., 247
Gleiter, R., 106, 145, 149, 218
Glending, E.D., 146
Gliaka, T., 331
Glish, G.L., 28
Gloaguen, B., 266
Godard, A., 201, 229
Goddard, R., 263
Godfrey, A.G., 283
Godfrey, C.R.A., 186
Göbel, T., 284, 285
Goghari, M.H., 29
Gogoll, A., 247
Goldberg, D.R., 141
Goldberg, I.B., 57
Goldschmidt, Z., 264
Goldstein, S., 72
Golebiowski, A., 107
Golik, J., 318
Gollins, D.W., 186
Golz, T., 80, 210
Gomez-Barrios, M.L., 313
Gomibuchi, T., 150
Gonon, K., 162
Gonzalez, J.M., 142, 158
Gonzalez, N.C., 141
Gonzalez-Lafont, A., 68
Gonzalez-Nunez, M.E., 182
Good, A.C.G., 69
Goodman, J.M., 66
Goodwin, C.J., 293
Goodwin, L., 57
Goralski, C.T., 130, 282, 294
Gorrod, J.W., 30
Goto, Y., 254
Gotor, Y., 304
Gottlieb, H.E., 264
Gottschalk, G., 156
Gou, D.-M., 302
Gould, I.R., 60
Gould, S.J., 317, 318, 325, 336
Goulet, R., 119
Goulet, S., 294
Gourdel, Y., 131, 277
Govil, G., 3
Gowri, G., 322
Grace, K.J.S., 235
Gräfe, U., 329
Graf, R., 138
Graff, C., 326
Graham, A.E., 289
Grainger, D.S., 9
Granberg, K.L., 247
Grandclaudon, P., 205
Grande, M., 183
Grandinetti, P.J., 16
Grandjean, D., 111
Grant, T.G., 156
Gras, J.-L., 116, 276
Gratti Comini, S., 78
Gray, J.L., 137
Green, D.V.S., 59, 60
Green, H., 15
Green, I.R., 153
Green, J.R., 254, 264
Greenspan, P.D., 250
Greeves, N., 193, 220
Grenier, L., 119, 294
Grey, A.G., 33
Grey, C.P., 12
Gribble, G.W., 147, 200
Grier, M.C., 270
Griesinger, C., 8
Griffin, D., 23
Griffin, R.G., 19
Griffiths, G.J., 300
Griffiths, O., 157
Griffiths, S.L., 248
Grigg, R., 112, 237, 240
Grill, V., 27
Grisenti, P., 299
Grissom, J.W., 93, 138, 149
Grogan, G., 306
Gronenborn, A.M., 11
Gronert, S., 66, 108
Gronowitz, S., 229, 234
Gross, M.L., 22, 24, 25
Grosshenny, V., 233
Grossman, R.B., 138, 220
Grotemeyer, J., 29
Groth, U., 125
Grubbs, R.H., 117
Gruber, B., 107
Grue-Sorensen, G., 327
Gruselle, M., 246
Grzesiek, S., 6, 7
Gschneider, D., 141
Gu, M., 24
Gu, Q.-M., 304
Guan, J.G., 57
Guan, L., 247
Gubatz, S., 323
Gudel, H.U., 56
Guella, G., 155
Günther, H., 4, 8
Guertin, K.R., 132, 289
Gueugnot, S., 234
Guibé, F., 252, 290
Guibe-Jampel, E., 300
Guijarro, A.G., 128, 167
Guillaumet, G., 206
Guillerm, G., 328
Guillon, T., 18
Guindon, Y., 97, 129
Guiry, P., 141
Guitian, E., 205
Gulevich, Y.V., 235
Guntha, S., 290
Guo, C., 118
Guo, J., 336
Guo, L., 135
Guo, Z.-X., 118, 179
Gupta, V., 94
Gurevich, A., 162
Gurney, K.A., 330
Gurski, A., 153
Gurthler, S., 185
Gurunmurthy, R., 162
Gusmeroli, M., 227
Guy, A., 304
Guy, L., 182
Guzik, H., 201
Guzzi, U., 108, 227
Gyoung, Y.S., 282

Haas, G.W., 24
Hachiya, I., 272, 278
Haddix, G.W., 13
Hadida, S., 108, 235
Hadjiarapoglou, L., 284
Haeffner, F., 249
Hafner, K., 79

Hagaman, E.W., 13
Hagemeyer, A., 16
Hagen, K.S., 136
Hager, L.P., 121, 306
Hai, A.K.M.A., 183
Hai, Y., 73
Haigh, D., 307
Hailes, H.C., 202
Haines, A.H., 179
Halcrow, M.A., 265
Hales, N.J., 257
Halevi, E.A., 65
Haley, M.M., 145
Hall, G.G., 147
Hall, S., 30
Hallberg, A., 227
Haller, B.-U., 124
Halley, K.A., 318
Hamada, Y., 230
Hamaguchi, M., 187
Hamasaki, T., 314
Hamelin, J., 294
Hamilton, I., 332
Hammerschmidt, F., 336
Hammes, S., 80, 210
Hammock, B.D., 304
Han, S., 13
Hanada, K., 302
Hanafusa, T., 23
Hananki, N., 272
Hanano, T., 119
Hanaoka, M., 245, 260, 262
Handy, N.C., 53, 57
Hanessian, S., 121, 286
Haning, H., 125
Hansen, J.A., 141
Hansen, J.J., 302
Hansson, S., 249
Hansson, T., 320
Hanzawa, Y., 136, 138, 290
Hara, R., 117, 243
Harayama, H., 119, 247
Hardick, D.J., 336
Harding, S.E., 322
Hardinger, S.A., 290
Harendza, M., 280
Harkema, S., 190
Harlow, R.L., 131
Harman, W.D., 142
Harmange, J.-C., 107
Harmon, K.M., 295
Harper, M.F., 324
Harran, P.G., 85
Harriman, A., 266
Harris, G.D., Jr., 239
Harris, K.D.M., 18, 19
Harris, M.C.J., 244
Harrison, A.G., 24, 25
Harrison, A.T., 108
Harrison, C.T., 282
Harrison, D.M., 330
Harrison, M.J., 322
Harrity, J.P.A., 223
Harrowven, D.C., 94, 194
Hart, D.J., 81
Hart, G.J., 216, 333
Hartung, J., 83, 284
Harusawa, S., 215
Harvey, D.F., 223
Harvey, R.G., 171
Hasegawa, E., 84
Hasegawa, M., 94, 159, 279
Hasegawa, Y., 125, 296
Hashimoto, T., 174
Hashimoto, Y., 159
Hashiyama, T., 284
Hashizume, M., 289
Haslego, M.L., 196
Hassan, A.A., 176
Hassner, A., 187
Hasumi, K., 318
Hatanaka, A., 312
Hattori, T., 163
Haubold, E.M., 149
Hauer, C.R., 31
Hauffe, K.D., 322
Haufler, R.E., 26
Hawker, C.J., 216, 333
Hawkins, A.R., 322
Hawkins, G.D., 57
Hayamizu, K., 13
Hayashi, A., 138
Hayashi, E., 276
Hayashi, N., 26
Hayashi, S., 13
Hayashi, T., 128, 242, 252
Hayashi, Y., 116, 264
Hayashizaka, N., 163
Hayward, C.M., 105, 269
He, P., 183
Healy, M.A.M., 235
Heaney, H., 284
Heathcock, C.H., 208, 220
Heaton, N., 267
Heaton, S.B., 264, 271
Heber, J., 222
Hechtberger, P., 305
Hedgecock, C.J.R., 267
Hedges, S.H., 328
Heeg, M.J.J., 74
Hefu, G., 240
Hegedus, L.S., 247
Heide, L.E., 322
Heidlas, J., 311
Heimer, N.E., 164
Heinemann, D., 57
Heinrich, N., 23
Heinze, J., 178
Heizmann, C.W., 31
Helal, C.J., 101, 287
Helene, C., 105
Hellberg, L.H., 295
Hellendahl, B., 219
Helmchen, G., 253, 271
Helmick, J.S., 155
Helquist, P., 198
Hemming, K., 76
Hemscheidt, T., 328, 329
Henderson, E., 13
Henderson, G.L., 254
Henderson, R.A., 31
Hendrickson, J.B., 283
Henkel, T., 317
Henning, K., 203
Henriquez, R., 169
Henry, J.R., 286
Henshilwood, J.A., 74, 263
Herberg, C., 78
Herbert, R.B., 310, 325, 328
Herdewijn, P.A., 228
Herges, R., 107
Herman, F., 333, 335
Herman, Z., 27
Herndon, J.W., 223
Herr, R.J., 239
Herrmann, W., 106
Hertel, I.V., 26
Hervai, B.H., 238
Herzfeld, J., 19
Heshmati, P., 246
Hesketh, A.R., 320
Hetzenegger, J., 65, 72
Hezroni-Langerman, D., 264
Hickey, E.R., 138
Hidai, E., 116, 276
Hidris, M.S., 77
Hiemstra, H., 254
Hierstetter, T., 65, 72
Higaki, C., 224
Higashi, Y., 301
Higashihara, T., 266
Higuchi, H., 172, 174
Hilbers, C.W., 4
Hill, C.L., 84
Hill, R.A., 311
Hillenkamp, F., 28
Hillers, S., 242
Hillier, I.H., 59, 60
Hilvert, D., 105
Hinkle, R.J., 108
Hino, T., 141
Hira, Y., 150
Hirai, N., 286
Hirako, K., 250
Hirama, M., 121, 138, 150
Hiramatsu, K., 235
Hirao, T., 252
Hirashima, T., 156
Hirata, K., 159
Hirayama, N., 132
Hirohara, Y., 136
Hiroi, K., 77
Hirose, Y., 271
Hirsch, A., 25, 107
Hirst, J., 162
Hirvela, L., 154
Hiyama, T., 111, 241
Hjortso, M.A., 313
Ho, J., 264
Ho, W.B., 206
Hobbs, G.R., 309
Hobson, A.D., 179

Hoch, U., 309
Hockey, S., 125, 296
Höft, E., 284
Hörnfeldt, A.-B., 229
Hoffman, A., 30
Hoffman, R.W., 235
Hoffmann, H.M.R., 134, 248
Hofmann, M., 12
Hogan, J.D., 28
Hohenberg, P., 51
Hoioki, J., 172
Hojo, M., 136
Hoke, S.H., II, 29
Holden, I., 324
Holkar, A.G., 158
Holladay, J.E., 141
Holland, H.L., 270
Hollander, J., 259
Holle, A., 28
Holliman, C.L., 22
Hollis, W.G., 137
Holmes, A.B., 290
Holmes, E.F., 295
Holmes, J.L., 22
Holms, H., 332
Holoboski, M.A., 102
Holwerda, R.A., 266
Hommann, K., 181
Hong, C.Y., 226
Hong, Y.P., 149
Honig, B., 57
Hoogsteen, K., 72, 136, 219
Hoorn, J.A., 135
Hopwood, D.A., 317
Horak, R.M., 305, 330
Hore, P.J., 9
Hormi, O.E.O., 154
Horz, K.H., 323
Hosaka, T., 141
Hoshino, M., 132
Hosokawa, M., 162
Hosokawa, T., 119
Hosomi, A., 136
Hosono, K., 266
Hosoya, N., 287, 288
Hotta, H., 163
Houk, K.N., 63, 65, 66, 71, 77, 98, 107, 108
Houpis, I.N., 232
Hovestreydt, E.R., 271
Hoveyda, A.H., 106
Howard, J.A.K., 266
Howard, K.J., 205
Howard, K.P., 11
Howdle, S.M., 217
Howell, J.A.S., 264
Hoye, T.R., 221
Hrusak, J., 22, 27
Hu, L., 255
Huang, D.F., 197
Huang, S.F., 22
Huang, T.-Y., 126
Huang, W., 137
Huang, X., 237
Huang, Y., 93
Huang, Z., 77, 290
Huang, Z.-T., 163
Huber, P., 10
Huber, R.S., 264
Huber, U., 319
Hudecek, M., 220
Hudlicky, T., 305
Hudson, C.E., 22
Huff, S., 29
Hughes, D.J., 191
Hughes, D.L., 196
Hugo, V.I., 153
Huhn, M.M., 57
Huhn, T., 125
Huisgen, R., 65
Hulmes, D.I., 131, 294
Hulst, R., 293
Hummel, C.W., 149
Humphreys, V.M., 267
Hung, S.M., 73
Hunt, A.R., 108, 166, 224
Hunt, D.F., 31
Hunziker, P., 31
Hur, C.U., 100
Hurd, R.E., 11
Hussoin, M.S., 283
Hutchens, W.T., 29
Hutchinson, C.R., 318
Hutchinson, D.W., 309, 336
Huttenloch, M.E., 138, 236
Huval, C.C., 94
Hvistendahl, G., 27
Hwang, B.K., 304
Hwu, J.R., 118
Hynes, J.T., 66

Ibata, T., 162
Ibers, J.A., 287
Ichihara, O., 293
Ichikawa, H., 74
Ichikawa, Y., 309
Ichinose, K., 335
Ida, T., 141, 153
Idrees, K., 266
Ihara, M., 107, 136, 200
Ihle, N.C., 220
Iida, H., 213
Iida, K., 121
Ikari, Y., 307
Ikeda, H., 78
Ikeda, I., 252
Ikeda, M., 83, 183
Ikeda, S., 228
Ikekawa, N., 318
Ikenaga, K., 235
Imada, Y., 250, 254
Imai, R., 95
Imamoto, T., 277
Imi, K., 136
Imoto, M., 323
Imperiali, B., 302
Imrie, C., 266
Inada, K., 281
Inagaki, A., 156
Inami, H., 251
Indolese, A.F., 136
Ingemann, S., 24
Ino, Y., 172
Inoue, H., 127, 130, 270
Inoue, Y., 177
Invernizzi, A.G., 72
Iqbal, J., 287
Ireland, R.E., 287
Irie, H., 132
Irie, R., 287, 288
Iseki, K., 125
Ishibashi, H., 83, 183
Ishida, M., 108
Ishihara, K., 141, 272, 279, 291
Ishii, M., 188, 302
Ishili, T., 175
Ishitani, H., 278
Ishiyama, T., 235
Isobe, M., 246
Isobe, S., 249
Isoe, S., 141
Ito, H., 136, 138, 290
Ito, K., 74, 279
Ito, N., 250
Ito, S., 169, 257
Ito, T., 202, 251
Ito, Y., 122, 235, 288, 292
Itoh, K., 158, 160, 235, 244, 282
Itoh, N., 108, 128, 273
Itoh, T., 167, 244
Iwamoto, S., 266
Iwao, M., 229
Iwaoka, T., 74
Iwasawa, N., 220
Iwata, C., 255
Iwata-Reuyl, D., 333
Iyer, R.P., 106
Iyer, V.S., 273
Iyoda, M., 95
Izatt, R.M., 105, 214

Jackson, G., 162
Jackson, R.F.W., 128, 235, 245
Jacob, L.A., 158
Jacobs, P.A., 159, 245
Jacobsen, E.N., 121, 122, 123, 288, 306
Jacobson, D.B., 27
Jaeger, E., 41
Jain, M., 162
Jakobsen, H.J., 13
Jandu, K.D., 180
Jang, M.-K., 304
Janson, M., 252
Janssens, B., 159
Jaouen, G., 246
Jardine, D.R., 47
Jardine, I., 48
Jarvi, E.T., 295
Jayaraman, H.C., 214

Jayasinghe, L.Y., 309
Jaynes, B.S., 84
Jean, T.-S., 138
Jedicka, B., 243, 266
Jeenes, D.J., 31
Jeffery, T., 108, 230
Jegorov, A., 331
Jenneskens, L.W., 171
Jennings, K.R., 35, 46, 47
Jensen, M.S., 135
Jensen, N.J., 24
Jensen, P., 57
Jensen, S.R., 319
Jeong, K.-S., 119, 284, 285
Jeong, N., 140, 220, 221, 246
Jerome, K.S., 217
Jeschke, T., 234
Jevnaker, B., 265
Ji, J., 240
Jia, X.Q., 162
Jiaang, W.-T., 86
Jiang, B., 229
Jiang, J., 74, 292
Jiang, S., 248
Jiang, Y., 73
Jimbo, M., 175
Jin, F., 137, 229
Jin, M.-J., 256
Jin, Z., 108, 255
Jing, N., 78
Joe, G.H., 90
John, B.K., 11
Johner, M., 185
Johnson, B.G., 53
Johnson, C.D., 107
Johnson, C.S., 5
Johnson, D.D., 105, 269
Johnson, G., 208
Johnson, J.V., 48
Johnson, R.E., 166, 225
Johnson, R.S., 34
Johnston, M.V., 23, 41
Jones, D.K., 280
Jones, D.W., 78
Jones, G.B., 264, 271
Jones, G.C., 42
Jones, J.A., 9
Jones, J.B., 105, 299
Jones, J.L., 46
Jones, M.Z., 35
Jones, P.G., 171
Jones, R.C.F., 205
Jones, R.G., 188
Jones, R.O., 56
Jones, R.V.H., 287
Jones, W.M., 145
Jordan, A.D., 72
Jordan, P.M., 315
Jorgensen, K.A., 80
Jorgensen, W.L., 58, 71
Joshi, S.K., 158
Jou, D.-C., 236
Joullié, M.M., 292
Ju, S., 325
Jubert, C., 113
Juhl-Christensen, J., 200
Jumnah, R., 250
Jun, J.-G., 206
Jun, W.S., 282
Jung, D.K., 137, 230
Jung, G., 29
Jung, J.C., 290
Jung, M.E., 90, 141, 154
Jung, S.-H., 133
Junga, H., 117, 138
Junggebauer, J., 280
Juntunen, S.K., 134
Jurczak, J., 107
Jursic, B.S., 294
Jutand, A., 244

Kabuto, C., 136
Kacan, M., 155
Kacun, M., 193
Kadam, S.M., 290
Kado, N., 226
Kählig, H., 336
Kaercher, G.R., 28
Kagan, H.B., 266
Kagechika, K., 243
Kageyama, M., 117, 243
Kagoshima, H., 159
Kahn, M., 105
Kahne, D., 94
Kajbaf, M., 30
Kajita, S., 249
Kajtar-Peredy, M., 185
Kakindii, O., 184
Kakiuchi, K., 175, 178
Kambe, N., 101, 128
Kambuchi, A., 141
Kameoka, C., 183
Kameyama, Y., 128
Kamigata, N., 95
Kamochi, Y., 283
Kampf, J.W., 128, 171, 259
Kanai, G., 153
Kanai, M., 46
Kaneda, T., 23
Kaneko, C., 74
Kane-Maguire, L.A.P., 258
Kanemasa, S., 126
Kann, N., 116, 275
Kannagara, G.S.K., 235
Kannan, K.R., 25
Kant, J., 183
Kantam, M.L., 290
Kaplan, L., 315
Kapoor, I.P.S., 162
Karanewsky, D.S., 291
Karas, M., 28, 37, 40, 42, 322
Karcher, T., 65, 72
Kardos, N., 250
Karikomi, M., 213
Karimi, B., 290
Karp, G.M., 187
Karr, D.E., 28
Karthikeyan, M., 165
Kasar, R.A., 141
Kassel, D.B., 39
Kassir, J.M., 150
Kasumov, T.M., 158
Kasuya, Y., 30
Kataoka, O., 245
Katritzky, A.R., 160, 165, 191
Katsoulos, G., 165
Katsuki, T., 279, 287, 288, 292
Katz, T.J., 138, 236
Kaufmann, R., 41
Kawabata, T., 132, 188
Kawada, A., 159
Kawaguchi, M., 178
Kawajiri, S., 65, 72
Kawakami, T., 184
Kawakita, K., 319
Kawamura, T., 235, 240
Kawarmine, K., 132
Kawasaki, H., 125, 296
Kawasaki, I., 188
Kawasoe, S., 286
Kawecki, R., 102
Kay, L.E., 7
Kayganich, K.A., 36
Kayser, F., 3, 167
Keck, G.E., 127, 133, 270
Keefe, L.J., 42
Keefer, P.K., 295
Keglevich, G., 200
Keim, W., 220
Kelderman, E., 190
Keller, M., 214
Kelley, J.A., 37
Kelly, M.T., 313
Kelly, S.C., 148
Kelly, S.M., 322
Kelly, T.R., 201
Kemp, T.J., 27
Kende, A.S., 132, 289
Kenedy, E., 324
Kennewell, P.D., 76, 237
Kenny, P.T.M., 33, 46
Kent, K.D., 45
Kent, S.B.H., 39
Kenttamaa, H.I., 21, 22, 23
Keough, T., 37, 42
Kerns, M.L., 155
Kerr, M.A., 235
Kerr, W.J., 223
Kerrick, S.T., 108, 166
Kerrigan, J.E., 141
Kerssebaum, R., 11
Kesselheim, C., 83
Kessler, H., 106
Kesura, G.M., 185
Khan, M.A., 253, 265
Khan, S., 97
Khan, S.I., 57, 141
Khemani, K.C., 25
Khosla, C., 317
Khrushchova, N., 266
Kido, M., 262

Kieffer-Higgins, S., 336
Kihara, N., 159
Kiho, T., 35
Kikuchi, N., 119
Kikuchs, I., 172
Kilbee, G.W., 324
Kilburn, J.D., 90
Kilby, G.W., 32
Kilgore, J.L., 321
Kim, B.H., 76, 106
Kim, C.K., 66
Kim, C.S., 208
Kim, H.J., 66
Kim, I.J., 260, 262
Kim, J., 135
Kim, J.N., 160
Kim, J.S., 135, 266
Kim, K., 141
Kim, K.D., 149
Kim, K.S., 76
Kim, M., 171
Kim, M.-J., 301
Kim, S., 89, 90, 135
Kim, S.H., 206
Kim, S.W., 247
Kim, S.Y., 100
Kim, Y., 155, 193
Kim, Y.C., 135
Kim, Y.H., 290
Kim, Y.S., 247
Kimarathasan, R., 74
Kiminkinen, L.K.M., 22
Kimowski, A., 164
Kimura, M., 247, 292
Kimura, T., 171, 236
Kimura, Y., 314
Kin, H., 273
Kinder, J.D., 233
King, F.L., 46
King, G., 59
King, P.M., 58
King, R.M., II, 158
Kingsmill, C.A., 21
Kingston, J., 266
Kinny, D., 27
Kinoshita, T., 326
Kinsel, G.R., 29
Kiplinger, J.P., 34
Kirilenko, A.G., 304
Kiriyama, Y., 232
Kirpekar, F., 37
Kirsch, D., 41
Kirschner, S., 80
Kiryu, T., 132
Kiselyov, A.S., 292
Kishi, Y., 244
Kitagawa, H., 159
Kitagawa, T., 145
Kitamori, Y., 172
Kitamura, Y., 326
Kitigawa, O., 119
Kiuchi, F., 333, 335
Klärner, F.-G., 80, 210
Klamt, A., 61
Klatt, M., 271
Kleanthous, C., 322
Klein, D.P., 248
Kleintop, B.L., 49
Klempier, N., 305
Kline, D.N., 77
Klohr, S.E., 318
Klots, C.E., 26
Kluge, H., 329
Klunder, A.J.H., 73
Knackmuss, H.-J., 156
Knaggs, A.R., 328, 331
Knani, M., 319
Knapp, R., 266
Knauer, M., 107
Kniaz, K., 25
Knight, D.W., 188, 193, 307
Knight, J., 90
Knight, S.D., 208, 269
Knobler, C.B., 174
Knochel, P., 106, 113, 138, 238, 245, 262, 271
Knölker, H.-J., 138, 222, 257
Kobatake, Y., 242
Kobayashi, H., 264, 266
Kobayashi, R., 57
Kobayashi, S., 159, 271, 272, 273, 278
Kobayashi, Y., 125
Kobbing, S., 136
Koch, A., 25
Kochi, J.K., 287
Kochling, H.J., 42
Kock-van Dalen, A.C., 304
Kočovský, P., 250, 272
Kodama, K., 183
Kodama, M., 232
Kodera, M., 335, 336
Koertvelyesi, T., 78
Kövér, K.E., 7
Koga, G., 292
Koga, K., 125, 157, 252, 293, 296
Koga, N., 67
Kogai, I., 158
Kogut, O.V., 304
Kohlpaintner, C.W., 106
Kohn, W., 51
Kohnke, F.H., 171
Koike, N., 163
Koja, K., 106, 125
Kol, M., 161, 286
Kolb, H.C., 119, 284, 285
Kolter, T., 105
Komatsu, K., 177
Komatsu, N., 277, 289
Komatsu, Y., 95
Komiya, N., 287
Kondakov, D.Y., 117, 243
Kondo, K., 243
Kondo, S., 286
Kondo, T., 254
Kondo, Y., 108, 225, 226, 228, 235, 264
Konig, B., 178
Konig, W., 309
Konings, R.N.H., 4
Konkoli, Z., 148
Kono, M., 30
Konopelski, J.P., 141
Konovalov, V.V., 168
Konshi, Y., 45
Koot, W.J., 254
Koprowski, M., 225
Koreeda, M., 80, 171
Korn, S.R., 295
Kornberg, B.E., 79
Korning, J., 171
Korsmeyer, K.K., 42
Kosemura, S., 321
Koskinen, A.M.P., 276
Koskinen, P.M., 276
Koster, C., 43, 44
Kostov, G.K., 75
Kosugi, M., 236
Kovacik, V., 35
Kowalewski, T., 18
Koyama, K., 160
Koyuncu, D., 155, 193
Kozlova, E.V., 304
Kozmin, A.S., 158
Kraebel, C.M., 123
Krafft, M.E., 140, 220
Kraka, E., 148
Krakowiak, K.E., 105, 214
Krasowski, M., 57
Kratky, C., 243, 266
Kratz, D., 106, 149
Kraus, G.A., 86
Krayushkin, M.M., 160
Kren, V., 309
Kress, M.H., 244
Kreutzberg, J., 259
Krishna, A., 165
Krishnamurthy, D., 270
Krishnamurthy, N., 306
Krishnamurthy, R., 81
Krishnamurthy, V., 88, 138
Krishnan, B., 167, 232, 318
Kristainsen, K., 37
Krogsgaard-Larsen, P., 302
Krohn, K., 141
Krow, G.R., 107
Krowiak, S.A., 203
Krstenansky, J.L., 33
Krueger, A.C., 74, 213
Krüger, C., 107, 262
Kruse, C.G., 163, 271
Krysan, D.J., 153, 241
Kubota, H., 252
Kucera, D.J., 239
Kuck, D., 23
Kudaka, I., 32
Kudo, T., 283
Kuehne, M.E., 208
Kümmerlen, J., 18
Kündig, E.P., 263, 265
Kuhn, N., 259

Kuichi, F., 336
Kukhar, V.P., 304
Kukolich, S.G., 254
Kulik, W., 23
Kulkarni, D.G., 107
Kulkarni, G.U., 25
Kulkarni, S.J., 159
Kumar, B.S.A., 17
Kumar, P., 290
Kumashiro, K.K., 148
Kumbhar, P.S., 159
Kumobayashi, H., 282
Kundu, N.G., 232, 234, 237
Kunec, E.K., 310
Kunisada, H., 286
Kuno, H., 281
Kunz, H., 106
Kuo, D.L., 251
Kurihara, A., 94, 279
Kurihara, H., 291
Kurihara, T., 215
Kurita, J., 188, 209
Kuroda, C., 142
Kuroda, T., 199
Kurono, M., 46
Kurosawa, H., 252
Kurth, M.J., 304
Kurumatani, S., 244
Kustanovich, I., 14
Kusumoto, N., 128
Kuthan, J., 79
Kutney, J.P., 311
Kuwajima, I., 128
Kux, U., 171
Kuzmicrkiewicz, W., 160
Kuznikowski, M., 225
Kvarnström, I., 252
Kvasnicka, V., 69, 75
Kvita, V., 200
Kwan, T.K., 36
Kwast, A., 206
Kwong, H.-L., 119, 285

Laasonen, K., 56
Labiad, B., 294
Labroue, D., 265
Lacey, M.P., 37, 42
Ladd, P., 69
Lage, M., 174
Lai, C., 169
Lai, G., 159
Lai, J.-S., 171
Lai-Hing, K., 47
Laine, R.A., 35, 36
Lal, G.S., 295
Lallemand, J.Y., 79
Lallemant, M., 162
Lam, K.S., 318
Lamb, C.J., 322
Lamb, J.H., 30
Lambalot, R.H., 314
Lambert, N., 31
Lambusta, D., 302
Laming, G.J., 57
Lammert, S.A., 47
Lamont, H.M., 330
Lamont, R.B., 121
Lamothe, S., 72
Lampard, C., 82
Lampe, E.-M., 259
Lamture, J.B., 291
Lan, X., 165, 191
Lanati, S., 72, 308
Landry, C.J.T., 14
Landuyt, L., 79
Lang, H., 165
Lange, C., 31
Langer, F., 245
Lantos, I., 199
Lappenbusch, W.C., 137
Larchevêque, M., 232
Larhed, M., 227
Larock, R.C., 222, 227, 237, 249
Larsen, B.S., 26
Laschat, S., 262
Lassaletta, J.-M., 295
Laswell, W.L., 183
Laszlo, P., 159, 287
Latajka, Z., 57
Latouche, R., 294
Lattman, E.E., 42
Laude, D.A., Jr., 28, 43, 44, 45
Laumann, C., 105
Lautens, M., 106
Lawrence, J.G., 336
Lawrence, N., 115, 277
Lawson, A.M., 34
Lay, J.O., Jr., 37
Laycock, M.V., 30
Layzell, T.P., 131, 294
Leadbetter, M.G., 29
Leary, J.A., 34
Lebeau, E., 202
Leblanc, D., 22
Leblanc, Y., 160
Lebrilla, C.B., 35
Lecea, B., 65, 75
Leclaire, M., 79
Le Corre, M., 131, 277
Lectka, T., 278
Lee, A.S., 278
Lee, B.S., 66
Lee, B.Y., 220, 221, 246
Lee, C., 52, 57, 86, 169, 190
Lee, C.T., 53
Lee, C.Y., 56
Lee, D.G., 286
Lee, E., 86, 100, 169, 190
Lee, G.-H., 254, 256
Lee, H., 171
Lee, H.-C., 126
Lee, I., 66
Lee, I.S., 301
Lee, J.-S., 265
Lee, N.H., 122
Lee, P.H., 135
Lee, P.W., 30
Lee, S., 108, 336
Lee, S.-G., 246
Lee, S.-H., 140
Lee, S.J., 221
Lee, S.M., 246
Lee, S.S., 265
Lee, S.W., 108
Lee, V., 333
Lee, V.J., 287
Lee, W.K., 188
Lee, Y.K., 16, 16
Leeper, F.J., 216, 333, 335, 336
Lees, W.J., 309
Leeson, P.D., 190
LeFloch, P., 208
Lehmann, R., 106
Le Hyaric, M., 287
Leibfritz, D., 11
Leigh, D.A., 162
Leimbacher, W., 31
Leistner, E.W., 322, 323
Lellouche, J.-P., 256
Lemaire, M., 281
Lemaire-Audoire, S., 252
Lemal, D.M., 78
LeMeillour, S., 46
Lemeune, S., 252
Lemmerz, R., 176
Lemoine, J., 35
Lempot, K., 185
Lentz, R., 138
Leong, W., 249
Leresche, J., 263
Lerman, O., 158
Lerner, R.A., 105, 180
Lessmann, K., 280
Levart, M., 287
Levin, J.I., 227
Levine, B.H., 111, 238
Lewis, J.D., 216, 333
Lewis, N., 82, 155, 193, 307
Lewis, S., 42
Ley, S.V., 188, 235, 291
Li, C.-J., 106, 250
Li, C.M., 29, 38
Li, C.-S., 236
Li, K.S., 86, 169
Li, L., 32, 42, 246
Li, W., 156
Li, W.-R., 292
Li, X., 24
Li, X.H., 29
Li, X.-Y., 160
Li, Y., 44, 63, 71, 111, 232, 325
Li, Z., 123, 288
Li, Z.-T., 95, 169
Liang, Y., 265
Liao, C.L., 23
Liao, Y.-H., 266
Liao, Y.-L., 254
Licandro, E., 263
Liebeskind, L.S., 136, 153, 229, 241, 254
Light-Wahl, K.J., 38

Lii, F.-L., 325
Lilly, M.D., 309
Lim, D., 71
Lim, J.A., 243
Lim, K.M., 89
Lim, S., 128, 243
Limbach, P.A., 44
Lin, F.-T., 133
Lin, G., 302
Lin, H.Y., 28
Lin, S., 291
Lin, S.-H., 254
Lin, W.-Y., 138
Lindbeck, A.C., 128
Lindell, S.D., 188, 235
Linden, A., 254
Lindh, F., 35
Lindner, H.J., 79, 83
Lindner, J., 29
Lindon, J.C., 11, 12
Lindström, M., 293
Ling, L., 300
Linnane, P., 162
Linnert, H.V., 146
Linstrumelle, G., 167, 232, 233, 234
Liotard, D.A., 57
Liotta, D.C., 280
Liou, C.C., 47
Lipniunas, P., 35
Lippard, S.J., 37
Lipshutz, B.H., 167, 232, 235, 290
Liso, F.-L., 236
Litrico, A., 150
Liu, D.H., 79
Liu, H.J., 73
Liu, L., 287
Liu, Q.-Y., 108
Liu, R.-S., 254, 256
Liu, S.-H., 302
Liu, Y., 248
Liu, Y.-C., 162, 302
Liu, Y.P., 67, 68, 79
Livinghouse, T., 124
Llenes, C.F., 29
Llera, J.-M., 295
Lloyd, M.D., 333
Lo, H.-C., 126
Loake, G.L., 322
Lobo, A.M., 180, 288
Locke, J.S., 74
Löfström, C., 134
Lohray, B.B., 285, 287
Londry, F.A., 48
Long, T.E., 14
Longvialle, P., 24
Loo, J.A., 38, 39, 45
Look, G.C., 105, 299
López, C.J., 284
Lopez, R., 302
Lopez, X., 65
Lorenzo, M.A., 196
Lothian, A.P., 158
Loubser, C., 266
Louey, J.P., 243
Loupy, A., 294
Lovey, R.G., 201
Lowe, G., 31
Lowen, G.T., 141
Lowis, D.R., 58
Lu, D.H., 67, 68
Lu, R., 57
Lu, X., 118, 220, 237, 240
Lu, Y.-F., 141
Lubman, D.M., 47
Lucas-Lokhorst, G., 36
Lünning, J., 290
Lugtenburg, 171
Luheshi, A.B.N., 76
Luibrand, R.T., 108
Lukacs, C., 310
Lumb, K.J., 31
Luna, E., 171
Luna, H., 305
Lunazzi, L., 147
Lunell, S., 57
Luo, C.W., 37
Luo, W., 79
Luteyn, J.N., 163
Luthra, R., 320
Lutjens, H., 113
Lutz, G., 278
Luu, B., 36
Lykke, K.R., 26
Lynch, G.C., 57, 67
Lynch, J.E., 183
Lynch, S.V., 135
Lyngh, G.C., 79
Lytle, F.W., 26

Ma, S., 76, 220, 237, 290
Ma, W., 30
Maas, P.J., 204
McAdoo, D.J., 22
McArdle, P., 256, 264
McCabe, R.W., 266
McCague, R., 300
McCallum, J.S., 253
McCarley, T.D., 22
McCarrick, M.A., 63
McCarthy, J.R., 295
McCauley, J.P., 25
Macciantelli, D., 147
McClain, R.D., 32
McCloskey, J.A., 37
McClure, K.F., 108, 227
McCombie, S.W., 106
McCormack, A.L., 47
McDaniel, K.F., 259
McDaniel, R., 317
McDonald, C.E., 286
Macdonald, C.J., 29
Macdonald, F., 30
McDonald, F.E., 247
McDougal, P.G., 141
McElvany, S.W., 25
McEwen, C.N., 26
McGarrity, J.F., 300
McGarry, P.F., 80
McGeady, P., 319
McGibbon, G.A., 21, 23
McGlinchey, M.J., 246
McGrane, P.L., 124
McGrath, K.J., 19
McGready, P., 161
McGuigan, P., 181
Machiguchi, T., 65, 72
Machinek, R., 150, 317
Maciel, G.E., 13
McIlwaini, D.B., 333
McIntosh, A., 49
McIntyre, C.R., 315, 316
McIntyre, L., 10
McIver, R.T., Jr., 44
McKay, R.G., 26
MacKeith, R.A., 300
McKennon, M.J., 157, 283
McKervey, M.A., 181
McKie, J.A., 281
McKillop, A., 155, 163, 193
McLafferty, F.W., 45
McLaren, L., 155, 193
McLean, D.B., 35
McLuckey, S.A., 28, 37, 47, 48, 49
McMahon, K., 168
McMillen, H.A., 93, 149
McMurray, B.T., 305
McNab, H., 86, 211
McNally, J.J., 79, 80
McNeill, A.H., 127
McNelis, E., 158
MacPherson, A.D., 86
McQuire, L., 208
McWherter, C.A., 18
Macy, T.S., 259
Madsen, U., 302
Maeda, K., 251
Maerk, T.D., 27
Maes, R.A.A., 38
Maestro, M.C., 271
Maeta, H., 117
Magnin, D.R., 250
Magnus, P., 208
Magriotis, P.A., 149
Mahajan, J.R., 124
Maharaj, V.J., 330
Mahr, N., 244
Maier, M.E., 124
Maier, W., 214
Maiorana, S., 262, 263
Mair, F.S., 125, 296
Maisa, W., 73
Maity, G., 290
Majdoub, M., 294
Majetich, G., 160
Majumdar, A., 11
Majumdar, T.K., 22
Mak, C.C., 230
Makin, H.L.J., 36

Makino, T., 138
Makita, K., 136
Mako, M., 254
Makosza, M., 164, 206
Malbuk, T.E., 169
Malenfant, E., 271
Malik, S.S., 156
Mallart, S., 250
Mallet, M., 201
Malm, J., 229
Malone, J.F., 305
Malyszko, J., 330
Mambu, L., 232
Mamos, P., 228
Manako, K., 108
Mandelbaum, A., 26
Mangeney, P., 281
Manning, H.W., 138
Mano, S., 131
Manolache, N., 151
Manoury, E., 287
Manriquez, J.M., 266
Mantell, S.J., 191
Mantle, P.G., 330
Marcaccini, S., 187
March, R.E., 48
Marchi, M., 237
Marcinow, Z., 169
Marcos, C.F., 248
Marcos, E.S., 60
Marcuzzi, A., 254
Marder, M., 162
Marek, I., 135, 138, 245
Maresca, L., 197
Margolin, A.L., 299, 304
Marguet, A., 328
Marinelli, F., 226, 233, 237
Marko, I.E., 141
Marko, L., 221
Marlow, G.E., 58
Marques, C.A., 198
Marquis, R.W., Jr., 208
Marr, J.C., 30
Marrs, C.M., 252
Marsais, F., 201, 229
Marsh, A., 141
Marshall, A.G., 43, 44
Marshall, D.R., 167, 232
Marshall, J.A., 190
Marson, C.M., 179
Martens, J., 271, 280
Martin, A., 141
Martin, A.R., 232
Martin, C.J., 114, 253, 271
Martin, D., 192
Martin, E., 141
Martin, N., 183
Martin, P., 200
Martin, R., 323
Martin, S.A., 32
Martin, T., 38
Martin Cabrejas, L.M., 271
Martin-Cantelejo, Y., 186
Martinez, J.P., 141
Martinez, L.A., 286
Martorell, G., 108
Martynow, J., 141
Maruoka, E., 196
Maruoka, K., 132, 278
Maruyama, A., 266
Maruyama, K., 270
Maruyama, T., 250
Maryanoff, C.A., 75, 196
Mascaretti, O.A., 107, 291
Mashima, K., 282
Masson, S., 165
Masters, J.J., 230, 247
Mastrorilli, E., 304
Masuda, H., 57
Masuda, K., 244
Mata, E.G., 107, 291
Mataka, S., 177
Matha, V., 331
Mathew, L., 291
Mathews, C.K., 26
Mathews, D.P., 295
Mathey, F., 208
Mathis, J.R., 66
Mathre, D.J., 280
Matsui, S., 165
Matsukawa, S., 77, 127
Matsuki, K., 130
Matsumoto, J.-I., 175
Matsumoto, K., 128, 186
Matsumoto, R., 314
Matsumoto, S., 235
Matsumura, S., 149
Matsunaga, K., 321
Matsunaga, S., 277
Matsuo, T., 39, 220
Matsushita, H., 281
Matsuura, T., 243
Matsuzaki, K., 316
Matthias, C., 23
Mattson, M., 134
Matusik, J.E., 29
Maxwell, C.A., 322
May, M.A., 43
Maye, J.P., 243
Mayer, F., 28
Mayer, H.K., 36
Mayer, M., 313, 314, 328
Mayo, J.E., 211
Mayr, H., 257
Mazzieri, M.R., 254
Mebane, R.C., 193
Medebielle, H., 163
Medici, A., 307
Meerholz, K., 178
Meffre, P., 286
Mehler, T., 271
Mehta, G., 170
Meinhardt, K.P., 79
Melikyan, G.G., 106
Melis, S., 165
Mella, M., 168
Mello, R., 182
Mellon, C., 271
Mellon, F.A., 30
Melnyk, P., 108, 207, 226
Méndez, N.Q., 248
Mereyala, H.B., 290
Merlic, C.A., 152
Merlini, L., 316
Merritt, A., 208
Merritt, K.D., 333
Merz, K.M., 58, 145
Merzouk, A., 252, 290
Meske, M., 137
Messmer, R.P., 146
Meth-Cohn, O., 204, 305
Metzenthin, T., 57
Metzger, J.W., 29
Metzler, M.R., 267
Meyer, C., 138
Meyer, S.D., 105, 269
Meyers, A.I., 156, 157, 167, 188, 283
Mi, A.Q., 73
Michael, J.P., 106, 108, 226
Michael, S.M., 47
Michel, H., 31
Michoud, C., 141, 230
Middlemiss, D., 223
Middleton, S., 138
Mieglo, A., 75
Miehlich, B., 52
Miertus, S., 59
Migita, T., 236
Mihn, B.J., 76
Mijoule, C., 57
Mikami, K., 77, 127, 128
Milan, S., 75
Miller, J.A., 295
Miller, J.M., 25
Miller, S.C., 295
Miller, S.J., 278
Miller, W.H., 244
Milliet, A., 22
Milne, D., 125, 296
Milot, G., 188
Milstein, D., 224, 227
Milzer, J., 270
Min, C., 333, 335
Minabe, M., 171
Minami, T., 111, 241
Minami, Y., 102, 150
Minamikawa, J.-I., 158
Minato, T., 65, 72
Minghetti, A., 330
Mink, J., 57
Min-Min, G., 138, 236
Minowa, N., 105, 231, 269
Mirza, U.A., 39
Mirzaei, Y.R., 188
Mischitz, M., 305
Mishina, F., 171
Misiti, D., 147, 197
Mitamura, S., 159
Mitarai, A., 169
Mitchell, M.B., 193
Mitchell, R.H., 273

Mitsudo, T.-a., 254
Miura, H., 326
Miura, M., 158, 235
Miura, T., 295
Miyamoto, H., 293
Miyano, S., 163
Miyashi, T., 78
Miyashita, A., 282
Miyashita, M., 132
Miyata, H., 321
Miyaura, M., 235
Miyaura, N., 141, 153, 232
Miyazawa, A., 175
Mizuno, K., 168
Mizuno, Y., 149
Mocek, U., 324, 326, 329, 331
Möller, U., 290
Moffre, P., 121
Mohan, V., 58
Mohr, B.J., 238
Moïse, C., 264
Molander, G.A., 128, 281
Molenaar-Langeveld, T.A., 24
Molina, A., 232
Molina, P., 196, 206
Molitor, K.D., 322
Moller, M., 174
Momose, T., 301
Monderhack, D., 182
Mongomery, D., 134
Moniz de Sa, M., 322
Monnig, C.A., 42
Mons, H.-E., 4, 8
Monstel, R.F., 141
Montagnani, R., 61
Montanari, F., 198
Montanari, S., 162
Montaufier, M.-T., 287
Moore, A.J., 266
Moore, B.S., 324, 328
Moore, H.W., 153
Moore, J.A., 327
Moore, J.S., 108
Moore, R.E., 314, 328
Moorlag, M., 167
Morand, K.L., 47, 48, 49
Moreau, J.J.E., 227
Moreno-Manas, M., 199
Moret, E., 106
Moret, M., 254, 262
Morey, J., 156
Morgan, J., 47
Mori, M., 243
Mori, T., 126, 158
Mori, Y., 145
Moriarty, R.M., 135
Moriguchi, S., 224
Moriguchi, T., 177
Morikawa, K., 284
Morimoto, T., 252, 302
Morinaga, Y., 95
Morisaki, M., 318
Morita, N., 257
Moritani, M., 141
Moriya, M., 301
Morizur, J.P., 24
Morman, R.J., 78
Morodome, M., 127, 270
Morokuma, K., 67
Morris, J., 332
Morris, K.F., 5
Morris, K.G., 151, 222
Morrow, G.W., 155
Morrow, J.C., 23
Morton, T.H., 22, 24
Morwick, T., 138
Morzherin, Y., 190
Mosandl, T., 146
Mosher, M.D., 247
Moskowitz, N., 81
Mosleh, A., 244
Moss, J.R., 266
Motallebi, S., 264
Motoki, K., 138
Motoki, Y., 138
Motoyama, I., 266
Moulines, F., 266
Mouloud, H.A.H., 287
Mourad, A.-F.E., 176
Mousdale, D., 332
Mowbray, C.E., 92
Mowrey, C.D., 41
Mowrey, R.C., 25
Moyake, R., 264
Moyano, A., 132
Muchowski, J.M., 165
Muci, A.R., 130, 281, 293
Mueller, C., 8
Müller, D.G., 312
Müller, K.-H., 257
Mueller, L., 7
Mueller, M.D., 30
Mueller, M.J., 330
Müller, P., 264
Müller, U., 246
Mueller-Westerhoff, U.T., 273
Muenster, H., 39
Muhammad, F., 115, 277
Mukai, C., 245, 260, 262
Mukaiyama, T., 159, 287
Mukata, T., 250
Mulder, P.P.J., 171
Mulheim, C., 31
Munasinghe, V.R.N., 182
Munoz, B., 304
Muñoz, S., 291
Munson, M.S.B., 28
Muraglia, E., 95, 193
Murahashi, S.-i., 119, 250, 254, 286, 287
Murai, S., 128, 153, 218, 250
Murakami, M., 122, 235
Murakami, N., 302
Murakami, Y., 224
Murase, N., 132, 278
Murashima, T., 158
Murata, K., 32
Murayama, K., 106
Murphree, S.S., 77
Murphy, C.K., 295
Murphy, J.A., 82, 103, 104
Murphy, M.M., 238
Murphy, P.J., 125, 296
Murphy, R.C., 36
Murray, C.W., 53, 57
Murray, J.S., 57
Musco, A., 249
Musselman, B.D., 34, 39
Musser, S.M., 37
Mussmann, L., 287
Mutter, M., 42
Muzart, J., 107, 286, 290
Myers, A.G., 128, 149
Mynett, D.M., 291

Nabela, K., 319
Nadelman, N.S., 162
Naemura, K., 175, 178
Nagai, K., 35
Nagai, T., 187
Nagata, T., 42
Nagatsu, A., 302
Nagra, D.S., 32
Nagy, P.I., 58
Naidu, B.N., 204
N'Ait Ajjou, A., 286
Naito, M., 128
Nakada, M., 149
Nakadaira, Y., 254
Nakafuku, K., 172, 174
Nakagawa, A., 326
Nakagawa, M., 141
Nakai, T., 127
Nakajima, H., 314
Nakajima, M., 252
Nakamura, D., 292
Nakamura, H., 122, 235
Nakamura, K., 108
Nakamura, N., 83
Nakamura, T., 128, 149
Nakanishi, J., 250
Nakanishi, K., 168
Nakanishi, Y., 281
Nakashima, H., 266
Nakata, M., 149
Nakatani, H., 83
Nakatani, K., 141, 233
Nakayama, H., 318
Nakayama, K., 264
Nakazawa, M., 66
Nallin-Omstead, M., 315
Namchuk, M., 29
Nandi, M., 263
Naota, T., 286, 287
Napolov, D.V., 60
Narasaka, K., 116, 276
Narasimhan, T.S.L., 26
Narayana, M., 13
Narto, S., 46

Naruta, Y., 106, 270
Nasaka, N., 235
Nash, J.J., 146
Nasini, G., 316
Natale, N.R., 188
Natile, G., 197
Nayak, S.K., 290
Naylor, A., 277
Naylor, S., 30
Ndibwami, A., 72
Nédélec, J.Y., 167, 235
Needham, J., 313
Neeleman, E., 286
Negishi, E.-i., 117, 154, 219, 235, 243, 244
Negri, J.T., 138
Neidlein, R., 171
Neil, D.A., 223
Neild, G.H., 12
Nelson, C.C., 37, 38
Nelson, J.A., 295
Nelson, R.W., 29, 37, 40
Nelson, S.G., 130, 281
Nelson, T.P., 167
Nestler, B., 122
Nestor, N.B., 286
Neto, C.C., 265
Neuheiser, F., 31
Neumann, W.P., 159, 280
Neunhoeffer, H., 201
Ng, C.Y., 23, 57
Nguyen, B.V., 232
Nguyen, J.T., 254
Nguyen, M.T., 24, 79
Nguyen, S.T., 117
Nguyen, T., 58, 331
Nguyen, V., 24
Ni, Z., 137
Nibbering, N.M.M., 24, 66
Nice, L.E., 286
Nicholas, J.B., 57
Nicholas, K.M., 245, 253
Nichols, M.A., 108
Nicholson, J.K., 12
Nicholson, L.W., 294
Nicholson, N.H., 332
Nickel, S., 271
Nickisch, K., 112
Niclas, H.-J., 291
Nicolaides, A., 146
Nicolaou, K.C., 105, 149, 231, 269
Nicolosi, G., 302
Nieder, M., 30
Nieger, M., 176
Nielsen, B., 302
Niessen, W.M.A., 35, 36
Nigawara, Y., 251
Nijssen, W.P.M., 69
Nikolski, M., 246
Nilsson, B., 35
Nilsson, M., 112
Nilsson, Y.I.M., 249
Nishida, A., 158
Nishida, H., 316
Nishide, K., 292
Nishigaichi, Y., 106, 270
Nishiguchi, H., 160
Nishiguchi, I., 156
Nishijima, K.-i., 235
Nishikawa, Y., 138
Nishima, S., 326
Nishimura, A., 111, 238
Nishimura, H., 242
Nishimura, K., 172
Nishimura, R., 171
Nishino, A., 301
Nishino, E., 251
Nishio, K., 271, 273
Nishiyama, H., 282
Nitta, M., 172
Niyazymbetov, M.E., 106
Njamela, O.L., 203
Nobel, D., 163
Noda, K., 288
Noda, Y., 235
Node, M., 292
Noe, M.C., 119, 121, 285
Noe, R., 262
Noguchi, H., 296, 320
Nohira, H., 282
Nohmi, T., 28
Nomura, M., 158, 235
Nonhebel, D.C., 169
Nordhaus, P., 237
Nordhoff, E., 37, 42
Normant, J.-F., 135, 138, 245
Norrby, P.-O., 249
Norris, D.J., 254
North, M., 271, 309
Nourbakhsh, S., 23
Nourooz-Zadeh, J., 304
Nourse, B.D., 49
Novak, M., 155
Novi, M., 163
Noyori, R., 270
Nozaki, K., 131
Nuber, B., 218, 267
Nucci, L., 155
Nugent, H.M., 266
Nugent, W.A., 131
Nukui, S., 113, 243
Nusair, M., 53
Nuss, J.M., 111, 238
Nuyken, O., 168

Oak, O.Z., 156
Oba, S., 254
Oberlies, N., 155
Obora, Y., 235, 240
O'Callaghan, N.M., 332
O'Connor, S.J., 239
Oda, H., 236
Oda, M., 292
Oda, Y., 287
Odagaki, Y., 141
Oelckers, B., 266
O'Ferrall, R.A.M., 148
Ogasawara, K., 213, 301
Ogata, M., 292
Ogawa, A., 94, 101, 128
Ogawa, S., 251
Ogawa, T., 302
Ogbu, C.O., 74
Ogden, M., 266
Ogino, Y., 284
Ogoshi, S., 250
Oguntimein, G.B., 302
Ogura, K., 108
O'Hagan, D., 300, 311, 331
O'Hara, K., 30
Ohara, Y., 111, 241
Ohashi, A., 149
Ohe, K., 250
Oh-e, T., 232
Ohga, Y., 132
Ohira, M., 286
Ohira, S., 141
Ohkita, H., 244
Ohkohchi, M., 26
Ohnishi, M., 136
Ohnishi, Y., 142
Ohno, T., 156
Ohno, Y., 271
Ohsaku, M., 67
Ohshima, T., 243
Ohsuka, A., 252
Ohta, H., 209
Ohta, K., 149
Ohta, S., 188
Ohta, Y., 116, 276, 302
Ohtani, I., 328
Ohtsuka, A., 149
Ohuchi, K., 215
Oishi, M., 132, 196
Oishi, S., 125
Oishi, T., 121, 307
Ojima, J., 172, 174
Okamoto, A., 158
Okamoto, S., 118, 284
Okamura, K., 108, 141, 273
Okazaki, M., 83
Okuda, S., 316
Okumoto, H., 111, 183, 238
Okuyama, H., 319
Olah, G.A., 107, 160
Oliver, P.A., 291
Oliveros, E., 293
Olivo, H.F., 300
Olivucci, M., 66, 67, 75, 145
Olson, E.R., 147
Olsson, T., 112
O'Malley, R.M., 29
Omura, A., 158
Omura, S., 316, 326
Ong, C.W., 258
Ono, N., 118
Opdenbusch, K., 111
Opella, S.J., 17
Oppolzer, W., 249
Ori, A., 75

Oriyama, T., 292
Orlando, R., 46, 47
Orlinkov, A., 159
Orozco, M., 58
Orr, J., 322
Osaki, H., 215
Osamura, Y., 108
Osella, D., 246
Oshima, K., 186
Oshio, A., 128
Osman, R., 62, 108
Osterhout, M.H., 130
Ostrowski, S., 164
Ostwald, R., 113
Ota, T., 302
Otani, T., 102, 150
Otera, J., 299
Otsuji, V., 168
Ott, W.R., 314
Otte, A.R., 134, 248
Otto, C., 278
Otto, H.-H., 185, 302
Ottow, E., 270
Oturan, M.A., 163
Ourison, G., 319
Overberg, A., 42
Overman, L.E., 208, 226, 239, 243, 269
Owen, D.A., 258
Owen, T.J., 33
Ozaki, M., 156
Ozaki, S., 300, 333
Ozawa, F., 242
Ozlu, Y., 93

Pace, P., 226
Padwa, A., 77, 137, 150
Paillaud, J.L., 148
Pairaudeau, G., 208
Pais, M., 31
Paiva, N.L., 322
Pal, M., 234, 237
Palacios, S.M., 162
Pale, P., 111
Paley, R.S., 232
Palin, M.G., 264
Palmer, C.F., 300
Palmer, I.J., 67, 145
Palmer, K.W., 78
Palmer, R.T., 263
Palmieri, G., 197
Palomo, C., 65, 75
Palovich, M., 86
Palumbo, G., 164
Pan, Y., 40
Panaye, A., 69
Pandey, B., 142, 290
Pandey, G., 165
Panetta, C.A., 164
Pang, Z., 320
Pannecouke, X., 36
Pannek, J.B., 257
Papagni, A., 263
Papai, I., 57
Papayannopoulos, I., 34
Papousek, D., 57
Pappalardo, R.R., 60, 61, 75
Paquette, L.A., 138
Paradisi, C., 47, 48, 162
Paraudeau, G., 269
Parchment, O.G., 59
Pardo, L., 62, 108
Paris, A., 24
Paris, M.F., 171
Park, C.M., 190
Park, J., 284
Park, J.-S., 336
Park, K.B., 289
Park, M.A., 28
Park, S.-B., 128, 282
Park, W.M., 76
Park, Y.C., 100
Parkanji, L., 185
Parker, A.R., 327
Parker, C.E., 30
Parker, D.H., 26
Parkinson, J.A., 12
Parlier, A., 223
Parmar, V.S., 302
Parodi, F.J., 313
Parr, G.R., 42
Parr, J., 282
Parr, R.G., 51, 52
Parrain, J.-L., 240
Parrinello, M., 56
Parry, D.E., 266
Parry, R.J., 325
Parsons, E.J., 217
Parsons, P.J., 93, 277
Parvez, M., 123
Pascualahuir, J.L., 60
Pasto, D.J., 74
Pastor, A., 206
Patel, U.R., 211
Paterson, I., 66, 180, 276
Pati, H.N., 302
Patil, V., 275
Patiño, R., 286
Pattenden, G., 90, 92, 106, 230
Patti, A., 302
Patzelt, H., 313
Pauson, P.L., 220
Pavlova, S.D., 75
Payne, A.N., 290
Payne, G., 48
Pearson, A.J., 106, 256, 257
Pearson, W.H., 128
Pedder, R.E., 48
Pedersen, M.J., 38
Pedragosa-Moreau, S., 305
Pedrini, P., 307
Pedro, J.R., 291
Pedrosa, R., 271
Peeran, M., 72
Peersen, O.B., 14
Pegg, N.A., 108, 227
Pegg, S.J., 324
Pellissier, H., 138
Pellon, P., 131, 277
Pelter, A., 155
Peltier, J.M., 35
Penco, S., 237
Peng, S.-M., 254, 256
Peng, T.-S., 248
Peng, Z., 58
Percas, M.A., 132
Pereira, I.A.C., 332
Pereira, M.M., 180, 288
Perellino, N.C., 330
Perez, A.J., 162
Perez, C., 302
Pérez-Encabo, A., 263
Pergola, F., 155
Periasamy, M., 156
Périchon, J., 167, 235
Perkins, J.R., 30
Perrio, S., 78, 263
Perrone, E., 185
Peschke, M., 47
Petasis, N.A., 244
Petersen, P.M., 216, 333
Petersen, T.H., 78
Peterson, M.J., 273, 291
Peterson, P.E., 264
Petit, A., 294
Petitou, M., 105
Petrillo, G., 163
Petrusiewicz, K.M., 225
Pettitt, B.M., 58
Pews, R.G., 158
Pfaltz, A., 106, 114, 198, 253, 271, 272
Pfeifer, K.H., 145
Pfenning, D.R., 168
Pflästerer, G., 218
Pflug, T., 237
Phan, L.T., 219
Phan, T.N., 23
Philippides, A., 216, 333
Phythian, S.J., 153
Piattelli, M., 302
Piazza, M.G., 270
Pickering, J., 179
Piers, E., 111
Pietra, F., 155
Pietruszka, J., 235
Pindur, U., 278
Pine, S.H., 117, 269
Pinel, C., 281
Pines, A., 16
Pinkston, J.D., 49
Pinson, J., 163
Pipeng, Y., 146
Pirrung, M.C., 306
Piscopio, A.D., 105, 231, 269
Pitsinos, E.N., 149
Plant, D., 11
Plaquevent, J.-C., 117, 277
Plater, M.J., 170
Platzer, N., 135
Plaziak, A.S., 37

Ple, N., 200, 201
Pleasance, S., 30
Pleixats, R., 199
Plummer, B.F., 171
Plzak, K.J., 125
Poch, M., 132
Pochapsky, S., 11
Poirot, P., 206
Poitier, P., 138
Poli, S., 307
Poliakoff, M., 217
Politzer, P., 57
Pommier, A., 107
Ponec, R., 71
Pons, J.-M., 107
Pontellini, R., 249
Pook, L.M., 81
Poon, D.J., 243
Pople, J.A., 53
Portnoy, M., 224, 227
Pospical, J., 69, 75
Postel, M., 288
Postema, M.H.D., 142
Poulter, C.D., 320
Poulter, G.T., 108
Powers, R., 11
Powers, T.S., 73
Prabhakar, S., 180, 288
Prabhakaran, P.C., 314
Pradeep, T., 25
Prakash, G.K.S., 160
Prakash, S., 160
Pramanik, B.N., 39
Pramatorova, V., 308
Prasad, A.K., 302
Prein, M., 158
Prescott, M.C., 36
Press, J.B., 79, 80
Prestegard, J.H., 11
Preston, P.N., 287
Preuss, H., 52
Price, K.R., 30
Price, N.C., 322
Prieba, G., 310
Prime, J.C., 267
Prins, T.J., 302
Proudfoot, G., 324
Proudfoot, J.R., 211
Prout, K., 267
Prusiner, S.B., 31
Puertas, S., 304
Punniyamurthy, T., 287
Puranik, D.B., 259
Purdon, M., 42
Pyun, H.-J., 319

Quallich, G.J., 130, 280
Quéguiner, G., 200, 201, 229
Quijano, L., 313
Quilliam, M.A., 30
Quinkert, G., 108
Quinn, J.P., 45
Quintard, J.P., 240

Raabe, G., 136, 271
Rabideau, P.W., 169
Rabinovitz, M., 156
Rabinowitz, J., 108
Rabinowitz, J.R., 62
Radanovich, L.J., 263
Rader, J., 241
Radford, S.E., 31
Ragazos, I.N., 66, 67, 145
Ragg, E., 316
Rahman, N.A., 186
Raithby, P.R., 125, 246, 296
Rajapohal, D., 113
Ram, B., 283
Rama, M., 266
Ramaiah, P., 160
Ramamoorthy, P.S., 97
Rama Rao, A.V., 159
Ramdani, M., 294
Ramesar, N., 203
Ramkumar, D., 289
Ramon, D.J., 167
Ramsden, C.A., 158
Ramsey, M.J., 48
Ramsey, U.P., 325
Rancourt, J., 97, 129
Rane, A.M., 120, 191, 284
Rangappa, K.S., 155
Ranu, B.C., 294
Rao, B.R., 283
Rao, C.N.R., 25
Rao, D., 24
Rao, P.R.V., 26
Rao, R., 158, 295
Rao, S.N., 148
Raphael, R.A., 202
Rappe, C., 30
Rapta, P., 168
Rashid, M.A., 111, 238
Raskin, I., 323
Rassat, A., 57
Rastelli, A., 75
Rastogi, V.K., 3
Ratanunga, T.D., 23
Rathke, E., 35
Ratovelomanana, V., 167, 235
Ratovelomanana-Vidal, V., 281
Rau, D., 257
Rauchschwalbe, G., 106
Rauhut, G., 61
Rautenstrauch, V., 293
Rauth, T., 27
Ravikumar, K., 170
Ravi Kumar, K.S., 294
Ravinovitz, M., 178
Rawal, V.H., 88, 91, 138, 141, 230
Raymo, F.M., 171
Reagan, A.C., 224
Reamer, R.A., 183, 280
Reau, R., 187
Rebière, F., 266
Rebolledo, F., 304
Reddy, G.S., 30
Reddy, J.P., 130
Reddy, M.M., 287
Reddy, R.S., 290
Redhouse, A.D., 76
Reese, W.G., 171
Reetz, M.T., 106, 125, 127
Reeves, W.P., 158
Regan, A.C., 197
Reguero, M., 61, 75
Rehahn, M., 266
Reich, H.J., 107, 128, 141, 269
Reich, I.L., 141
Reichling, J., 323
Reimann, C.T., 41
Reimes, U., 73
Rein, T., 116, 275
Reinhold, T.L., 106
Reinhoudt, D.N., 105, 190
Reinold, S., 322
Reis, L.V., 180, 288
Reisch, J., 237
Reiser, O., 178, 242, 253
Reissig, H.-U., 181
Reitsam, K., 107
Reitz, O., 106
Rempel, D.L., 22, 43
Ren, H., 7
Rennels, R.A., 111, 238
Resck, I.S., 124
Resnati, G., 107, 295
Restituyo, J.A., 183
Reszka, K.J., 168
Rettie, A.B., 128
Reusch, W., 36
Rexroth, A., 8
Reynolds, C.A., 58, 59, 69
Reynolds, J.H., 148
Reynolds, K., 134, 324
Rhee, H., 261
Rhee, Y.H., 100
Rheingold, A.L., 138, 150, , 269278
Riant, O., 266
Ricard, L., 208, 266
Ricca, D.J., 239
Ricca, T.L., 44
Rice, J.E., 227
Richards, W.G., 58, 59, 69
Richter, N., 125
Riddell, F.G., 18
Riddoch, A., 46
Ridge, D.P., 28
Ridlehuber, R.W., 162
Riedel, M., 262
Rieger, D.L., 111, 232
Rielly, M.D., 79
Riera, A., 132
Riesinger, S.W., 229
Rigby, J.H., 74, 106, 213, 263
Righetti, P.P., 72, 78
Riley, T.A., 37
Rinaldi, D., 60
Rink, H., 42
Ripoll, J.-L., 107

Risbridger, G.D., 332
Ritter, K., 106, 167
Riva, R., 227
Rivail, J.L., 60
Rivero, I.A., 295
Riveros, J.M., 146
Robb, M.A., 61, 66, 67, 75, 145
Robert, A., 77
Roberts, S.M., 105, 300, 306
Roberts, W.M., 152
Robins, D.J., 326, 327
Robinson, C., 332
Robinson, J.A., 313
Robinson, R.A., 290
Robison, T.W., 266
Rocca, P., 201, 229
Rockwood, A.L., 28, 39
Rodewald, U., 174
Rodriguez, J., 119
Rodríguez, R., 271
Rodriguez-Lopez, J., 186
Roehrig, M.A., 254
Roell, B.C., Jr., 259
Roepstorff, O., 27, 37
Roessner, C.A., 333, 335
Rogers, D.H., 128
Rogerson, M., 18
Rohmer, M., 319
Rohr, J., 311, 317
Roitman, E., 321
Rolfe, J., 44
Román, E., 266
Romanens, P., 263
Romanow, W.J., 25
Romberger, M.L., 133
Romero, R.H., 220
Romero, S.H., 140
Romesberg, F.E., 125, 296
Romo, D., 105
Romo, L.K., 34
Romon, D., 269
Roof, M.B., 290
Roper, T.D., 279
Rosankiewicz, J.R., 34
Rosch, N., 61
Rosen, A., 57
Rosenberg, F.E., 108
Rosenberg, R., 35
Rosenblum, M., 266
Roskamp, E.J., 183, 292
Ross, C.W., III, 44
Ross, M.M., 25
Ross, P.L., 23
Rossi, R.A., 162, 235
Rossiter, J., 324
Rost, B., 70
Roth, G.P., 167, 232, 290
Roth, J.R., 336
Roth, K.-D., 246
Rouden, J., 305
Rouillard, L., 271
Rousseau, G., 300
Rouvray, D.H., 69
Row, T.N., 25
Roxburgh, C.J., 144, 209
Roy, S.C., 290
Royer, A.C., 193
Rozema, M.J., 113, 245
Rozen, S., 158, 161, 286
Rubello, A., 148
Rubenfield, M., 336
Rubino, F.M., 29
Rubio, A., 118
Rudewicz, P., 46
Rudler, H., 223
Ruechardt, C., 78
Rück, K., 106
Ruel, R., 244
Ruf, S., 185
Ruiz-Lopez, M.F., 60
Ruiz-Montes, J., 249
Runsink, J., 141
Ruscic, B., 23, 25
Russel, R.G., 318
Russell, A.T., 276
Ryan, W.J., 264
Rybakova, A.V., 304
Ryu, E.K., 160
Ryu, I., 94, 101, 128, 129, 279
Ryu, S.I., 45
Rzepa, H.S., 61, 75

Saa, J.M., 108, 156
Sabat, M., 142
Saberi, S.P., 151, 222, 266
Sabol, J.S., 295
Sacki, M., 150
Sadakane, M., 183
Sadler, I.H., 12, 211
Saeki, M., 102
Saeki, N., 247
Saha, M., 156, 294
Saha-Moller, C.R., 309
Sahm, H., 319
Saigo, K., 159
Saiki, S., 175
Saimoto, H., 149
Sainsbury, M., 107
Saint-Clair, J.-F., 165
Saito, K., 74
Saito, T., 128
Saito, Y., 26
Sakaguchi, K., 31
Sakai, H., 266
Sakai, K., 106, 138
Sakai, N., 131
Sakaki, J.-i., 282
Sakakibara, J., 302
Sakamoto, T., 108, 225, 226, 228, 235
Sakamoto, Y., 66, 215
Saksena, A.K., 201
Sakuda, S.-i., 128
Sakuma, K., 125, 296
Sakurai, T., 39
Salahub, D.R., 57
Salman, S.R., 11
Salomon, C.J., 107, 291
Salter, M., 151
Salter, M.M., 248
Salter, M.W., 222
Salvador, J.M., 300
Salvino, J.M., 296
Salzner, U., 108
Samuel, O., 266
Samuelsson, B.E., 46, 252
Sancéau, J.-Y., 121, 286
Sandanayaka, V.P., 263
Sander, C., 70
Sanders, J.C.P., 57
Sanders, M., 324
Sanderson, P.N., 11
Sandford, G., 295
Sanfilippo, C., 302
Sangwan, R.S., 320
Sankararaman, S., 289, 291
Sankawa, U., 320
Sannicolo, F., 196
Sano, H., 266
Sano, S., 336
Santander, P.J., 333
Santaniello, E., 299
Santelli, M., 138
Santelli-Rouvier, C., 249
Santhakumar, V., 240
Santhi, P.L., 290
Santiago, J., 266
Santos, P.P.O., 180, 288
Sanz-Cervera, J.F., 331
Sappa, E., 221
Saquat, M., 165
Sarakinos, G., 279
Sarkar, A., 263, 283
Sarrazin, L., 246
Sarshar, S., 119, 285
Sasai, H., 108, 114, 128, 243, 273
Sasaki, H., 287
Sasaki, K., 169
Sathe, K.M., 263
Sato, F., 118, 251, 284
Sato, H., 26
Sato, K., 108
Sato, M., 162, 266
Sato, T., 83, 183
Satoh, J.Y., 142
Satoh, S., 163
Satoh, T., 158, 235
Sattlet, I., 330
Sauer, J., 65, 72
Sauerwein, M., 326
Saulnier, M.G., 200
Saunders, A.C.G., 69
Saunders, J.K., 141
Saunders, J.K.M., 105
Sauvage, J.-P., 230
Saveant, J.-M., 163
Savignac, M., 252
Savin, A., 52
Sawada, M., 23
Scaiano, J.C., 80, 148

Schaaf, P.M.M., 322
Schack, C.J., 57
Schaefer, H.F., III, 80
Schaefer, J., 18
Schäfer, M., 244
Schaeffer, M.J., 118
Schaer, M., 42
Schaller, T., 14
Schelder, D.J.A., 283
Scherer, H.J., 141
Scheuer, P.J., 311
Scheuplein, S.W., 111, 150, 233
Schieltz, D., 37
Schimpf, R., 210, 226
Schimpff, D.G., 254
Schink, H.E., 252
Schioett, B., 80
Schirmeister, T., 302
Schleyer, P.von R., 108
Schlingloff, G., 286
Schlögl, K., 266
Schlosser, M., 106, 115, 165, 275
Schmalz, H.-G., 259
Schmid, R.D., 302
Schmidt, A., 18
Schmidt, J., 23
Schmidt, T., 252
Schmidt, W., 157
Schmittberger, T., 301
Schmittel, M., 73
Schnatter, W.F.K., 151
Schneider, B., 330
Schneider, G., 70
Schneier, A., 29
Schnell, A., 28
Schoellkopf, K., 270
Schoenewolf, M., 317
Schoenmakers, P.J., 69
Schöttler, M., 311
Schofield, C.J., 332, 333
Schols, H.A., 35, 36
Schreier, P., 309
Schrell, A., 310
Schreurs, J., 33
Schrieber, S.L., 105, 269
Schrimpf, M.R., 273
Schrobilgen, G.J., 57
Schroeder, D., 22, 27
Schuchardt, U., 106
Schulman, J.M., 146
Schultz, A.G., 102
Schultz, M., 323
Schultz, P.A., 146
Schultz, P.G., 105
Schulz, D.J., 90
Schumann, G., 317
Schuster, M., 304
Schuurmann, G., 61
Schwab, J.M., 327
Schwalbe, H., 8
Schwarm, M., 283
Schwartz, B.L., 32
Schwartz, H., 107
Schwartz, J.C., 48
Schwarz, H., 22, 23, 27
Schwarz, M., 311
Schwarzhans, K.E., 27
Schweikert, E.A., 28
Schweikhard, L., 43
Schweizer, W.B., 282
Schwenkreis, T., 287
Schwiebert, K.E., 248
Scopes, D.I.C., 247
Scorrano, G., 162
Scott, A.I., 107, 214, 333, 335
Scott, I.L., 140, 220
Scott, L.T., 145, 196
Scrocco, E., 59
Searle, P.A., 107
Seaton, N., 324
Sebald, A., 14, 18
Sebek, P., 79
Sedmera, P., 79, 331
Seebach, D., 188, 282
Seely, F.L., 81
Sehata, M., 138
Sehra, M., 198
Seibt, W., 41
Seitz, G., 73
Sekiguchi, S., 162
Selig, H., 25
Selivanov, B.A., 168
Selzle, H.L., 23
Semenov, V.V., 292
Seminario, J.M., 57
Semmelhack, M.F., 261, 262
Semones, M.A., 150
Sengupta, S., 166, 240
Senko, M.W., 45
Seo, S., 320
Seres, P., 141
Servi, S., 308
Seto, H., 320
Setti, E.L., 162
Sevilir, N., 31
Sewell, T.J., 333
Seyler, J.W., 248
Shabanowitz, J., 31
Shackleton, C., 321
Shackleton, C.H.L., 30
Shalaby, M.L., 30
Shan, C.-K., 138
Shaozu, W., 266
Share, A.C., 307
Sharghi, H., 159
Sharma, N.D., 305
Sharma, N.K., 302
Sharma, P.K., 171
Sharma, S., 252
Sharma, S.K., 302
Sharp, K., 57
Sharpless, K.B., 119, 284, 285
Shaw, A.W., 286
Sheil, M.M., 32
Sheldon, R.A., 286, 284, 304
Sheldrake, G.N., 305
Shen, B., 318
Shen, C.-Y., 138
Shen, G.J., 309
Shen, T.Y., 197
Shen, W., 88, 93
Shenggang, Y., 240
Sherbine, J.P., 133
Sheu, B.-A., 255
Shevlin, C.G., 180
Shi, Y., 138, 219, 238
Shi, Y.-J., 280
Shiau, C.-Y., 332
Shibasaki, H., 30
Shibasaki, M., 108, 113, 114, 128, 168, 243, 273, 281
Shibata, O., 250, 254
Shibata, T., 284
Shibayama, K., 149
Shida, Y., 30
Shidori, K., 224
Shieh, W.-C., 121
Shiel, M.M., 46
Shigekuni, M., 175
Shigiwara, A., 74
Shihua, W., 240
Shim, J.H., 319
Shimizu, I., 250
Shimomura, K., 326
Shin, H.S., 206
Shin, W., 265
Shindo, M., 157
Shinhama, K., 158
Shinkai, I., 183, 280
Shinkai, S., 105
Shinohara, H., 26
Shioiri, T., 106, 230, 295
Shiraki, T., 150
Shiratori, S., 209
Shiro, M., 264
Shirori, T., 202
Shishkina, I.P., 304
Shizuma, M., 23
Shon, Y.S., 282
Shook, C.A., 133
Short, K.M., 74, 263
Shreeve, J.L., 12
Shtah, S.R., 170
Shteingarts, V.D., 168
Shui, X., 199
Shushan, B., 29
Siangouri-Feulner, I., 71
Sibille, S., 167, 235
Sicking, W., 65
Sidduri, A., 245
Sieburth, S.McN., 165
Siedem, C.S., 154
Siegbahn, K., 41
Siegel, J.S., 146, 147, 149
Siegel, M.G., 215
Siegmann, K., 167
Siegmund, E., 49
Siehl, D.L., 322
Sierra, M.A., 186
Siesel, D.A., 232
Sieskind, O., 159
Sih, C.J., 303, 304

Silla, E., 60
Silverman, L.S., 333
Simard, M., 108
Simion, C., 205
Simms, J.R., 49
Simon, V., 69
Simonet, J., 171
Simonin, P., 319
Simonsen, S.H., 135
Simpkins, N.S., 84, 125, 138, 296
Simpson, E.S.C., 287
Simpson, T.J., 315, 316
Singaram, B., 130, 282, 294
Singer, M., 283
Singer, R.D., 106, 271
Singh, B.K., 322
Singh, G., 162, 287
Singh-Sangwan, N., 320
Singleton, D.A., 94, 141
Singleton, T., 262
Sinigaglia, M., 307
Sirisoma, U.N., 278
Sirois, M., 22
Sironi, A., 254
Sivaraman, N., 26
Sjögren, M.P.T., 249
Skabara, P.J., 266
Skelton, B.W., 271
Skibsted, J., 13
Sklenak, S., 69, 75
Slawin, A.M.Z., 151, 248, 263
Smaill, J.B., 180, 276
Smalley, R.K., 76
Smallridge, A.J., 309
Smith, A.B., III, 25
Smith, A.L., 149
Smith, B.J., 54
Smith, D.A., 58, 65
Smith, D.M., 77
Smith, D.R., 147
Smith, L.M., 42
Smith, R.D., 37, 38, 39, 45
Smith, R.W., 35
Smith, S.O., 14
Smits, J.R.M., 69
Smyth, M.S., 102
Snaith, R., 125, 296
Snatzke, G., 308
Sneden, A.T., 105
Snijman, P.W., 153
So, S.-S., 69
Soai, K., 128, 271
Sobey, W.J., 332
Sobukawa, M., 125
Sodano, G., 311
Sodeoka, M., 113, 114, 168, 243
Soderquist, J.A., 120, 191, 284
Sokolov, V., 266
Soldenko, V.A., 304
Solé, D., 102
Solladie, G., 118
Soloshonok, V.A., 304
Somanathan, R., 295
Somogyi, A., 47
Son, J.C., 191
Sonada, K., 128
Sonawane, H.R., 107
Sonderholm, L., 26
Sonoda, N., 94, 101, 279
Sorey, S.D., 317
Sorgi, K.L., 79, 80, 196
Soria, A., 75
Sorokin, V.D., 158
Sosa, C., 54, 57
Sotiriou-Leventis, C., 156
Souay, P., 72
Souizi, A., 77
Southworth, B.A., 295
Spaltenstein, A., 310
Sparks, M.A., 310
Spassov, G., 308
Speckamp, W.N., 254
Speir, J.P., 41
Spencer, D.M., 105
Spencer, I.D., 328, 327, 329
Spencer, J.B., 315, 333, 335
Spencer, N., 147
Spencer, P., 266
Spengler, B., 41
Sphon, J.A., 29
Spielmann, H.P., 8
Spiess, H.W., 16
Spinace, E.V., 106
Spraul, M., 12
Sprinz, J., 253, 271
Squires, R.R., 146
Srdanov, G., 25
Srebnik, M., 235
Sridharan, V., 112, 237, 240
Srinivas, R., 22
Srinivasan, K., 257
Srinivasan, T.G., 26
Stacel, S.J., 138
Stadtmüller, H., 138, 238
Staempfli, A., 34
Stafford, J.A., 291
Stageland, E.L., 130
Stahl, B., 42
Stahl, D.C., 30
Stahl, N., 31
Staley, S.W., 232
Stallman, J.B., 140
Stamford, N.P.J., 333
Standing, K.G., 40
Stang, P.J., 108, 141, 158
Stangeland, E.L., 282
Stará, I.G., 250, 272
Stark, W.M., 216, 333
Starý, I., 250, 272
Stasko, A., 168
Staunton, J., 202
Stave, M.S., 57
Stec, D., 158
Steding, A., 37
Steffen, L.K., 171
Stein, N., 107
Steiner, V., 42
Steinke, T., 61
Stempf, I., 125
Stenstroem, Y., 27
Stephan, C., 218
Stephenson, G.R., 257, 258
Stepowska, H., 267
Sterbenz, J.T., 253
Sterin, S., 182
Sterner, O., 320
Steurnagel, S., 323
Stevenson, P., 134
Stewart, J.J.P., 61
Stewart, S.K., 108, 166, 224
Stille, J.R., 159
Stiller, L., 58
Stinson, S.C., 269
Stirk, K.G., 22
Stoddart, J.F., 147
Stoll, H., 52
Stolowich, N.J., 333, 335
Stone, G.B., 118
Stout, R.L., 38
Strange, J.H., 18
Stratmann, K., 312
Street, L.J., 203
Streitwieser, A., 146
Strekowski, L., 292
Strife, R.J., 49
Stringer, M.B., 24
Strnad, M., 71
Strongin, R.M., 25
Stryker, J.M., 248
Stüdemann, T., 138
Stuerga, D., 162
Stürmer, R., 293
Stumpf, R., 124
Stupczewski, J.T., 164
Subba Rao, Y.V., 159
Subrahmanyam, M., 159
Subramaniam, R., 322
Subramanian, R., 72
Suchen, S.D., 168
Suemune, H., 106
Suffert, J., 111, 150, 233
Suga, T., 319
Sugai, T., 309
Sugathapala, P.M., 263
Sugihara, T., 154, 219
Suginome, M., 122, 235
Sugita, T., 277, 289
Sugiura, Y., 132, 150
Sulsky, R., 250
Sumiya, T., 292
Sun, B.Q., 16
Sun, C.-M., 315
Sun, S., 99, 129
Sun, Y., 14, 254
Sundaresan, J., 201
Sundermeyer, J., 106
Sundqvist, B.U.R., 41
Suriano, J.A., 221
Susfalk, R.B., 235
Sustmann, R., 65, 71
Sutherland, A.G., 299

Sutter, B., 319
Suziki, Y., 165
Suzuki, A., 141, 153, 232, 235
Suzuki, F., 199
Suzuki, H., 107, 158
Suzuki, I., 273
Suzuki, K., 117, 159
Suzuki, N., 117, 243, 244
Suzuki, T., 25, 108, 128, 162, 163, 273
Svedas, Y.K., 304
Svensson, P., 148
Svensson, S.C.T., 252
Svoboda, I., 83
Swafford, A.M., 193
Swanson, S., 258
Swarbick, T.M., 141
Sweeley, C.C., 36
Sweigart, D.A., 264, 265
Swenton, J.S., 155
Swestock, J., 166, 225
Sygula, A., 169
Sygula, R., 169
Szalontai, G., 7
Szántay, C., 232
Szarek, W.A., 35
Szczeciński, P., 263
Sziraky, P., 304
Szymoniak, J., 264

Tabata, H., 316
Tabata, Y., 323
Taber, D.F., 138, 243
Tabet, J.-C., 22, 24, 46, 328
Tacconi, G., 78
Tachdijian, C., 183
Tachiro, M., 177
Tada, T., 188
Tadano, K., 251
Tadokoro, T., 111, 238
Tae, J.S., 86, 169, 190
Taga, T., 116, 276
Tagliani, A., 308
Tagliavini, E., 270
Taguchi, T., 119, 125, 136, 138, 290
Takada, K., 141
Takagi, I., 251
Takagi, K., 167
Takahashi, K., 74, 111, 241, 281
Takahashi, T., 66, 117, 150, 243, 244
Takahashi, Y., 78
Takahori, T., 272
Takai, M., 215
Takai, T., 287
Takai, Y., 23, 172
Takano, M., 136
Takano, S., 213, 301
Takao, K., 251
Takasaki, T., 78
Takase, K., 172
Takashima, M., 224
Takaya, H., 131, 282
Takayama, M., 26
Takazawa, N., 108, 235
Takeda, K., 138, 320
Takeda, M., 130
Takehara, H., 229
Takemoto, S., 114
Takemoto, T., 243
Takemoto, Y., 255
Takemura, S., 215
Taketomi, T., 282
Takeuchi, H., 160
Takeuchi, J., 255
Takeuchi, K., 132, 177
Takigiku, R., 42
Takuwa, A., 106, 270
Tamaru, Y., 247, 254
Tamás, J., 185, 232
Tamura, J., 34, 39
Tamura, Y., 119
Tan, C.-W., 126
Tan, W., 314
Tanabe, K., 141
Tanahashi, A., 252
Tanaka, A., 254
Tanaka, H., 150, 183
Tanaka, K., 108, 116, 165, 184, 273, 276, 293
Tanaka, S., 246, 247, 250
Tang, K., 42
Tang, S.C., 13
Tani, K., 118, 284
Tani, Y., 322
Taniguchi, N., 136
Taniguchi, T., 136
Tanino, K., 128
Taniseki, Y., 254
Tanya, M.J., 171
Tao, C., 257
Tarbet, K.H., 127, 270
Tarbit, B., 181
Tashiro, M., 175
Tasuaka, M., 129
Tatematsu, T., 295
Tatsukawa, A., 126
Tatsumi, A., 158
Tatsuta, K., 154
Tau, S.-I., 255
Tauwa, A., 106
Tavani, C., 163
Taylor, D.L., 204
Taylor, M.K., 292
Taylor, N.J., 254
Taylor, P.J., 59
Taylor, R.J.K., 118, 155, 179, 193, 203
Teasdale, A.J., 237
Tebbe, M., 198
Teesch, L.M., 27
Tejeda, J., 187
ten Hoeve, W., 163, 293
Teplow, D.B., 31
Terada, M., 127
Terashima, S., 233
Terlouw, J.K., 21, 22, 23
Terrett, N.K., 84, 138
Tessier, C.A., 233
Tester, R.W., 193
Tews, H., 280
Texier-Boullet, F., 294
Thakur, R.S., 320
Thal, C., 108, 207, 226
Theurig, M., 235
Thibault, P., 30
Thibaut, D., 333, 335, 336
Thiecke, J.R.G., 163
Thiel, W., 266
Thiericke, R., 311, 313, 314, 328, 329
Thom, S.M., 230
Thomas, B.E., IV, 65, 77
Thomas, E.J., 127
Thomas, G., 99
Thomas, P.J., 158
Thomas, R., 37
Thomas, S.E., 151, 222, 248, 263
Thomas-dit-Dumond, L., 201
Thompson, C.F., 264
Thompson, D.F., 92
Thompson, J.R., 76
Thomson, N., 309
Thorat, T.S., 159
Thorimbert, S., 250
Thorneley, R.N.F., 29
Thuong, N.T., 105
Thurman, E.M., 38
Tietze, L.F., 107, 210, 226
Tillyer, R.D., 180
Timofeeva, L.M., 60
Tinachoe, S.W., 90
Titman, J.J., 16
Tius, T.A., 235
Tkaczyk, M., 25, 48
Tobe, Y., 175, 178
Toda, F., 106, 184
Toda, T., 213
Todd, J.F.J., 47, 48
Todoroki, R., 307
Todres, Z.V., 263
Togni, A., 253
Toia, R.F., 315
Tokuda, M., 150
Tokunaga, M., 132, 282
Tolman, J.R., 11
Tomasi, J., 59, 61
Tomaszewski, M.J., 81, 280
Tom Dieck, H., 218
Tomer, K.B., 30
Tomioka, K., 157, 296
Tomita, K., 136
Tomoda, H., 316
Toneva, R., 283
Tonnies, S.D., 133
Topol, I.A., 57
Torchio, M., 263
Torii, S., 111, 183, 238
Torisawa, Y., 141
Torode, J.S., 193, 220

Toromanoff, E., 138
Torroba, T., 187
Tortajorda, J., 24
Toru, T., 100, 129
Toshima, K., 149
Tóth, M., 311
Toupet, L., 138
Touzin, A.-M., 250
Towne, T.B., 247
Townsend, C.A., 333
Toyooka, N., 301
Toyota, M., 138
Trafford, D.J.H., 36
Traldi, P., 47, 48
Tran, V.D., 239
Treiber, K.D., 247
Tressl, R., 311
Trew, S.J., 336
Trewhella, M.A., 309
Trombini, C., 270
Trost, B.M., 136, 138, 218, 219, 238, 249, 250, 252
Truhlar, D.G., 57, 67, 68, 79
Truong, T.N., 67, 79
Tsai, Y.-M., 86
Tsarbopoulos, A., 39
Tsay, S.C., 149
Tschaen, D.M., 183, 280
Tschudin, R., 11
Tseng, W.-H., 138
Tso, H.-W., 118
Tsubota, M., 171
Tsuchiya, T., 188, 209
Tsuchuja, R., 331
Tsuda, N., 132
Tsuge, A., 177
Tsuji, K., 39
Tsuji, R., 177
Tsuji, Y., 235, 240, 249, 250
Tsukiyama, T., 246
Tsunoi, S., 94, 279
Tsusaka, M., 100
Tsvetkov, V.D., 168
Tsyganov, D.V., 160
Tu, C.-C., 249, 304
Tubul, A., 138
Tucker, C.E., 138, 238
Tully, B., 171
Tunde, P., 198
Tunon, I., 60
Turchi, I.J., 75, 79, 80
Turck, A., 200, 201
Turecek, F., 24
Turner, I., 205
Turner, M.M., 309
Turner, N.J., 300, 309
Turner, R.M., 188, 235
Turos, E., 248
Tyler, A.N., 34
Tyler, J.W., 332
Tyrrell, E., 246

Uchida, S., 247
Uchiyama, M., 226
Udvarnoki, G., 317
Ueda, M., 292
Ueda, R., 183
Ueda, Y., 85
Uejima, A., 165
Uemetsu, T., 304
Uemura, J., 85
Uemura, M., 242, 264
Uemura, N., 83
Uemura, S., 277, 289
Ueno, K., 169
Ueno, Y., 100, 129
Ugalde, J.M., 65
Uggerud, E., 27
Ugi, I., 107, 199
Uguen, D., 301
Uhrin, D., 7
Ukai, S., 35
Ulmer, C.W., 65
Ulmer, G., 26
Umani-Ronchi, A., 270
Umemura, M., 77
Umezawa, K., 323
Underwood, D.J., 24
Undheim, K., 200, 250, 265
Uneyama, K., 85
Uno, H., 107
Unverzagt, C., 105
Uomori, A., 320
Uozumi, Y., 128, 252
Urabe, H., 251
Urbanos, F., 265
Urry, D.W., 107
Urushibara, S., 171
Utimoto, K., 186
Uwakwe, P.U., 162

Vadecord, J., 117, 277
Vaillancourt, V., 112
Vainiotalo, P., 21
Vaissermann, J., 223, 246
Valdes, C., 141
Valentine, B.P., 332
Vallee, Y., 107
Valvano, N.L., 291
Van Aerschot, A.A., 228
Van Berkel, G.J., 28, 37
van Boeckel, C.A.A., 105
Van Bramer, S.E., 23
Van Bruggen, N., 324
van den Berg, K.J., 193
Vanderbilt, D., 56, 57
van der Gen, A., 271
Van der Greef, J., 35, 36
Van der Hoeven, R.A.M., 35, 36
van der Kerk-van Hoof, A., 35
van der Made, A.W., 160
van der Made, R.H., 160
VanDerveer, D., 141
Van Dongen, W.D., 33
Van Dorsselaer, A., 36
Van Dyke, G., 171
van Eck, E.R.H., 16
Van Epp, J.E., 324
Van Es, D.S., 176
Vanest, Q.C., 69
van Hummel, G.I., 190
van Leeuwen, M., 163
van Leusen, A.M., 193
van Lier, J.E., 232
Van Nieuwenhze, M.S., 285
Van Pelt, C.E., 140, 220
Vanquickenborne, L.G., 79
van Rantwijk, F., 304
Van Rooyen, P.H., 266
Van Seggern, H., 73
Van't Land, C.W., 329
Vardhan, A., 302
Vargas, D., 313
Varma, M., 291
Varma, R.S., 105, 291
Vath, J.E., 32
Vaugeois, J., 108
Vaughan, G.B., 25
Vaughan, J.G., 186
Vaughan, W.S., 259
Vaupel, A., 138, 238
Vázquez, J., 295
Vázquez-Tato, M.P., 294
Vedejs, E., 273, 291
Veeman, W.S., 16
Veggaar, R., 330
Veitch, J.A., 318
Venanzi, L.M., 253, 290
Venkateswara Rao, B., 81
Venturni, A., 75
Vera, W.J., 141
Verbroom, W., 190
Verentchikov, A., 40
Verhoeven, T.R., 280
Verweij, J., 182
Vessières, A., 246
Vestal, M.L., 39
Vettori, U., 47
Vey, A., 31
Vicens, J., 105
Vicker, N., 208
Vidal, J., 182
Vijn, R.J., 204
Vilaplanu, M.J., 206
Villani, C., 147
Villemin, D., 294
Vincent, M.A., 59
Viola, A., 74
Virgili, A., 75
Vishwakarma, R.A., 333
Vitagliano, A., 249
Vlahov, R., 308
Vogel, P., 148
Vogel, R., 168
Vogt, D., 287
Vogtle, F., 176
Volante, R.P., 183
Volk, K.J., 318
Vollhardt, K.P.C., 146
Volpin, M., 159

Vomero, S., 210
von Matt, P., 114, 198, 253, 271
Von Philipsborn, W., 254
Voragen, A.G.J., 35, 36
Vosko, S.H., 53
Vouros, P., 30
Vuilhorgne, M., 333, 335
Vuister, G.W., 9, 11
Vulpetti, A., 66

Waas, J., 245
Wada, Y., 39
Wadsworth, H.J., 77
Wagner, C., 317
Wagner, P.J., 158
Wagschal, K.C., 319
Wahl, J.H., 45
Waizumi, K., 57
Walker, A.J., 179
Walker, J.A., 325
Walkup, R.D., 247
Wallace, D.W., 254
Wallace, E.M., 167, 244
Walls, F.C., 30
Wally, H., 266
Walsh, C.T., 336
Walsh, E.J., 192
Walsh, K.A., 34
Walsh, P., 284
Walter, C.J., 141
Walters, I.A.S., 293
Walther, B.W., 57
Walton, N.J., 326, 327
Wandless, T.J., 105
Wang, A.C., 9
Wang, A.P.L., 32, 42
Wang, J., 136, 241
Wang, J.T., 57
Wang, K.-T., 304
Wang, K.K., 93
Wang, M., 77
Wang, P., 324
Wang, Q., 160
Wang, S., 159
Wang, S.-L., 236, 254
Wang, W.-B., 183
Wang, Y., 138, 248
Wang, Z., 286
Wang, Z.E., 141
Wang, Z.-M., 284
Wang, Z.-X., 266
Ward, J.M., 309
Ward, M.F., 151, 222
Ward, S.E., 290
Warkentin, J., 81, 280
Warner, K., 324
Warren, M.J., 333
Warren, S., 121
Warshel, A., 58
Wartenberg, F.-H., 138, 219
Wasylishen, R.E., 13, 14
Watanabe, H., 85
Watanabe, M., 229, 266
Watanabe, Y., 100, 129, 254
Watling, R., 316
Watrelot, S., 240
Watson, A.B., 326, 327
Watson, C.H., 28
Watson, J.T., 33, 36
Watson, R.J., 155, 193
Watterson, S.H., 137
Wazeer, M.I.M., 197
Webb, H.M., 25
Webb, T.H., 105
Weber, J., 56, 57
Weber, M., 49
Weickhardt, K., 286
Weinhold, F., 146
Weinreb, P.H., 138
Weinreb, S.M., 239, 286, 291
Weinstein, H., 62, 108
Weiss, E., 108
Weiss, J., 105
Weiss, R.G., 19
Weissensteiner, W., 243, 266
Weitz, I.S., 156
Welch, J.T., 102
Wellauer, T., 178
Welle, R., 322
Wellington, E.M.H., 336
Welpy, J.K., 39
Welzel, P., 290
Wemmer, D.E., 8
Wender, P.A., 149
Wensbo, D., 234
Wentrup, C., 146
Werner, H., 244
Wessel, T., 174
West, F.G., 193, 204
West, G.M.J., 69
West, R., 86
Westermann, J., 112
Westmann, A., 41
Westwood, R., 76
Wetterich, F., 97, 98, 99, 129
Wey, H.G., 263
Weyns, N.J., 228
Whang, H.S., 86, 169
Whang, Z., 271
Wheatley, J.R., 191
Whitby, R.J., 244
White, A.H., 271
White, J.D., 135
White, P.S., 255
Whitehouse, C.M., 40
Whitesides, G.M., 309, 310
Whiting, A., 108, 166, 224
Whiting, D.A., 324
Whittaker, M., 267
Whitten, W.B., 48
Wider, G., 11
Widhalm, M., 243, 266
Wiegelmann, J.E.C., 241
Wier, M.R., 48
Wiermann, R., 323
Wierschke, S.U., 146
Wiersum, U.E., 171
Wierzchleyski, A.T., 263
Wijaya, N., 290
Wijsman, G.W., 176
Wilcox, C.S., 105
Wilde, A., 134, 248
Wilen, S.H., 270
Wilk, L., 53
Wilkins, C.L., 40, 42, 43, 44
Wilkinson, B., 310
Willard, P.G., 108
Willem, R., 3
Willets, A.J., 306, 309
Willey, K.F., 47
Williams, A.C., 250
Williams, D.J., 147, 151, 248, 263
Williams, D.R., 130
Williams, E.R., 42
Williams, F., 57, 322
Williams, I.H., 62
Williams, J.D., 48
Williams, J.M.J., 114, 163, 198, 250, 253, 271
Williams, K.W., 310
Williams, P., 37
Williams, R.M., 331
Williamson, B.L., 141
Williard, P.G., 264, 296
Willis, C.L., 77
Willker, W., 11
Willoughby, C.A., 281
Wilmesmeier, S., 323
Wilson, C., 108, 226
Wilson, I.D., 12
Wilson, K.E., 295
Wilson, S.R., 107, 269
Wilson, W.W., 57
Wilt, J.W., 72
Winchester, W.R., 135
Winger, B.E., 38
Wink, M., 326
Winkler, J.D., 215
Winter, S.B.D., 122
Wipf, P., 106, 128, 155, 193, 243
Wirnsberger, G., 305
Wishart, N., 235, 245
Wison, P.D., 138
Wittekind, M., 7
Wnuk, S.F., 106
Wojcicki, A., 254
Woleben, C.M., 137
Wolf, J., 244
Wolf, R.E., 107
Wolfe, J.J., 69
Wolfsberger, M., 65
Wolkow, C., 42
Wollowitz, S., 107, 269
Wong, C.-H., 105, 299, 304, 309
Wong, N.R., 171
Wong, T., 111
Woo, S.H., 153
Wood, J.M., 29
Wood, T.D., 43

Woodall, T.M., 130, 280
Woodcock, S., 59
Woodley, J.M., 309
Woods, A., 42
Wooley, J.G., 326
Woolsey, N.F., 263
Wormald, P.C., 328
Wrede, P., 70
Wrigglesworth, R., 179
Wright, A.D., 21, 46, 47
Wright, B., 45
Wright, J.D., 69
Wright, J.L.C., 325
Wrinn, M., 57
Wrobel, Z., 206
Wu, F.-C., 302
Wu, G., 13, 14
Wu, H.F., 48
Wu, J., 12
Wu, K.J., 37
Wu, X., 13, 14
Wu, X.-M., 138
Wu, Y., 13, 108
Wu, Y.-C., 138
Wu, Y.D., 63
Wu, Y.-J., 136
Wu, Y.-S., 171
Wu, Z., 25, 324
Wu, Z.J., 183
Wudl, F., 25
Wuest, J.D., 108
Wuethrich, K., 11
Wulff, W.D., 73, 138, 150, 151, 222, 278
Wurster, J.A., 192
Wurthner, F., 193
Wurz, P., 26
Wyler, H., 207
Wynberg, H., 163, 293
Wysocki, V.H., 46, 47
Wythes, M.J., 235, 245

Xavier, L., 111, 232
Xhang, X., 44
Xi, M., 167
Xia, X.F., 58
Xiao, M., 133
Xie, L., 160
Xinquan, H., 240
Xu, D., 152, 284
Xu, F., 208
Xu, L.H., 263
Xu, S.L., 150
Xu, W., 201
Xu, Y., 137, 229
Xu, Y.-C., 150, 202
Xu, Z., 108

Yabe, M., 153
Yada, Y., 319
Yadav, G.D., 159
Yajima, T., 254
Yakubov, A.P., 160
Yalpani, N., 323
Yamabe, S., 65, 72
Yamada, H., 23, 150
Yamada, N., 250
Yamada, T., 287
Yamaguchi, H., 26
Yamaguchi, K., 138
Yamaguchi, M., 138
Yamaguchi, S., 282
Yamaguchi, Y., 80
Yamamoto, A., 250
Yamamoto, G., 172, 174
Yamamoto, H., 127, 132, 141, 196, 270, 272, 278, 291
Yamamoto, Y., 106, 128, 132, 273
Yamamura, S., 321
Yamanaka, H., 108, 225, 226, 228, 235
Yamashita, M., 188
Yamato, M., 141
Yamato, T., 175
Yamazaki, A., 252
Yamazaki, H., 101
Yamazaki, Y., 264, 266
Yamori, S., 213
Yan, T.-H., 126
Yanagisawa, A., 127, 270
Yanai, M., 136, 219
Yang, A.S., 57
Yang, C., 171
Yang, D.-L., 162
Yang, G.K., 248
Yang, H., 23, 253
Yang, T.K., 73
Yang, W., 51, 52, 53
Yang, Y., 232
Yang, Z.-Y., 232
Yao, Q., 89
Yashunsky, D.V., 138
Yasuda, N., 111, 232
Yasuhara, A., 228
Yasui, K., 254
Yasuike, S., 188, 209
Yasukata, T., 293
Yasunami, M., 172
Yau, P.Y., 41
Ye, J., 112
Yeh, M.-C.P., 255
Yelm, K.E., 141
Yeske, P.E., 77
Yeung, B., 30
Yeung, E.S., 41
Yeung, K.-S., 276
Yin, H.M., 23
Yin, J., 145
Yip, T.T., 29, 38, 40, 42
Yip, Y.C., 132
Yohannes, D., 105, 269
Yokokawa, F., 230
Yokota, K., 292
Yokoyama, Y., 224
Yoneda, N., 163
Yoneda, R., 215
Yoo, R.K., 25
Yoo, S.-e., 140
Yoon, E., 35
Yoon, N.M., 282
Yoon, T.-S., 265
Yorozu, K., 287
Yoshida, I., 35
Yoshida, K., 46, 102, 150
Yoshida, M., 95
Yoshida, N., 138
Yoshida, W., 250
Yoshihara, K., 132
Yoshii, E., 138
Yoshikawa, A., 183
Yoshimura, T., 292
Yoshimura, Y., 320
Yoshizumi, T., 282
Yost, R.A., 48, 49
You, Z., 80
Young, P., 60
Young, R.J., 171
Youngs, W.J., 233
Yu, H., 34
Yu, M.S., 136, 241
Yu, R.H., 136
Yu, S.J., 22
Yu, W., 32
Yu, Y., 255
Yuan, W., 304
Yuki, Y., 286
Yulan, Z., 266
Yum, E.K., 249
Yus, M., 128, 167
Yvanaeff, C., 294

Zachara, J., 263
Zaia, J., 31
Zair, T., 249
Zaltsgendler, I., 160
Zamojski, A., 267
Zare, R.N., 42
Zargarian, D., 284
Zdravkovski, Z., 294
Zecca, L., 29
Zecchi, G., 213, 262
Zeech, A., 330
Zefirov, N.S., 158
Zeitz, H.-G., 97
Zeller, L., 21
Zeltz, H.-G., 129
Zelvelder, E., 162
Zeng, Z., 331
Zenglu, L., 266
Zenk, M.H., 330
Zercher, C.K., 149
Zettlmeier, W., 270
Zevaco, T., 288
Zhand, S.-W., 254
Zhang, B., 133
Zhang, D., 320
Zhang, H.-C., 228
Zhang, J.Y., 32

Zhang, Q., 128, 201
Zhang, W., 90, 106, 279
Zhang, W.-Y., 199
Zhang, X., 282, 287
Zhang, X.-L., 284
Zhang, X.-M., 162
Zhang, X.-P., 115, 275
Zhang, Y., 160
Zhang, Z., 191
Zhao, S.-H., 284
Zhao, W.-Y., 163
Zhdankin, V.V., 158
Zheng, L., 19
Zheng, Q., 232
Zheng, W., 245
Zheng, Z.B., 211
Zhoa, H., 32
Zhong, H.M., 91
Zhou, J., 40
Zhou, M., 273
Zhou, P., 295, 331
Zhou, Q.-L., 114, 272
Zhou, T., 254
Zhou, Y., 248
Zhou, Z., 34
Zhu, G., 9, 220
Zhu, J., 94
Zhuo, J.-C., 207
Ziegler, C.B., Jr., 101
Ziegler, F.E., 85
Ziegler, T., 54, 57, 66
Ziessel, R., 233
Zijlstra, R.W.J., 293
Ziller, J.W., 12
Zilm, K.W., 13, 14, 148
Zimmer, R., 107, 181
Zimmerman, C.N., 141
Zink, T., 79
Zora, M., 223
Zou, X., 162
Zuiderweg, E.R.P., 11
Zumbulyadis, N., 14
Zupan, J., 69
Zwannenberg, B., 73
Zych, K., 171